THE PHYSICS OF STRUCTURALLY DISORDERED MATTER:

AN INTRODUCTION

Graduate Student Series in Physics
Other books in the series

Weak Interactions
D Bailin

Gauge Theories in Particle Physics
I J R Aitchison and A J G Hey

Collective Effects in Solids and Liquids
N H March and M Parrinello

Hadron Interactions
P D B Collins and A D Martin

Superfluidity and Superconductivity
D R Tilley and J Tilley

Introduction to Gauge Field Theory
D Bailin and A Love

Mechanics of Deformable Media
A B Bhatia and R N Singh

Supersymmetry, Superfields and Supergravity
Prem P Srivastava

GRADUATE STUDENT SERIES IN PHYSICS

Series Editor: Professor Douglas F Brewer, M.A., D.Phil.
Professor of Experimental Physics, University of Sussex

THE PHYSICS OF STRUCTURALLY DISORDERED MATTER:

AN INTRODUCTION

N E CUSACK
School of Mathematics and Physics
University of East Anglia

ADAM HILGER, BRISTOL AND PHILADELPHIA

British Library Cataloguing in Publication Data

Cusack, N.E.
The physics of structurally disordered matter: an introduction.—(Graduate student series in physics)
1. Matter—Properties
I. Title II. Series
530.4 QC173.3

ISBN 0-85274-591-5
ISBN 0-85274-829-9 Pbk

First published 1987
Reprinted 1988

Published under the Adam Hilger imprint by IOP Publishing Ltd
Techno House, Redcliffe Way, Bristol BS1 6NX, England
242 Cherry Street, Philadelphia, PA 19106, USA

Typeset by Datapage International Ltd, Dublin, Ireland

Printed in Great Britain by J W Arrowsmith Ltd, Bristol

CONTENTS

Preface ix

Acknowledgments xi

Abbreviations and acronyms xiii

Common symbols xv

1 Order and disorder: an introductory discussion **1**

1.1 Ordering rules 1
1.2 Objects in space 3
1.3 Order parameters 5
1.4 Physical systems: ensembles 6
1.5 Structural order: crystalline and non-crystalline matter 6
1.6 Crystal imperfections 7
1.7 Some varieties of disorder 8
1.8 Disorder and entropy: quenched materials 9
1.9 Examples of structurally disordered materials 12
1.10 Quasicrystals 14
1.11 Order–disorder transformation. Ising model 17
1.12 Some problems presented by disordered matter 22
References 24

2 Describing disordered structures **25**

2.1 Sphere packings. The Bernal model 25
2.2 Voronoi and other polyhedra 27
2.3 Distribution functions 30
2.4 Extension to binary systems 33
2.5 Models grown from seeds 36
2.6 Models of chemical short-range order 37
2.7 Short- and long-range order parameters 40
2.8 Curved space and the structure of glasses 43
2.9 Fractals 47
References 53

3 The experimental investigation of disordered structures **55**
3.1 Scattered intensity 55
3.2 X-rays 56
3.3 The average $S(Q)$ for a real sample 58
3.4 The importance of the structure factor 60
3.5 Formulae for binary mixtures 61
3.6 Incoherent and inelastic scattering 62
3.7 Neutron and electron scattering 62
3.8 Some experimental considerations. Pure samples 63
3.9 From $S(Q)$ to $g(r)$ 65
3.10 Binary systems and isotopic substitution 67
3.11 Examples of structure investigations 68
3.12 Different kinds of structure factor. Fluctuations 74
3.13 Other clues to structure. Triplet distribution 79
References 81

4 Monte Carlo and molecular dynamics methods **83**
4.1 The Monte Carlo method 83
4.2 Molecular dynamics 87
4.3 Some results for the hard-sphere model 90
4.4 Some results for the Lennard-Jones model 93
4.5 Some results for a liquid metal potential 95
4.6 Quenching and computer glasses 97
References 99

5 The static properties of simple liquids **100**
5.1 Intermolecular potentials 100
5.2 Connections between thermodynamic quantities and $g(r)$ 102
5.3 Attempts to derive $g(r)$ from $\varphi(r)$ 103
5.4 The superposition approximation and the BGY equation 110
5.5 Density expansions. Direct correlation function 111
5.6 The HNC and PY equations 114
5.7 More recent theories, MSA and others 118
5.8 Perturbation and variation methods 121
5.9 Extracting $\varphi(r)$ from $g(r)$ 124
5.10 Cell and allied theories 125
5.11 Specimen comparisons of theory and experiment 126
References 129

6 Simple liquid metals **131**
6.1 Experimentation with liquid metals 131
6.2 The general description of a metallic liquid 132

6.3 Pseudopotentials and model potentials 133
6.4 Dielectric screening functions 137
6.5 Input data for NFE calculations 140
6.6 Energy of a liquid metal 140
6.7 Effective interionic potential 144
6.8 Thermodynamic properties 146
6.9 Thermodynamic properties of liquid alloys 147
6.10 The one-component plasma 152
6.11 Electron transport properties; introduction 153
6.12 Electron transport properties and NFE theory 156
6.13 Successes and limitations of the theory 158
6.14 The Hall effect and the thermal conductivity 160
6.15 Optical properties 161
6.16 Modifications to the resistivity formula 163
Appendix Hard-sphere formulae for binary systems 165
References 166

7 Electrons in disordered matter 168
7.1 The Green and spectral functions 168
7.2 The coherent potential approximation 172
7.3 Liquid metals 175
7.4 Clusters and the recursion method 179
7.5 Other methods and miscellaneous points 183
7.6 Some relevant experiments 185
7.7 Electron transport 189
Appendix 195
References 196

8 Thermal motion 198
8.1 $G(r, t)$, $S(Q, \omega)$ and density correlations 199
8.2 Neutron inelastic scattering 202
8.3 Specimen experiments 204
8.4 Some theoretical approaches: fluids 211
8.5 Some theoretical approaches: solids 217
8.6 Other systems and techniques 223
References 225

9 Percolation and localisation 227
9.1 Percolation—an introduction 227
9.2 Site and bond percolation. Clusters 229
9.3 A few more general problems 233
9.4 Electron transport: effective-medium theory 235
9.5 Localisation—an introduction 238
9.6 Some simple, if not compelling, ideas 241

9.7 Localisation in 1-D 242
9.8 The Anderson model 244
9.9 Scaling arguments 246
9.10 Mobility edges and gaps 249
9.11 Experimental tests—some preliminaries 251
9.12 Experiments—a few examples 252
9.13 Hopping conductivity 257
References 263

10 Insulating glasses **266**
10.1 Empirical description of the glass transition 267
10.2 Some thermodynamic and statistical considerations 269
10.3 Can all materials be vitrified? 273
10.4 A glass phase transition? Models 273
10.5 Constraints and stability 276
10.6 The structure of oxide glasses 279
10.7 Low-temperature properties and two-level systems 286
References 293

11 Amorphous and liquid semiconductors **295**
11.1 Amorphous semiconductors—an introduction 296
11.2 Preparation 297
11.3 Composition and structure 299
11.4 Defects and dangling bonds in a-Si 302
11.5 Defects in chalcogenide glasses 305
11.6 ESR and photoluminescence 308
11.7 Density of states in the gap 314
11.8 Bands, tails and optical gaps 320
11.9 Some electron transport properties 328
11.10 A few more properties of amorphous semiconductors 340
11.11 Compound-forming liquid alloys 341
11.12 Expanded liquid metals 350
References 356

12 The structure and electronic properties of metallic glasses **360**
12.1 Formation of metallic glasses 361
12.2 Structure and vibrational spectrum 368
12.3 Electronic spectrum 372
12.4 Electron transport properties I: simple metal alloys 377
12.5 Electron transport properties II: non-simple cases 381
12.6 Other important properties 388
References 392

Index **395**

PREFACE

In this book I have set out to introduce the physics of structurally disordered matter to those who are starting research in some aspect of this vast field, or considering whether to start. I have tried to write the book I wish had been available for my research students—not to speak of myself—when I took up experiments on liquid metals many years ago.

Those who have already begun research in, say, thermal motion in metallic glasses will be rapidly acquiring much more expert knowledge than this introduction offers. I have assumed, however, that this means they will be more, not less, likely to be interested in some other (but possibly related) aspects of some other (but not wholly dissimilar) systems; say, electronic motion in liquid alloys.

I also hope the book might be a useful source for lecturers in universities and polytechnics who like the idea of introducing more about disordered systems into their courses on condensed matter than is usual at present.

It would not be difficult to compile an 'anti-index' of subjects which might have been in the book but are not: liquid crystals, liquid helium, polymers, ionic melts and other things of which I regret the absence. Selection was dictated by the ratio of subject matter to available time—a parameter which showed an alarming tendency to diverge and needed a somewhat arbitrary cut-off.

There should be a word about the references. For the most part they are simply those I found useful myself. However, they are numerous and I should be surprised if they do not help a reader new to the field to become rapidly involved in the literature. This I took to be one of the functions of an Introduction.

I am not particularly attracted to formal dedications but would like to add that while writing the book I often thought of my research students and collaborators, and of the pleasure of working with them.

N E Cusack
March 1986

ACKNOWLEDGMENTS

It is a pleasure to thank those who let me consult some of their writings before publication: Dr N Cowlam, Professor J Hafner, Professor F Hensel, Professor J Jäckle, Dr M Itoh, Professor U Mizutani, Professor G J Morgan, Professor G C Shephard, Dr M Silbert and Professor W H Young. My friend and colleague Professor W H Young read the typescript and I am very grateful to him for constructive remarks. Although they are in no way responsible for the outcome, a number of people generously spent time advising me on sections within their own special fields of interest. This was very valuable and it is a pleasure to thank Dr D C Champeney, Dr G Everest, Dr T Gaskell, Dr D Gaunt, Dr P Gray, Dr D Greig, Dr J Hirsch, Dr M Itoh, Professor A E Owen, Mr J Reeve and Professor G C Shephard.

The task of typing the manuscript was long and tedious and I am greatly indebted to four secretaries who at various times have given their skill and patience: Mrs Sue Brodie, Mrs Fiona Kelly, Mrs Jenny Rivett and Mrs Anne Steven.

Mr Dick Fuller spent much time and photographic expertise preparing diagrams and I am very grateful to him too.

A great many publishers and authors gave permission to reproduce published diagrams and graphs and this help is acknowledged appropriately elsewhere.

N E Cusack
March 1986

ABBREVIATIONS AND ACRONYMS

Abbreviation or acronym	Meaning	First mentioned or defined in section
ACAR	Angular correlation of annihilation radiation	7.6
ARUPS	Angle-resolved ultraviolet photoemission spectroscopy	7.6
BIS	Bremsstrahlung isochromat spectroscopy	11.8
BGY	Born–Green–Yvon	5.3
CA	Cluster aggregation	2.9
CPA	Coherent potential approximation	7.2
CRN	Continuous random network	2.6
CS	Carnahan–Starling	4.3
CSRO	Chemical short-range order	1.7
CVD	Chemical vapour deposition	11.2
DKP	Disordered Kronig–Penney (model)	9.7
DLA	Diffusion-limited aggregation	2.9
DLTS	Deep-level transient spectroscopy	11.7
DOS	Density of states	8.5
DRP	Dense random packing	2.1
EMA	Effective-medium approximation	7.3
EMT	Effective-medium theory	9.4
ESR	Electron spin resonance	11.6
EXAFS	Extended x-ray absorption fine structure	3.13
EXP	Exponential (liquid state model)	5.7
FE	Field effect	11.7
GD	Glow discharge	11.2
GFA	Glass forming ability	12.1
HNC	Hypernetted chain	5.6
HS	Hard sphere	4.3
ICTS	Isothermal capacitance transient spectroscopy	11.7
ID	ideal	6.9
IY	Ishida–Yonezawa	7.3
KKR	Kohn–Korringa–Rostoker	7.2

KP	Kronig–Penney	
LESR	Light-induced electron spin resonance	9.7
LEXP	Linearised exponential	5.7
LJ	Lennard-Jones	4.4
LRO	Long-range order	1.7
MAS–NMR	Magic-angle spinning–nuclear magnetic resonance	10.5
MC	Monte Carlo	4.1
MD	Molecular dynamics	4.2
MHNC	Modified hypernetted chain	5.6
MNM(T)	Metal–non-metal (transition)	9.10
MSA	Mean spherical approximation	5.7
NFE	Nearly free electron	Just before 6.1
OCP	One-component plasma	6.10
OCT	Optimised cluster theory	5.7
ODS	Optical density of states	7.6
ORPA	Optimised random-phase approximation	5.7
OZ	Ornstein–Zernike	5.5
PD	Photodarkening	11.10
PDS	Photothermal deflection spectroscopy	11.7
PECVP	Plasma-enhanced chemical vapour deposition	11.2
PL	Photoluminescence	11.6
PS	Photostructural (change)	11.10
PY	Percus–Yevick	5.6
QCA	Quasicrystalline approximation	7.1
RDF	Radial distribution function	2.3
RKKY	Rudermann–Kittel–Kasuya–Yoshida (interaction)	12.6
RPM	Random-phase model	7.7
SCLC	Space-charge-limited currents	11.7
SO	Spin–orbit	6.14
SRO	Short-range order	1.8
SWE	Staebler–Wronski effect	11.10
TBA	Tight-binding approximation	7.2
TCR	Temperature coefficient of resistivity	12.4
TEM	Transmission electron microscopy	9.4
TLS	Two-level system	10.6
TPA	Transient photoabsorption	11.9
TPC	Transient photocurrent	11.9
TTT	Time temperature transformation (curves)	12.1
UPS	Ultraviolet photoelectron spectroscopy	7.6
VCA	Virtual-crystal approximation	7.2
WCA	Weeks–Chandler–Andersen	5.8
XPS	X-ray photoelectron spectroscopy	11.8

COMMON SYMBOLS

Symbol	Meaning	Introduced in section
a	Thermal diffusivity	8.4
a_i	Activity	6.9
b	Scattering length	3.7
B	Tight-binding band width	9.6
B_n	nth virial coefficient	4.3
c_i	Concentration of i	
$c_{ij}(r)$, $c(\boldsymbol{r})$	Direct correlation function	5.5
C, C_v, C_p	Heat capacity	
d, d_c	Density, critical density	
d_f	Fractal dimension	2.9
D, $D_{\mu\nu}$	Diffusion coefficient	7.7
D^0, D^-, D^+	Dangling bond energies	11.4
E	Energy	
E_c, E_v	Mobility edges	9.10
E_F	Fermi level	
$\boldsymbol{f}$, $\boldsymbol{F}$	Force	
f, $f^{(0)}$, $f(E, T)$	Fermi function	6.11
$f(\theta)$, $f(\boldsymbol{Q})$, f_A	Scattering amplitude	3.1
F	Helmholtz free energy	
$F(\boldsymbol{Q})$	Total structure factor	3.5
g	g-factor (density of states)	7.7
g	g- or splitting factor	11.6
g, $g^{(2)}$, $g^{(3)}$ ...	Pair, triplet, ..., distribution function	2.3
g_T	Total pair distribution function	3.10
g_{ij}	Partial pair distribution function	2.4
g_{ij}	Conductance i to j	9.13
$g(\varepsilon)$	Density of states in energy	6.15, 7.1
$g(L)$	Non-dimensional conductance	9.9
G, $G^{(0)}$	Gibbs free energy	
G	Shear modulus	8.4
G	Conductance	9.4
G, $G(E)$, $G(\boldsymbol{r}, \boldsymbol{r}', E)$	Green operator, Green function	7.1

$G(\boldsymbol{r}, t)$	Space–time correlation function	8.1
$h(\boldsymbol{r})$	Pair correlation function	5.2
H	Hamiltonian	
H	Enthalpy	6.9
$I(\theta)$, $I(\boldsymbol{Q})$	Scattered intensity	3.1
$\boldsymbol{j}$	Current density	
$\boldsymbol{k}$	Wavevector	3.1
$k(\omega)$	Optical extinction coefficient	6.15
l_e, l_i	Elastic, inelastic scattering length	9.11
L_i	Diffusion length	9.11
L	Lorenz number	6.14
$L_{\mu\nu}^{\alpha\beta}$	Onsager coefficient	7.7
m^*	Effective mass	
M	Magnetisation	
n, $n(\boldsymbol{r})$	Particle number density	
n_0	Average number density	
$n^{(1)}, n^{(2)}, \ldots$	Particle distribution function	2.3
n_1, n_{ij}	Coordination number	2.4
$n_e(\boldsymbol{r})$	Electron number density	3.2
$n(\omega)$	Refractive index	6.15
$n(\varepsilon)$, $N(E)$	Particle energy distribution	7.6
N	Number of particles in system	
p	Pressure	
$\boldsymbol{p}$	Momentum	
p_c	Percolation threshold	9.2
p_c	Critical pressure	
P	Probability	
Q	Configurational partition function	4.1
$\boldsymbol{Q}$	Wavevector	3.1
$\boldsymbol{r}$	Position vector	
r_s	Radius of spherical volume per electron	6.6
$\boldsymbol{R}$, $\boldsymbol{R}_i$	Position vector	
R	Resistance	
R_H	Hall coefficient	6.11
S	Order parameter	1.3
S	Entropy	
S_F	Fermi surface area	6.11
$S(\boldsymbol{Q})$	Structure factor	3.2
$S(\boldsymbol{Q}, \omega)$	Dynamic structure factor; scattering law	3.7 8.1

S_{NN}, S_{Nc}, S_{cc}, S_{ij}, $S_{ij}^{(AL)}$	Partial structure factors	3.5, 3.12
$S^{\rho}(Q)$	Resistivity structure factor	6.16, 12.4
t_i, T	Scattering amplitude matrix	7.2
T	Absolute temperature	
T_c	A critical temperature	
T_{CR}	Crystallisation temperature	12.1
T_G	Glass transition temperature	10.1
T_L	Liquidus temperature	
T_M	Melting point	
T_{RG}	Reduced glass transition temperature	12.1
U	Total energy	
U_{ps}, $u(\boldsymbol{r})$, $u(\boldsymbol{Q})$	Pseudopotential	6.3
v_i	Partial molar volume	3.12
$\boldsymbol{v}$	Velocity	
V	Volume	
V	Tight-binding transfer integral	9.8
V, v_i, $V(\boldsymbol{r})$	Potential energy operator, function	7.1
V_G	Gate voltage	11.7
W	Debye–Waller factor	12.4
z	A complex energy	7.1
z	Valency	6.6, 6.8
Z	Canonical partition function	
$\mathscr{Z}$	Grand canonical partition function	
α, $\alpha(T)$	Thermoelectric power	6.11
α, $\alpha(\omega)$	Optical, ultrasonic absorption coefficient	6.15, 10.7
α^{-1}	Localisation length	9.5
α_p	Thermal expansivity	
β	$(k_BT)^{-1}$	
γ_{el}	Electronic specific heat coefficient	12.3
Γ	One-component plasma parameter	6.10
Γ	Acoustic attenuation coefficient	8.4
ε; ε_{ij}	Energy; strain component	
$\varepsilon(\boldsymbol{Q})$	Dielectric screening function	6.4
$\varepsilon(\omega)$	Permittivity	6.15
ζ	Electrochemical potential	6.11
η	Viscosity	

η_{ij}, η_{ij}^0	Short-range order parameter	2.7
θ_D	Debye temperature	
κ_T	Isothermal compressibility	
λ	Thermal conductivity	6.11
Λ	Mean free path	6.11
μ	Absorption coefficient	3.12
μ, $\mu(E)$	Carrier mobility	11.9
μ_d	Drift mobility	11.9
μ_i	Chemical potential	6.9
μ_H	Hall mobility	9.4
ν	Frequency	
$\nu^{(1)}$, $\nu^{(2)}$, . . .	Density function	2.3
ξ	Thermoelectric coefficient	12.4
ρ, $\rho(T)$	Electrical resistivity	6.11
$\rho(r)$, $\rho_{ij}(r)$	Radial distribution function	2.3, 2.4
$\rho(\boldsymbol{r})$	Electric charge density	12.3
$\rho(E)$, $\rho(\boldsymbol{k}, E)$	Spectral operator, function	7.1
σ	Molecular diameter	4.3
σ, $\sigma(\omega)$	Electrical conductivity	6.11
σ_{ij}	Stress component	12.6
$\sigma_{\mu\nu}$	Conductivity component	7.7
Σ, $\Sigma(\boldsymbol{k}, E)$	Self-energy operator, function	7.3
τ	Relaxation or collision time	6.11, 9.11
φ, Φ	Potential energy function	4.1, 4.3
χ, $\chi(\boldsymbol{Q})$	Susceptibility, admittance	6.4, 7.7
$\psi(t)$	Velocity autocorrelation function	8.3
ω	Angular frequency	
Ω	Volume	7.1
$d\Omega$	Element of solid angle	

1

ORDER AND DISORDER: AN INTRODUCTORY DISCUSSION

It would be businesslike to open with clear definitions of order and disorder and then to proceed at once to introduce the physics of disordered matter. This is not so easy. The word 'order' in English, and even within science, has many meanings. Like other verbal borrowings by science, it carries a number of associations from everyday life. It has been argued that it is something of a waste of time to seek a definition of order: everyone knows more or less what is meant, precise definitions can always be given when it comes to detailed scientific problems and that is what really matters. This is somewhat reminiscent of the words of Robert Herrick in a theological context:

> God is above the sphere of our esteem
> And is the best known not defining Him.

There is a great deal to be said for this opinion and certainly it appears to be widely held in so far as there are many cogent writings on matters of order and disorder in physics which manage perfectly well without any discussion of the general concept. Nevertheless, while accepting that order might be defined in various ways using language from more than one discipline, it seems worth spending a little space trying to formulate at least one definition.

What is required is a definition of order that covers all that is needed in physics but does not exclude things in other spheres, such as nature or art, which common intuition would also regard as ordered. Disorder then follows as absence of, or a detraction from, order. This immediately raises the question: must order be either present or absent, or is it a matter of degree? It is necessary to have the concept of perfect, complete or ideal order and perfection is indeed either present or not; it is equally necessary to recognise that greater or smaller departures from perfection are conceivable and that numerical measures of the amount of order, called order parameters, will be indispensible.

1.1 Ordering rules

The first proposition is that a set of things is ordered or not in respect of a particular property that all the things possess. Since an object can have n

properties the set might be ordered in m properties at once, $m \leqslant n$; thus molecules might be ordered in position and in chemical nature but perhaps not in orientation. Let us consider a collection of things whose positions are defined by that of some specified point in each, say its centre of mass. If some prescription exists which allots to every member of the collection a unique placing with respect to all other members of the collection by the repeated application to one member after another of a rule, and if every member of the collection actually occupies its unique placing, the collection would be called *perfectly ordered in position*. The rule repeatedly applied will be called the *ordering rule*. Motifs in a straight frieze could be so ordered by prescribing that each was 50 cm to the right of its neighbour. Orientation of the motif, if this is significant, can be made the subject of another ordering rule. Similar ideas would apply to ordering in time—say of pulses of sound—or to events positioned in both space and time. A straight frieze immediately recalls a one-dimensional crystal lattice. However, a mural painter might decorate a wall by drawing an Archimedean spiral and placing identical sunburst symbols, identically oriented, every 50 cm along it. It would be difficult to find an analogue in physics, but hard to deny that it was a completely ordered system.

Repeated application is an important ingredient in the idea of the ordering rule. If a collection were finite, it would be possible to generate a unique arrangement according to an elaborate rule which specified exactly where to put every member of the collection. The result might be highly irregular and in no sense ordered. The essence of perfect ordering is that one can go from one member of the collection to another by repeated application of the same rule.

Position is defined by a vector quantity. When vector properties are in question their values can be represented as points in an appropriate space such as momentum space. If these points are subject to an ordering rule in that space, then by an extension of the preceding ideas it would be reasonable to describe the objects as, for example, 'ordered in momentum'. The momenta of electrons in a perfect dense electron gas subject to periodic boundary conditions on a cube at absolute zero form a simple cubic lattice in momentum space and could be regarded as ordered in this sense. It is conceivable for vectors, or other quantities with components, to be perfectly ordered in some components and not in others.

Suppose objects have scalar properties, say lights with brightnesses or atoms with atomic numbers. Under what circumstances would we say that lights are ordered in brightness? We would certainly say so if they were presented simultaneously along a straight line ranked in increasing brightness. However, is position along the line necessary to the ordering in brightness? If the lights had serial numbers 1, 2, 3, . . . we would say they were ordered in brightness if no light had a brightness greater than one of lower serial number. This correlates the serial number with the brightness and, in principle, any two scalar properties could be perfectly correlated without reference to positions in space or time. However, in physics, when the ordering of a scalar property is in question, position in space or time is usually also involved. Examples are the chemical nature (atomic number) in relation to position in an alloy; or the loudness in relation to timing in a sequence of sound pulses.

After this preliminary discussion we may now be in a position to formulate a definition of perfect order. A set of objects, $\{O_i\}$, each with properties $\{p_n\}$, will be perfectly ordered in respect of property p_g if there exist (i) a rule and (ii) at least one object O_h such that, starting with a knowledge of p_g for O_h, p_g for each of the other subjects taken one at a time can be specified uniquely relative to that of O_h by successive applications of the rule.

This definition calls for a number of comments. The word 'uniquely' implies that the rule has no probabilistic elements: it has to produce the same result every time it is used to order the set. The set may be finite or infinite in number and the objects may be physical such as atoms and mosaic tiles or abstract such as points or point events. The set of properties $\{p_n\}$ may, but is not obliged to, include spatiotemporal coordinates. In physics it may well include the orientation coordinates, atomic number, magnetic or electric dipole moments of atoms or molecules. While the definition covers linear friezes or perfect crystals as simple cases it allows also for ordering along an Archimedean spiral, for the brightness of ordered lights with respect to their serial numbers and many more complex examples. Readers may find it interesting to devise alternative definitions of order.

1.2 Objects in space

One of the simplest examples of order is that of the positions of identical objects. There is no necessity that the positions are *equally* spaced. The ordering rule might produce arrays like those of figure 1.1, and many others which are more complex. The essence of the ordering rule is simply that by using it a unique array can be constructed by going from one object by the prescribed rule to any other. Alternatively, the rule can be used to check whether any object is in or out of position. The rule must be precise and quantitative; we would say that the objects were ordered along a straight line if each one were, say, 1 mm to the left of its neighbour but not if the prescription were 'a little way to the left', or even, 'any distance to the left between 1 and 2 mm'. Such imprecise criteria would not allow a unique array to be arrived at. This point should perhaps be emphasised. It is quite possible to lay down rules for constructing an array of points which do not however, produce a *perfectly ordered* array. The two procedures suggested immediately above are cases in point. A rule only becomes an ordering rule when it generates a *unique* array by repeated application.

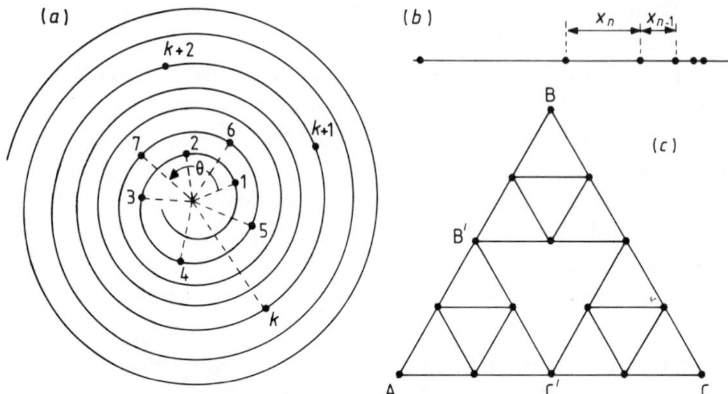

Figure 1.1 Ordered sets of points. (*a*) $\theta_k = (k-1)\theta$ for constant θ in a given spiral. (*b*) $x_n = 2x_{n-1}$ on a straight line. (*c*) The points shown are on a hexagonal Bravais lattice, but if the figure were extended by stacking three units ABC together in the same way that ABC is derived from three units AB′C′ (and so on recursively) there would still be an ordered set of points but not a Bravais lattice (see also §2.9).

If the objects are not all the same they can of course still be ordered by a position ordering rule. However, they may also be ordered in respect of whatever other properties they possess in common. If two or more properties are ordered at once we could call it a case of *multiple ordering*.

In physics a simple example of multiple ordering occurs in the crystalline binary alloy. Here there are two kinds of atom of which the equilibrium positions are perfectly ordered. However, the two different species may alternate regularly or may be randomised (see figure 1.2). Transition from the ordered to the chemically disordered state is known to metallurgists as an

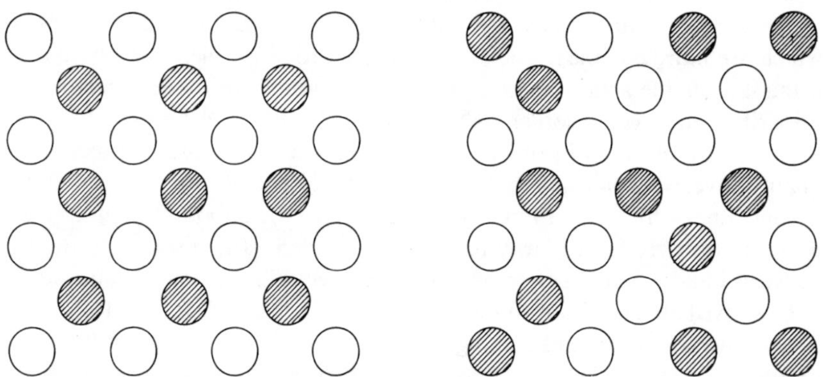

Figure 1.2 Ordered and disordered forms of a binary AB alloy shown schematically in 2-D.

order–disorder transformation and is discussed further in §1.11. Particles can vary in ways other than chemical nature, e.g., isotopic nature, orientation of magnetic moment, degree of ionisation, etc, and this leaves plenty of scope for multiple ordering criteria. Many examples may occur to the reader from physics and everyday life.

1.3 Order parameters

Insofar as a collection of things is not in conformity with an ordering rule, the collection is *disordered* with respect to that rule. However, given that it is not perfectly ordered, the kind and amount of disorder can show great variety. Clearly there is a difference in disorder between a collection of atoms of which 99.9 % conform to a position ordering rule (a crystal with defects) and another collection in which none do (a perfect gas). Moreover, where disorder in spatial position is concerned, ordering may be evident locally but not over long distances in which case the radial extent of local ordering is open to question. It will therefore be necessary later to discuss *modes* and *different degrees* of disorder.

Granted that there are degrees of disorder, it is not surprising that they are referred to in phrases like 'partially ordered' or 'having some detectable vestige of order'. Contexts usually make clear what is meant. However, what numerical measure is there for a 'vestige' or any other amount of order? The appropriate quantity, called the *order parameter*, has to be devised for each problem because it depends on the phenomenon under discussion. Order parameters frequently come into discussions of transitions between more and less disordered phases. Two examples will be given here. The first is the magnetisation, M, of a set of electron spins which may point 'up' or 'down' as in ferromagnetism. Saturation, M_{sat}, corresponds to complete order in alignment; randomly oriented spins give $M = 0$. The second example refers to the binary alloy mentioned in the previous section. Suppose c_A is the fraction of A-type atoms and r is the fraction of those that are in the positions they would occupy if perfect chemical order prevailed. Then the quantity $S = (r - c_A)/(1 - c_A)$ will be 1 for complete order and 0 when there are no more A's in correct positions than there would be by chance. M and S interpolate between the completely ordered and completely disordered extremes and are examples of order parameters. They give meaning to expressions like 'partially ordered'.

A rule which produced a partially ordered collection could be called a *partial ordering rule*. Rules with probabilistic elements could do this. They would not generate unique outcomes and each array produced would be partially ordered; only by chance would perfect order result.

Some order parameters are used in later chapters. Devising an order parameter may be quite a difficult or subtle matter.

1.4 Physical systems: ensembles

Hitherto we have been asking to what extent a particular collection is disordered, e.g., a particular array of atoms at specified places. In physics it is usually more pertinent to ask this question of a physical system in a given thermodynamic state. The implication is that the ensemble average of the order parameter must be calculated, there being no special significance attached to any one of the vast number of configurations that are compatible with the prescribed thermodynamic variables. Taking advantage of the equivalence in statistical mechanics of the ensemble average and the time average, we may say alternatively that the variables of the system, and therefore any order in them, fluctuate from moment to moment and we require the time average.

In the ensemble the positions, orientations, etc, whose order is in question have probability distributions and any set of values chosen at random from the distributions would be perfectly ordered only by an extremely remote statistical accident. Suppose a coordinate has a range of possible values each of which is equally probable. Suppose further that the existence of an object with one value of the coordinate has no influence on the value attached to any other object. An example would be the position coordinates of the molecules in a classical perfect gas of infinite volume: knowing where one molecule is gives no information at all about the positions of the others. Another example is the axial direction of molecules which in certain crystals can point with equal probability in any one of a number of directions, each molecule being independent of the others. Introducing correlations and constraints in the values of the variables will decrease the disorder or, if preferred, will introduce or increase the partial order. For example, giving the gas molecules finite diameters and attractive forces correlates their positions somewhat; increasing the number density of finite-sized molecules will do so more still. Similarly, coupling magnetic moments together or subjecting them to external aligning fields will decrease disorder in their orientation. A major statistical problem in each case is to devise an appropriate order parameter and see how it varies with strength of coupling, external influences, temperature and other things.

1.5 Structural order: crystalline and non-crystalline matter

Most of the preceding discussion refers to order in a somewhat general sense. In the physics discussed in this book the order or disorder most commonly in question is that of the position of atoms or molecules in space. We refer to this as *structural* order or disorder.

The fundamental order concept in the field is that of the Bravais lattice which is an array of points in space such that, if any one is taken as the origin for position vectors, $\boldsymbol{R}$, all the other points are given by

$$\boldsymbol{R} = n_1\boldsymbol{a}_1 + n_2\boldsymbol{a}_2 + n_3\boldsymbol{a}_3 \tag{1.1}$$

where $\boldsymbol{a}_1$, $\boldsymbol{a}_2$, and $\boldsymbol{a}_3$ are fixed primitive lattice vectors and the n's are the positive and negative integers.

Equation (1.1) is the ordering rule for the position of lattice points. Readers will be familiar with the facts that there are five different kinds of Bravais lattice in two dimensions and fourteen in three. The kinds differ from one another in the varieties of rotational and mirror symmetry they display. The parallelepiped formed by $\boldsymbol{a}_1$, $\boldsymbol{a}_2$ and $\boldsymbol{a}_3$ is the primitive unit cell.

If every lattice point has associated with it an identical group of atoms called a basis and every basis is identically placed and orientated with respect to its lattice point the resulting array of atoms is a perfect crystal. The ordering rule (1.1) then applies to every set of corresponding atoms in the bases. The elaboration and analysis of these concepts and the experimental determination of the structures of crystals are the main constituents of the elegant science of crystallography and for such subjects the reader is referred to crystallography texts.

It need hardly be emphasised that the perfect crystal is the acme of structurally ordered systems nor that it is a somewhat ideal concept. Real crystals have a certain degree of disorder because there are many kinds of structural imperfection some of which are always present. Provided however that the vast majority of atoms conform to the ordering rule it is common and natural to call the system a single crystal and to refer to matter composed of a single crystal, or a conglomeration of them, as *crystalline matter*.

Non-crystalline matter is all other matter and includes liquids, gases and some solids which in recognition of their non-crystallinity are called *amorphous solids*.

There are systems intermediate between crystals and liquids in which there is some order but not enough to produce crystalline rigidity. These are called *liquid crystals*. The order may be positional in one or two but not three dimensions or it may be in the axial alignment of rod-like molecules.

1.6 Crystal imperfections

More for completeness in this introductory chapter than because they are to be extensively discussed for their own sakes, we recall here some of the important structural imperfections in crystals. Since the reader will be familiar with them some well known ones are simply listed by name—vacancy and divacancy, interstitial and substitutional impurity, stacking fault, dislocation, F-centre. In general the prevailing crystalline order is necessary for the definition of departures from it.

Most of the properties studied in solid state physics which are exploited in technology result from the presence and mutual interactions of imperfections, as witness the colour of gems, the electrical properties of semiconductors, the ductility and other mechanical properties of metals.

With the exception of the chemical impurity atom the crystal imperfections are only marginally relevant to non-crystalline matter or, more precisely, if concepts like vacancy, dislocation and stacking fault are to be used for non-crystalline systems, their exact meaning and application ought to be defined with some care. However, a collection of impurity atoms in an otherwise perfect crystal may, because of its random distribution, constitute a disordered array of atoms embedded in a crystalline host and be a system of interest in its own right—an example is the array of donor atoms in a doped semiconductor. Defects in amorphous solids will be defined in Chapters 10 and 11.

1.7 Some varieties of disorder

We first distinguish between long- and short-range order denoted by LRO and SRO. As the names imply the distinction is in the spatial extent over which the ordering rule applies. If this distance is macroscopic or even over the width of a colloidal particle of 10^4 atoms the order is long range. If it applies only over a few atomic radii it is short range.

Since the surface atoms of a solid particle are under forces different from those in the interior, the interatomic spacings near the surface are not typical of the bulk. In general therefore we do not expect perfect lattice order to exist throughout a particle so small that a significant fraction of its atoms are in the surface—a particle of, say, a few hundred atoms. A few tens of atoms in the interior may however show SRO. A multiparticle aggregate of such small groups would thus have local SRO at best. Some people have suggested that liquids and amorphous solids have small regions with SRO and certainly chemical bonding requirements might impose some structural organisation on nearest neighbours. Whether this happens, and to what extent it is reasonably described as SRO, have proved difficult questions which will be taken up later in connection with particular systems. The concept of medium- or intermediate-range order is also assuming importance especially in amorphous solids. Medium-range order will probably turn out to be even more difficult to deal with theoretically than SRO or LRO (see, e.g., §10.5).

A macroscopic structure possessing crystalline LRO overall is subject to equation (1.1) and is therefore sometimes said to have translational order. Nevertheless it may still incorporate some disorder of various kinds which we will now list and exemplify in some important cases.

(i) Short-range disorder associated with localised defects such as a vacancy or an impurity atom near which the neighbouring host atoms are somewhat displaced.

(ii) Chemical disorder, sometimes called *compositional* disorder, as in the binary alloy example in §1.3.

(iii) Disorder in spin orientation as in a ferromagnet, ferrimagnet or antiferromagnet above its critical point.

(iv) Disorder in molecular orientation which sometimes occurs when, as in CCl_4, the molecules can rotate into alternative directions while remaining at their sites.

In the last three cases perfect order will prevail at $T = 0$ and disorder above some critical temperature, T_c. How the disorder enters at $T > 0$ and increases towards T_c is a major question for such systems; it is the problem of the temperature dependence of the order parameter.

The preceding four cases are examples of *cellular disorder*. This is any disorder describable by reference to the properties of particles situated at long-range-ordered sites. The properties include intrinsic ones like spin direction or chemical nature; the presence or absence as with point defects in moderate or low concentrations; and vibratory displacement as in thermal motion. To enlarge on the latter somewhat: in samples at finite temperature the atoms or molecules will have one or more kinds of motion; translatory, vibratory or rotatory. Looking at the motion classically, the ith molecule has a position vector $\boldsymbol{R}_i(t)$ and the set of positions, $\{\boldsymbol{R}_i(t)\}$, is highly unlikely to have LRO even in a single crystal or, in other words, thermal motion is an agent of disorder. The set of averaged positions, $\{\langle \boldsymbol{R}_i(t)\rangle_t\}$, where $\langle\rangle_t$ denotes a time average, will have LRO in the case of a single crystal in which the atoms are oscillating about lattice positions but even the time-averaged positions will be disordered in an amorphous solid. In fluid systems translatory motion is a major aspect and it is sensible to consider the distribution of relative positions of molecules; a terminology for this will be introduced later (§2.3).

A macroscopic sample, such as a fluid or glass, to which equation (1.1) has no application is said to have *topological* disorder. Such a system, if it is a binary mixture or alloy of species A and B, will also have chemical disorder in the long range. If, however, A atoms attract B atoms more than other A atoms, there may be preferred local formations such as an A atom surrounded by four B atoms and this phenomenon could be called *partial chemical ordering* in the short-range or chemical short-range order (CSRO). Corresponding remarks apply to repulsion between A and B atoms or to mixtures of more than two components. Topologically disordered structures may have ordered spins if the latter are all aligned—as in a saturated amorphous ferromagnet.

1.8 Disorder and entropy: quenched materials

In statistical mechanics entropy is frequently described as a measure of disorder and the reader will be familiar with the common instances of this such as the way in which the increase in disorder when a crystal melts is accompanied by a gain in entropy given by L/T_m where L and T_m are the latent heat and melting temperature respectively.

Recognising that this connection is a matter of some depth, hardly susceptible of summary in a few sentences, we can nevertheless give it some formal basis by quoting Boltzmann's equation for entropy, S, viz,

$$S = k_B \ln W \tag{1.2}$$

where the connection with disorder can be found in the meaning of W.

W is the number of quantum states of a macroscopic system compatible with the thermodynamic variables prescribed for the system. Alternatively expressed, it is the number of microstates compatible with the macrostate. In the language of the microcanonical ensemble theory, W is the number of quantum states compatible with the prescribed volume (V), total particle number (N) and total energy (E) of the system under discussion. In canonical ensemble theory W is the number of quantum states contributing significant weight to the computed average energy of a system with prescribed V, N and T. In the latter case, for the extremely large values of N required for macroscopic samples, all the quantum states in W have energies so close to the average energy that the differences can be ignored.

An increase in entropy therefore corresponds to the increase in the number of microstates compatible with the final thermodynamic state over that for the initial state. By a basic postulate of statistical mechanics these quantum states are equiprobable so there is a greater uncertainty after the entropy increase as to which of the possible quantum states the system is in. This increase in uncertainty connects with increases in disorder and requires some discussion with the help of examples.

The first example is the introduction of local disorder in the form of vacancies by heating a crystal. The advent of more vacancies (for a given N and V) allows more alternative equiprobable arrangements of the vacancies among the particles which adds to W and therefore contributes entropy. This contribution is the *configurational* entropy, so called because of its origin in the spatial arrangements. From this example and others, such as the configurational entropy of spins partially aligned by a magnetic field, we see that the increase in disorder results in an increase in entropy.

Another example is the increase of $R \ln V_2/V_1$ per mole in the entropy of a perfect Maxwell–Boltzmann gas of point particles expanded at constant temperature from V_1 to V_2. The formula follows from thermodynamics; statistically W increases because the number of allowed states of particles in a box rises with V. However, it is not immediately apparent how or whether to relate the rise in entropy to an increase in disorder because one might think the disorder in the gas at V_1 was already as great as it could be. So far as particle momentum is concerned this is true; in any instantaneous configuration, there is no rule leading from knowledge of the momentum of one particle to anything about the momentum of any other particle. However, the positions are also part of the description of the configuration and if we know the position of one particle we know that another is not further away than the walls of the container. This constraint represents a vestige of order and the density could be regarded as an order parameter, decreasing to zero as the gas goes from a finite volume to a state of complete disorder in particle position at infinite volume.

As a final example consider the mixing of two fluids and let us assume first that this is done without change in total volume and with no interaction. In that case the greater volume throughout which any one molecule can now diffuse increases the uncertainty in its position. The corresponding entropy increase over the entropy of the separated fluids is called the entropy of mixing, ΔS, and arises when each microstate of one fluid is multiplied by the total number of those of the other to compute W for the mixture. W is greater than in the initial state and ΔS is therefore positive. However, if there were chemical interactions leading to clustering or compound formation partial chemical ordering would occur, possibly accompanied by a decrease in total volume. Both the gain in order and the shrinkage in volume through such causes work against the effect of shared volume and may even make ΔS negative.

Uncertainity and therefore entropy is decreased by additional information about the system—which makes the connection between statistical thermodynamics and information theory. This interesting connection can be pursued in Brillouin (1962) and Levine and Tribus (1978). Disorder and entropy, though closely connected, are qualitatively different kinds of thing. This can be seen from the fact that entropy is essentially statistical—state probabilities in an ensemble of systems are part of its definition and it does not make sense to ask for the entropy of a single configuration. The amount of disorder in a single configuration is, on the other hand, quite conceivable. In summary one could say that insofar as particular configurations of the system can show disorder the ensemble of systems allows a calculable finite entropy; the latter is a measure of the number of configurations permitted by the extent of the disorder.

Since fluids can be in thermodynamic equilibrium, the calculation of their entropy and other thermodynamic properties, though technically very difficult, is based on established procedures. Other disordered matter is not in thermodynamic equilibrium because it has been quenched. Quenching is cooling so rapid that the atoms do not have time to adopt configurations appropriate to equilibrium at the final temperature. Ordinary and metallic glasses are examples of quenched systems, another example would be an alloy containing a random array of atoms with magnetic moments. Quenching does not inhibit thermal motion; the atoms will vibrate about their sites and realignments of magnetic moments occur. What is precluded by lack of sufficient thermal excitation is the rearrangement of the atomic positions.

This may not be prevented in an absolute sense, simply reduced drastically in speed; it is well known that glass does not recrystallise in a great many years.

It is not obvious how to apply equilibrium thermodynamics or statistical mechanics to quenched systems, and this is a profound problem. One approach to this is to divide the variables needed to describe the microscopic state into two sets, $\{X\}$ and $\{Y\}$. $\{X\}$ are the quenched variables, e.g., those that specify atomic sites but do not participate in the thermal motion. $\{Y\}$ are needed to describe the thermal motion, e.g., the vibrations and alignments. $p(\{X\})$ is the probability that the $\{X\}$ have certain values and is supposed to be known *a priori*—it describes the system and depends on the method of preparation. For a given $\{X\}$, a thermodynamic quantity, say the free energy, can be calculated using the variables $\{Y\}$. The free energy is then averaged over the distribution, $p(\{X\})$, of quenched variables. This method is that of Brout (1959). Others have tried to devise ways of constructing a fictitious system whose thermodynamic properties obtained by standard procedures would be those of the real quenched system (see, e.g., Huber (1981)). To both of these procedures it may be objected that dividing the variables into $\{X\}$ and $\{Y\}$ may be difficult or arbitrary. Presumably as the temperature rises, quenched variables gradually become active—a process known as annealing—and $p(\{X\})$ is changed. These matters will be referred to again in connection with particular materials, notably glasses, in Chapter 10 (see also Jäckle (1981)).

1.9 Examples of structurally disordered materials

In this section some of the disordered systems will be introduced briefly with remarks about their occurrence and with typical examples of the different categories.

In principle the subject includes the whole class of liquids. The fundamental problem in the theory of liquids, as in all condensed matter, is to understand the macroscopic properties and the structure in terms of the interparticle interactions. Since the latter fall into various categories, different aspects assume importance depending on the system under discussion. Some important cases are:

(i) liquid argon, typifying the simplest liquids characterised by van der Waals' molecular forces;

(ii) liquid sodium chloride as a representative of molten salts in which electrostatic ionic forces dominate;

(iii) liquid sodium to exemplify metallic liquids with a Fermi gas of conduction electrons;

(iv) liquid semiconductors which are not very easy to typify but for which selenium serves as an example.

In the cases of metallic and semiconducting elements and alloys interest was first focused on their electronic properties and on how and why these were different in the liquid and crystalline phases. Only in the 1970s was substantial progress made in understanding the thermodynamic properties. With liquid argon on the other hand the equation of state and thermodynamic properties have always been centres of interest and in all classes of liquid the problem of structure still remains incompletely solved. The more complex cases of liquid helium, organic molecular liquids, and polymers must be regarded as outside our scope except for occasional references.

Liquids can be maintained in thermodynamic equilibrium. Amorphous solids on the contrary have to be manufactured by special methods which prevent the atoms from falling into the ordered arrangements characteristic of the ground or equilibrium states. These methods amount to very fast cooling and examples are the condensation of metal vapour on to very cold substrates and rapid quenching from the melt. Among amorphous solids recently studied are: certain metallic elements, metallic alloys especially magnetic ones of technological potential such as Fe–Ni–P–B, germanium, silicon and other semiconductors which, apart from their scientific interest, have possibilities for device design, and glasses.

Glasses form a subset of amorphous solids. Not all authorities would take this view and would instead regard 'amorphous solid', 'glass', 'non-crystalline solid' as synonyms. Others following a more traditional concept of glass would use the word to signify an amorphous solid obtained by supercooling from the liquid state continuously, i.e., without the first-order phase transition of crystallisation. A convention is then required for distinguishing the liquid from the solid. This is a matter of viscosity and the dividing line is 10^{12} Pa s (10^{13} P), a viscosity so high that such a substance would certainly be called a solid in everyday life. Near this viscosity a glass transition occurs (see Chapter 10). A *vitreous solid* is another name for a glass in this sense and well known examples are SiO_2 (vitreous silica), B_2O_3 and *chalcogenide glasses* which are those, like As_2S_3 and As_3Se, in which the chalcogenide elements S, Se and Te play a vital role. Ordinary domestic glass is a mixture of SiO_2, Na_2O and B_2O_3.

Yet another way of specifying a glass is through the microscopic structure; sometimes a glass is defined as an amorphous solid in which there is sufficient flexibility in the lengths and angles of the chemical bonds between neighbours that the constraints imposed by chemical bonding requirements can be satisfied without LRO being a necessary consequence. Whether the macroscopic and microscopic definitions select out the same systems is then an empirical question to which at the time of writing the answer seems not entirely clear though it is probable that many systems that satisfy one definition satisfy the other also. Amorphous metallic elements, however, are not glasses by either definition. In any detailed discussion the context ought to make clear what aspects of the definitions really matter to the argument.

1.10 Quasicrystals

Figure 1.1(*a*) is a simple example of an ordered array of points on a spiral. Spiral tiling of a plane is also conceivable as shown in figure 1.3. There is an infinity of such possibilities making ordered systems without translational periodicity. Readers interested in the mathematical or aesthetic properties of tilings, whether periodic or not, will probably be aware of the famous designs of M C Escher, of an illustrated discussion by Gardner (1977) and perhaps of the underlying theorems comprehensively expounded by Grünbaum and Shephard (1986).

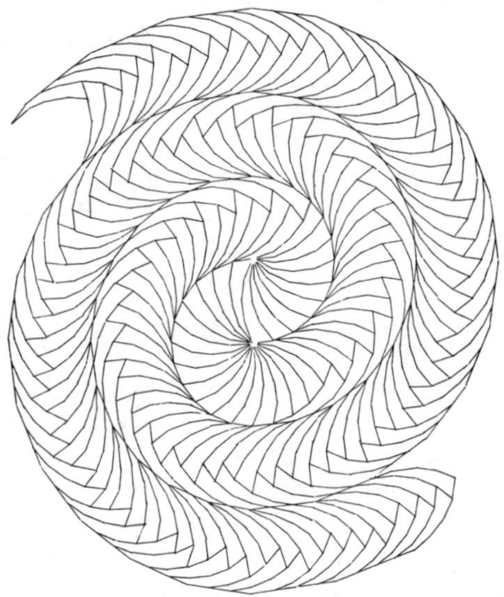

Figure 1.3 One of the numerous spiral tilings discussed by Grünbaum and Shephard (1986).

Of all the remarkable tilings in two dimensions some named after Penrose are of special interest (Penrose 1974). They consist of two distinct unit cells which jointly fill all space in spite of being present in numbers which have an irrational ratio and of having areas in the same ratio. In a frequently described example the two tiles are the rhombuses of figure 1.4(*a*) and one pattern they make is figure 1.5. This is generated by fitting the two tiles together so that both single- and double-arrowed sides always match. The ratio of acute to obtuse tiles is $\tau \equiv (1+\sqrt{5})/2$ in both area and frequency. Another pair of tiles which will do this is the 'dart' and 'kite' of figure 1.4(*b*) (see Gardner (1977) and Grünbaum and Shephard (1986) for beautiful illustrations). The appearance in the Penrose pattern of local pentagonal symmetry is clear. By starting at a vertex where either five kites or five darts meet and by building outwards to infinity two patterns with perfect pentagonal symmetry can be made (Gardner 1977).

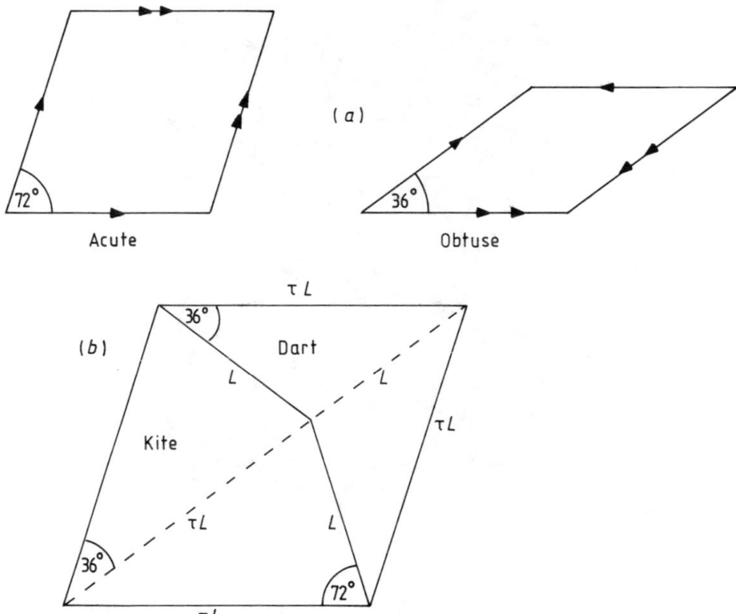

Figure 1.4 (*a*) The two tiles leading to figure 1.5; all sides are the same length. (*b*) The kite and dart tiles; $\tau = (1 + \sqrt{5})/2$.

The arrays of vertices do not form crystal lattices. The reader may find it interesting to consider whether any property of the tilings is ordered in the sense of §1.1 or, if not, whether it would be according to some more general definition of order. There is certainly not a unique Penrose tiling; it has been proved that their number is uncountable and includes an infinite number that have bilateral symmetry. The angles in the rhombuses are all integral multiples of 36°, or 1/10 of a revolution, consequently all their sides are parallel to one or other side of a regular decagon and the pattern preserves this orientational consistency over any distance—a property called long-range orientational order. Gardner (1977) remarks: 'most patterns . . . are a mystifying mixture of order and unexpected deviations from order. As the patterns expand they seem to be always striving to repeat themselves but never quite managing it'. The remarkable thing about Penrose tiles is not that they can tile aperiodically (even a square can do that) but that they cannot tile otherwise.

Penrose tilings in three dimensions can also be made by filling space with two kinds of polyhedron (Mackay 1982, Levine and Steinhardt 1984). As in two dimensions the vertices do not make a crystal lattice but there is long-range orientational order of which the symmetry is that of a regular icosahedron with six fivefold, ten threefold and fifteen twofold axes. Arrays of points like this have come to be called quasicrystalline and Levine and

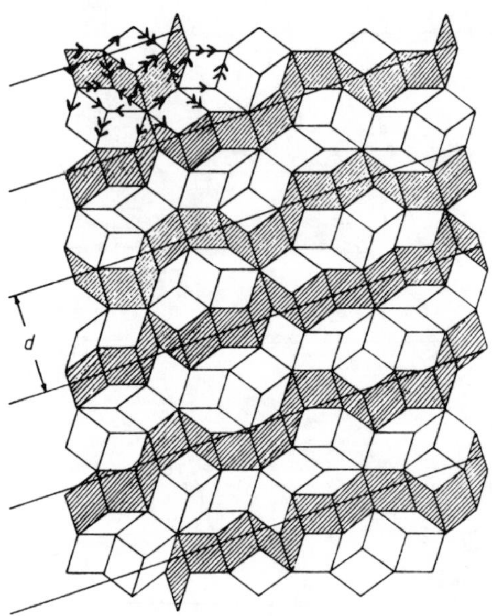

Figure 1.5 A Penrose tiling in 2-D. In the top left-hand corner some arrows are drawn to show the operation of the fitting rules. Some rows of tiles are shaded to show how the pattern implies a grid of equally spaced lines leading to Bragg reflections. The common sides of the shaded rhombuses are parallel (after Nelson and Halperin 1985).

Steinhardt (1984) attribute the following essential properties to them: there is a minimum distance between any two points; every point lies within some distance $R > 0$ of some other point; it is possible to remove a subset of points such that those left form another quasicrystalline array with nearest-neighbour (NN) distance multiplied by a constant factor (the self-similarity property also possessed by Bravais lattices); there is perfect long-range orientational order; the spacings of the points are quasiperiodic.

The simplest illustration of the meaning of quasiperiodic is the generation of a sequence of intervals along a straight line by the following procedure. Its generalisation to two and three dimensions (2-D, 3-D), and to $k > 2$ incommensurate intervals, is given by Levine and Steinhardt (1984). Let L and S stand for long and short intervals such that $L = \tau S$. Start with a two-interval sequence of S followed by L, written SL; create a second-generation sequence by substituting for S and L according to the rule

$$S \rightarrow L \qquad L \rightarrow SL.$$

This creates LSL from SL. Using the same substitution rule the next three generations are $SLLSL$; $LSLSLLSL$; $SLLSLLSLSLLSL$. The length of the nth generation sequence is $\tau^{n+1}S$. Obviously such a sequence of intervals can be infinitely extended; less obvious are some of the remarkable properties of the infinite sequence such as that any finite sequence can be found infinitely often and that, in spite of the formation of the sequences by repeated application of the same rule, there is no translational periodicity. There exist subtle and highly interesting connections between such sequences and Penrose tilings (Grünbaum and Shephard 1986). In 2-D (and 3-D?) *any* patch of Penrose tiles can be found infinitely often in the same tiling and, indeed, in any Penrose tiling using the same tiles.

In 1-D, 2-D, or 3-D each of the two kinds of tile could contain its own characteristic array of atoms making the pattern into a macroscopic solid but it would not be a crystal. Would it nevertheless give rise to sharp Bragg peaks in its diffraction pattern? As the shaded lines in figure 1.5 suggest it does so, and it was the detection of diffraction patterns clearly displaying the fivefold, threefold and twofold axes belonging to icosahedral symmetry that constituted the discovery of quasicrystals in 1984 (Shechtman *et al* 1984). The material was a rapidly quenched alloy of 14 at. % Mn in Al. It is metastable and remains quasicrystalline even at about 300 °C but crystallises quite rapidly at 400 °C. The material is neither a crystal nor a glass and has opened a new chapter in structure studies. Other quasicrystals are being discovered (Poon *et al* 1985) and the reader's attention has been drawn to a field which at the time of writing is new and under intensive study. New materials and new theorems in mathematics are very likely to emerge.

Long-range orientational order in a quasicrystal suggests that before the system solidified it might have had short-range orientational order. Local icosahedral symmetry has been suspected in liquids for at least thirty years and the discovery of quasicrystals, as well as the advent of computer simulations (Chapter 4), are stimulating more study of it (Frank 1952, Steinhardt *et al* 1983, Sachdev and Nelson 1985, see also §2.1). Another possibility is a solid with long-range orientational order but without perfect quasiperiodicity, e.g., a random packing of oriented icosahedral clusters. Quite sharp but not true Bragg peaks are possible with such systems (Stephens and Goldman 1986).

Whether the properties of a quasicrystal are in general closer to those of the crystal, or of the glass, of identical composition is an interesting question. In some cases the vibrational and electronic densities of states, and some magnetic resonance features, make quasicrystals appear closer to glasses than to crystals but the question remains open.

1.11 Order–disorder transformation. Ising model

Some materials, such as amorphous Si, are created with disorder; others achieve disorder by undergoing a thermodynamic phase transition. The reader will be familiar with the distinction between first- and higher-order phase transitions. Despite its very great interest it will not be attempted in this book to deal with phase transition theory. However, at this point an elementary account of order–disorder transformations in a binary alloy will be given in terms of the Ising model since this introduces a number of ideas that will recur. The phenomenon to be discussed is the gradual randomisation of the positions of A and B atoms in an alloy as the temperature is raised from absolute zero (figure 1.2).

It is of the essence that interparticle interactions are invoked. This immediately causes substantial difficulties in condensed matter theory and it is necessary to choose simple interactions to avoid insuperable problems. Preferably this should lead to a rigorous partition function and if several order–disorder phenomena could be handled with the same model then so much the better. A model that, simultaneously, is interacting, simple in concept, open to rigorous treatment and versatile is the Ising model. It dates from 1925 but reference to appropriate journals will show that it still remains a source of challenging problems. The Ising model is set up as follows.

We consider a lattice of sites $\{i\}$ and we attribute to the particle at i a variable σ_i which has two possible values $+1$ and -1. For the alloy problem of figure 1.2, $\sigma_i = +1$ or -1 means an A, or B, atom at site i, respectively. The distribution of values of σ_i describes the configuration of the alloy. We may say at once that $\sigma_i = +1$ or -1 could equally well indicate electron spin up and down in which case we should be discussing spin alignment or ferromagnetism. A third possibility is that σ denotes the presence or absence of a single kind of atom so that different concentrations of $+1$ and -1 values imply different number densities. Let V_{11}, V_{22} be the nearest-neighbour (NN) interaction energies when both neighbours have $\sigma = +1$ or -1 respectively; $V_{12} = V_{21}$ is the energy when neighbouring σ are unlike. Longer-range interactions are excluded. Multiple occupation of sites is forbidden. Fowler and Guggenheim called systems like this regular assemblies; they look simple to handle but are not (Domb 1960, Stanley 1971).

To proceed further we need the total energy in terms of the V in order to form a partition function. We must also specify the V for the problem in hand and devise an order parameter. For these purposes we define the following numbers: z = number of nearest neighbouring sites; N_{11}, N_{22} = number of NN pairs with $\sigma = 1$ or $\sigma = -1$ respectively; N_{12} = number of NN pairs with one of each σ; N_1, N_2 = number of sites with $\sigma = 1$, $\sigma = -1$ respectively; $N = N_1 + N_2$ = total number of sites. These numbers are not independent and a little reflection shows that

$$zN_1 = 2N_{11} + N_{12} \tag{1.3a}$$

$$zN_2 = 2N_{22} + N_{12}. \tag{1.3b}$$

Suppose N and z are specified and that we choose the values of N_1 and N_{12}. The other numbers then follow from equation (1.3) as

$$N_{11} = \tfrac{1}{2}(zN_1 - N_{12}) \tag{1.4a}$$

$$N_{22} = \tfrac{1}{2}(zN_2 - N_{12}) = \tfrac{1}{2}z[(N - N_1) - N_{12}] \tag{1.4b}$$

$$N_2 = N - N_1. \tag{1.4c}$$

The total energy is

$$E = N_{11}V_1 + N_{22}V_{22} + N_{12}V_{12}$$

and substituting from equation (1.4) we find

$$E(N, N_1, N_{12}) = \tfrac{1}{2}zNV_{22} + \tfrac{1}{2}zN_1(V_{11} - V_{22}) + N_{12}[V_{12} - \tfrac{1}{2}(V_{11} + V_{22})]. \quad (1.5)$$

Many different alloy configurations, say $g(N, N_1, N_{12})$ of them, would be compatible with the same values of N, N_1, N_{12} so we may write the partition function

$$Z(T, N) = \sum_{N_1, N_{12}} g(N, N_1, N_{12}) \exp[-\beta E(N, N_1, N_{12})]$$

where $\beta \equiv (k_B T)^{-1}$.

The evaluation of $g(N, N_1, N_{12})$ is the first of numerous major problems in the statistical mechanics of disordered matter which will arise in ensuing chapters. Although it emerges from a simple model containing only cellular and not topological disorder the combinatorial problem is formidable. It is in fact a celebrated problem of which the rigorous solution in two dimensions by Onsager in 1944 was a landmark in phase transition theory. Although other methods of handling it have been derived since, it remains too complex to solve here. The corresponding solution in 3-D is still sought and even in 1-D the matter is not trivial (Domb 1960, Stanley 1971). The approximation we shall shortly use is typical of the expedients adopted in many theories in disordered matter—it will replace a distribution by its average. This is sometimes unavoidable, often reasonable, but sometimes dangerous depending on the problem in hand, the shape of the distribution and the definition of the average.

The appropriate order parameter for the alloy problem has already been given in §1.3:

$$0 \leqslant S \equiv \frac{r - c_A}{1 - c_A} \leqslant 1 \quad (1.6)$$

where $c_A \equiv N_1/N$; r = fraction of A atoms on the sites they would occupy if LRO prevailed. It is convenient to call the latter α-sites; β-sites are those that are 'wrong' for A atoms and 'right' for B atoms. It is common to relable V_{11}, V_{22} and V_{12} as $-V_{AA}$, $-V_{BB}$ and $-V_{AB}$ respectively and they are then the interatomic potentials. $V \equiv V_{AB} - \frac{1}{2}(V_{AA} + V_{BB})$ is called the ordering potential and large V favours CSRO, the ordered state being the stable one at low temperature. (The analogous problem arises later with liquid alloys—see §11.11.) We now define the partition function $Z(T, S)$ for a given S

$$Z(T, S) = \sum_{N_{12}} g_S(N, N_1, N_{12}) \exp[-\beta E_S(N, N_1, N_{12})].$$

This involves the sum over all configurations with a given S and a specified concentration which is defined by N_1. The Helmholtz free energy is

$$F(S) = -k_B T \ln Z(T, S)$$

and if this is minimised with respect to S there results an expression $S = S(T)$ which would show how the order depends on T for a given c_A. This calculation is impeded by the problem of g_S so we introduce a simplification.

The configurations for a given S will have a variety of E_S. The approximation is to put

$$E_S \simeq \langle E \rangle = -(\langle N_{11} \rangle V_{AA} + \langle N_{22} \rangle V_{BB} + \langle N_{12} \rangle V_{AB})$$

where $\langle\rangle$ denotes an average over the configurations. To find $\langle N_1 \rangle$ etc, we specify c_A by choosing an alloy of formula AB such as CuZn. Then

$$S = \frac{r - \frac{1}{2}}{1 - \frac{1}{2}}$$

$$r = \tfrac{1}{2}(1 + S) \qquad \text{and} \qquad 1 - r = \tfrac{1}{2}(1 - S).$$

There will be $Nr/2$ or $N(1 + S)/4$ A atoms on α-sites and each of them has z nearest neighbours on β-sites. Of these z sites, $z(1 - S)/2$ will have A (i.e. 'wrong') atoms. Thus the average number of A–A pairs is

$$\langle N_{11} \rangle = \tfrac{1}{4}N(1 + S) \times \tfrac{1}{2}z(1 - S) = \tfrac{1}{8}Nz(1 - S^2).$$

Similarly

$$\langle N_{22} \rangle = \tfrac{1}{8}Nz(1 - S^2) \qquad \langle N_{12} \rangle = \tfrac{1}{4}Nz(1 + S^2).$$

Substituting in the expression for $\langle E \rangle$ we find

$$\langle E \rangle = E_0 - \tfrac{1}{4}NzS^2V$$

where E_0 is the value of E for $S = 0$. $Z(T, S)$ now becomes

$$Z(T, S) = g(S) \exp(-\beta \langle E \rangle)$$

whence

$$F(S) = -k_B T \ln g(S) + E_0 - \tfrac{1}{4}NzS^2V.$$

We now meet the problem of $g(S)$ in a much simplified form. Using table 1.1, $g(S)$ is the number of ways of putting the tabulated numbers of atoms

Table 1.1

Atoms \ Sites	On α-sites	On β-sites
Number of A atoms	$\frac{1}{4}N(1 + S)$	$\frac{1}{4}N(1 - S)$
Number of B atoms	$\frac{1}{4}N(1 - S)$	$\frac{1}{4}N(1 + S)$

on the $\frac{1}{2}N$ α- and $\frac{1}{2}N$ β-sites and

$$g(S) = \frac{(N/2)!}{[N(1+S)/4]![N(1-S)/4]!} \times \frac{(N/2)!}{[N(1-S)/4]![N(1+S)/4]!}.$$

Then, using Stirling's theorem,

$$\ln g(S) \simeq -\tfrac{1}{2}N[(1+S)\ln(1+S) + (1-S)\ln(1-S) - 2\ln 2].$$

Substituting this into $F(S)$ and putting $\mathrm{d}F/\mathrm{d}S = 0$ we find the minimum F when

$$S = \tanh\frac{zVS}{2k_\mathrm{B}T}. \tag{1.7}$$

The solution of this is in figure 1.6 where the value of T for the limiting tangent is

$$T_\mathrm{c} = \frac{zV}{2k_\mathrm{B}}.$$

The reader will recognise the identity of this with the Weiss or mean-field theory of ferromagnetism due to electron spins. In that case T_c is the Curie temperature and $V \to 2J$ which stands for the exchange energy difference between antiparallel and parallel neighbouring spins, $J > 0$. If $J < 0$ and the lattice consists of two equal interpenetrating sublattices, the Ising model leads to antiferromagnetism with a maximum value of N_{12} in the ground state.

The main purpose of this section was to illustrate an approach to the calculation of the degree of disorder in a particular phenomenon. The transition from order to disorder occurs over a range $0 < T < T_\mathrm{c}$ of temperature

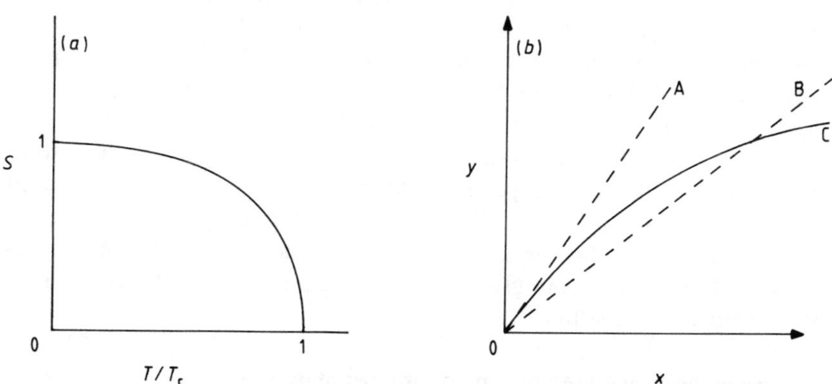

Figure 1.6 Schematic solution of equation (1.7). In (*b*) the lines are: A, $y = 2k_\mathrm{B}T_\mathrm{c}x/zV$; B, $y = 2k_\mathrm{B}Tx/zV$; C, $y = \tanh x$.

and is of second order. In some alloys such as FCC A_3B the transition is first order. There are other transitions, e.g., the glass transition (Chapter 10), where the very nature of the transition is in question and it is not obvious what to take as an order parameter.

Before leaving the Ising model, its richness in possibilities should be emphasised. In the case that $\sigma = \pm 1$ means the presence or absence of an atom, the Ising model is representing the number density in atoms per unit cell, N_1/N, as a function of T or, in other words, the equation of state of a fluid in which V_{11} is an attractive potential and $V_{12} = 0 = V_{22}$. It is a special model of a fluid in which the topological disorder is represented by the irregular distribution of the N_1 atoms on the occupied sites; it is called a lattice gas. The order parameter corresponding to S in the alloy or to the magnetic moment per particle, $\mu_m(N_2 - N_1)/N$, in the ferromagnet, is the number density N_1/N. It turns out that in the lattice gas the partition function corresponding to the canonical one used above has to be the grand canonical, and the quantity equivalent to the Helmholtz free energy per particle is the fluid pressure. These parallels and the general theory of the lattice gas were elucidated by Lee and Yang (1952). Even in the approximate treatment given above the partition function leads to critical points and accompanying specific heat anomalies (peaks), in qualitative agreement with experiment (Domb 1960, Stanley 1971).

The Ising model can be exploited further by including next-nearest-neighbour (NNN) interactions. Suppose in the magnetic case the NN interaction, $J_1 > 0$, would lead to a ferromagnetic ground state but the NNN interaction, $J_2 < 0$, favours antiferromagnetism. When both are present in the 2-D square lattice, ferromagnetism exists in the ground state down to $J_2/J_1 = -\frac{1}{2}$ and antiferromagnetism beyond this. However, further phenomena can be explored if the atoms carrying magnetic moments are supposed to occupy randomly only a fraction of the sites making a diluted magnet. Then a variety of magnetic behaviours are possible depending jointly on the concentration of magnetic atoms and the value of J_2/J_1. Elaborations of the Ising model like this have been used to discuss spin glasses. These are magnets in which randomness in J_1 and J_2 interactions due to dilution cause the magnetic moments to freeze into random directions at low temperatures. This leads to characteristic magnetic behaviours such as peaks in the susceptibility of which the shape depends on the time of observation. A few per cent of Mn in crystalline Cu or of Fe in Au produce spin glasses, and concentrated amorphous alloys with topological disorder can show similar phenomena. The study of spin glasses has become a major and difficult branch of magnetism (Chowdhury 1986).

1.12 Some problems presented by disordered matter

In this section we introduce a selection of the problems which make disordered matter a difficult subject to study.

In the first place it is not easy to find a rational mathematical way of describing disordered structures, especially topologically disordered ones. When the particles are jumbled like marbles in a bag, there is no condensed summary, like the unit cell in a crystal, that contains all the information. Either all the coordinates of all the particles are needed or some statistical extracts of this such as the distribution of numbers of nearest neighbours. Even the definition of nearest neighbour requires some thought.

Secondly, in fluids, the molecular motion continually changes the configuration of the particles and experiments must be, and theories should be, concerned with the time or configuration averages of properties. If, say, the density of states of conduction electrons in a liquid metal is required, it should be calculated for a particular configuration and averaged over all configurations. Taking these averages properly has proved to be quite a subtle task in some cases. It is part of the problem of solving Schrödinger's equation when the potential energy term varies irregularly from point to point. As we shall see in later chapters, what is useful here is the statistical information referred to in the previous paragraph, such as the distribution of interparticle distances.

Here a third problem arises because it is difficult to obtain these statistical data for real systems. Diffraction experiments supply knowledge of interparticle distances but not of spatial arrangements. For example, the shapes and sizes of triangles formed by particles three at a time is virtually unknown. When it is needed it has to be calculated approximately from the interparticle distance distribution and these approximations are another stumbling block.

On the experimental side there is the problem of sample characterisation. Some important disordered systems, e.g., amorphous semiconductors, are not thermodynamic equilibrium systems, i.e., they are not uniquely defined by their composition, temperature and pressure. Their history and method of preparation can play a considerable part in determining their properties and in generalising about them there is a problem of comparing like with like.

The preceding problems are sufficiently serious to have made progress in the field quite slow. Simplifying principles, such as crystalline order leading to Bloch's theorem in solid state physics or almost perfect randomness leading to the kinetic theory of gases, are not available for disordered systems generally and it is commonplace to remark how relatively backward the theory of liquids is compared with that of crystals and gases. Nevertheless, progress has been made and the object of the ensuing chapters is not to show how the problems have been solved—because in many cases they have not been—but to illustrate some of the ideas that anyone beginning in the field would have to acquire. Theoretical discussion of many of the subjects can be pursued in Ziman (1979), Zallen (1983), Elliott (1984), Careri (1984), Parsonage and Stavely (1978), Mott and Davies (1979), and Yonezawa and Ninomiya (1983) which cover a wide selection of topics from the various authors' different points of view.

References

Brillouin L 1962 *Science and Information Theory* (New York: Academic)

Brout R 1959 *Phys. Rev.* **115** 824

Careri G 1984 *Order and Disorder in Nature* (Benjamin Cummings)

Chowdhury D 1986 *Spin Glasses and other Frustrated Systems* (New York: Wiley)

Domb C 1960 *Adv. Phys.* **9** 149

Elliott S R 1984 *Physics of Amorphous Materials* (London: Longman)

Frank F C 1952 *Proc. R. Soc.* A **215** 43

Gardner M 1977 *Sci. Am.* **236** 110

Grünbaum B and Shephard G C 1986 *Tilings and Patterns* (New York: Freeman)

Huber A 1981 *Recent Developments in Condensed Matter Physics* vol 2, ed J T Devreese (New York: Plenum) p 139

Jäckle J 1981 *Phil. Mag.* B **44** 533

Lee T D and Yang C N 1952 *Phys. Rev.* **87** 410

Levine D and Steinhardt P J 1984 *Phys. Rev. Lett.* **53** 2477

Levine R D and Tribus M (ed) 1978 *The Maximum Entropy Formalism* (Cambridge, Mass., MIT)

Mackay A L 1982 *Physica* **114A** 609

Mott N F and Davies E A 1979 *Electronic Processes in Non-Crystalline Materials* (Oxford: Clarendon)

Nelson D R and Halperin B I 1985 *Sci.* **229** 233

Parsonage N G and Stavely L A K 1978 *Disorder in Crystals* (Oxford: Oxford University Press)

Penrose R 1974 *Bull. Inst. Math. Appl.* **10** 266

Poon S J, Drehman A J and Lawless K R 1985 *Phys. Rev. Lett.* **55** 2324

Sachdev S and Nelson D R 1985 *Phys. Rev.* B **32** 1985

Shechtman D, Bleck I, Gratias D and Cahn J W 1984 *Phys. Rev. Lett.* **53** 1951

Stanley H E 1971 *An Introduction to Phase Transitions and Critical Phenomena* (Oxford: Clarendon)

Steinhardt P J, Nelson D R and Ronchetti M 1983 *Phys. Rev.* B **28** 784

Stephens P W and Goldman A I 1986 *Phys. Rev. Lett.* **56** 1168

Zallen R 1983 *The Physics of Amorphous Solids* (New York: Wiley)

Ziman J M 1979 *Models of Disorder* (Cambridge: Cambridge University Press)

Yonezawa F and Ninomiya (ed) 1983 *Topological Disorder in Condensed Matter* (Berlin: Springer)

2

DESCRIBING DISORDERED STRUCTURES

This chapter tackles the problem of how to describe a disordered structure short of listing the coordinates of all the atomic nuclei. Such a list, apart from being impossible to compile, is a clear case of enumerating the trees rather than looking at the wood. What is needed is a set of numbers which give statistical information sufficient to characterise the structure or at least go some way towards doing so. Such statistics would include the distribution of the number of nearest neighbours and the number of interparticle distances of given length. If the system is in thermodynamic equilibrium, then the appropriate numbers will be statistical mechanical averages.

There is also the question of intuitive appreciation. Many people have found it useful to construct models which can be looked at, looked through or pictorially represented. In this chapter we will describe a selection of such models as well as defining some of the most widely used statistical measures of disordered structures.

2.1 Sphere packings. The Bernal model

The densest packing of equal spheres occurs when each is touched by twelve nearest neighbours and the system is long-range ordered, either face-centered cubic (FCC) or close-packed hexagonal (CPH). The spheres then occupy 74.05 % of the space. The twelve neighbours are not uniformly distributed over the surface of the thirteenth sphere they all touch. A uniform distribution is possible but then the twelve do not touch each other and each is surrounded on the surface of the thirteenth by five neighbours whose centres make a regular pentagon (figure 2.1). The centres of the twelve neighbours are the vertices of a regular icosahedron and the structure is said to have icosahedral symmetry. Its energy is lower than that of alternative thirteen-atom clusters. Groups of thirteen like this, if stacked closely together, cannot have LRO and will occupy less than 74 % of space.

Long-range icosahedral orientational order in a quasicrystal suggested that there might be short-range orientational order in the melt (§1.10). This suggestion arose in the study of liquids long before the discovery of quasicrystals. If a few clusters like that of figure 2.1 are adjacent in a liquid they could confer local orientational order with icosahedral symmetry. Papers stemming from the important work by Frank in 1952 have often suggested that icosahedral arrangements occur in ordinary and especially in supercooled liquids (see §1.10 for references).

Figure 2.1 Icosahedral close packing of spheres showing pentagonal surface coordination.

It is a seemingly simple question to ask: what is the densest disordered packing of equal spheres? There is apparently no mathematical answer. However, empirical studies show that in such an arrangement the spheres occupy about 64 % of the space. Typical of these empirical studies were the seminal investigations, started by J D Bernal in the late fifties (Bernal 1959, 1960, Bernal and Mason 1960). In one of these a few thousand ball bearings were jammed closely but irregularly together and soaked in black paint. When the paint was drained off it adhered because of surface tension to the points of contact or near contact. These points could then be counted after the mass was dried and carefully picked apart. Displaying great ingenuity and patience Bernal and his assistant J Mason established a number of results such as the distribution of the number of close contacts, of which the average was 8.5, and the proportion of space occupied which was about 64 %.

The study of irregular sphere packings could be attempted for its own sake—Bernal called it an example of 'an almost unknown subject, statistical geometry'—but the main stimulus was the hypothesis that a dense random packing (DRP) of spheres is a good representation of the structure of simple liquids. By simple liquids is meant monatomic liquids of which the atoms have spherically symmetrical interatomic force fields; examples are molten alkali metals and liquified rare gases. This hypothesis is supported by a good deal of evidence (see, e.g., §2.2) and one example is that the increase in volume in going from FCC to DRP is about 16 %, which is also the expansion when crystalline argon melts. Accordingly, the DRP of spheres is sometimes referred to as the Bernal model of a liquid. Bernal's own formulation of the liquid state was 'a homogeneous, coherent and essentially irregular assemblage of molecules containing no crystalline regions or holes large enough to admit another molecule'. It has not been possible to construct homogeneous sphere packings occupying between 74 % (LRO) and 64 % (DRP) of space, and it is widely believed that this is intrinsically impossible in three dimensions. According to the Bernal model, the discontinuity is an expression of the first-order nature of the melting transition and the entropy of melting results from the large number of configurations possible in the disordered state as well as from the associated change in volume.

One of the ways in which the Bernal model has been studied is through its Voronoi polyhedra.

2.2 Voronoi and other polyhedra

Consider any array of sites. Figure 2.2 illustrates in two dimensions the following construction which could be executed in three dimensions: from any site P draw straight lines to all other sites; perpendicularly bisect these lines with planes; locate the smallest convex polyhedron made by these planes about P. This is the Voronoi polyhedron for the site P and encloses all the space which is nearer to P than to any other site.

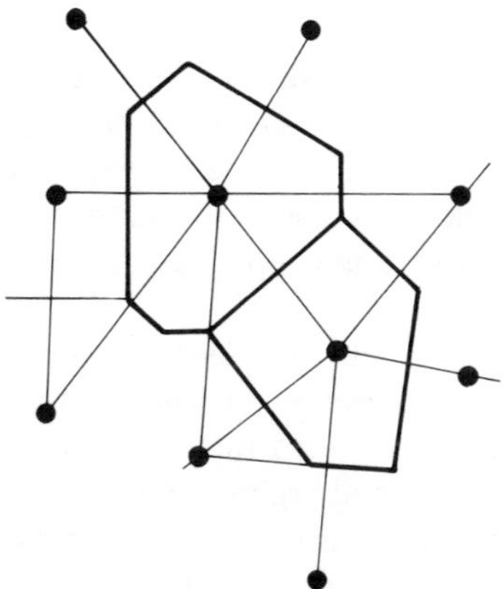

Figure 2.2 Atomic sites, construction lines and two Voronoi polygons.

A similar construction could be made for every other site resulting in a space-filling set of polyhedra, one per site, with shared faces, edges and vertices.

If the original array of sites were a Bravais lattice all polyhedra would be the same and would have form and symmetry characteristic of the lattice.

The most familiar such shape is probably the truncated octahedron belonging to the body-centred cubic lattice; this is the Wigner–Seitz cell of the BCC structure and the Brillouin zone in reciprocal space of the FCC structure.

Disordered arrays generate a distribution of polyhedra or cells with various shapes and sizes. The vertices and edges of the set are sometimes called a Voronoi froth and the froth forms a network threading the original sites which, in physics, will be thought of as atomic positions. The statistical distribution of quantities such as cell volume, number of faces per cell and number of edges per face constitutes a description of the disordered structure and that is the value of the concept. In one of Bernal's empirical studies referred to above, the 65 polyhedra characterising a random packing of spheres showed 32 different combinations of polygonal faces among which pentagonal faces were predominant. Bernal concluded (Bernal 1959, 1960, Bernal and Mason 1960) that 'irregular dense packing and pentagonal arrangements are necessarily connected'. Such insights can be backed up by detailed statistical studies and we will now describe one of these to exemplify the use of Voronoi polyhedra.

The work is that of Finney (1970) and its starting point was the set of spatial coordinates of 7994 ball bearings in a dense random packing constructed and then measured by Bernal *et al* (1970) following a similar but smaller study by Scott (1960). From the coordinates the Voronoi polyhedra could be constructed by computer calculations which generated the coordinates of the vertices and the centres of each cell, the number and geometrical characteristics of the faces, the surface area and volume of each cell. From this vast array of data many statistics could be inferred. Examples are: the spheres occupied 63.66 ± 0.0004 % of space; the average number of faces per polygon was 14.251 and of edges per face, 5.158; there was a spread of about 25 % in cell volumes. These numbers are examples of the statistics referred to in the first paragraph of this chapter as desiderata for characterising a disordered structure.

To relate these results to physical systems Finney considered not real liquid argon, but a simulated argon constructed at an appropriate density in a computer. The machine calculated the configurations of 108 atoms interacting through a well established argon–argon interatomic potential. This is an exploitation of the Monte Carlo method (see §4.1). The Voronoi polyhedra were calculated from the atomic coordinates and their statistics bore a strong resemblance to those of the ball bearing model. When we add that the thermodynamic properties calculated for the simulated argon are in very good agreement with those of real argon we can infer that the structure of real liquid argon is well represented by the dense random packing of spheres.

Voronoi polyhedra are general in concept and not restricted in use to dense random packing nor to simple liquids. The interested reader will find other applications of which one further example is the attempt by Srolovitz *et al* (1981) to relate the geometry of the polyhedra to local stresses in amorphous solid metals. This paper illustrates the general point that however useful the polyhedra are for description and visualisation, it is difficult to relate them to observable physical properties.

The polyhedra also have a drawback when they are used for mixtures of different sized spheres. In a binary system of large and small spheres the bisecting planes which form the cell faces may intersect the larger spheres and allocate an implausibly large cell volume to the smaller ones. An alternative construction is to replace the bisecting planes by radical planes, i.e., by those planes which are the loci of points from which the tangents to the two spheres are of equal length. Such planes will pass through the point of contact of touching spheres and there will be bigger cells for bigger spheres. These polyhedra, like Voronoi cells, fill space and divide it uniquely between the spheres. At the time of writing the radical plane polyhedra are just coming into use (Gellatly and Finney 1982).

Another idea introduced into this area by Bernal is that of the interstitial hole (Bernal 1959, 1960, Bernal and Mason 1960). If the atomic sites are treated as the vertices instead of the centres of the cells, then the latter enclose the interstitial spaces instead of the atoms. The statistics of the interstitial cells would be an alternative description of the structure. In CPH

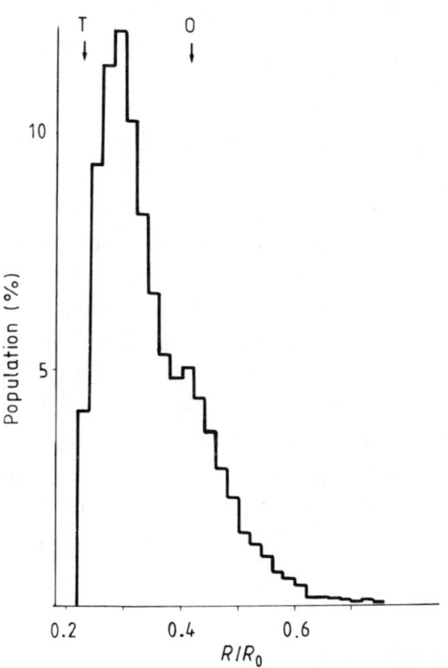

Figure 2.3 Frequency of occurrence of different sized interstices from a model of 4000 hard spheres; the maximum corresponds to slightly distorted tetrahedral holes. T, tetrahedral; O, octahedral (from Finney and Wallace 1981).

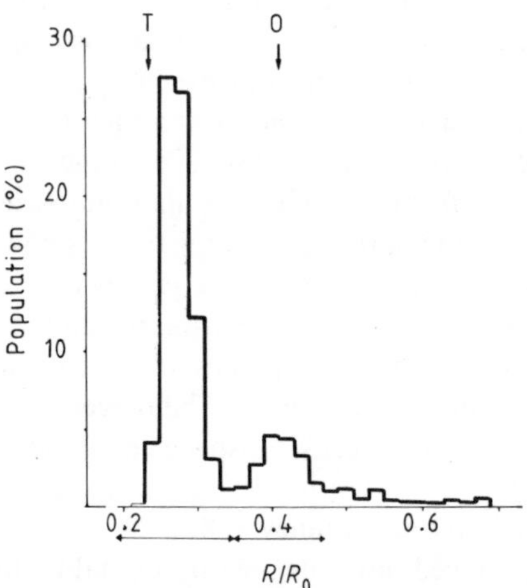

Figure 2.4 As figure 2.3 but for a relaxed model of 999 spheres with a Lennard-Jones potential. T and O interstices are now much more individualised in this more realistic model (for 'relaxation' see §2.5) (from Finney and Wallace 1981).

or FCC structures the interstitial cells are regular tetrahedra or octahedra. In the DRP structures, tetrahedra and octahedra, somewhat distorted from regular shape, also occur along with more complex and less regular figures. One way of illustrating this has been to calculate the radius, R, of the largest inscribed sphere which occupies the interstice and touches the surrounding spheres of the model. If the latter have radius R_0, then $R/R_0 = 0.225$ for a perfect tetrahedral interstice and 0.414 for an octahedral one. Figure 2.3 shows the distribution of R/R_0 from a ball bearing model of 4000 spheres. It appears that distorted tetrahedral interstices predominate with a considerable variety of larger holes such as dodecahedra (Finney and Wallace 1981).

The analysis of interstitial holes can be pursued in the work of Finney and Wallace (1981). They add the interesting idea that if realistic interatomic forces were allowed to determine the interparticle distances, i.e., the spheres are not hard, then the interstitial distribution would be radically different. In fact, the less regular interstitial cells almost disappear and tetrahedral and octahedral ones remain, though they are distorted, as befits a disordered structure (figure 2.4). Finney and Wallace argue that interstitial statistics have distinct advantages over Voronoi polyhedron data, being more sensitive reflectors of the disorder in the structure.

2.3 Distribution functions

The modelling and visualisation methods in the preceding and later sections have a definite value but by far the most common way of describing structures is by way of distribution functions. These can be obtained from models by measurement and arithmetic, from real materials by experiment or from first principles by theory. We now define some of the most frequently used functions.

Consider first N fixed identical particles in a volume V. The position vectors of their centres are $\{\boldsymbol{R}_i\}$. The average number density will be denoted by

$$n_0 \equiv N/V. \tag{2.1}$$

We shall assume that only one particle centre can occupy one point. All points in space are then either singly occupied sites or empty. In terms of the Dirac δ-function the single-particle density function is

$$\nu^{(1)}(\boldsymbol{r}) \equiv \sum_{i=1}^{N} \delta(\boldsymbol{r} - \boldsymbol{R}_i) \tag{2.2a}$$

and

$$\int \nu^{(1)}(\boldsymbol{r})\, \mathrm{d}\boldsymbol{r} = N. \tag{2.2b}$$

The two-particle density function is

$$\nu^{(2)}(\boldsymbol{r}_1, \boldsymbol{r}_2) \equiv \sum_{i}^{N} \sum_{j \neq i}^{N-1} \delta(\boldsymbol{r}_1 - \boldsymbol{R}_i)\delta(\boldsymbol{r}_2 - \boldsymbol{R}_j). \tag{2.3a}$$

This is zero unless two particles are at $\boldsymbol{R}_i$ and $\boldsymbol{R}_j$. If we integrate this over $\boldsymbol{r}_2$ there is a contribution every time $\boldsymbol{r}_2 = \boldsymbol{R}_j \neq \boldsymbol{R}_i$, and thus

$$\int \nu^{(2)}\, \mathrm{d}\boldsymbol{r}_2 = (N-1) \sum_{i}^{N} \delta(\boldsymbol{r}_1 - \boldsymbol{R}_i) = (N-1)\nu^{(1)}(\boldsymbol{r}_1). \tag{2.3b}$$

If the particles are atoms or ions in physical systems they are not fixed. Time or ensemble averages must be taken and we denote $\langle \nu^{(1)}(\boldsymbol{r}) \rangle$ by $n^{(1)}(\boldsymbol{r})$. $n^{(1)}(\boldsymbol{r})\, \mathrm{d}\boldsymbol{r}$ then means the average number of particle centres in $\mathrm{d}\boldsymbol{r}$ and, on the assumption that this cannot be more than unity because particles will not coincide, $n^{(1)}(\boldsymbol{r})\, \mathrm{d}\boldsymbol{r}$ is also the probability of finding a particle in $\mathrm{d}\boldsymbol{r}$. Thus,

$$\int_V n^{(1)}(\boldsymbol{r})\, \mathrm{d}\boldsymbol{r} = N \tag{2.4}$$

where $n^{(1)}(\boldsymbol{r})$ is called the number density function or *one-particle distribution function*.

If we average equation (2.3*b*) we find

$$\mathrm{d}\boldsymbol{r}_1 \int_V n^{(2)}(\boldsymbol{r}_1, \boldsymbol{r}_2)\, \mathrm{d}\boldsymbol{r}_2 = (N-1) n^{(1)}(\boldsymbol{r}_1)\, \mathrm{d}\boldsymbol{r}_1. \tag{2.5}$$

Here $n^{(2)}(\boldsymbol{r}_1, \boldsymbol{r}_2)\, \mathrm{d}\boldsymbol{r}_1\, \mathrm{d}\boldsymbol{r}_2$ means the probability of finding particles simultaneously in $\mathrm{d}\boldsymbol{r}_1, \mathrm{d}\boldsymbol{r}_2$. Both sides of equation (2.5) mean the probability of finding a particle at $\boldsymbol{r}_1$, namely, $n^{(1)}(\boldsymbol{r}_1)\, \mathrm{d}\boldsymbol{r}_1$, multiplied by the number of particles summed as $d\boldsymbol{r}_2$ traces out the volume V, namely $(N-1)$. $n^{(2)}$ is the *two-particle distribution function.*

$n^{(2)}(\boldsymbol{r}_1, \boldsymbol{r}_2) \neq n^{(1)}(\boldsymbol{r}_1)n^{(1)}(\boldsymbol{r}_2)$ because the probability of occupation of $\boldsymbol{r}_2$ may be affected by that of $\boldsymbol{r}_1$. This is to be expected if there are significant interparticle forces over the distance $\boldsymbol{r}_1 - \boldsymbol{r}_2$. A *pair-distribution function*, $g^{(2)}(\boldsymbol{r}_1, \boldsymbol{r}_2)$, may be defined by

$$n^{(2)}(\boldsymbol{r}_1, \boldsymbol{r}_2) \equiv n^{(1)}(\boldsymbol{r}_1)n^{(1)}(\boldsymbol{r}_2)g^{(2)}(\boldsymbol{r}_1, \boldsymbol{r}_2) \tag{2.6a}$$

and, if $\boldsymbol{r}_1 - \boldsymbol{r}_2 \rightarrow \infty$ we expect $g^{(2)} \rightarrow 1$ because the interparticle influences vanish.

If the system is homogeneous, $n^{(1)}(\boldsymbol{r}_1) = n^{(1)}(\boldsymbol{r}_2) = n_0$; consequently

$$n^{(2)}(\boldsymbol{r}_1, \boldsymbol{r}_2) = n_0^2 g^{(2)}(\boldsymbol{r}_1, \boldsymbol{r}_2). \tag{2.6b}$$

It is often convenient to put $\boldsymbol{r}_2 - \boldsymbol{r}_1 = \boldsymbol{r}$ and choose the origin at particle 1. Equations (2.5) and (2.6*b*) then give

$$n_0 \int g^{(2)}(\boldsymbol{r})\, d\boldsymbol{r} = N - 1. \tag{2.7}$$

The interpretation of equation (2.7) is that $n_0 g^{(2)}(\boldsymbol{r})\, d\boldsymbol{r}$ gives the probability of finding a particle at a point $d\boldsymbol{r}$ which is a distance $\boldsymbol{r}$ from another particle at the origin.

Sometimes it is useful to have an integrand that *counts in* the particle at the origin. It is

$$z(\boldsymbol{r}) \equiv n_0 g^{(2)}(\boldsymbol{r}) + \delta(\boldsymbol{r}) \tag{2.8a}$$

$$\int_V z(\boldsymbol{r})\, d\boldsymbol{r} = N. \tag{2.8b}$$

Many of the disordered materials are not only homogeneous but isotropic. In that case

$$g^{(2)}(\boldsymbol{r}) \rightarrow g^{(2)}(|\boldsymbol{r}|) \rightarrow g(r). \tag{2.9}$$

In the last term the superscript has been dropped: $g(r)$ conventionally denotes the pair distribution function.

Of course three-particle and higher functions can be defined by extension of these formulae. Thus $n^{(3)}(\boldsymbol{r}_1, \boldsymbol{r}_2, \boldsymbol{r}_3)\, d\boldsymbol{r}_1\, d\boldsymbol{r}_2\, d\boldsymbol{r}_3$ means the probability of there being three particles simultaneously in $d\boldsymbol{r}_1$, $d\boldsymbol{r}_2$, $d\boldsymbol{r}_3$. Let

$$n^{(3)} \equiv n^{(1)}(\boldsymbol{r}_1)n^{(1)}(\boldsymbol{r}_2)n^{(1)}(\boldsymbol{r}_3)g^{(3)}(\boldsymbol{r}_1, \boldsymbol{r}_2, \boldsymbol{r}_3) \tag{2.10a}$$

where $g^{(3)}$ is the *triplet distribution function* and in amorphous matter $^{(3)} \rightarrow 1$ if the separations are all large. For a homogeneous system

$$n^{(3)} = n_0^3 g^{(3)}. \tag{2.10b}$$

It follows from the definitions that

$$\int n^{(3)}\,\mathrm{d}\boldsymbol{r}_3 = (N-2)n^{(2)}(\boldsymbol{r}_1, \boldsymbol{r}_2) \tag{2.11a}$$

$$\iint n^{(3)}\,\mathrm{d}\boldsymbol{r}_3\,\mathrm{d}\boldsymbol{r}_2 = (N-1)(N-2)n^{(1)}(\boldsymbol{r}_1) \tag{2.11b}$$

$$\iiint n^{(3)}\,\mathrm{d}\boldsymbol{r}_3\,\mathrm{d}\boldsymbol{r}_2\,\mathrm{d}\boldsymbol{r}_1 = N(N-1)(N-2). \tag{2.11c}$$

It is clear that each higher function expresses more about the structure than the lower ones. Equation (2.11a) shows that an infinite number of arrangements of particles by threes is compatible with the *same* pair distribution function. This is important because it shows how little information about the details of a disordered structure is expressed by $g(r)$. Unfortunately $g(r)$ is often all the information experiments will provide.

The average number of particles in a spherical shell centred on a particle at the origin, and dr thick, is

$$\int_{\text{shell}} n_0 g(r)\,\mathrm{d}r = 4\pi r^2 n_0 g(r)\,\mathrm{d}r = n_0\rho(r)\,\mathrm{d}r \tag{2.12}$$

where $\rho(r) \equiv 4\pi r^2 g(r)$ is called the *radial distribution function* or RDF for short.

Figure 2.5 exemplifies typical pair distribution functions, and illustrates one way in which computer models like those of §2.5 can be compared with experiment. Peak shapes, splittings, positions and heights can be subjects of comparison.

From equation (2.12) we infer that the average number of particles surrounding the origin particle out to a distance r_1 is

$$n_1 \equiv n_0 \int_0^{r_1} \rho(r)\,\mathrm{d}r.$$

If r_1 just takes in the first peak of $\rho(r)$, n_1 may be called the *number of nearest neighbours* or sometimes the *coordination number*. Figure 2.6 shows that n_1 is not uniquely defined so that the number of nearest neighbours is subject to uncertainty. In a simple liquid near its freezing point it is typically 10 to 11.

2.4 Extension to binary systems

For describing alloys, solutions and mixed systems generally, we need generalisations of the formulae in §2.3. These now follow for homogeneous isotropic binary systems only.

Let us suppose the volume V contains N_A A-type atoms and N_B B-type

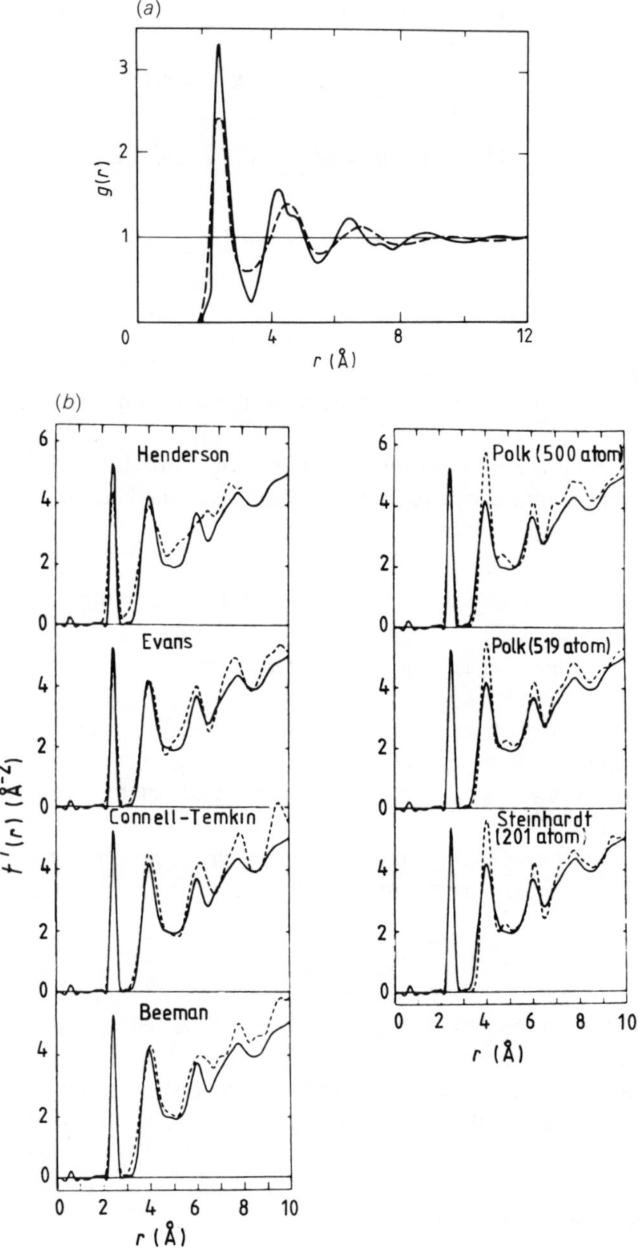

Figure 2.5 (*a*) $g(r)$ for liquid and amorphous Ni (from Waseda 1980). (*b*) Experimental correlation function (full curves) for amorphous Ge compared with various proposed models. The ordinate is $4\pi r n_0 g(r)$ derived with a factor $M(Q)$ as in §3.9 (from Wright 1982b).

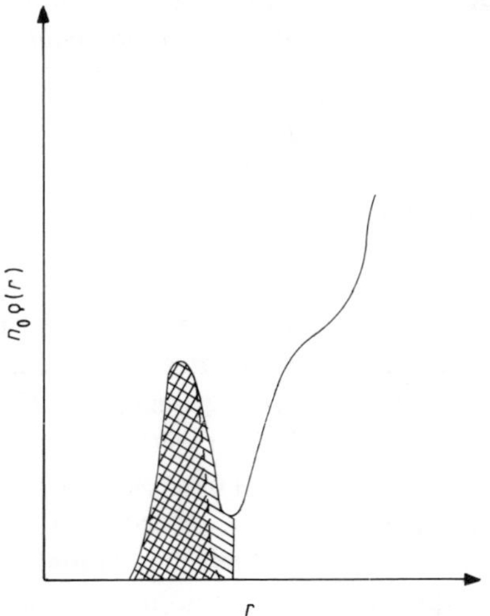

Figure 2.6 Schematic diagram to show two ways of integrating over the first peak. One selects out a symmetrical peak, the other integrates to the first minimum. The reader will see other possibilities.

atoms. The concentrations expressed as atomic fractions are

$$c_A \equiv N_A/N \qquad c_B \equiv N_B/N \qquad N \equiv N_A + N_B. \tag{2.13}$$

Let $n_{ij}(r)$ be the average number density of type-j atoms at a distance r from a type-i atom. Thus, for example, $n_{BB}(r)$ would give the number of B atoms per unit volume at a distance r from a type-B atom taken as the origin.

Define

$$g_{ij}(r) \equiv \frac{n_{ij}(r)}{n_0 c_j} = \frac{n_{ij}(r)}{n_j} \tag{2.14a}$$

where $n_0 \equiv N/V$ as in equation (2.1), and $n_j = N_j/V$. Since the numbers of $A \rightarrow B$ and $B \rightarrow A$ interparticle distances must be identical:

$$g_{AB} = g_{BA}. \tag{2.14b}$$

$g_{ij}(r)$ is the generalisation of $g(r)$ in equation (2.9) and is called a *partial pair distribution function.* For large r, the i–j interaction is negligible so $n_{ij} \rightarrow n_j$ and $g_{ij} \rightarrow 1$. For very small r, $g_{ij} = 0$ because atoms do not interpenetrate. The *partial radial distribution function*, ρ_{ij}, is defined as in equation (2.12) by

$$4\pi r^2 g_{ij}(r) \equiv \rho_{ij}(r). \tag{2.15a}$$

Thus, for example, the average number of type-B atoms within a distance r_1 of a type-A atom is

$$n_B \int_0^{r_1} \rho_{AB}(r)\, dr.$$

In analogy with the definition of n_1, the coordination number, in the previous section we now have four *partial coordination numbers*, n_{ij}:

$$n_{ij} = n_j \int_0^{r_1} \rho_{ij}(r)\, dr. \qquad (2.15b)$$

2.5 Models grown from seeds

It is not surprising that some workers have substituted computer calculations for the manipulation of ball bearings and plastic spheres. The object is the same however: to obtain a set of coordinates of the centres of spheres with DRP. The method was clearly defined by Bennett (1972). The calculation starts with a seed cluster which could be a triangle of three or a tetrahedron of four mutually touching spheres, often of equal diameter, σ, but not necessarily so. An additional sphere is then placed by the computer in hard contact with three already present and left in position. The latter process is repeated indefinitely, typically a few thousand times. A dense irregular model gradually grows and is characterised by the coordinates. It can be reproduced on the bench with plastic balls if desired.

The structure of the model clearly depends on the nature of the seed and on the criteria according to which the positions of added spheres are acceptable or not. Examples of these criteria will now be taken from Ichikawa (1975) whose seed was a tetrahedron. The procedure was:

(i) choose three spheres on the surface of the cluster so that their separations are less than $k\sigma$, with $k > 1$;

(ii) find the position of another sphere which would lie in contact with the chosen three without overlapping any other sphere present—this position is called a pocket;

(iii) locate all pockets presented by the cluster;

(iv) put the new sphere in the pocket closest to the centre of the cluster and keep it there;

(v) recalculate all the pockets of the new cluster and proceed as before. The whole calculation can be repeated for a series of k-values.

Choosing the pocket closest to the centre roughly simulates reducing the energy due to long-range interactions of the particles. Bennett called models based on this principle 'global'. An alternative would be to choose the pocket which allowed the arriving sphere to locate itself closest to the plane of its three contacts. This was called a 'local' model and roughly simulates the minimisation of short-range interaction energy.

Many constructions of this kind are described in the literature and they use seeds and aggregation criteria suitable for their author's purpose. This has often been to model the structure of amorphous metals. A commonly used test of whether the model is satisfactory is to compare it with what can be learned of the structure of real amorphous metals by x-ray or neutron scattering. This usually means the radial distribution function (figure 2.5).

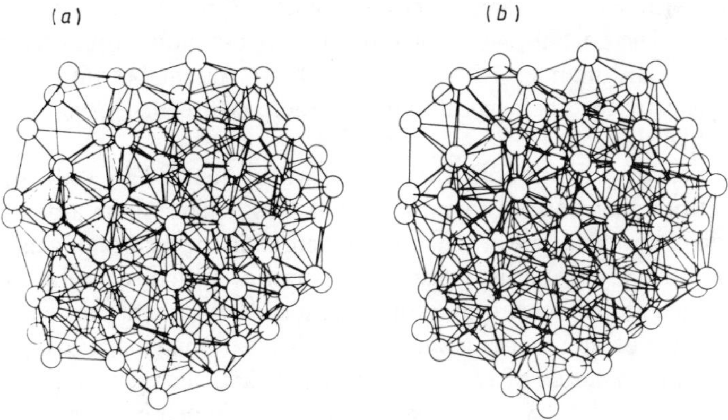

Figure 2.7 100 atoms in a model (*a*) before and (*b*) after relaxation (from Barker *et al* 1975).

A random packing of hard spheres is obviously an idealised, perhaps unrealistic, representation of atomic packing. It overemphasises the exclusion of one atom from another by the repulsive forces and does not minimise the energy. Attempts to remove these objections can be made by a process called *relaxation*. A further effort of computing is needed which starts with the coordinates of, say, 1000 spheres in the Bernal model and calculates E_0 and ΔE, where E_0 is the potential energy of the original configuration under some assumed interatomic potential, say Lennard-Jones 6:12, and ΔE is the departure from E_0 resulting from slight adjustments of position. The computer seeks energy minima: in effect it allows the spheres to shake down into lower energy configurations not far from the original. Figure 2.7 (taken from Barker *et al* 1975) shows a 100-atom region before and after relaxation. The process reduced the energy by almost 5 % and made small but significant changes in the radial distribution function and Voronoi statistics. In particular, the split second peak of the radial distribution function often observed in amorphous metals had its two subpeaks interchanged in height—which improved the agreement with experiment. Relaxation also sharpened the first peak and deepened the first minimum. Packings grown from seeds on computers have been subjected to various versions of relaxation. The choice of interatomic potential to be used is important. It should presumably be appropriate to the class of material, e.g., metal or insulator, and compatible with whatever is known about the elastic properties.

2.6 Models of chemical short-range order

Dense random packings of hard spheres represent one extreme of structural disorder. The short-range configuration is dominated by the geometry of spheres. It is quite conceivable that the SRO could be dominated by chemical bonding with definite valencies, bond lengths and bond angles exerting their influences. The partial pair, triplet and higher functions would have to express this. The number of nearest neighbours might be quite small and determined by the chemical affinities of the constituents. Common glasses behave like this.

The idea of a continuous random network, denoted by CRN, based on covalent bonding was put forward in 1932 by Zachariasen in a paper which has been regarded as one of the most influential in the history of glass science (Zachariasen 1932). Let us consider as a unit of structure a tetrahedron of four oxygen atoms bonded to a four-valent silicon atom at the centre. Crystalline forms of SiO_2 contain such a unit. We can imagine an oxygen atom as the common vertex of two such tetrahedra and therefore bonded to two Si atoms (see figure 2.8). If many tetrahedra are joined by common vertices in this way a network will be built up which will not have LRO if some randomness is built in. Randomness could enter by distributing a range of Si–O–Si bond angles about a mean, by stretching or shrinking bond lengths, by rotating adjacent tetrahedra somewhat about the line of the Si–O bonds, or some combination of these.

In the 1960s some ball-and-spoke models were built in this way with several hundred tetrahedral units (Evans and King 1966, Bell and Dean 1966). Typically the Si–O–Si angle spread $\pm 20°$ about its mean of 140°. In a similar approach by Polk (1971) to modelling amorphous Si and Ge, about 12 % of the bonds were stretched by 10 % on the average. These models are in the Bernal tradition of building an ‘essentially irregular assemblage’ but they are in the Zachariasen tradition of determining the SRO by maintaining the chemical bonding.

It appears possible to build such CRN to indefinite size without including unsatisfied bonds. They may be held to represent ideal glassy structures. Real materials like vitreous silica may approximate to the ideal but contain defects just as crystals depart from perfection. As with DRP sphere models, CRN models can be measured for their coordinates, RDF, densities and other properties. The number of four-, five- and n-membered closed rings of bonds can be counted for use in the theoretical discussion of glasses. Allowing for the statistical imprecision resulting from the relatively small number of atoms and for the experimental error in measuring the RDF (§3.9), the agreement between models and experiment is moderately good (figure 2.5). Perfect agreement is rarely if ever achieved. It must be remembered that the models embody definite conditions on the triplet and higher distributions about which the RDF can reveal little or nothing (Gaskell 1979a,b).

By contrasting the CRN having chemical SRO with a DRP of spheres, we do not mean to imply that the latter could not contain chemical SRO (Gaskell 1979). In a binary system it is quite conceivable that A and B atoms might preferentially cluster together with definite implications for g_{AA}, g_{AB} and g_{BB}. Such might be the case in some amorphous alloys and modelling them by the

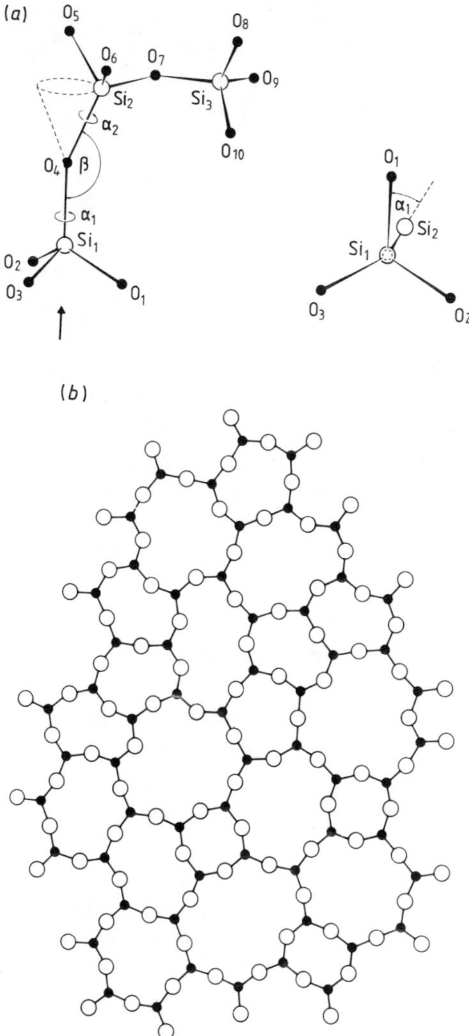

Figure 2.8 (*a*) SiO_4 tetrahedra joined at common oxygen atoms. In a disordered system the angles α and β have distributions about mean values. (*b*) Zachariasen's original diagram of a CRN in 2-D (from Wright 1982b).

seed method of §2.5 is quite difficult because each added sphere must *simultaneously* contribute appropriately to all three partial pair distributions and it will be hard to know which kind to add at which point. Nevertheless such seed growth methods have been applied with moderate success to generate models describing the structure of binary amorphous systems of metal and metalloid such as Fe–P and Pd–Si. (See §12.2 for an example and reference.)

It might be thought that a binary model of a system like Pd–Si might be envisaged with the Si atoms occupying some of the interstices of the DRP structure (Polk 1970). However, it does not appear that there is really enough room for this and probably any such insertions would seriously reorganise the basic packing.

A way of building chemical SRO into a dense packing would be to devise a basic unit or seed which expresses the chemical requirement, and then build one unit onto another in a way similar to building a CRN glass model. A plausible initial seed could be inferred from local structures in related crystalline compounds. For example, in cementite (Fe_3C) the C atom is inside a trigonal prism of Fe atoms, and there are other similar cases. Gaskell (1979) inferred that trigonal prismatic seeds might well be useful for building models by making such prisms share edges, choosing at random the edges to be shared while preventing the overlap of the spheres and requiring the model to be as dense as possible. Such representations of structures can be built by hand or computer and can then be relaxed (§2.5). Gaskell's model for metal–metalloid glasses had a density and a RDF compatible with the observed quantities for real amorphous Pd_4Si (see also §§12.1, 12.2).

2.7 Short- and long-range order parameters

Two more examples of the various ways of describing or characterising structure are the short- and long-range order parameters. We deal first with the short-range one. This is a number, η, of which the magnitude indicates the degree of chemical SRO in a binary system; $\eta = 0$ for complete chemical disorder.

It is clear that the degree to which one type of atom clusters round another is expressed by the average partial coordination numbers n_{AA}, n_{BB} and n_{AB}. Let n^*_{ij} be the value of these numbers when complete chemical disorder prevails and now define:

$$\eta_{ij} \equiv \frac{n_{ij}}{n^*_{ij}} - 1. \tag{2.16}$$

The η_{ij} are not independent and the following relations will be shown to hold between them:

$$\eta_{AA} = -\eta_{AB}\frac{n_B c_B}{n_A c_A} \tag{2.17a}$$

$$\eta_{BB} = -\eta_{AB}\frac{n_A c_A}{n_B c_B} \tag{2.17b}$$

$$\eta_{AB} = \eta_{BA} \tag{2.17c}$$

$$\eta^2_{AB} = \eta_{AA}\eta_{BB}. \tag{2.17d}$$

η_{AB} therefore suffices to characterise the degree of chemical SRO. It is zero for total chemical disorder ($n_{AB} = n^*_{AB}$), $\eta_{AB} > 0$ if there is a preference for AB associations and $\eta_{AB} < 0$ if there is a bias against AB associations.

The total number of atoms, including both kinds, round an A atom is called the total coordination number n_A and

$$n_A \equiv n_{AA} + n_{AB} \qquad n_B \equiv n_{BB} + n_{BA}. \tag{2.18}$$

For a given composition and given values of n_A and n_B, the quantity η_{AB} will have a maximum value, η^{max}_{AB}, and our final definition will be

$$\eta^0_{AB} \equiv \eta_{AB}/\eta^{max}_{AB}. \tag{2.19}$$

This quantity is 0 or 1 for, respectively, complete chemical disorder and maximum preference for A–B neighbours. If A–A and B–B neighbours are preferred, $\eta^0 < 0$.

η^0_{AB} is the short-range order parameter introduced in 1981 for binary amorphous or liquid systems by Cargill and Spaepen (1981). It could be calculated from the various models discussed in this chapter or from experimental partial coordination numbers obtained by the methods in the next chapter. Amorphous $Co_{0.8}P_{0.2}$ has $\eta^0_{AB} \simeq 1$ and $Cu_{0.57}Zr_{0.43}$ is a disordered example with $\eta^0_{AB} = 0.01$ (Cowley 1950a).

We now reproduce some of the arguments used by Cargill and Spaepen leading to equation (2.17). The main point is to get an expression for n^*_{ij} by maximising the configurational entropy subject to the necessary relations in equation (2.18) and to the further constraint that

$$N_A n_{AB} = N_B n_{BA}. \tag{2.20}$$

The latter follows because both products are equal to the total number of A–B nearest-neighbour pairs. Let W_A be the number of different arrangements of n_A neighbours round an A atom. W_B is similarly defined and W is the total number of distinct nearest-neighbour arrangements in the whole sample. We now write the configurational entropy as $S = k_B \ln W$ where $W = W_A^{N_A} W_B^{N_B}$, $W_A = n_A!/(n_{AB}!n_{AA}!)$ and $W_B = n_B!/(n_{BA}!n_{BB}!)$. The necessary assumption here is that all the arrangements can be regarded as independent and this is really an oversimplification.

The Lagrangian method of undetermined multipliers serves to maximise S and leads quite easily to

$$n^*_{AB} n^*_{BA} = n^*_{AA} n^*_{BB} \tag{2.21a}$$

and

$$n^*_{ij} = c_j n_i n_j / \langle n \rangle \tag{2.21b}$$

where

$$\langle n \rangle \equiv c_A n_A + c_B n_B. \tag{2.21c}$$

It follows from equations (2.16) and (2.21*b*) that

$$\eta_{ij} = \frac{n_{ij}\langle n\rangle}{c_j n_i n_j} - 1. \tag{2.21d}$$

The four relations (2.17) follow quite easily from equations (2.21*d*) and (2.18).

The quantity η was originally introduced to describe amorphous solid metals with topological disorder. Three partial pair distributions are needed to obtain the η_{ij} and a minimum of three scattering experiments are needed (§3.10). A similar problem of characterising SRO occurs in binary alloys with cellular disorder and there are various ways of defining suitable parameters. A well known one is named after Cowley (1950a). As before c_A and c_B are the fractions of A and B atoms. Let us consider the degree of order in the kth shell, consisting of n_k atoms, round a B-type atom. If there were no order, the average number of A atoms in the shell would be $c_A n_k$ but suppose it is in fact $n_k^{(A)}$. The SRO parameter for the shell is then

$$\alpha_k \equiv 1 - \frac{n_k^{(A)}}{c_A n_k}. \tag{2.22}$$

$\alpha_k = 0$ in a completely disordered alloy. For perfect order α_k has a maximum absolute value and varies from shell to shell; $\alpha_k^{\max}$ may be unity or fractional, positive or negative and depends on the type of lattice (Cowley 1950a). In a simple case, that of β-brass, CuZn, at 50:50 composition, equation (2.22) gives $\alpha_1^{\max} = -1$, $\alpha_2^{\max} = +1$, $\alpha_3^{\max} = -1$ etc. For Cu_3Au the values are $-\frac{1}{3}$, $+1$, $-\frac{1}{3}$ etc. Experimentally α_k may be obtained from the diffuse x-ray scattering (Cowley 1950b) and only one scattering experiment is necessary which also gives the lattice sites from the sharp peaks.

The quantities η_{AB}^0 and α_k, being SRO parameters, refer to the composition of coordinate shells, not to a lattice. A LRO parameter is relevant to cellular disorder and must refer to the lattice sites. A common LRO parameter was mentioned in §1.3 and is

$$S \equiv \frac{r - c_A}{1 - c_A} \tag{2.23}$$

where r is the fraction of A-type atoms that are occupying the lattice positions they would have if complete chemical order prevailed. Clearly $1 \geqslant S \geqslant 0$. If no LRO exists there can be SRO at most and $\alpha_k \to 0$ as $k \to \infty$. Conversely if $\alpha_k \to$ a finite limiting value, α_k^{∞}, as $k \to \infty$, this is a sign of LRO. α_k^{∞} can be shown to be related to S (Cowley 1950a). S can also be measured by x-ray diffraction (Cowley 1950b).

The use of scattering experiments to determine distribution functions and other descriptive measures of topologically disordered structures is the subject of the next chapter.

2.8 Curved space and the structure of glasses

Even to describe the structure of metallic or CRN glasses has not proved easy. Very local order—like neighbours at the corners of a tetrahedron in amorphous Si—is not hard to envisage; likewise statistical measures such as $g(r)$, n_1. However, neither of these is a very good starting point for building up a picture of medium-range order. Connected with this is the question of what, in any disordered structure, is usefully identified as a 'defect'?

A radical approach to this problem was introduced by Sadoc and his collaborators in about 1979. What it does is to start with a long-range ordered—and by that token completely understood—structure in a space of n dimensions and then to map it by a systematic procedure on to a region of $(n-1)$-dimensional space. In the space of lower dimension, the structure will in general not have LRO. If the original structure has been suitably chosen, and the mapping procedure is appropriate, the new structure may be like that of real glasses. If so, some general understanding of glass structure may follow from that of the structure in the higher dimension. Naturally if the glasses are in 3-D the starting structure will be in 4-D and the subject, like that of quasicrystals in §1.10, has considerable topological depth which will only be pointed to here, not plumbed.

It is a geometrical commonplace, though perennially interesting, that a plane cannot be tiled with regular pentagons whereas the surface of a sphere can. Since we are concerned with glasses, the vertices of the twelve pentagons which cover a sphere will be called atoms. The regular polyhedron of which these atoms are vertices is the dodecahedron. Sadoc has emphasised the formal extension of this: it is not possible to fill 3-D space with regular tetrahedra but if the 3-D space is curved in 4-D space to make a hypersphere, then tiling, or tessellation, can be achieved. Just as the pentagons and atoms on the ordinary sphere are associated with a regular polyhedron, the tetrahedra and atoms on the hypersphere have a corresponding polytope, where a polytope is to 4-D what a polyhedron is to 3-D. There are six regular polytopes which fit into the hypersphere in the way the five Platonic solids (which include the dodecahedron) fit into the ordinary sphere. They are all of mathematical interest but not all equally relevant to glass structures. This emerges when we treat the vertices as atoms and ask how many neighbouring vertices there are (Klémen and Sadoc 1979).

Regular polytopes can be identified by a numerical symbol, $\{p, q, r\}$, and polyhedra by $\{p, q\}$, where p = symmetry order of a face ($p = 4$ for a square), q = number of faces round a vertex ($q = 3$ for a tetrahedron), r = number of $\{p, q\}$ cells that share a common edge (the Schafli notation). The six regular polytopes are $\{3, 3, 3\}$, $\{4, 3, 3\}$, $\{3, 3, 4\}$, $\{3, 4, 3\}$, $\{5, 3, 3\}$ and $\{3, 3, 5\}$. The polytope $\{3, 3, 5\}$ is a tessellation by tetrahedra (600 of them) and has 120 vertices each surrounded uniformly with twelve neighbours. This promises well for constructing 3-D structures of the DRP type which have large coordination numbers. In $\{5, 3, 3\}$, on the other hand, dodecahedra replace tetrahedra, there are 600 vertices, each with four neighbours, and this is a possibility for CRN with low coordination. Incidentally $\{5, 3, 3\}$ can be derived from $\{3, 3, 5\}$ by placing vertices for the former at the midpoint of each cell of the latter; $\{3, 3, 5\}$ can be similarly derived from $\{5, 3, 3\}$.

Ignoring all the less immediately appealing polytopes and all the formal geometry, we have now to map the atoms on the hypersphere onto three dimensions. Just as flat maps distort geographical features, this mapping may alter nearest-neighbour distances, bond angles and coordination numbers. Since there is an infinity of possible mappings, those can be chosen which minimise changes in coordination number and NN distance even if bond angles alter in consequence—this choice would help to minimise the strain energy.

Mapping is difficult both in concept and practice. In this context the disclination is an important idea. It is possible to define a disclination without reference to atomic structure and to differentiate it from a dislocation. Consider a hollow square of material like a picture frame, and imagine it cut through one limb (see figure 2.9). If the opposing faces of the cut are now supposed to move relative to one another, the motions fall into two possible classes: if the faces translate parallel to one another, edge or screw dislocations result; if the relative motion is rotational, wedge and twist disclinations form. When the movements open up gaps, we imagine them filled by sealing in added substance. When the movements would potentially force the faces to interpenetrate, this is prevented by removing substance. After any necessary removal or addition, the faces are notionally sealed together and any strain distributed throughout the now distorted square. For crystalline matter, the movements, and the addition and subtraction, must be compatible with the translational and rotational symmetries of the lattice so that the ultimate knitting together preserves the lattice lines though they will in general be strained (figure 2.9). Sixfold symmetry would for example require a minimum of 60° rotation. Both disclinations and dislocations are line defects in 3-D.

Very large strain energies would result from preserving rotational symmetry in disclinations so they do not occur in ordinary 3-D crystals. They are common in surface structures, thin skins and liquid crystals and lead to much interesting physics. An illustrated introduction to their geometry is given by Harris (1977). To concentrate solely on the matter in hand, we note that disclinations introduce curvature into otherwise flat systems. The most common illustration of this is in the 2-D hexagonal lattice. The reader may like to confirm that if a 60° wedge of lattice is cut out and the network then reconstituted, one hexagon will have become a pentagon—the site of a point defect—and the sheet will have become a cone. Adding a 60° wedge transforms one hexagon to a heptagon and the sheet into a warped saddle-like surface.

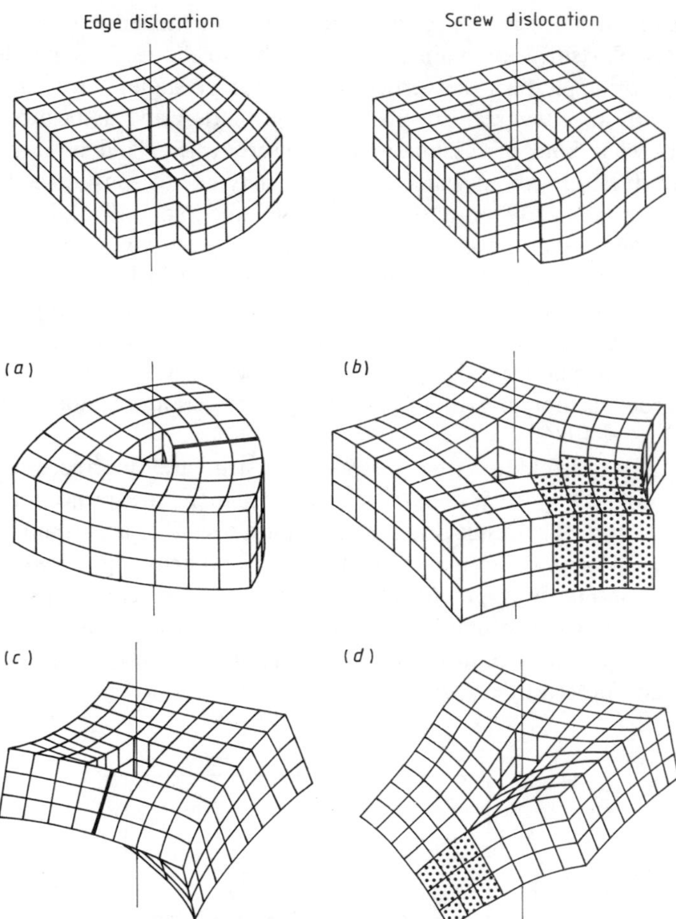

Figure 2.9 Dislocations and disclinations introduced into a hollow square in a cubic lattice. Note that positive- and negative-wedge disclinations are made by removing and inserting extra material respectively. In the twist disclinations rotations about two different axes are shown, one parallel, and one perpendicular, to the axis of the square. (*a*) and (*b*) are wedge disclinations; (*c*) and (*d*) are twist disclinations (from Harris 1977).

The relevance of this to amorphous structures is that if a space is already curved it can be flattened, or at any rate, decurved to some extent by inserting wedges or appropriate sections of material. If the curved space is perfectly tessellated—as in the hypersphere tiled with tetrahedra—this decurving process will introduce disclinations in the tessellation. In principle disclinations could be introduced until the curvature was removed, if not absolutely then at least on the average so that a wrinkled or corrugated but otherwise flat surface remains. Applying this to the {3, 3, 5} polytope one could in principle turn the vertices into a non-crystalline array in flat 3-D space with approximately twelvefold coordination and associated strains. Exactly what kind of disclinations to insert, where, and how many, are difficult questions about which the reader may like to consult the literature (Venkataraman and Sahoo 1985a,b, Mosseri and Sadoc 1985). For example, Mosseri and Sadoc have devised a method of decurving the {3, 3, 5} and {5, 3, 3} sets of atoms, not by introducing one disclination at a time, but by an iterative procedure which inserts a whole network of disclinations at each step. The resulting flat 3-D structures are said to be more suitable for the discussion of quasicrystals (§1.10) than glasses because they are still somewhat ordered, though without periodicity (Mosseri and Sadoc 1985, Sadoc and Mosseri 1985).

It is also possible to flatten the polytope structure by projecting the vertices onto 3-D space just as cartographical maps are projections of the earth's surface. The result is a model of finite size—120 atoms in the case of {3, 3, 5}—which can be enlarged by fitting several mappings together with as little discontinuity as practicable. Perhaps fitting several maps of Africa together is not unacceptable as an analogy. A 651-atom aggregate of this kind has a pair distribution like that in figure 2.10 and the details of the generation procedure are given by Sadoc (1981).

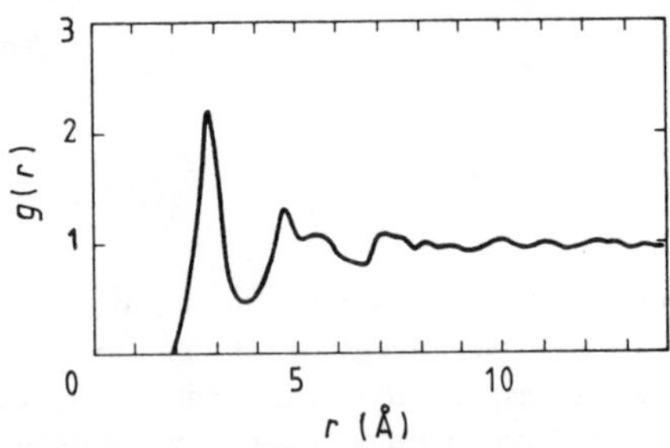

Figure 2.10 Pair distribution of a 651-atom model projected into 3-D space from the {3, 3, 5} polytope (from Sadoc 1981).

These researches have brought a flood of new thought about glass structure in general, and a systematic method of computing structures from a well characterised starting point—the chosen polytope. However, at the time of writing it is probably premature to determine whether they hold the key to understanding properties of particular glasses. It must be added that a number of alternative approaches on the same general lines are being explored (see Venkataraman and Sahoo (1985a,b) for an introduction and other references). Furthermore, the reader's attention is particularly drawn to the theory of glass developed by Rivier (Rivier and Duffy 1982, Rivier 1983, 1984) with its original concept of odd lines—linear defects similar to, but not identical with, disclinations. Too complex to attempt to outline here, this work is not simply an attempt to describe glass structure, but a theory of why the structure should have the character it does and why it leads to physical properties such as the entropy (§10.2) and the two-level phenomena (§10.7).

2.9 Fractals

Since the recent influential work of Mandelbrot (1982), the description of some disordered physical systems has been more and more frequently couched in fractal language. In the ensuing chapters the fractal concept has not been systematically used, though it may well be that in the coming years fractal descriptions will become routine. In this section no more will be attempted than to introduce the fractal dimension and to give some examples of things to which it is relevant. It is a subject which lends itself to pictorial illustration and the reader is strongly recommended to consult Mandelbrot (1982), Pynn and Skjeltorp (1985), and Pietronero and Tosatti (1985), and other references therein, not only for breadth and depth but also for visual impact and pleasure.

Lines, surfaces and solids may be subdivided into parts by sets of points, lines and surfaces respectively, and are therefore commonly attributed the intuitive topological dimensions $d = 1, 2, 3$. Without trying to give a formal general definition of self-similarity let us consider the subdivision of a geometrical object such that the parts are identical with each other, and are geometrically similar to, and collectively constitute, the original shape. Suppose there are N subdivisions scaled down from the original by a factor $r < 1$ in linear dimensions. For example, with $r = \frac{1}{3}$, a parallelogram could be partitioned in this way into $n = 9$ subparallelograms. We now define

$$d_f \equiv \ln N/\ln(1/r) \qquad \text{or } Nr^{d_f} = 1 \tag{2.24}$$

and consider the quantity d_f. In the parallelogram and any comparable example, $d_f = d$, whatever N and r.

Suppose now that N of the identical subdivisions can be displaced or rotated relative to one another and reunited according to some rule. d_f can then differ from d and a new class of geometrical objects appear.

Two famous and much discussed examples are the Koch lines and the Sierpinski gasket. A straight line may be reconstructed into, say, four segments ($N = 4$), each one third of the original length ($r = \frac{1}{3}$). Two successive stages of this are shown in figure 2.11. In this case, $d_f = \ln 4/\ln 3 \simeq 1.2618$ and the Koch line can evolve indefinitely, tending to an infinitely long curve, continuous but without a definite tangent at every point because it has 'a

Figure 2.11

corner almost everywhere' (Mandelbrot 1982). Small sections of the nth generation are similar to the $(n-1)$th generation. Suppose the nth generation ($n \gg 1$) were observed with a finite resolving power. If a fraction $1/q$ of it is magnified by a factor q so that previously unresolved detail can be seen, it will look like the nth generation itself. The length $L(\eta)$ clearly depends on the step size (η) of the dividers used for measuring it. In this case, $L(\frac{1}{3}\eta) = \frac{4}{3}L(\eta)$. Assuming $L(\eta) \propto \eta^{\text{const}}$, it follows from equation (2.24) that

$$L(\eta) \propto \eta^{1-d_f} \tag{2.25}$$

The Sierpinski gasket was approached in figure 1.1. The first three generations are shown in figure 2.12. The rule is to retain the three outer equilateral triangles of the four into which the black triangles of the previous generation can be divided; the white triangles are holes. $d_f = \ln 3/\ln 2 \simeq 1.585$. It too can be envisaged with $n \gg 1$, ultimately proceeding to a limit in which there is no black area—only an infinitely contorted line (see Mandelbrot (1982) for more detail).

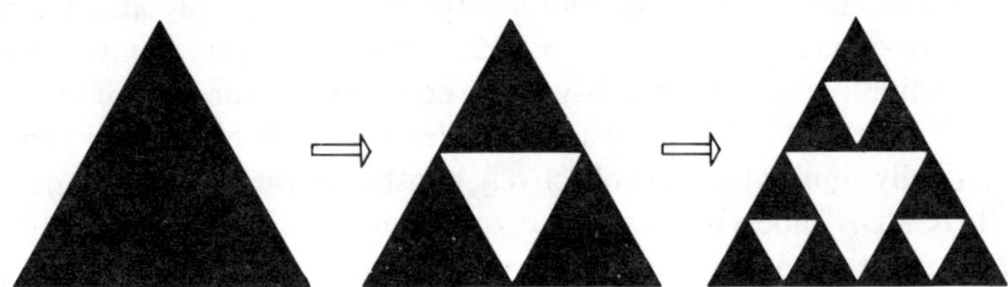

Figure 2.12

Constructs like this, of which there are infinitely many, may reasonably be said to have ordered structures but do not have translation invariance. They do have scale invariance: as $n \to \infty$, the magnification of a small part can proceed without limit until it looks like a larger part. Such objects are called self-similar fractals—self-similar because of the scale invariance and fractals because $d_f > d$. d_f is called a fractal dimension. Fractals have a large mathematical literature and are fascinating occasions for computer graphics.

The fractal dimension also emerges from other considerations. For example, consider the mass or amount of material. For uniform non-fractal objects of dimension d, the mass within a radius r rises as r^d and the density is constant. After many stages of generation, the Sierpinski gasket would be seen to contain less material (i.e. have more holes) the larger the magnification with which it was inspected and thus to have a density dependent on the scale of the observation. Suppose a uniform triangular lamina of side L has area $A(L)$ and mass $M(L)$. When $L \to 2L$, $A(2L) \to 2^d A(L)$ and $M(2L) \to 2^d M(L)$ where $d = 2$. The density, $\rho = \text{mass/area}$, is unchanged. Because of its mode of generation, the corresponding statements for a Sierpinski gasket are: $L \to 2L$, $A(2L) \to 4A(L)$, $M(2L) \to 3M(L)$ where $4 = 2^d$, $3 = 2^{d_f}$. So $\rho(2L) = 2^{(d_f - d)}\rho(L)$. In general, fractal objects scaled by any positive factor b have $M(bL) \propto b^{d_f} M(L)$ or, putting $b = L^{-1}$,

$$M(L) \propto L^{d_f} \tag{2.26}$$

It follows that

$$\rho(L) \propto L^{d_f - d}. \tag{2.27}$$

Another useful relation concerns the density correlation function, $C(L) = \langle \rho(\boldsymbol{r} + \boldsymbol{L})\rho(\boldsymbol{r}) \rangle$ where $\langle\rangle$ indicates $C(L)$ is averaged over all points $\boldsymbol{r}$. The density at distance L from an origin point in a space of dimension d will be the ratio of mass to volume in a shell of size proportional to L^{d-1}. From equation (2.26), it follows that

$$C(L) \propto L^{d_f - d}. \tag{2.28a}$$

In Chapters 3 and 8 it will be shown that density correlations like $C(L)$ are related by Fourier transformation to quantities called structure factors, $S(Q)$, which can be derived from scattered intensities in diffraction experiments, $Q \equiv (4\pi/\lambda) \sin\theta$ where 2θ is the scattering angle and λ the wavelength of the light, x-ray or neutron beam. For fractal structures obeying equation (2.28a):

$$S(Q) \propto Q^{-d_f}. \tag{2.28b}$$

Measurements of $M(L)$, $\rho(L)$, $C(L)$ or $S(Q)$ can therefore be used for the experimental determination of d_f.

The preceding paragraphs refer to geometrical objects which are exactly self-similar. When it is noticed that a Koch line is a model, albeit rather idealised and regular, of an indented coastline, it immediately looks plausible that various macroscopic objects in nature might be self-similar in some sense (Mandelbrot 1982). In addition to coastlines, examples are silhouettes of leafy trees, mountain range horizons, dendrites, jets of high-pressure water injected into liquid of higher viscosity, coral, the powdery distribution of stars, oscilloscope traces of noise, the system of channels in porous rock and many others. There are microscopic examples too: the track of a Brownian particle; the infinite percolating cluster on a lattice, or its backbone, or a very large finite cluster near the threshold (§9.2); the pattern of spin-up and spin-down sites in the Ising model near T_c (§1.11); the clustered aggregate of molecules deposited from a solution or vapour; the wavefunction of an electron in a one-dimensional disordered chain (§9.7); a many-branched polymer. Many of these examples, macroscopic and microscopic, are very hard to parametrise and have shapes suggesting words like ramified, wispy or tortuous.

Such things are not exactly self-similar. The nature of their self-similarity can be seen by photographing (notionally) a section S_1 of one of them and then enlarging a small subsection S_2 to the size of S_1; the original and the enlargement might be mistaken for each other. They are said to be statistically self-similar if they have identical statistical properties such as the distribution of angles between adjacent segments, the distributions of voids or particle sizes or whatever is an appropriate measure. Formally, a set S is statistically self-similar if it is composed of N distinct subsets, each a factor r smaller than the original, and all identical in all statistical respects with the set S scaled down by r (see, e.g., the article by Voss in Pynn and Skjeltorp (1985)).

In principle the Koch lines and Sierpinski gasket could be indefinitely generated to larger or smaller scales. For the fractal objects in nature which are statistically self-similar, there are usually upper and lower bounds to the scales it is reasonable to consider. The lower bound, a, in microscopic objects is commonly the molecular size or lattice spacing. The upper bond, ξ, could be a size limit imposed by a boundary condition or by the actual size to which a system has grown, or it might be a correlation or connectedness length such as determines the size of a finite cluster in percolation (§9.2) or a density, magnetisation or concentration fluctuation near a critical point (§1.11). In some cases, as at critical points, $\xi \to \infty$.

d_f is only one of a considerable number of fractal dimensions defined for various purposes. d_f parametrises complicated fractal shapes and describes the scaling of mass, density etc with length. Of the ensemble of possible random walks on, say, a square lattice some will be fractals and some not. Suppose $L(t)$ is the distance as the crow flies from beginning to end of a random walk of t steps. Then the ensemble average of $L(t) \propto t^{1/2}$ according to random-walk theory. If the random walker paid out a thread, like Theseus in the labyrinth, the mass of thread (assumed uniform) would be proportional to t, so we could write it $M(L) \propto \langle L \rangle^2$. This looks like a case of equation (2.26) with $d_f = 2$ but is in principle different because it expresses an average property of an ensemble, not of an individual fractal. Stanley (who discusses this example in Pynn and Skjeltorp (1985)) therefore writes $M(L) \propto L^{d_w}$ where $d_w = 2$ in this case and is another kind of fractal dimension. This holds for $a \ll L \ll \xi$.

The random walks in this ensemble are self-intersecting, i.e. a particular site may be revisited during one walk. Another problem is to find the shortest, albeit circuitous, path between sites a distance L apart given that a substantial proportion, randomly distributed, of possible steps on the lattice have been rendered impassable. In this case $M(L) \propto L^{d_{min}}$. d_w and d_{min} make the point that a variety of fractal dimensions with different significances have been devised and to some extent investigated theoretically and experimentally. Ten different ones are referred to by Stanley (Pynn and Skjeltorp 1985). More new numbers are required to express the density of states of excitations on fractal structures (see, e.g., Orbach in Pynn and Skjeltorp (1985)).

The rest of this section selects a few examples of objects thought to be fractal in order to show the use of fractal dimensions in describing properties it would otherwise be hard to characterise. First we take the path of a molecule in liquid argon. This cannot be seen but it can be simulated by molecular dynamics (§4.2). If it has any similarity to a Koch curve we expect its length $L(\eta)$ to behave like equation (2.25). Measurements on the computed trajectory showed that this was so with $d_f = 1.65$ for a sixteenfold variation in the length (Powles and Quirke 1984, Powles 1985). At large η, $L(\eta)$ tends to the end-to-end length which varies as $(6Dt)^{1/2}$ where D is the self-diffusion coefficient and t the duration. At small η, $L(\eta)$ tends to the total path length, $\sim \langle v \rangle t$, where $\langle v \rangle$ is the mean speed. The significance of thus finding a fractal dimension to a molecular trajectory was made somewhat obscure by the later discovery that d_f was a function of t, approaching two as $t \to \infty$ (Powles and Quirke 1984, Powles 1985).

Diffusion-limited aggregation, abbreviated DLA, is another occasion for fractal studies. Metal particles of size 30 to 40 Å, derived by pulsed evaporation from a heated filament, agglomerate in the ambient gas (He) and can be collected on an electron microscope grid (Forrest and Witten (1979), Witten and Sander (1981), see also Meakin *et al* in Pietronero and Tosatti (1985)). The aggregate looks like figure 2.13 and can be simulated on a lattice by the following rules: (i) a seed particle is placed on any central site; (ii) particle 2 is added at a random distant site; (iii) particle 2 randomly walks until either it reaches a site next to the seed and sticks, starting a cluster; or it reaches the lattice boundary and is removed; (iv) particles 3, 4, . . . proceed like particle 2, successively. The probability, S, of sticking is put at 1. Why is the cluster not much more compact? Apparently because a growing branch screens the centre from approaching particles and grows preferentially; after a time it becomes very improbable that a random walking particle would find its way up a channel between two branches. The model simulates DLA (Forrest and Witten 1979, Witten and Sander 1981).

By counting particles on electron micrographs of real aggregates, or by suitable programs in computer simulations, equations (2.26) to (2.28) could be tested. Typical values of d_f from Zn, Fe and SiO_2 particles, obtained from equations (2.26) and (2.27) respectively, are 1.60 ± 0.04 and 1.67 ± 0.02. The simulation result for either square or triangular lattices is about 1.7 (Forrest and Witten 1979, Witten and Sander 1981). d_f is higher in higher Euclidean dimensions, e.g., about 2.5 for $d = 3$ (Meakin 1983). The article by Sander in Pynn and Skjeltorp (1985) carries these ideas considerably further but

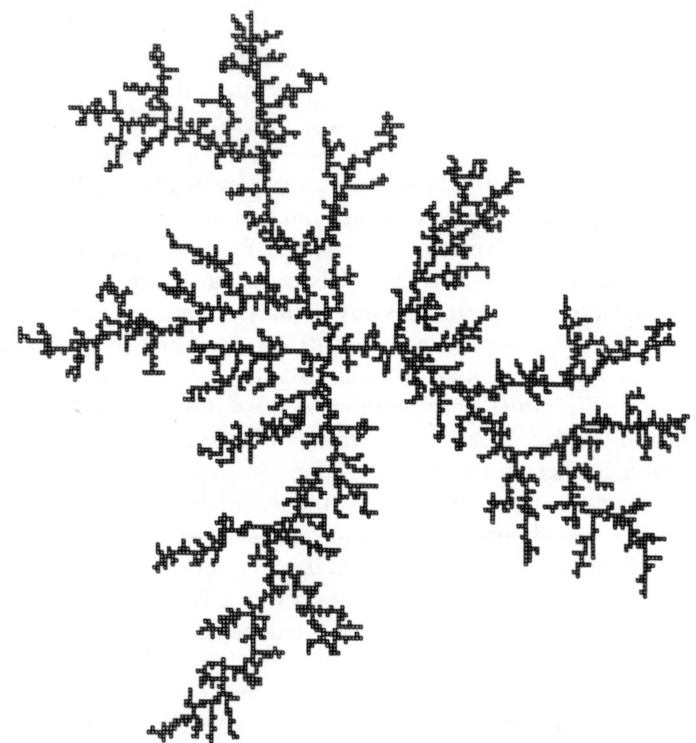

Figure 2.13 A DLA structure obtained by computer simulation with 4047 particles; $d_f \simeq 1.65$. The sticking coefficient was 1.0 but as it is decreased below ~ 0.1 the structure becomes more compact and d_f rises. (Private communication from M Grimson.)

there is much theoretical development required. It does seem clear, however, that smoke particle aggregates are fractals and that d_f is a valuable parameter in describing them. Other growth clusters, e.g., of electrodeposited ions or colloidal particles, may be similar. Indeed, clusters of colloidal particles of diameter 27 Å aggregating in 3-D have been studied by light and x-ray scattering and equation (2.28*b*) found to be obeyed with $d_f = 2.12 \pm 0.05$ (Schaefer *et al* 1984). This is less than the $d_f \sim 2.5$ from the DLA model described above and greater than $d_f \sim 1.8$ from the cluster aggregation (CA) model in which pairs, triplets, etc of particles cluster together first and then the clusters stick to each other. Colloidal aggregation is scientifically interesting and technologically important in many industries and further studies continue (Pynn and Skjeltorp 1985, Pietronero and Tosatti 1985).

As a final example we note that fractal geometry relates to percolation. In Chapter 9, percolation clusters will be introduced; they are illustrated in figures 9.1 and 9.2 and are indeed ramified structures. In an influential study, Kapitulnik *et al* (1983) used a computer to generate percolation clusters on a 2-D lattice. Only finite lattices can be used but the characteristics of the largest cluster on a finite sample approximate closely to those of a section of a true infinite cluster if the site occupation probability (p) is significantly greater than the percolation threshold (p_c). In each of many such large clusters which happened to include the central site O, a count was made of the number, $M(L)$, of sites connected to O in a square of side L centred on O. It appeared that $M(L) \propto L^{d_f}$ with $d_f \simeq 1.90$; or the density $\rho(L) = M(L)/L^2 \propto L^{d_f-2}$. This established the statistical self-similarity of the percolating clusters and discovered their fractal dimension. d_f can be connected with the percolation indices introduced in §9.2 (Schaefer *et al* 1984) and the fractal nature of percolation structures is now much studied (Pynn and Skjeltorp 1985, Pietronero and Tosatti 1985).

This section has tried to show that fractal dimensions are useful parameters for characterising various structures which, though of highly irregular form, have properties independent of scale—at least between upper and lower scale limits. The deeper problem of why the structures have the particular fractal dimensions they do will no doubt demand continuing attention.

References

Barker J A, Finney J L and Hoare M R 1975 *Nature* **257** 120
Bell R J and Dean P 1966 *Nature* **212** 1354
Bennett C H 1972 *J. Appl. Phys.* **43** 2727
Bernal J D 1959 *Nature* **183** 141
—— 1960 *Nature* **185** 68
Bernal J D, Cherry I A, Finney J L and Knight K R 1970 *J. Phys. E: Sci. Instrum.* **3** 388
Bernal J D and Mason J 1960 *Nature* **188** 910
Cargill G S and Spaepen F 1981 *J. Non-Cryst. Solids* **43** 91
Cowley J M 1950a *Phys. Rev.* **77** 669
—— 1950b *J. Appl. Phys.* **21** 24
Evans D L and King S V 1966 *Nature* **212** 1353
Finney J L 1970 *Proc. R. Soc.* A **319** 479, 495
Finney J L and Wallace J 1981 *J. Non-Cryst. Solids* **43** 165
Forrest S R and Witten T A 1979 *J. Phys. A: Math. Gen.* **12** L109
Gaskell P H 1979a *J. Phys. C: Solid State Phys.* **12** 4337
—— 1979b *J. Non-Cryst. Solids* **32** 207
Gellatly B J and Finney J L 1982 *J. Non-Cryst. Solids* **50** 313
Harris W F 1977 *Sci. Am.* **237** 130
Ichikawa T 1975 *Phys. Status Solidi* A **29** 293
Kapitulnik A, Aharony A, Deutscher G and Stauffer D 1983 *J. Phys A: Math. Gen.* **16** L269
Klémen M and Sadoc J 1979 *J. Physique Lett.* **40** L569
Mandelbrot B B 1982 *The Fractal Geometry of Nature* (San Francisco: Freeman)
Meakin P 1983 *Phys. Rev.* A **27** 1495
Mosseri R and Sadoc J F 1985 *J. Non-Cryst. Solids* **77/78** 179

Pietronero L and Tosatti E (ed) 1985 *Fractals in Physics* (Amsterdam: North-Holland)

Polk D E 1970 *Scripta Metall.* **4** 117

—— 1971 *J. Non-Cryst. Solids* **5** 365

Powles J G 1985 *Phys. Lett.* **107A** 403

Powles J G and Quirke N 1984 *Phys. Rev. Lett.* **52** 1571

Pynn R and Skjeltorp A 1985 (ed) *Scaling Phenomena in Disordered Systems, NATO ASI Series B* vol 133 (New York: Plenum)

Rivier N 1983 in *Topological Order in Condensed Matter* ed F Yonezawa and T Ninomiya (Berlin: Springer)

—— 1984 *Theory of Glass, Institute For Theoretical Physics, Santa Barbara* NSF-ITP-84-133

Rivier N and Duffy D M 1982 *J. Physique* **43** 293

Sadoc J 1981 *J. Non-Cryst. Solids* **44** 1

Sadoc J and Mosseri R 1985 *J. Physique* **46** 1809

Schaefer D W, Martin J E, Wiltzius P and Cannell D S 1984 *Phys. Rev. Lett.* **52** 2371

Scott G D 1960 *Nature* **188** 908

Srolovitz D, Maida K, Takeuchi S, Egami T and Vitek V 1981 *J. Phys. F: Met. Phys.* **11** 2209

Waseda Y 1980 *Structure of Non-Crystalline Materials* (New York: McGraw-Hill)

Witten T A and Sander L M 1981 *Phys. Rev. Lett.* **47** 1400

Wright A C 1982a *J. Non-Cryst. Solids* **48** 265

—— 1982b *J. Non-Cryst. Solids* **49** 63

Venkataraman G and Sahoo D 1985a *Contemp. Phys.* **26** 579

—— 1985b *Contemp. Phys.* **27** 3

Zachariasen W H 1932 *J. Am. Chem. Soc.* **54** 3841

3

THE EXPERIMENTAL INVESTIGATION OF DISORDERED STRUCTURES

In order to discover the distribution functions of §2.3 or to test the applicability of the models in §§2.5 and 2.6 the structure of real matter must be studied experimentally. This is almost always done by scattering experiments which are interpreted with diffraction theory. There are many standard works and valuable articles concerning this field (James 1962, Guinier 1963, Waseda 1980, Guinier and Fournet 1955, Lovesey 1984, Pings 1968, Bacon 1975, *International Tables for X-ray Crystallography* 1962, Howe 1978) and this chapter is intended to introduce the main ideas.

3.1 Scattered intensity

Some essential quantities can be defined with the help of figure 3.1. $\hbar\boldsymbol{k}_i$, $\hbar\boldsymbol{k}_f$ are the momenta of the photons, electrons or neutrons scattered by scattering centres inside the sample. One of these is chosen as the origin O; another is at P, distant $\boldsymbol{R}_\mathrm{P}$ from O. We immediately make some assumptions, namely, the scattering is elastic so $|\boldsymbol{k}_i| = |\boldsymbol{k}_f|$; more than one scattering event in the sample is so improbable that only single scattering need be considered. These assumptions by no means always hold but the formulae resulting from them are indispensible starting points. We define as follows the momentum transfer vector, $\boldsymbol{Q}$, and the connection with the scattering angle, θ:

$$\hbar\boldsymbol{Q} \equiv \hbar(\boldsymbol{k}_i - \boldsymbol{k}_f) \qquad |\boldsymbol{Q}| = 2k \sin \tfrac{1}{2}\theta. \tag{3.1}$$

We assume the incident waves are unpolarised.

If there is an isolated scattering centre at O the scattered wave has the form $(f(\theta)/r)\exp[\mathrm{i}(kr - \omega t)]$ where r is the distance in the θ-direction. Waves from P and other scattering centres arrive at the distant detector with different phases because of their various path lengths. The phase difference for waves from O and P is $\boldsymbol{Q} \cdot \boldsymbol{R}_\mathrm{P}$, consequently the intensity at the detector is

$$I(\boldsymbol{Q}) = A\left|\sum f_\mathrm{P}(\boldsymbol{Q}) \exp(\mathrm{i}\boldsymbol{Q} \cdot \boldsymbol{R}_\mathrm{P})\right|^2 \tag{3.2}$$

where the sum is over all the scattering centres such as P, and $\boldsymbol{Q}$ is used as a variable instead of θ; A is a proportionality constant. The quantity $|f(\theta)|^2$ is a cross section and will depend on the nature of the incoming waves and of the scattering centre; $f(\theta)$ is a *scattering amplitude*.

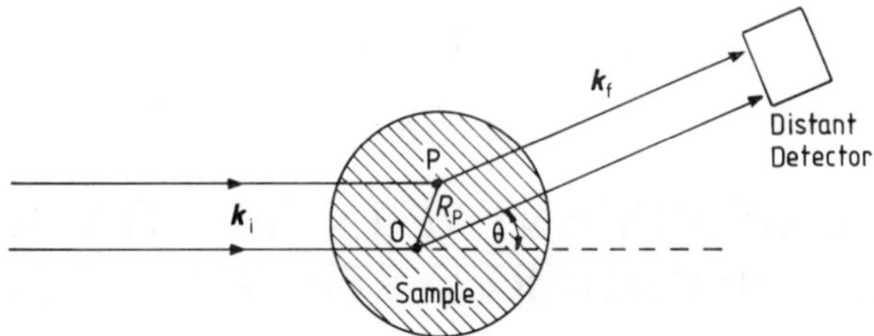

Figure 3.1 Definitions of quantities for the coherent scattering calculation.

3.2 X-rays

X-ray radiation is scattered by electrons and it is conventional to express $I(\boldsymbol{Q})$ in terms of the intensity scattered by a single electron. A single electron at O would give

$$I_e(\boldsymbol{Q}) = A|f_e(\boldsymbol{Q})|^2 \tag{3.3}$$

where $|f_e(\boldsymbol{Q})|^2$ is the known cross section for an electron.

What, for x-rays, is to be regarded as a 'scattering centre'? Since electron wavefunctions give a continuous distribution of position probability it is reasonable to take a volume element of sample, $d\boldsymbol{r}$, containing electron charge $en_e(\boldsymbol{r})\,d\boldsymbol{r}$, as the scattering centre and replace the sum in equation (3.2) by an integral over the sample. The scattering from $d\boldsymbol{r}$ then has amplitude proportional to $f_e(\boldsymbol{Q})n_e(\boldsymbol{r})\,d\boldsymbol{r}$ and, using equation (3.3), we have

$$I_{eu}(\boldsymbol{Q}) \equiv I(\boldsymbol{Q})/I_e(\boldsymbol{Q}) = \left|\int n_e(\boldsymbol{r}) \exp(i\boldsymbol{Q}\cdot\boldsymbol{r})\,d\boldsymbol{r}\right|^2 \tag{3.4}$$

where $n_e(\boldsymbol{r})$ is the total electron number density and the quantity in equation (3.4) is called the intensity in the Q-direction in electron units.

Even in metals, where the conduction electron wavefunctions permeate the entire sample, the charge density still resembles that of a set of separate atoms to a very good approximation. We therefore divide expression (3.4) into contributions from N identical fixed spherical atoms with centres at $\{\boldsymbol{R}_i\}$. Then

$$n_e(\boldsymbol{r}) \to \sum_{i=1}^{N} n_{ea}(\boldsymbol{r} - \boldsymbol{R}_i) \tag{3.5}$$

where $n_{ea}(\boldsymbol{r} - \boldsymbol{R}_i)$ is the electron density in the ith atom at a distance $(\boldsymbol{r} - \boldsymbol{R}_i)$ from its centre and

$$I_{eu}(\boldsymbol{Q}) = \left|\int \sum_{i=1}^{N} n_{ea}(\boldsymbol{r} - \boldsymbol{R}_i) \exp(i\boldsymbol{Q}\cdot\boldsymbol{r})\,d\boldsymbol{r}\right|^2. \tag{3.6a}$$

If we multiply and divide under the summation by $\exp(\mathrm{i}\boldsymbol{Q}\cdot\boldsymbol{R}_i)$, this becomes

$$I_{\mathrm{eu}}(\boldsymbol{Q}) = \left|\sum_{i=1}^{N} \exp(\mathrm{i}\boldsymbol{Q}\cdot\boldsymbol{R}_i)\right|^2 |f_{\mathrm{a}}(\boldsymbol{Q})|^2 \tag{3.6b}$$

where

$$f_{\mathrm{a}}(\boldsymbol{Q}) \equiv \int_{\mathrm{atom}} n_{\mathrm{ea}}(\boldsymbol{r}) \exp(\mathrm{i}\boldsymbol{Q}\cdot\boldsymbol{r})\, \mathrm{d}\boldsymbol{r}. \tag{3.6c}$$

$f_{\mathrm{a}}(\boldsymbol{Q})$ is an important quantity known as the atomic scattering factor in electron units. It is known and tabulated (*International Tables for X-ray Crystallography* 1962) and will not be discussed in detail (see figure 3.2).

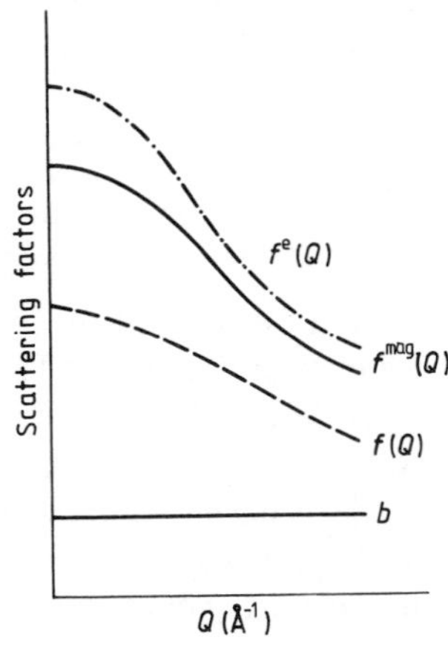

Figure 3.2 Single-centre scattering factors for: electrons by atoms (f^{e}), neutrons by magnetic ions (f^{mag}), x-rays by atoms (f), and neutrons by nuclei (b) (from Waseda 1980).

Equation (3.6) achieves a separation of the intensity into two factors one characteristic of the atoms and the other depending on where they are. The latter factor may be written as

$$\left|\sum_{i=1}^{N} \exp(\mathrm{i}\boldsymbol{Q}\cdot\boldsymbol{R}_i)\right|^2 \equiv NS(\boldsymbol{Q}) \tag{3.7}$$

which defines $S(\boldsymbol{Q})$, *the structure factor*. $S(\boldsymbol{Q})$ will claim a lot of attention. Let us first relate it to equation (2.2*a*) for $\nu^{(1)}(\boldsymbol{r})$. Since $\nu^{(1)}$ is the density function we expect, in analogy with equation (3.4), that

$$I_{\mathrm{eu}}(\boldsymbol{Q}) = |f_{\mathrm{a}}(\boldsymbol{Q})|^2 \left|\int \nu^{(1)}(\boldsymbol{r}) \exp(\mathrm{i}\boldsymbol{Q}\cdot\boldsymbol{r})\, \mathrm{d}\boldsymbol{r}\right|^2 \tag{3.8a}$$

and from equations (3.6*b*) and (3.7)

$$NS(\boldsymbol{Q}) = \int v^{(1)}(\boldsymbol{r}_1) \exp(\mathrm{i}\boldsymbol{Q}\cdot\boldsymbol{r}_1)\, \mathrm{d}\boldsymbol{r}_1 \times \int v^{(1)}(\boldsymbol{r}_2) \exp(-\mathrm{i}\boldsymbol{Q}\cdot\boldsymbol{r}_2)\, \mathrm{d}\boldsymbol{r}_2. \quad (3.8b)$$

By substituting equation (2.2*a*) into (3.8*b*), equation (3.7) is readily recaptured. If we now rewrite equation (3.8*b*) as

$$NS(\boldsymbol{Q}) = \int v^{(1)}(\boldsymbol{r}_1) v^{(1)}(\boldsymbol{r}_2) \exp[\mathrm{i}\boldsymbol{Q}\cdot(\boldsymbol{r}_1 - \boldsymbol{r}_2)]\, \mathrm{d}\boldsymbol{r}_1\, \mathrm{d}\boldsymbol{r}_2$$

and change the variables so that $\boldsymbol{r}_2 \to \boldsymbol{r}'$, $\boldsymbol{r}_1 - \boldsymbol{r}_2 \to \boldsymbol{r}$ we have

$$NS(\boldsymbol{Q}) = \int P_{\mathrm{a}}(\boldsymbol{r}) \exp(\mathrm{i}\boldsymbol{Q}\cdot\boldsymbol{r})\, \mathrm{d}\boldsymbol{r} \quad (3.9a)$$

where

$$P_{\mathrm{a}}(\boldsymbol{r}) \equiv \int v^{(1)}(\boldsymbol{r} + \boldsymbol{r}') v^{(1)}(\boldsymbol{r}')\, \mathrm{d}\boldsymbol{r}'. \quad (3.9b)$$

$P_{\mathrm{a}}(\boldsymbol{r})$ is called the autocorrelation function of the atomic positions.

There is no statistical mechanics in this so far. The structure, represented by $S(\boldsymbol{Q})$, is that of a fixed array. We could calculate $S(\boldsymbol{Q})$ from $\{\boldsymbol{R}_i\}$, taking the $\{\boldsymbol{R}_i\}$ from one of the models in the previous chapter. Model $S(\boldsymbol{Q})$ have been found this way. If the model is big enough it will contain many different local configurations and its $S(\boldsymbol{Q})$ will approximate more or less to the configuration average which a scattering experiment would obtain for a real material.

3.3 The average $S(\boldsymbol{Q})$ for a real sample

To find the ensemble or configuration average of $S(\boldsymbol{Q})$ we need that of $P_{\mathrm{a}}(\boldsymbol{r})$ for use in equation (3.9*a*). We now show that $\langle P_{\mathrm{a}}(\boldsymbol{r})\rangle$ is related to the $g(r)$ of equation (2.7).

Now, from equations (3.9*b*) and (2.2*a*) we have

$$P_{\mathrm{a}}(\boldsymbol{r}) = \int \sum_i \delta(\boldsymbol{r} + \boldsymbol{r}' - \boldsymbol{R}_i) \sum_j \delta(\boldsymbol{r}' - \boldsymbol{R}_j)\, \mathrm{d}\boldsymbol{r}' = \sum_{i,j} \delta(\boldsymbol{r} + \boldsymbol{R}_j - \boldsymbol{R}_i).$$

Put $\boldsymbol{R}_j = \boldsymbol{R}_1$ and sum over i using $\boldsymbol{R}_1$ as origin (see figure 3.3). This gives $\Sigma_i\, \delta(\boldsymbol{r} - \boldsymbol{R}_i)$ which is the density function at a distance $\boldsymbol{r}$ away from the point $\boldsymbol{R}_1 - a$ at which there is also a particle. The average of this is exactly the function $z(\boldsymbol{r})$ in equation (2.8). Similarly for $\boldsymbol{R}_j = \boldsymbol{R}_2$, $\boldsymbol{R}_j = \boldsymbol{R}_3$ etc, giving N

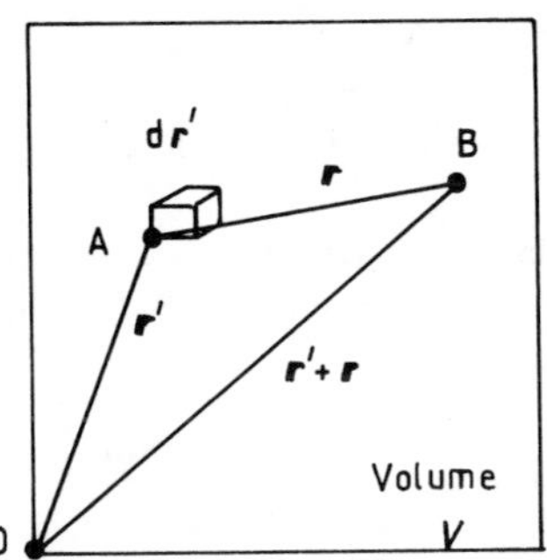

Figure 3.3 Definitions of quantities.

similar terms. Thus finally,

$$\langle P_a(\boldsymbol{r})\rangle = N(n_0 g(\boldsymbol{r}) + \delta(\boldsymbol{r}))$$

$$\begin{aligned}\langle S(\boldsymbol{Q})\rangle &= N^{-1}\int_V \langle P_a(\boldsymbol{r})\rangle \exp(\mathrm{i}\boldsymbol{Q}\cdot\boldsymbol{r})\,\mathrm{d}\boldsymbol{r}\\ &= \int [n_0(g(\boldsymbol{r})-1) + n_0 + \delta(\boldsymbol{r})]\exp(\mathrm{i}\boldsymbol{Q}\cdot\boldsymbol{r})\,\mathrm{d}\boldsymbol{r}\\ &= n_0\int_V (g(\boldsymbol{r})-1)\exp(\mathrm{i}\boldsymbol{Q}\cdot\boldsymbol{r})\,\mathrm{d}\boldsymbol{r} + (2\pi)^3 n_0\delta(\boldsymbol{Q}) + 1.\end{aligned} \tag{3.10}$$

To reach equation (3.10) we have used the properties of the δ-function, notably

$$\delta(\boldsymbol{Q}) = (2\pi)^{-3}\int \exp(\mathrm{i}\boldsymbol{Q}\cdot\boldsymbol{r})\,\mathrm{d}\boldsymbol{r}.$$

There is also a tacit assumption that the volume V is a representative subsection of an infinite volume with uniform properties rather than a finite sample with a bounding surface. The $\delta(\boldsymbol{Q})$ term implies a strong scattering intensity in the forward direction with $|\boldsymbol{Q}| = 0 = \theta$. This peak would fall in the incident beam and not be observed in practice but should in any case be replaced by another term which arises as follows. Strictly, when $\mathrm{d}\boldsymbol{r}'$ goes over the sample volume during integration, $v^{(1)}(\boldsymbol{r}'+\boldsymbol{r})$ falls to zero outside the body and the exact value of integral depends on the shape and size of the sample. In Guinier (1963) it is shown that the second term of equation (3.10) becomes

$$n_0 V^{-1}\left|\int \sigma(\boldsymbol{r})\exp(\mathrm{i}\boldsymbol{Q}\cdot\boldsymbol{r})\,\mathrm{d}\boldsymbol{r}\right|^2$$

where $\sigma(\boldsymbol{r}) = 1$ inside and 0 outside the sample. This term does not contain information about the structure and it is negligible unless Q is very small in which event θ would be too near the incident direction for observation. Thus to a very good approximation

$$\langle S(\boldsymbol{Q})\rangle = 1 + n_0 \int (g(\boldsymbol{r}) - 1)\exp(\mathrm{i}\boldsymbol{Q}\cdot\boldsymbol{r})\,\mathrm{d}\boldsymbol{r}. \qquad (3.11a)$$

For an isotropic material—which most disordered substances are—$\langle S(\boldsymbol{Q})\rangle \to \langle S|\boldsymbol{Q}|\rangle$ which is often written $S(Q)$ with the averaging bracket tacitly assumed, $g(\boldsymbol{r}) \to g(r)$, and

$$S(Q) = 1 + n_0 \int_0^\infty (g(r) - 1)\frac{\sin Qr}{Qr}\,4\pi r^2\,\mathrm{d}r. \qquad (3.11b)$$

Finally,

$$I_{\mathrm{eu}}(Q) = N|f_{\mathrm{a}}|^2 S(Q) \qquad (3.11c)$$

and

$$S(Q) \to 1 \text{ as } Q \to \infty. \qquad (3.11d)$$

3.4 The importance of the structure factor

In crystals, $S(\boldsymbol{Q})$ gives the Bragg peaks and is at the foundations of crystallography. For materials without LRO, equation (3.11*b*) enables $g(r)$ to be found if $S(Q)$ is inferred from the observed intensity—though this inference is not without difficulty (§3.8). In any event, the Fourier transformation of equation (3.11*b*) gives

$$g(r) = 1 + \frac{1}{8\pi^3 n_0}\int_0^\infty (S(Q) - 1)\frac{\sin Qr}{Qr}\,4\pi Q^2\,\mathrm{d}Q. \qquad (3.12)$$

The structure factor is therefore the route to the pair distribution, the RDF and the coordination number.

$S(Q)$ and $g(r)$ have strong connections with many theoretical functions in statistical mechanics, e.g., in liquid state theory the pressure is

$$p = n_0 k_{\mathrm{B}} T - \tfrac{1}{6} n_0^2 \int g(r)\frac{\mathrm{d}\varphi(r)}{\mathrm{d}\ln r}\,\mathrm{d}\boldsymbol{r} \qquad (3.13)$$

where $\varphi(r)$ is the interatomic potential (§4.3).

As we have already seen $S(Q)$ can be calculated for computer or other models in order to compare with observed $S(Q)$. In theories of electronic and many other properties of liquid metals or semiconductors, $S(Q)$ is almost invariably required and its measured value is used for numerical calculations.

In all these applications $S(Q)$ or $g(r)$—which we have seen to be interconvertible—carry the structural information. Because of the single-scattering assumption and the use of diffraction theory only the interparticle distance vectors were involved in the intensity expression, equation (3.2). Because there are no periodicities or preferred directions in the structure, only the distribution of interparticle distances, i.e. $g(r)$, can be inferred. Scattering experiments do not determine $g^{(3)}$, $g^{(4)}$ etc. This is a considerable limitation.

Before giving examples of the practical use of these formulae two generalisations are required, firstly to mixtures of different atomic species and secondly to waves other than x-rays.

3.5 Formulae for binary mixtures

Returning to the amplitude summation in equation (3.2) and treating the atoms as scattering centres we may rewrite it for a binary system as

$$I(\boldsymbol{Q}) = A\left|\sum_k f_{\rm A}(Q)\exp(\mathrm{i}\boldsymbol{Q}\cdot\boldsymbol{R}_k^{(\rm A)}) + \sum_l f_{\rm B}(Q)\exp(\mathrm{i}\boldsymbol{Q}\cdot\boldsymbol{R}_l^{(\rm B)})\right|^2 \quad (3.14)$$

where two types of atom, A and B, each have their characteristic scattering factors. The two sums are over the $N_{\rm A}$ A atoms and $N_{\rm B}$ B atoms respectively. It is clear that the intensity will have terms in $f_{\rm A}^2$, $f_{\rm B}^2$ and $2f_{\rm A}f_{\rm B}$ associated with A-type, B-type and AB-type structure factors. These terms introduce *partial structure factors* S_{ij}, which are related by a Fourier transformation like equation (3.12) to the partial pair distributions introduced in §2.4. The results for a homogeneous isotropic binary material follow from the same reasoning as that of §§3.2 and 3.3 and are:

$$I_{\rm eu}(Q) = NF(Q) \quad (3.15a)$$

$$F(Q) \equiv f_{\rm A}^2(c_{\rm A}c_{\rm B} + c_{\rm A}^2 S_{\rm AA}(Q)) + f_{\rm B}^2(c_{\rm A}c_{\rm B} + c_{\rm B}^2 S_{\rm BB}(Q)) + 2f_{\rm A}f_{\rm B}c_{\rm A}c_{\rm B}(S_{\rm AB}(Q) - 1) \quad (3.15b)$$

$$S_{ij} \equiv 1 + n_0\int_0^\infty (g_{ij}(r) - 1)\frac{\sin Qr}{Qr}\,4\pi r^2\,\mathrm{d}r \quad (3.15c)$$

$$g_{ij}(r) = 1 + \frac{1}{8\pi^3 n_0}\int (S_{ij}(Q) - 1)\frac{\sin Qr}{Qr}\,4\pi Q^2\,\mathrm{d}Q \quad (3.15d)$$

$$S_{\rm AB} = S_{\rm BA} \quad (3.15e)$$

$$n_0 \equiv (N_{\rm A} + N_{\rm B})/V \qquad c_i \equiv N_i/N \qquad i, j = \text{A or B}. \quad (3.15f)$$

When $c_{\rm A} = 1$, $c_{\rm B} = 0$, equation (3.15) reduces to the formulae of §3.3. Equation (3.15*c*) is the definition of the partial structure factor in terms of the partial pair distribution and leads to the expression (3.15*b*) for the intensity. Other definitions of the partial structure factors are conceivable with corresponding intensity formulae (see §3.12). Equation (3.15*e*) results from equation (2.14*b*).

3.6 Incoherent and inelastic scattering

Scattering can be coherent or incoherent, elastic or inelastic. Coherent scattering is that for which the angular dependence of the intensity is determined by the interference of scattered wavelets from different scattering centres. It depends on the structure factors. The incoherent scattering has an angular dependence affected only by the intrinsic cross sections of the scattering centres and not by their relative positions. Let us label EC, IC, EI and II the pairings elastic–coherent, inelastic–coherent, elastic–incoherent and inelastic–incoherent, respectively.

The x-ray scattering in equation (3.11*c*) is type EC. In equation (3.15) for the binary system there are EC-type and EI-type terms. The latter may be written $c_A c_B(f_A - f_B)^2$ or $\langle\langle f^2 \rangle\rangle - \langle\langle f \rangle\rangle^2$ where $\langle\langle x \rangle\rangle \equiv c_A x_A + c_B x_B$. X-rays are subject to Compton scattering which can be shown to be of type II.

For measuring structure factors, incoherent and inelastic scattering have a certain amount of nuisance value though they are interesting in their own right. For example, Compton scattering has to be subtracted from the observed intensity before processing the data. This point recurs in neutron scattering.

3.7 Neutron and electron scattering

An aspect of nuclear reactor technology is the provision of monoenergetic neutron beams in wavelengths suitable for crystallography. The formulae of the preceding sections are used except that the atomic scattering factor $f_a(Q)$ is replaced by an appropriate quantity representing neutron scattering by the nucleus. This quantity is the scattering length, b, which is related to the bound-atom cross section, σ_s, by $\sigma_s = 4\pi b^2$. Only s-wave scattering is important at the energies involved (Bacon 1975) and this is isotropic (figure 3.2).

Neutron scattering is considerably more complicated than x-ray scattering, however. A chemically pure sample is not necessarily isotopically pure. If there were two isotopes with scattering lengths b_1, b_2, we should expect an equation like (3.15*b*) to apply. However, there would be only one structure factor to consider since the atomic distribution is determined by chemical not nuclear effects. The three S-terms therefore reduce to $\langle\langle b \rangle\rangle^2 S$ and the incoherent term to $\langle\langle b^2 \rangle\rangle - \langle\langle b \rangle\rangle^2$. In other words a weighted mean scattering length deals with isotopic mixtures. However, the incoherent term will be present even in an isotopically pure sample because the system (nucleus + neutron) can exist in two spin states with different b. An incoherent term of the form $\langle\langle b^2 \rangle\rangle - \langle\langle b \rangle\rangle^2$ arises from the relative weight of the two spin states. The meaning of b, written without brackets, will therefore be extended to stand for a mean scattering length weighted for both isotopic mixture and spin states. $\langle\langle b \rangle\rangle$ will mean $c_A b_A + c_B b_B$ as in §3.6.

More serious still is the inelasticity of neutron scattering. The important quantity is not really $S(\boldsymbol{Q})$ but $S(\boldsymbol{Q}, \omega)$ where $\hbar\omega$ is the energy lost by the neutron in its collision. Mechanics gives

$$Q^2 = 2k_i^2\left[1 - \frac{m\omega}{\hbar k_i^2} - \left(1 - \frac{2m\omega}{\hbar k_i^2}\right)^{1/2} \cos\theta\right] \tag{3.16}$$

where $\hbar k_i$ is the initial momentum of the neutron. If $\omega = 0$ and $k_f = k_i$, equation (3.16) reduces to equation (3.1), and this is valid for x-radiation because the energy transfers are very small. In neutron practice, however, θ alone does not define Q uniquely and a detector receiving at an angle θ records a coherent intensity, *not* $Nb^2S(Q)$ in analogy with equation (3.11*c*), but

$$I(\theta) = Nb^2 \int_{-\infty}^{E_i/\hbar} D(k_f)\frac{k_f}{k_i} S(\boldsymbol{Q}, \omega)\, d\omega \tag{3.17}$$

where the integral has the following meaning. It integrates over all the energies of the incoming neutrons with a weighting factor, $D(k_f)$, expressing the efficiency of the detector as a function of neutron speed. Since the observable is the counting rate, the factor k_f/k_i is required because the rate is proportional to the speed of arrival. If the effective spread of ω were small enough for k_f to be essentially constant—the so called static approximation—the coherent intensity from an element becomes proportional to $Nb^2S(\boldsymbol{Q})$. This is not really good enough in practice and there are methods of correcting for inelastic scattering. In the terms of §3.6, neutron scattering has EC, IC, EI and II contributions. The underlying theory can be pursued in Lovesey (1984) and Howe (1978).

Because of its magnetic moment a neutron can also be scattered by a magnetic interaction with unpaired electron states or spins. The cross section for this resembles f_a in that it depends on the atomic structure (figure 3.2).

Electrons are scattered by the screened Coulomb fields of the atoms. The scattering factor, $f_{el}(Q)$, is related to the $f_a(Q)$ for x-rays because of the electrostatic connection between field and charge density (see figure 3.2). The relation is

$$f_{el}(Q) = \frac{8\pi^2 me^2}{h^2}\left(\frac{Z - f_a(Q)}{Q^2}\right). \tag{3.18}$$

Electrons are subject to inelastic and multiple scattering especially in higher-Z elements. They are suited to measurements on thin amorphous layers. An example with a technical description is given in Leung and Wright (1974).

3.8 Some experimental considerations. Pure samples

The preceding sections show that the single scattering cross section and the structure factors are two of the influences that determine the observed intensity. A third influence is a whole variety of extraneous phenomena that have to be corrected for as part of the data processing. Exactly how to do the corrections has to be thought out for each experiment and can be found in original papers. The problem is merely illustrated by the following discussion which is far from exhaustive.

Consider the following formulae for x-ray, neutron and electron observed intensities:

$$I^{\text{obs}}(Q)\ (\text{x-ray}) = \alpha\beta PN(f_a^2 S(Q) + f_c^2(Q) + \Delta) \tag{3.19a}$$

$$I^{\text{obs}}(Q)\ (\text{neutrons}) = \alpha\beta N\{[b_{\text{coh}}^2(S(Q) + p(Q)) + b_{\text{inc}}^2(1 + p(Q))] + \Delta\} \tag{3.19b}$$

$$I^{\text{obs}}(Q)\ (\text{electrons}) = \alpha\beta N(f_{\text{el}}^2 S(Q) + f_{\text{in}}^2 + \Delta) + I_{\text{substrate}}. \tag{3.19c}$$

The factor α corrects for the effect on the intensity of absorption in the sample and its container if any. It depends on the material and geometry and is well known for standard cases, e.g. cylindrical samples in cylindrical cells (Waseda 1980). P refers to the polarisation of x-rays and depends on whatever reflections are imposed on the rays by the instrumental design.

f_c refers to the inelastic incoherent Compton scattering and its known theoretical value can be inserted. f_{in}^2, referring to the inelastic scattering of electrons, must also be inserted. In both cases the effect is not necessarily very large and can be removed by extra instrumentation, e.g. by putting a velocity filter or monochromator in the scattered beam to reject electrons or photons which have lost energy.

Δ is for multiple scattering which was excluded from the discussion in §3.1. It is small and is sometimes neglected for x-rays, but it is more serious with electrons and neutrons. In neutron scattering Δ is isotropic and can be subtracted (Howe 1978). The factors $p(Q)$ in equation (3.19b) are the corrections for inelastic neutron scattering. These are small for heavy elements but are serious at low Z (Howe 1978, Dahlberg and Kunsch 1983). $I_{\text{substrate}}$ in equation (3.19c) is the intensity scattered from the foil or substrate on which the very thin samples required in electron diffraction are often supported.

There remains the product βN which converts the factors f^2 and b^2 which are expressed in absolute units into the observed counting rates. This is quite a difficult matter. $S(Q) \to 1$ as $Q \to \infty$ and it will be seen in §3.12 that $S(0) = n_0 \kappa_T k_B T$ where κ_T is the isothermal compressibility. If either or both these limiting values were inserted in equation (3.19) with the observed value of $I(Q)$ for very large or very small Q, βN could be calculated. This is somewhat unsatisfactory because high- and low-Q are the ranges in which $I(Q)$ is least accurate: with x-rays and electrons the scattered intensity is small at high Q. Nevertheless versions of this method have been used one of which is, from equation (3.19*a*),

$$\beta N = \int_{Q_1}^{Q_2} (f_a^2 + f_c^2 + \Delta)\, dQ \left((\alpha P)^{-1} \int_{Q_1}^{Q_2} I^{\mathrm{obs}}(Q)\, dQ \right)^{-1} \quad (3.20a)$$

where Q_1 is a fairly high value, say, 8 Å^{-1}, and Q_2 is the observed upper limit of Q which might be 15 Å^{-1}. $S(Q) \simeq 1$ in the numerator. Alternatively, use can be made of the fact that the Fourier transform of $S(Q)$ is $g(r)$ (equation (3.12)) and that for $r \to 0$, $g(r) = 0$ because of the mutual impenetrability of atoms. This leads to (Waseda 1980, Pings 1968)

$$\beta N = \int_0^{Q_2} [Q^2(f_a^2 + f_c^2 + \Delta)\, dQ - 2\pi^2 n_0] \left((\alpha P)^{-1} \int_0^{Q_2} I^{\mathrm{obs}}(Q)\, dQ \right)^{-1}. \quad (3.20b)$$

In neutron experiments this problem can be solved another way because the scattering from the element vanadium is over 99 % incoherent and is consequently isotropic. A solid vanadium replica substituted for the sample will give scattering according to equation (3.19*b*) with N and b^2 known and the coherent term negligible. This gives β which will apply also to the sample. N and b^2 for the latter must be known to extract $S(Q)$ from the sample scattering (North *et al* 1968).

These numerous corrections, and others not referred to, are applied to intensities observed in either reflection or transmission with diffraction equipment or, if $S(Q, \omega)$ is required, with neutron spectrometers (see figure 3.4). Typically, unless special measures are taken, $\sim 0.5 < Q < \sim 12$ Å^{-1}. Good measurements of $S(Q)$ might achieve an overall accuracy at the first peak of about ± 2 %: there are examples in §3.11. Modern radiation sources, like synchrotrons for x-rays and spallation neutron sources, are extending observable Q-values to 40 Å^{-1} and more (White and Windsor 1984, Biggin 1986).

3.9 From $S(Q)$ to $g(r)$

Assuming the best $S(Q)$ has been extracted from $I^{\mathrm{obs}}(Q)$, one use for it is to obtain $g(r)$ from equation (3.12). Computing a Fourier transform is a routine procedure but errors in $g(r)$ result from the limited range of Q over which $S(Q)$ can be measured and from the errors in $S(Q)$ itself. Probably the most serious matter is the truncation of $S(Q)$ at $Q_{\max}$. This means that equations like (3.12) and (3.15*d*) have in their integrands a factor $M(Q)$ which is a step function of magnitude unity which cuts off to zero at $Q = Q_{\max}$. The implication is that the transform will be a distorted $g(r)$ containing spurious ripples particularly noticeable below the first peak and

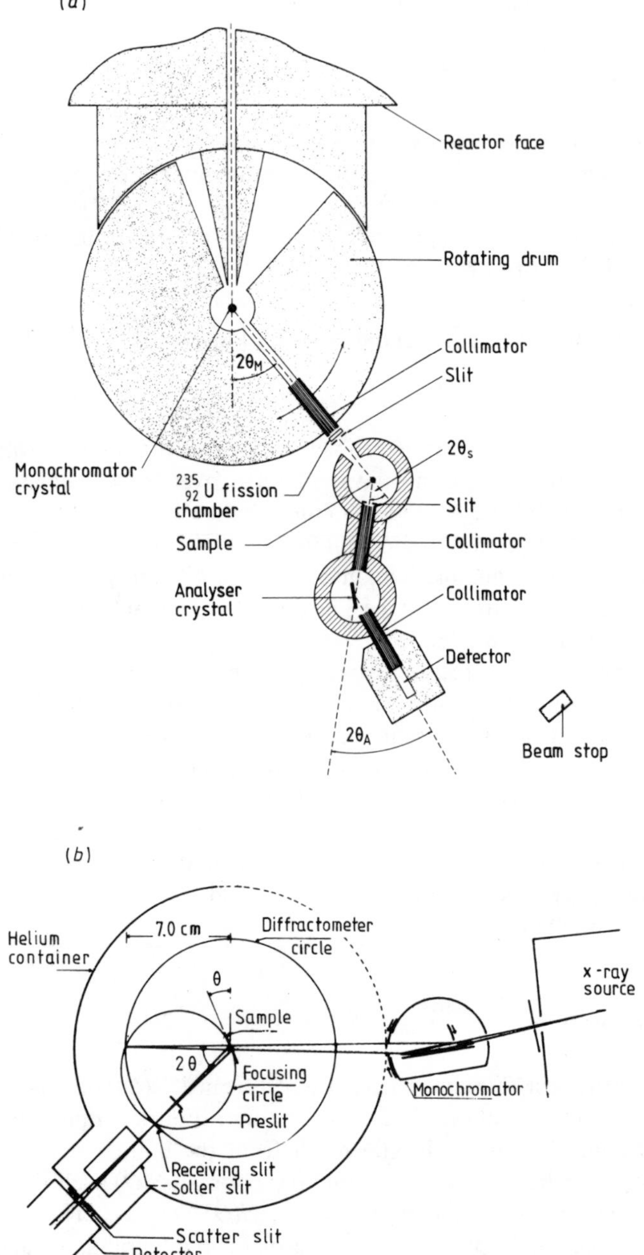

Figure 3.4 (a) Layout of a triple-axis neutron spectrometer for measuring $S(Q, \omega)$ (for details see White and Windsor 1984; Biggin 1986). (*b*) Layout of equipment for measuring $S(Q)$ by transmitted x-rays (from Van der Lugt 1979).

these can lead to erroneous structural inferences. Error analysis and correction techniques have been devised (Kaplow *et al* 1965, Wright 1974). One of these involves transforming $S(Q)$ and $g(r)$ cyclically into one another and removing obvious truncation errors from $g(r)$. This operation continues until $g(r)$ is zero and devoid of spurious ripples at small r, but will also transform into an $S(Q)$ which agrees with the observed one within experimental error. However, this is not the only approach and may be criticised on the grounds that $g(r)$ is not necessarily zero at small r if inferred from x-rays scattered by electrons. There is really no substitute for non-existent data and the removal of spurious ripples from $g(r)$ can only be achieved by sacrificing something elsewhere. Some workers prefer to do this by using an $M(Q)$ which is not a step function but a more gentle cut-off such as

$$M(Q) = \sin(\pi Q/Q_{\max})/(\pi Q/Q_{\max}).$$

This certainly reduces spurious ripples but broadens somewhat the peaks in the transform, i.e., it spoils the resolution.

3.10 Binary systems and isotopic substitution

It is possible to Fourier transform the intensity from the binary system to obtain a total pair distribution function, $g_T(r)$. From equations (3.15*b*) and (3.15*d*),

$$F(Q) - \langle\langle f^2 \rangle\rangle = \sum_{i,j} c_i c_j f_i f_j (S_{ij} - 1) \qquad (3.21a)$$

and if we Fourier transform this according to equation (3.15*d*) there results

$$\frac{1}{8\pi^3 n_0}\int (F(Q) - \langle\langle f^2 \rangle\rangle)\frac{\sin Qr}{Qr}\, 4\pi Q^2\, \mathrm{d}Q = \sum_{i,j} c_i c_j f_i f_j (g_{ij} - 1). \quad (3.21b)$$

Let us define $g_T(r)$ by

$$\langle\langle f^2 \rangle\rangle (g_T(r) - 1) \equiv \sum_{i,j} c_i c_j f_i f_j (g_{ij} - 1). \qquad (3.21c)$$

Equation (3.21*b*) then shows how $g_T(r)$ can be found from the observed $F(Q)$.

It is by no means impossible to make useful inferences from $g_T(r)$ obtained in this way but it would clearly reveal more if the three g_{ij}'s could be measured separately.

Since the three S_{ij} are independent, three separate determinations of $F(Q)$ are required to measure them with three distinctly different ratios of f_A to f_B. Varying c_A, c_B is not an alternative because in general S_{ij} varies with c. Except for the special case of anomalous dispersion (Waseda 1980), $f(Q)$ for x-rays is not itself open to variation though one could substitute one or both elements by others which do not affect the structure. This is 'isomorphous substitution', e.g., substitute Zr for Hf in amorphous NiHf alloys (Wagner 1980). Alternatively three different radiations (x-rays, electrons, neutrons) could be used on the same mixture. A third possibility is to alter the neutron scattering lengths by changing the isotopic abundances in a mixture of fixed chemical composition (isotopic substitution).

This kind of programme was first envisaged by Keating in 1963, and first implemented by Enderby, North and Egelstaff in 1966 (Enderby *et al* 1966). Since then the method of isotopic substitution has been exploited, especially by Enderby and co-workers, to find S_{ij} and therefore g_{ij} in liquid and amorphous alloys and semiconductors, molten salts and solutions. Sometimes a combination of neutron, with x-ray or electron, scattering experiments is suitable.

Clearly the isotopic method is limited by the availability of suitable isotopes. The $F(Q)$ resulting from the three experiments must be sufficiently different to make the solution of the three simultaneous equations for S_{ij} a practical possibility (Edwards *et al* 1975, Biggin and Enderby 1982). All the errors and corrections involved in neutron scattering will be present, enhanced by the necessity of combining three results. There are some interesting possibilities to exploit however. Since scattering lengths can be negative, b for an element can be negative or zero and to some extent disposable if isotopes can be mixed. An example is given in §3.11.

3.11 Examples of structure investigations

A large number of examples are given in Waseda (1980). In this section a few are selected from the literature to illustrate the achievements possible with the methods outlined in the previous sections.

X-rays in transmission were used by Greenfield *et al* (1971) to provide accurate tabulated structure factors for liquid Na and K. Points with Q-values as low as 0.3 Å^{-1} were observed and the result is shown in figure 3.5. Somewhat comparable work with neutrons on simple liquid metals is reported in North *et al* (1968) and these two papers provide an interesting comparison of two techniques. The x-ray work on a simple non-metallic liquid, Ar, over a large density range is described in Pings (1968) and part of this is shown in figure 3.6. This shows that as the density decreases the coordination number, defined as

$$2n_0 \int_0^{r(\text{peak})} \rho(r)\, \mathrm{d}r$$

falls significantly. This reveals the important point that expansion reorganises the SRO, it does not merely increase the interatomic distances as it would in a crystal.

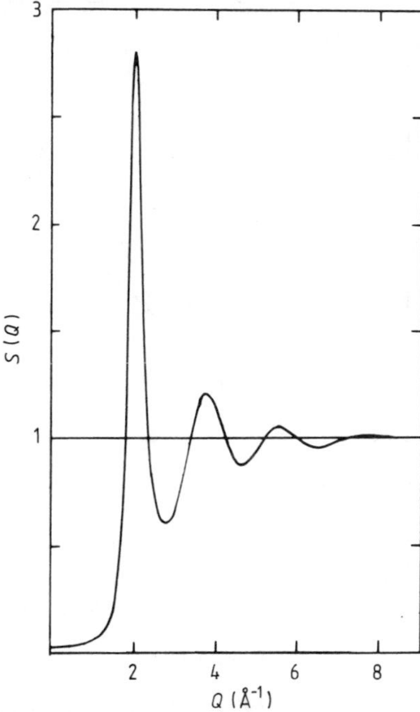

Figure 3.5 $S(Q)$ from Greenfield *et al* (1971) for liquid Na.

One type of non-simple liquid is molecular, like liquid CCl_4. If the scattering from one isolated molecule is known—which is often so for simple molecules—this information must supply part of the content of $F(Q)$. An assumption about the relative orientation of molecules in the liquid will supply more. These two items, plus a single measurement of $F(Q)$, go some way towards distinguishing the three partial structure factors without requiring isotopic substitution (Egelstaff *et al* 1971a,b). The element Ti has a small negative scattering length for neutrons and this circumstance enables $g_T(r)$ of equation (3.21*c*) to be useful. For $TiCl_4$

$$g_T = 1 + 0.004(g_{TiTi} - 1) + 0.588(g_{ClCl} - 1) - 0.101(g_{TiCl} - 1).$$

Figure 3.7 from Enderby (1978) shows the effect of the g_{TiCl} term clearly with a Ti–Cl intramolecular separation of 2.17 Å. The examples of CCl_4 and $TiCl_4$ illustrate inferences made without separate measurements of the partial structure factors in binary systems.

The study of molten salts was advanced by the application of the isotopic substitution method. Figure 3.8(*a*) shows $F(Q)$ of equation (3.15*b*) measured by Edwards *et al* (1975) for three liquid NaCl samples containing natural Na but three isotopically different Cl anions. Because of the different Cl scattering lengths, the curves are sufficiently different for the S_{ij} to be

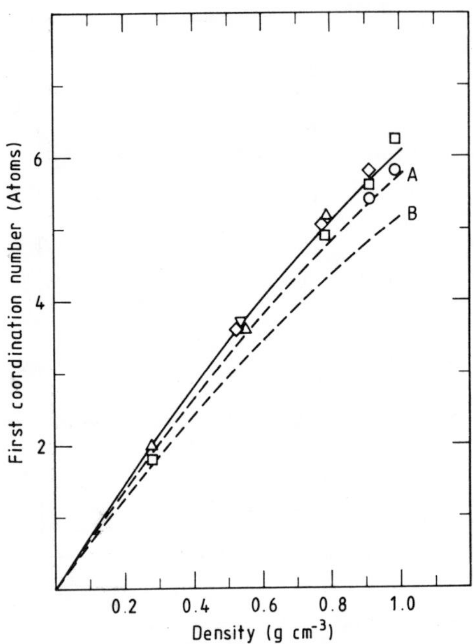

Figure 3.6 Data for Ar from Pings (1968) showing that the coordination number is a function of density. Broken curves are theoretical estimates from the LJ potential.

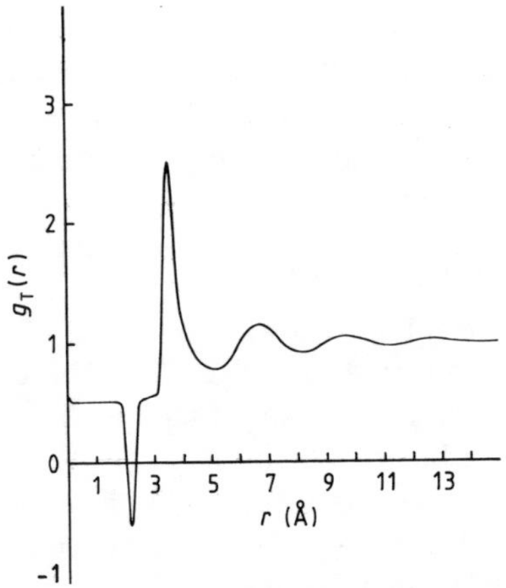

Figure 3.7 Enderby's total $g_T(r)$ for $TiCl_4$ (from Enderby 1978).

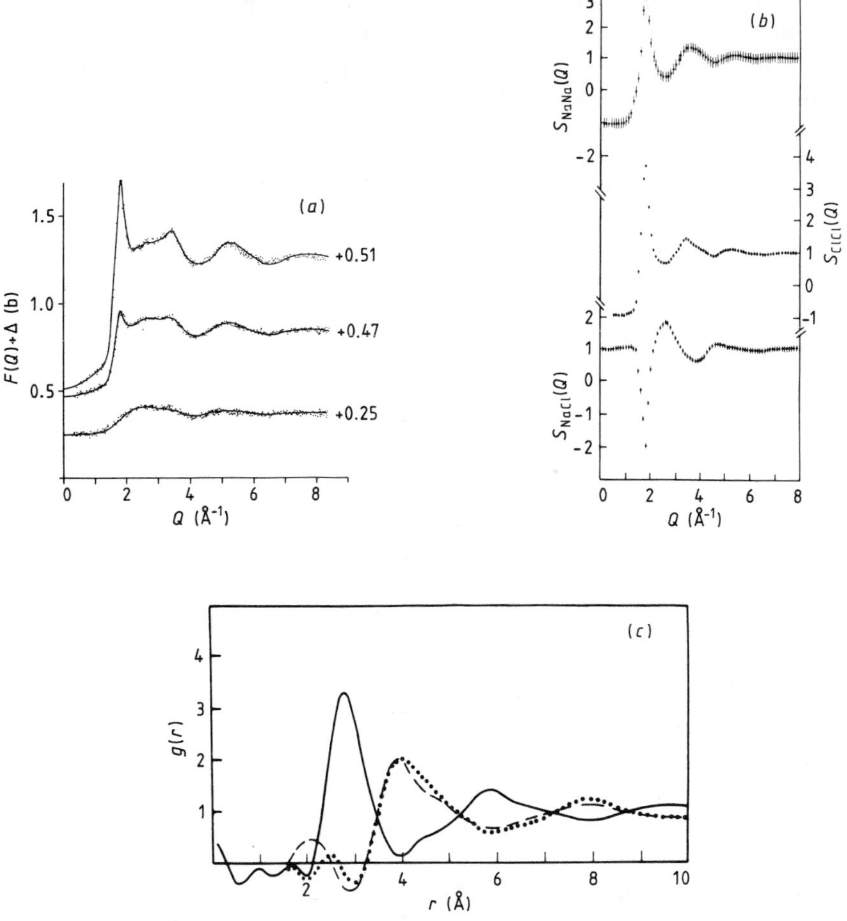

Figure 3.8 (*a*) $F(Q)$ from Edwards *et al* (1975) for neutron scattering by NaCl. Top curve, $Na^{35}Cl$; middle curve, NaCl; bottom curve $Na^{37}Cl$. The points are observed. The lines are derived from the S_{ij}. $\Delta = c_a b_a^2 + c_b b_b^2$. (*b*) Partial structure factors S_{ij} for NaCl (from Edwards *et al* 1975). (*c*) Partial g_{ij} (from Biggin and Enderby 1982) from the data of (*a*) and (*b*). Full curve, $g_{NaCl}(r)$; broken curve, $g_{NaNa}(r)$; dotted curve, $g_{ClCl}(r)$.

resolved (figure 3.8(*b*)) and the S_{ij} can be transformed into g_{ij} (figure 3.8(*c*)). It is evident that the first shell of neighbours, given by the peak in g_{NaCl}, is of opposite electrical sign to that of the ion at the origin; the coordination number is 5.8 ± 0.1. The g_{ClCl} and g_{NaNa} peaks coincide and contain 13.0 ± 0.5 second-nearest neighbours. Above 5 Å the charges cancel one another. Similarities to and differences from crystalline NaCl are obvious.

Clearly it would be impossible to infer such things from a single $F(Q)$. Figure 3.9 shows the outcome for $BaCl_2$ (Edwards *et al* 1978). It is distinguished from NaCl especially in that the g_{BaBa} and g_{ClCl} curves are quite different, partially because of the ionic charge difference but probably also because the ions are not equally mobile.

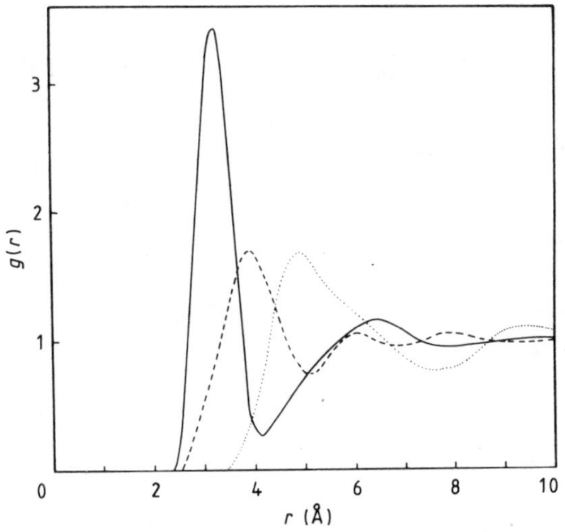

Figure 3.9 $g_{ij}(r)$ for $BaCl_2$ from Edwards *et al* (1978). Full curve, $g_{BaCl}(r)$; broken curve, $g_{ClCl}(r)$; dotted curve, $g_{BaBa}(r)$.

Consideration of amorphous alloys of Ni and Ti illustrates the way in which special expedients can be used to maximise the yield of information. First we note that both Ni and Ti have isotopes both with $b < 0$ and $b > 0$. In principle therefore isotopes could be mixed to give either $b_{Ni} = 0$ or $b_{Ti} = 0$; these are called null elements. With $b_{Ni} = 0$, equation (3.15*b*) shows that $F(Q)$ would give S_{TiTi} directly: similarly $b_{Ti} = 0$ would give S_{NiNi}. A zero alloy, i.e., one with $c_{Ni}b_{Ni} + c_{Ti}b_{Ti} = \langle\langle b \rangle\rangle = 0$, could also be prepared as could another alloy with $\Delta b = b_{Ti} - b_{Ni} = 0$. These last two alloys would yield $F(Q)$ which give directly the quantities called S_{cc} and S_{NN} which will be introduced in the next section and which, together with a third quantity, S_{Nc}, are altogether equivalent to the three S_{ij}. Between them, S_{TiTi}, S_{NiNi}, S_{cc}, S_{NN} give all the partial structure factors. At the time of writing this programme has not been realised but neutron and x-ray scattering from $Ni_{40}Ti_{60}$ has been observed (Wagner 1980). In this material, because of the relative sizes of the b, the x-ray $F(Q)$ was dominated by S_{NN} and the neutron $F(Q)$ by S_{cc}. S_{Nc} exerted so small an effect that it was sufficient to calculate it by an approximate theory. The result is shown in figure 3.10 and amounts to a very plausible solution of the partial structure factor

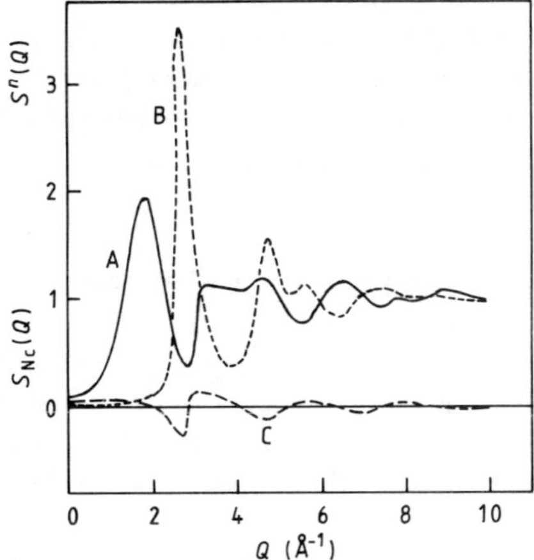

Figure 3.10 Bhatia–Thornton structure factors for amorphous $Ni_{40}Ti_{60}$. A, $S_{cc}(Q)/c_1c_2$; B, $S_{NN}(Q)$; C, $S_{Nc}(Q)_{HS}$ (from Wagner 1980).

problem in this alloy. From the corresponding g_{ij} it could be inferred that Ni had 9.3 Ti neighbours instead of the 7.7 it would have in a random mixture, i.e., SRO existed (see also §12.2).

As a final example we consider the archetypal glass, SiO_2. One of the more recent of many experiments used a time-of-flight neutron spectrometer and a pulsed neutron source (Misawa *et al* 1980). This took the Q-value up to about 45 Å^{-1}. Figure 3.11 shows the total radial distribution function obtained by Fourier transforming $(F(Q)/\langle\langle b^2\rangle\rangle - 1)$. The interpretation of the

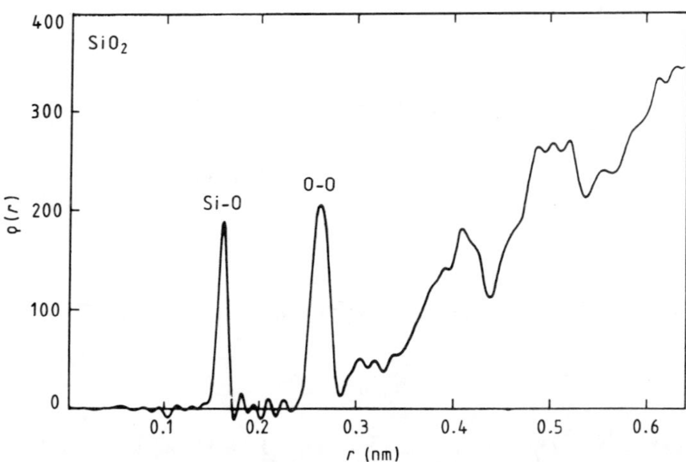

Figure 3.11 The radial distribution junction of vitreous silica derived from the scattering of neutrons from a spallation source (from Misawa *et al* 1980).

two prominent peaks is that they represent the Si–O and O–O distances of 0.1613 and 0.2628 nm respectively. The areas of the peaks lead to corresponding coordination numbers 4 and 2. The widths of the peaks indicate the variation in bond lengths. These results are clearly consistent with the Si–O tetrahedral unit and CRN of §2.6 (see also §10.6).

3.12 Different kinds of structure factor. Fluctuations

In §§3.5 and 3.11 it was mentioned in passing that the S_{ij} used so far are just one set of possible partial structure factors. In this section we introduce others and make the further points that when $Q = 0$ the structure factor is connected on the one hand with fluctuations and on the other with thermodynamic quantities.

It will be convenient to use previous notations except that the superscript will be dropped from the density function $\nu^{(1)}$. V now stands for a fixed volume within a much larger amount of binary mixture and we must differentiate between the fluctuating numbers N_A, N_B and N of atoms in V and their averages $\langle N_A \rangle$ etc. There will be corresponding instantaneous and average number densities ν_A, n_A etc.

Let

$$\Delta\nu_A(\boldsymbol{r}) \equiv \nu_A(\boldsymbol{r}) - n_A = \sum_k \delta(\boldsymbol{r} - \boldsymbol{R}_k^{(A)}) - n_A \tag{3.22a}$$

and define the Fourier component $\Delta\nu_A(\boldsymbol{Q})$ by

$$\Delta\nu_A(\boldsymbol{r}) = V^{-1} \sum_Q \nu_A(\boldsymbol{Q}) \exp(-\mathrm{i}\boldsymbol{Q} \cdot \boldsymbol{r}) \tag{3.22b}$$

then

$$\begin{aligned}\Delta\nu_A(\boldsymbol{Q}) &= \int \exp(\mathrm{i}\boldsymbol{Q} \cdot \boldsymbol{r})\, \Delta\nu_A(\boldsymbol{r})\, \mathrm{d}\boldsymbol{r} \\ &= \sum_k \exp(\mathrm{i}\boldsymbol{Q} \cdot \boldsymbol{R}_k^{(A)}) - n_A \int_V \exp(\mathrm{i}\boldsymbol{Q} \cdot \boldsymbol{r})\, \mathrm{d}\boldsymbol{r}.\end{aligned}$$

Similarly

$$\Delta\nu_B(\boldsymbol{Q}) = \sum_l \exp(\mathrm{i}\boldsymbol{Q} \cdot \boldsymbol{R}_l^{(B)}) - n_B \int_V \exp(\mathrm{i}\boldsymbol{Q} \cdot \boldsymbol{r})\, \mathrm{d}\boldsymbol{r}$$

and

$$\Delta\nu(\boldsymbol{Q}) \equiv \Delta\nu_A(\boldsymbol{Q}) + \Delta\nu_B(\boldsymbol{Q}) = \sum_{j=k,l} \exp(\mathrm{i}\boldsymbol{Q} \cdot \boldsymbol{R}_j) - n_0 \int_V \exp(\mathrm{i}\boldsymbol{Q} \cdot \boldsymbol{r})\, \mathrm{d}\boldsymbol{r}. \tag{3.22c}$$

The new partial structure factor $S_{NN}(\boldsymbol{Q})$ will now be defined by

$$S_{NN}(\boldsymbol{Q}) \equiv \frac{1}{\langle N \rangle} \langle \Delta v(\boldsymbol{Q})^* \Delta v(\boldsymbol{Q}) \rangle. \tag{3.23}$$

For a homogeneous isotropic binary mixture $\boldsymbol{Q} \to Q$. In the limit $\boldsymbol{Q} = 0$, it follows from equation (3.22*c*) that

$$S_{NN}(0) = \frac{1}{\langle N \rangle} \langle (N - \langle N \rangle)^2 \rangle \equiv \frac{\langle (\Delta N)^2 \rangle}{\langle N \rangle} \tag{3.24}$$

which is a statement of the mean-square fluctuation in the number N.

Before discussing S_{NN} let us introduce S_{cc} and S_{Nc}. For this we require the definition

$$\Delta c(\boldsymbol{r}) \equiv (V/\langle N \rangle)(c_B \Delta v_A(\boldsymbol{r}) - c_A \Delta v_B(\boldsymbol{r})). \tag{3.25a}$$

Δc is an expression of the concentration fluctuations within V, and the definition is reasonable because it ensures that, if the density fluctuations, Δv, of each component were proportional to their respective average concentrations (i.e., the latter remain unchanged), $\Delta c(\boldsymbol{r})$ would vanish. $\Delta c(\boldsymbol{Q})$ is introduced by

$$\Delta c(\boldsymbol{r}) = \sum_{Q} \Delta c(\boldsymbol{Q}) \exp(-\mathrm{i}\boldsymbol{Q} \cdot \boldsymbol{r}) \tag{3.25b}$$

and

$$\Delta c(\boldsymbol{Q}) = V^{-1} \int_V \Delta c(\boldsymbol{r}) \exp(\mathrm{i}\boldsymbol{Q} \cdot \boldsymbol{r}) \, \mathrm{d}\boldsymbol{r}. \tag{3.25c}$$

Two more new structure factors are now defined as follows:

$$S_{cc} \equiv \langle N \rangle \langle \Delta c(\boldsymbol{Q})^* \Delta c(\boldsymbol{Q}) \rangle \tag{3.26}$$

$$S_{Nc} \equiv \mathscr{R} \langle \Delta v(\boldsymbol{Q})^* \Delta c(\boldsymbol{Q}) \rangle. \tag{3.27}$$

Substitution into equations (3.26) and (3.27) from (3.22*c*) and (3.25*c*), taking the limit $Q \to 0$, leads to

$$S_{cc}(0) = \langle N \rangle \langle (\Delta c)^2 \rangle \tag{3.28a}$$

$$\Delta c = \langle N \rangle^{-1} [c_B(N_A - \langle N_A \rangle) - c_A(N_B - \langle N_B \rangle)] \tag{3.28b}$$

and

$$S_{Nc}(0) = \langle \Delta N \Delta c \rangle. \tag{3.29}$$

Inspection of equation (3.22*c*) shows that the three new structure factors contain the sum $\Sigma_{j=k,l} \exp(\mathrm{i}\boldsymbol{Q} \cdot \boldsymbol{R}_j)$ though it enters in different ways. Thus, the same positional information is carried that is in equation (3.14), leading to the structure factors S_{ij}. In fact, Bhatia and Thornton (1970), who introduced S_{cc}, S_{NN} and S_{Nc} in 1970, showed that the latter are linear combinations of the S_{ij}. The linear relations with the arguments Q omitted are

$$S_{NN} = c_A^2 S_{AA} + c_B^2 S_{BB} + 2c_A c_B S_{AB} \tag{3.30a}$$

$$S_{Nc} = c_A c_B[c_A(S_{AA} - S_{AB}) - c_B(S_{BB} - S_{AB})] \tag{3.30b}$$

$$S_{cc} = c_A c_B[1 + c_A c_B(S_{AA} + S_{BB} - 2S_{AB})] \tag{3.30c}$$

conversely

$$c_A^2 S_{AA} = c_A^2 S_{NN} + 2c_A S_{Nc} + S_{cc} - c_A c_B \tag{3.31a}$$

$$c_B^2 S_{BB} = c_B^2 S_{NN} - 2c_B S_{Nc} + S_{cc} - c_A c_B \tag{3.31b}$$

$$c_A c_B S_{AB} = c_A c_B S_{NN} + (c_B - c_A) S_{Nc} - S_{cc} + c_A c_B. \tag{3.31c}$$

The equivalence of these two sets of structure factors means that the scattering intensity in equation (3.15) could be re-expressed. Substituting equation (3.31) into (3.15*b*) leads to

$$F(Q) = \langle\langle f \rangle\rangle^2 S_{NN} + (\Delta f)^2 S_{cc} + 2\Delta f \langle\langle f \rangle\rangle S_{Nc} \tag{3.32a}$$

where

$$\Delta f \equiv f_A - f_B. \tag{3.32b}$$

Equation (3.32) justifies the remarks about S_{cc} and S_{NN} in §3.11.

It is probably most readily seen from equation (3.30) that, at large Q, S_{NN} oscillates about and tends to one, S_{Nc} behaves similarly about zero and S_{cc} about $c_A c_B$. $S_{cc} = c_A c_B$ if $S_{AA} = S_{BB} = S_{AB}$.

At $Q = 0$, often called the long-wavelength limit, equations (3.24), (3.28*a*) and (3.29) show the connection with fluctuations. In a pure material, S_{cc} and S_{Nc} vanish and $S_{NN} \to S$. Thus $S(0)$ is $\langle(\Delta N)^2\rangle / \langle N \rangle$ for a pure material.

The fluctuation expressions can be connected with thermodynamic quantities. We shall demonstrate the simplest relation and quote the rest in table 3.2. For these purposes the thermodynamic symbols in table 3.1 will be useful.

Table 3.1 List of thermodynamic symbols.

Name or definition	Symbol
Isothermal compressibility	κ_T
Gibbs free energy	G
Activity of ith component	a_i
Number of moles or particles of ith component according to context	n_i
Partial molar Gibbs free energy of ith component or chemical potential $\equiv (\partial G/\partial n_i)_{T,p,n_j}$	μ_i
Partial molar volume $(\partial V/\partial n_i)_{T,p,n_j}$	v_i
$n_0(v_A - v_B)$	δ
Boltzmann's constant	k_B
$n_0 k_B T \kappa_T$	θ
Grand canonical partition function	$\mathscr{X}$
Energy of jth state of a system	E_j

Table 3.2 The connection of fluctuation expressions with thermodynamic quantities. Note: $\langle\rangle$ = ensemble average: $\langle\!\langle x \rangle\!\rangle \equiv c_A x_A + c_B x_B$.

Fluctuation expression	Structure factor for $Q = 0$	Thermodyamic expression
$\dfrac{\langle N^2\rangle - \langle N\rangle^2}{\langle N\rangle} = \dfrac{\langle(\Delta N)^2\rangle}{\langle N\rangle}$	$S(0)$	$\theta \equiv n_0 k_B T \kappa_T$
$\dfrac{\langle(\Delta N)^2\rangle}{\langle N\rangle}$	$S_{NN}(0)$	$\theta + \delta^2 S_{cc}(0)$
$\langle N\rangle\langle(\Delta c)^2\rangle$	$S_{cc}(0)$	$\dfrac{RT}{(\partial^2 G/\partial c_A^2)_{T,p,N}} = \dfrac{k_B T(1-c_A)}{(\partial\mu_A/\partial c_A)_{T,p,N}}$
$\langle\Delta N \Delta c\rangle$	$S_{Nc}(0)$	$-S_{cc}(0)$
$1 + \langle N\rangle\left(\dfrac{\langle N_A^2\rangle - \langle N_A\rangle^2}{\langle N_A\rangle^2}\right)$	$S_{AA}(0)$	$\theta - \dfrac{c_B}{c_A} + \left(\dfrac{1}{c_A} - \delta\right)^2 S_{cc}(0)$
$1 + \langle N\rangle\left(\dfrac{\langle N_B^2\rangle - \langle N_B\rangle^2}{\langle N_B\rangle^2}\right)$	$S_{BB}(0)$	$\theta - \dfrac{c_A}{c_B} + \left(\dfrac{1}{c_B} + \delta\right)^2 S_{cc}(0)$
$1 + \langle N\rangle\left(\dfrac{\langle N_A N_B\rangle - \langle N_A\rangle\langle N_B\rangle}{\langle N_A\rangle\langle N_B\rangle}\right)$	$S_{AB}(0)$	$\theta + 1 - \left(\dfrac{1}{c_A} - \delta\right)\left(\dfrac{1}{c_B} + \delta\right) S_{cc}(0)$
	$F(0)$	$\theta\langle\!\langle f\rangle\!\rangle^2 + S_{cc}\langle\!\langle f\rangle\!\rangle^2\left(\delta - \dfrac{f_A - f_B}{\langle\!\langle f\rangle\!\rangle}\right)^2$

We now show that $S(0) = \theta$. The way V has been used in this section indicates that V, T and μ (but not N) are the specified variables and the system in V is accordingly a member of a grand canonical ensemble. For a pure system the partition function is therefore

$$\mathscr{Z} = \sum_N \sum_j \exp\left(\frac{\mu N - E_j}{k_B T}\right) \tag{3.33a}$$

and the probability of a system having N particles and total energy E_j is

$$p(N, E_j) = \mathscr{Z}^{-1} \exp \frac{\mu N - E_j}{k_B T}. \tag{3.33b}$$

From the latter the average, $\langle X \rangle$, of any property X is obtainable as $\Sigma_{N,j} Xp(N, E_j)$ and consequently

$$\begin{aligned}\langle(\Delta N)^2\rangle &\equiv \langle(N - \langle N \rangle)^2\rangle = \langle N^2 \rangle - \langle N \rangle^2 \\ &= (k_B T)^2 \left[\mathscr{Z}^{-1} \left(\frac{\partial^2 \mathscr{Z}}{\partial \mu^2}\right)_{T,V} - \left(\frac{\partial \ln \mathscr{Z}}{\partial \mu}\right)^2_{T,V} \right] \\ &= k_B T \left(\frac{\partial \langle N \rangle}{\partial \mu}\right)_{T,V} \end{aligned} \tag{3.34}$$

Manipulation of the last derivative brings the required result for, by the chain rule,

$$\left(\frac{\partial \langle N \rangle}{\partial \mu}\right)_{T,V} = \left(\frac{\partial \langle N \rangle}{\partial n_0}\right)_{T,V} \left(\frac{\partial n_0}{\partial p}\right)_{T,V} \left(\frac{\partial p}{\partial \mu}\right)_{T,V}$$

where $n_0 \equiv \langle N \rangle / V$. Now

$$\left(\frac{\partial \langle N \rangle}{\partial n_0}\right)_{T,V} = V \qquad \left(\frac{\partial n_0}{\partial p}\right)_{T,V} = -\frac{\langle N \rangle}{V^2}\left(\frac{\partial V}{\partial p}\right)_{T,N} \qquad \left(\frac{\partial p}{\partial \mu}\right)_{T,V} = n_0$$

of these, the third relation follows from the thermodynamic equations $V = (\partial G / \partial p)_T$ and $G = \langle N \rangle \mu$. From equation (3.34)

$$\frac{\langle(\Delta N)^2\rangle}{\langle N \rangle} = -k_B T \frac{n_0}{V}\left(\frac{\partial V}{\partial p}\right)_{T,N} = k_B T n_0 \kappa_T = \theta \tag{3.35a}$$

thus,

$$S(0) = \theta. \tag{3.35b}$$

Other connections between thermodynamic quantities and fluctuations can be demonstrated (Bhatia and Thornton 1970) (see table 3.2).

There are other structure factors—called the Ashcroft–Langreth—quite common in the literature and valuable for some purposes. Writing them as $S_{ij}^{(\mathrm{AL})}(Q)$, their relation with others is,

$$S_{ij}^{(\mathrm{AL})} = \delta_{ij} + (c_i c_j)^{1/2}(S_{ij} - 1) \tag{3.36a}$$

$$S_{11}^{(\mathrm{AL})} = c_1 S_{\mathrm{NN}} + S_{\mathrm{cc}}/c_1 + 2S_{\mathrm{Nc}} \tag{3.36b}$$

$$S_{22}^{(\mathrm{AL})} = c_2 S_{\mathrm{NN}} + S_{\mathrm{cc}}/c_2 - 2S_{\mathrm{Nc}} \tag{3.36c}$$

$$S_{12}^{(\mathrm{AL})} = (c_1 c_2)^{1/2} S_{\mathrm{NN}} - S_{\mathrm{cc}}/(c_1 c_2)^{1/2} + [(c_2/c_1)^{1/2} - (c_1/c_2)^{1/2}]S_{\mathrm{Nc}} \tag{3.36d}$$

$$S_{\mathrm{NN}} = c_1 S_{11}^{(\mathrm{AL})} + c_2 S_{22}^{(\mathrm{AL})} + 2(c_1 c_2)^{1/2} S_{12}^{(\mathrm{AL})} \tag{3.36e}$$

$$S_{\mathrm{cc}} = c_1 c_2 [c_2 S_{11} + c_1 S_{22}^{(\mathrm{AL})} - 2(c_1 c_2)^{1/2} S_{12}^{(\mathrm{AL})}] \tag{3.36f}$$

$$S_{\mathrm{Nc}} = c_1 c_2 [S_{11}^{(\mathrm{AL})} - S_{22}^{(\mathrm{AL})} + (c_2 - c_1) S_{12}^{(\mathrm{AL})}/(c_1 c_2)^{1/2}]. \tag{3.36g}$$

The $S_{ij}^{(\mathrm{AL})}$ may be defined by

$$S_{ij}^{(\mathrm{AL})} = \delta_{ij} + (c_1 c_2)^{1/2} n_0 \int (g_{ij}(r) - 1) \exp(-\mathrm{i}\boldsymbol{Q} \cdot \boldsymbol{r})\, \mathrm{d}\boldsymbol{r}. \tag{3.36h}$$

Since the new structure factors and the S_{ij} are mutually interconvertible it is a matter of convenience which is used. The Bhatia–Thornton ones connect more directly with fluctuations and the S_{ij}, often called Faber–Ziman structure factors, may be thought to arise more naturally in scattering discussions and in connection with the $g_{ij}(r)$. $S_{\mathrm{cc}}(0)$ has featured in much discussion of liquid mixtures because, if $S_{\mathrm{cc}}(0) \simeq 0$ for, say, $c_0 = c$, fluctuations are negligible and this signifies a stable compound with the composition c_0. On the other hand if $S_{\mathrm{cc}}(0)$ becomes very large then the large concentration fluctuations imply a tendency to phase separation or liquid immiscibility. It can be shown that $S_{\mathrm{cc}}(0) = c_{\mathrm{A}} c_{\mathrm{B}}$ if A and B mix randomly without volume change or heat of mixing (see §6.9). As table 3.2 shows, $S_{\mathrm{cc}}(0)$ is obtainable as a function of composition by thermodynamic measurements or from $F(0)$. Since scattering observations becomes impossible at very small Q, $F(0)$ has to be obtained by extrapolating the intensity to $Q = 0$. This should lead to the same values for $S_{\mathrm{cc}}(0)$ as thermodynamic measurements and where this point has been tested the results are satisfactory though not accurate.

3.13 Other clues to structure. Triplet distribution

Although scattering studies are the major source of structural information, clues can be picked up elsewhere. Chemical knowledge of molecular structure or preferred bond orientations, or crystallographic knowledge of related crystal structures, may suggest hypotheses about SRO in liquids or glasses. Even bulk properties, like density, may eliminate some structures from consideration and the composition dependence of thermodynamic variables such as the heat of mixing may indicate that SRO of some kind—as distinct from random mixing—may be occurring. However, ideally we require techniques in which the signal is affected by the number and disposition of the atoms round any one of them thought of as the origin. The reader's attention will now be drawn briefly to a few such techniques but it is one of the problems in the study of disordered matter that the inference of the structure from the signal is somewhat complex and indirect.

In x-ray absorption it is often noticed that on the high-energy side of the absorption edge the spectrum has a complex oscillatory form. This phenomenon is extended x-ray absorption fine structure, called EXAFS. X-ray absorption chiefly results from photoelectron emission and EXAFS stems from the fact that on its departure from its parent atom the photoelectron is backscattered by the neighbouring atoms. There is interference between the photoelectron wave and the backscattered wavelets and whether this is constructive or destructive depends on the photon energy. This is the origin of the ripples in the spectrum. The ripples must therefore depend also on how many neighbouring atoms there are, of what kind and how far away. In other words the partial radial distribution functions are involved. For a monatomic target, the theory (Lee *et al* 1981) leads to

$$\frac{\Delta\mu}{\mu_0} = -\left(\frac{3f(k,\pi)}{2k}\right)\int_0^\infty \frac{g(r)\exp(-r/L)}{r^2}\sin(2kr+\alpha(k))\,\mathrm{d}r \qquad (3.37)$$

where $\Delta\mu/\mu_0$ is the fractional change in the absorption coefficient represented by the ripples; k is the photoelectron wavenumber, $f(k,\pi)$ is the backscattering amplitude, L is the photoelectron mean-free path, $\alpha(k)$ is the scattering phaseshift. Each element i in a binary mixture will have its own absorption edge and two partial distributions, g_{ii} and g_{ij}, will control the backscattering. Despite the difficulties of extracting anything about $g_{ij}(r)$ from EXAFS, methods of analysis have been evolved (Lee *et al* 1981, Teo and Joy 1981) and applications to liquids and glasses can be found in Lee *et al* (1981), Teo and Joy (1981) and Bianconi *et al* (1983). For example the EXAFS of Ni in an amorphous Ni–Mo alloy could be fitted accurately with equation (3.37) if Ni were assumed to have 3.6 Mo neighbours at 2.59 Å and 8.4 Ni neighbours at 2.36 Å. EXAFS has made significant contributions to the understanding of glass structures (§10.6).

The fields from neighbouring atoms affect the magnetic, quadrupole and Mössbauer resonances of nuclei. Therefore, these phenomena can also furnish clues about SRO. For example, a given environment results in a characteristic Mössbauer spectrum. In a disordered structure a range of environments will result in a range of superposed spectra. It may be possible to devise models of SRO which would result in the observed spectra. As with scattering and EXAFS, there is a considerable distance in data processing and interpretation to be traversed before structural inferences can be reached. An application of NMR to glassy solids is in §10.6.

It has several times been remarked that scattering experiments measure only $g(r)$ or $g_{ij}(r)$ and that this is a serious limitation. It should be added however that the triplet distribution is theoretically related to the variation with pressure of the pair distribution. Since scattering experiments can be performed at more than one pressure, $(\partial S(Q)/\partial p)_T$ can in principle tell us something about $g^{(3)}$. The theory and an experimental application are given in Egelstaff *et al* (1971a).

References

Bacon G E 1975 *Neutron Diffraction* (Oxford: Clarendon)

Bhatia A B and Thornton D E 1970 *Phys. Rev.* B **2** 2004

Bianconi A, Incoccia L and Stipcich S (ed) 1983 *EXAFS and New Edge Structure* (Berlin: Springer)

Biggin S (ed) 1986 *Neutron Beams: Their Use and Potential in Scientific Research* (Swindon: SERC)

Biggin S and Enderby J E 1982 *J. Phys. C: Solid State Phys.* **15** L305

Dahlberg U and Kunsch B 1983 *Phys. Chem. Liq.* **12** 237

Edwards F G, Enderby J E, Howe R A and Page D J 1975 *J. Phys. C: Solid State Phys.* **8** 3483

Edwards F G, Howe R A, Enderby J E and Page E T 1978 *J. Phys. C: Solid State Phys.* **11** 1053

Egelstaff P A, Page D I and Heard C R T 1971a *J. Phys. C: Solid State Phys.* **4** 1453

Egelstaff P A, Page E I and Powles J G 1971b *Mol. Phys.* **20** 881

Enderby J E 1978 *The Metal Non-Metal Transition in Disordered Systems* (*St Andrews*) *1978* ed L R Friedman and D P Tunstall (Edinburgh: Scottish Universities Summer School in Physics Publications) p 435

Enderby J E, North D M and Egelstaff P A 1966 *Phil. Mag.* **14** 961

Greenfield A J, Wellendorf J and Wiser N 1971 *Phys. Rev* A **4** 1607

Guinier A 1963 *X-ray Diffraction* (San Francisco: W H Freeman)

Guinier A and Fournet G 1955 *Small Angle Scattering of X-rays* (New York: Wiley)

Howe R A 1978 *Progress in Liquid Physics* ed C A Croxton (Chichester: Wiley) p 503

International Tables for X-ray Crystallography 1962 (Chester: Union of Crystallography)

James R 1962 *The Optical Principles of the Diffraction of X-rays* (London: Bell)

Kaplow R, Strong S L and Averbach B L 1965 *Phys. Rev.* **138A** 1336

Lee P A, Citrin P M and Kincaid B M 1981 *Rev. Mod. Phys.* **53** 769

Leung P K and Wright J G 1974 *Phil. Mag.* **30** 185

Lovesey S W 1984 *The Theory of Neutron Scattering from Condensed Matter* (Oxford: Clarendon)

Misawa M, Price D L and Suzuki K 1980 *J. Non-Cryst. Solids* **37** 85

North D M, Enderby J E and Egelstaff P A 1968 *J. Phys. C: Solid State Phys.* **1** 784, 1075

Pings C J 1968 *Physics of Simple Liquids* ed H N V Temperley, J S Rowlinson and G S Rushbrooke (Amsterdam: North-Holland) p 387

Teo B and Joy D C (ed) 1981 *EXAFS Spectroscopy Techniques and Applications* (New York: Plenum).
Van der Lugt 1979 *Acta. Crystallogr.* **35** 431
Wagner 1980 *J. Non-Cryst. Solids* **42** 3
Waseda Y 1980 *The Structure of Non-Crystalline Materials* (New York: McGraw-Hill)
White J W and Windsor C G 1984 *Rep. Prog. Phys.* **47** 707
Wright A C 1974 *Adv. Struct. Res. by Diffraction Methods* **5** 1

4

MONTE CARLO AND MOLECULAR DYNAMICS METHODS

The complicated configuration of atoms in a disordered structure and our lack of detailed knowledge about it make the routine application of quantum mechanics and statistical mechanics very difficult. The partition function can be obtained only by approximate methods even if simplifications are used for the interparticle potentials. Approximate theoretical methods are not lacking but they have been supplemented since the 1950s by numerical calculations with computers. The two most important of these will be introduced in this chapter.

These numerical simulation methods may be a source of information about some things, such as the triplet distribution, which cannot be obtained from experiment. However their chief value probably lies in their ability to proceed from a well defined physical model to accurate numerical consequences. The latter can then be used to test approximate theories applied to the same physical model. In many ways this is better than testing the theory against observations on real matter because, even if there were no experimental errors, the assumed model might not suit the real sample.

Only the principles of these methods will be described. As with laboratory techniques, there are numerous tricks of the trade for which the specialist literature must be consulted.

The second object of this chapter will be to set out some results of the numerical methods which have particular relevance to the subjects in the rest of the book.

4.1 The Monte Carlo method

It will be convenient to recall first some formulae from statistical mechanics. If the systems in an ensemble have prescribed V, N and T the ensemble is canonical and the normalised probability that a system is in its ith quantum state with energy E_i is

$$P_i = \exp(-\beta E_i)/Z \tag{4.1a}$$

where

$$Z \equiv \sum_i \exp(-\beta E_i) \tag{4.1b}$$

and

$$\beta \equiv (k_B T)^{-1}. \tag{4.1c}$$

The ensemble average value of a mechanical variable, X, such as energy is then

$$\langle X \rangle = \sum_i X_i P_i = \frac{1}{Z} \sum_i X_i \exp(-\beta E_i) \tag{4.1d}$$

$\langle X \rangle$ is the theoretical outcome needed for comparison with X_{obs}.

By applying these formulae, it is shown in statistical mechanics texts that the classical version of equation (4.1) is quite adequate for handling fluid behaviour with a few exceptions such as H_2 and He at very low temperatures. The classical version is

$$P = \frac{\exp(-\beta E)}{\int \exp(-\beta E)\, d\boldsymbol{r}^{(N)}\, d\boldsymbol{p}^{(N)}} \tag{4.2a}$$

$$Z_{cl} = \frac{1}{N!(h)^{3N}} \int \exp(-\beta E)\, d\boldsymbol{r}^{(N)}\, d\boldsymbol{p}^{(N)} \tag{4.2b}$$

$$\langle X \rangle = \frac{\int X \exp(-\beta E)\, d\boldsymbol{r}^{(N)}\, d\boldsymbol{p}^{(N)}}{\int \exp(-\beta E)\, d\boldsymbol{r}^{(N)}\, d\boldsymbol{p}^{(N)}}. \tag{4.2c}$$

In equation (4.2): $E = E(\boldsymbol{r}^{(N)}, \boldsymbol{p}^{(N)})$ and $d\boldsymbol{r}^{(N)}\, d\boldsymbol{p}^{(N)}$ means the volume element of a $6N$-dimensional phase space with $3N$ position and $3N$ momentum coordinates; it is short for $d\boldsymbol{r}_1\, d\boldsymbol{r}_2 \ldots d\boldsymbol{r}_N\, d\boldsymbol{p}_1\, d\boldsymbol{p}_2 \ldots d\boldsymbol{p}_N$. Assuming that E can be divided into kinetic and potential parts, T and Φ, depending on the p and the r respectively, the integrals over $\boldsymbol{p}^{(N)}$ can be carried out to give

$$Z_{cl} = \frac{1}{N!}\left(\frac{2\pi m k_B T}{h^2}\right)^{3N/2} \int \exp(-\beta\Phi)\, d\boldsymbol{r}^{(N)}. \tag{4.2d}$$

The expression for the average potential energy, $\langle \Phi \rangle$, is then

$$\langle \Phi \rangle = \frac{1}{Q} \int \Phi(\boldsymbol{r}^{(N)}) \exp(-\beta\Phi)\, d\boldsymbol{r}^{(N)} \tag{4.3a}$$

where

$$Q \equiv \int \exp(-\beta\Phi)\, d\boldsymbol{r}^{(N)} \tag{4.3b}$$

and the probability, strictly the probability density, for a configuration with particle coordinates $\{\boldsymbol{r}^{(N)}\}$ is

$$P(\boldsymbol{r}^{(N)}) = Q^{-1} \exp(-\beta\Phi(\boldsymbol{r}^{(N)})). \tag{4.3c}$$

The total energy is

$$U \equiv \langle E \rangle = \langle T \rangle + \langle \Phi \rangle \tag{4.3d}$$

where

$$\langle T\rangle = \tfrac{3}{2}k_B T.$$

The Monte Carlo method is a technique for evaluating integrals or averages by randomly sampling them. It works for multiple integrals and has accordingly been put to great use for equations like (4.2*c*) or (4.3*a*). The most frequently used sampling method was introduced by Metropolis *et al* (1953). Let us think of its application to the evaluation of $\langle\Phi\rangle$. The idea is to generate a sequence of configurations, $C_1, C_2, \ldots, C_k, \ldots$, with corresponding potential energies $\Phi_1, \Phi_2, \ldots, \Phi_k, \ldots$. These are to have the property that they are chosen randomly from a distribution *such that the probability of choosing* C_k *is proportional to* $\exp(-\beta\Phi_k)$. Equation (4.3) shows that if Φ is now averaged over the members of the sequence, the result will approximate to $\langle\Phi\rangle$, the more so the longer the sequence.

Carrying this out requires a number of devices. The system is specified by the number of particles and to avoid excessive computation this is restricted to a few hundred or even a few tens. For many fluid studies this is apparently sufficient to give useful results (Barker and Henderson 1976). The particles are conceptually enclosed in a box, often but not necessarily cubic. With such small numbers of particles a high proportion would be in an unsymmetrical environment, namely near the walls. To avoid the distorting effect of this on the results for bulk properties, periodic boundary conditions are imposed, the effect of which is to make the cube one of a space-filling array of identical cubes each containing identical configurations of the particles. Every particle now has other particles on all sides (figure 4.1). The interparticle forces have to be prescribed by a potential function. It is then a matter of computation to find Φ_k, given C_k.

It remains to decide C_1 and to give rules for proceeding from C_k to C_{k+1} in such a way that the sequence has the property described above. C_1 is arbitrary and could be an ordered configuration; then Φ can be computed from the potential function. The random sampling now enters by the choice of a particle at random and the adjustment of its position from $\boldsymbol{r}$ to $\boldsymbol{r}+\delta\boldsymbol{r}$ where $\delta\boldsymbol{r}$ is also chosen randomly as a displacement to any point inside a small cube centred on the position $\boldsymbol{r}$. This produces a possible C_2 for which Φ_2 and $\exp(-\beta\Phi_2)$ can be computed, and $\exp[-\beta(\Phi_2-\Phi_1)]$, or Δ, also follows. The rule is: compare Δ with a random number, x, selected from the range (0, 1); if $\Delta > x$, C_2 is the new configuration; if not, the new one is rejected and C_2 is made the same as C_1. C_3, C_4 etc are successively specified by the same rule. Comparing Δ with x has the effect of accepting the altered configuration if the energy has been decreased since $\Delta > 1$; but, if $\Delta < 1$ because $\Phi_2 > \Phi_1$, then the randomness of x ensures that the new configuration will be accepted or rejected with probabilities Δ and $(1-\Delta)$ respectively. It is plausible but not obvious that this procedure generates a sequence in which, as required, the probability of C_k is proportional to

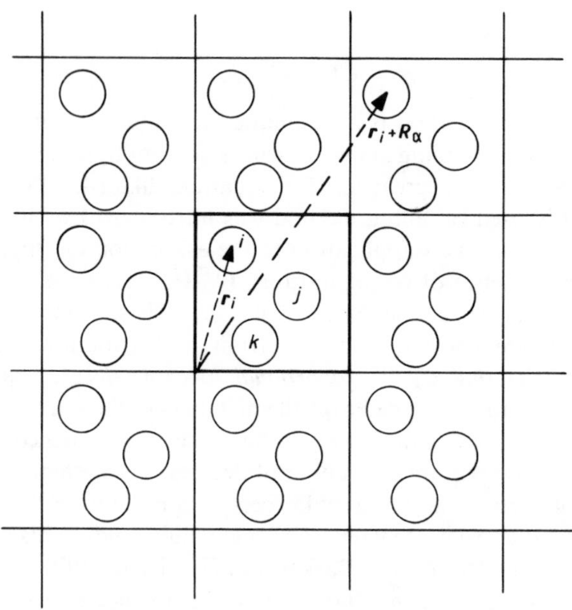

Figure 4.1 Schematic illustration in 2-D of periodic boundary conditions. Particles $i, j, k, \ldots$ in the box at the origin interact with all other particles, e.g.,

$$U = \sum_{\alpha} \sum_{i \leqslant j} \varphi(|\boldsymbol{r}_{ij} + \boldsymbol{R}_\alpha|)$$

where $\{\boldsymbol{R}_\alpha\}$ are the lattice vectors.

$\exp(-\beta\Phi_k)$. The proof that this is so is given in Metropolis *et al* (1953), Valleau and Whittington (1977) and Valleau and Torrie (1977) using the theory of Markov chains.

Practitioners have to consider questions such as the length of sequence to achieve a required accuracy in $\langle\Phi\rangle$, the optimum size of $|\delta\boldsymbol{r}|$, the harmful as opposed to useful effects of the periodic boundary conditions and the relation of the range of the interparticle potential to the size of the sample box. Leaving these technical points aside, we note that any $\langle X\rangle$ obtainable by equation (4.1c) is accessible by calculating it for each configuration and taking the average. In addition to $\langle\Phi\rangle$, $\langle\Phi^2\rangle$ could be evaluated and therefore the heat capacity from

$$C_v = \frac{\langle E^2\rangle - \langle E\rangle^2}{k_B T^2}. \tag{4.4}$$

By averaging the histograms of interparticle distances from all the configurations, $g(r)$ or the RDF could be found and the pressure then follows from equation (3.13). Entropy, S, and Helmholtz free energy, $F = U - TS$ cannot be evaluated from equations (4.1c) or (4.2c) because S cannot be calculated for one state at a time. Various indirect methods are available however. One is to integrate the thermodynamic relation $p = -(\partial F/\partial V)_T$ which leads to

$$F_2 - F_1 = N \int_1^2 n^{-2} p(n)\, \mathrm{d}n$$

where n is the number density. If F_1 is sufficiently well known for state 1—say at a low density where a theory is adequate—then F_2 is found by integrating $n^{-2}p(n)$. This involves a Monte Carlo evaluation of p at a series of densities.

The preceding paragraphs refer to translational motions in three-dimensional classical fluids using the canonical ensemble. It should be observed that the technique is not confined to translation, to 3-D, to fluids, to classical statistics, to canonical ensembles, nor indeed to statistical mechanics (Valleau and Whittington 1977, Valleau and Torrie 1977, Wood 1968).

4.2 Molecular dynamics

We have just seen that the Monte Carlo method evaluates ensemble averages. By contrast and as its name suggests, molecular dynamics, MD for short, follows the time evolution of a system of N particles by solving the N simultaneous equations of motion. Both translatory and rotatory motions can be included. N particles, with N of the order 10^3, are confined in a box with periodic boundary conditions. These ensure that when a particle passes outward through the wall another enters at the opposite side. The interparticle force is specified and the initial state, though arbitrary, might be conveniently chosen by taking a configuration from an equilibrium state reached in a former calculation. The particles will have an approximately Maxwell–Boltzmann velocity distribution and starting in this way speeds up the approach to equilibrium; otherwise time will be taken up while the system proceeds to equilibrium from some improbable arbitrary state.

There are numerous techniques for solving the equations of motion but all involve computing the force on the ith particle from

$$\boldsymbol{F}_i(t) = \sum_{j \neq i} \boldsymbol{f}_{ij}(r_{ij}(t)) \tag{4.5}$$

which assumes that the interparticle force $\boldsymbol{f}_{ij}$ depends only on the interparticle distance r_{ij}. Assuming $\boldsymbol{F}_i(t)$ is known for some time t, Newton's law of motion is used to find the positions $\{\boldsymbol{r}_i\}$ at a time Δt later. After this time step, $F_i(t + \Delta t)$ is computed as a preliminary to the next Δt. In this way the positions at successive steps are found and the particle velocities $\{\boldsymbol{v}_i\}$ are known by the same token. Thus the time trajectory of the system in the $6N$-dimensional phase space is traced. If a quantity X is calculated at each of ν successive time steps, each Δt long, then

$$\langle X \rangle \equiv \lim_{t_{\max} \to \infty} \frac{1}{t_{\max}} \int_{\nu}^{t_{\max}} X(t)\, dt \simeq \frac{1}{\nu} \sum_{m=1}^{\nu} X(m\Delta t). \tag{4.6}$$

Assuming the calculation has proceeded long enough for the system to be in equilibrium by $t = 0$, equation (4.6) gives an approximation to the static property $\langle X \rangle$ which might, for example, be kinetic energy, $\langle \Sigma_{i=1}^{N} \frac{1}{2} m v_i^2 \rangle$. Other equilibrium properties are the temperature given by $T = (m/3Nk_B)\, \Sigma_{i=i}^{N} v_i^2$ according to the equipartition law and the pressure which follows from the virial equation (Hirschfelder *et al* 1954, Erpenbeck and Wood 1977, Kushick and Berne 1977, Hockney and Eastwood 1981)

$$pV = Nk_B T + \frac{1}{3}\left\langle \sum_{i=1}^{N} \boldsymbol{r}_i \cdot \boldsymbol{F}_i \right\rangle. \tag{4.7}$$

$g(r)$ is also derivable by averaging the histogram of interparticle separations over all particles in many time steps.

However, it is clear that the calculations need not be confined to thermodynamic properties or even static ones. The velocity autocorrelation function defined by

$$z(\tau) \equiv \langle \boldsymbol{v}(t) \cdot \boldsymbol{v}(t + \tau) \rangle \tag{4.8}$$

is accessible in the equilibrium state. Like other correlation functions it is connected theoretically to transport properties and, in this particular case, to the self-diffusion coefficient $D = \frac{1}{3} \int_0^\infty z(\tau)\, d\tau$. Furthermore, the interest of this calculation may be not in the equilibrium state but the way it is approached from an arbitrary initial condition. Molecular properties such as the velocity distribution function and the mean-square displacement of a particle in time Δt are also calculable.

It is clear that a great deal of information could be acquired once the pioneers of the subject had solved the problems of implementation. The details of technique can be found in original papers or reviews (e.g., Erpenbeck and Wood 1977, Kushick and Berne 1977, Hockney and Eastwood 1981, Alder and Wainwright 1960, Rahman 1964, Verlet 1967). One example of the rule for reaching $\boldsymbol{r}_i(t + \Delta t)$ can be derived by expanding this, and also $\boldsymbol{r}_i(t - \Delta t)$, to second order by Taylor's theorem and subtracting one from the other. The result (Verlet 1967) is

$$\boldsymbol{r}_i(t + \Delta t) = -\boldsymbol{r}_i(t - \Delta t) + 2\boldsymbol{r}_i(t) + \boldsymbol{F}_i(t)(\Delta t)^2/m_i \tag{4.9a}$$

where m_i is the mass of particle i and

$$\boldsymbol{v}_i(t) = \frac{\boldsymbol{r}_i(t + \Delta t) - \boldsymbol{r}_i(t - \Delta t)}{2\Delta t}. \tag{4.9b}$$

The time step in a calculation for Ar might be 10^{-14} s; a few hundred steps might establish equilibrium and many hundreds of steps then follow in the equilibrium state. This could be realised for several hundred particles. No doubt with the continual development of computing power longer and more complex operations become possible all the time.

In the preceding paragraph it has been tacitly assumed that the interparticle force is a continuous function of position. In some cases, of which the hard-sphere model is the most important, the forces are impulsive and there is free flight between collisions. In that case, starting with an initial set of positions and velocities, the computer must find out when and where the first collision will occur. The known laws of collisions then determine the post-collision velocities of the two particles and the computation then seeks the next collision—and so on. The time step is then comparable with the mean-free time between collisions (Erpenbeck and Wood 1977, Kushick and Berne 1977, Hockney and Eastwood 1981, Alder and Wainwright 1960).

It is interesting to note that in MD computations with fixed total energy and momentum as well as fixed N and V the temperature shows fluctuations about a mean. In a canonical ensemble investigated by the MC method, N, V and T are fixed and the energy fluctuates. In either case the pressure can be computed and in Erpenbeck and Wood (1977), Kushick and Berne (1977), Hockney and Eastwood (1981) it is shown that the two pressures differ by a factor $(1 - N^{-1})$ which is close to unity for typical N.

As a matter of fact the pressure or temperature can be specified for a calculation and the volume allowed to vary. It was a significant development in 1980 when Andersen showed how to do this (Andersen 1980). The point is that the phenomenon of interest—e.g. phase transitions—may be associated with density or temperature changes and a simulation with N, V and E fixed and p and T fluctuating may give misleading results. MD calculations have therefore been developed which hold p or T or both constant and allow V to vary while retaining periodic boundary conditions. Averages are then over isothermal–isobaric (N, T, p), isobaric–isenthalpic (N, p, H) or canonical (N, V, T) ensembles instead of microcanonical (N, V, E).

To implement the (N, p, H) calculation in which V becomes a variable, the quantities $V, \boldsymbol{r}_i$ and $\boldsymbol{p}_i$ in the usual method are replaced by $Q(\equiv V)$, $\boldsymbol{\rho}_i(\equiv \boldsymbol{r}_i V^{1/3})$ and $\boldsymbol{\pi}_i(\equiv \boldsymbol{p}_i Q^{1/3})$. A MD calculation then has to be performed on the system described by $(Q, \boldsymbol{\rho}_i, \boldsymbol{\pi}_i)$ and it can be shown that the trajectory of this system in its phase space corresponds to one in the system $(V, \boldsymbol{r}_i, \boldsymbol{p}_i)$ with V now become a function of time. Andersen showed that the average of a structural or thermal property calculated along this trajectory (cf equation (4.6)) is the ensemble average of that property in the (N, p, H) ensemble. The overall strategy is to write a Lagrangian in terms of the new variables including a term representing pV where p is the prescribed external pressure. The equations of motion come from the Langrangian. The equilibrium system has been reached when the internal pressure calculated from the internal variables equals p. There are corresponding arguments for (N, p, T) and (N, V, T) ensembles.

The cell volume V can be written $\boldsymbol{a} \cdot \boldsymbol{b} \wedge \boldsymbol{c}$ where $\boldsymbol{a}$, $\boldsymbol{b}$ and $\boldsymbol{c}$ are the cell edges. If the nine components of these vectors are treated as variables not only V but also the cell shape can respond to system changes at constant pressure. This further generalisation of MD is appropriate for studying structural phase changes in crystals (Parinello and Rahman 1981).

These methods will be referred to again in connection with glass transitions (§4.6). Here we note a metallic example—an appropriate application because the volume-dependent aspects of the energy of a metal (see Chapter 6) can be explicitly incorporated into the MD as V changes. Using the extended version of MD, Barnett *et al* (1985) derived several thermodynamic and transport properties of liquid Mg, including the equilibrium density, at three temperatures in moderate agreement with observations.

Evolution has not stopped here and at the time of writing a method is being developed for combining MD with the density functional theory used for solving the Schrödinger equation for valence electron motion in metals and semiconductors. Simultaneous solutions look possible for the connected problems of what the electronic properties are, what interatomic forces they imply, and how the atoms respond (Car and Parinello 1985).

4.3 Some results for the hard-sphere model

The hard-sphere or HS potential may be written

$$\begin{aligned} \varphi(r) &= +\infty \qquad && r < \sigma \\ &= 0 && r \geqslant \sigma. \end{aligned} \tag{4.10}$$

This is the interparticle potential energy; r is the separation of centres, σ the diameter. The same form applies in two- or one-dimensional systems to hard discs or rods. It could be described as an unrealistic potential but its great value derives from its simplicity which makes some exact or numerically accurate results accessible. These in turn can be used to test deductions from analytical theories or as starting points in perturbation treatments. When physical phenomena are governed largely by repulsive interparticle forces, the HS potential has some direct application to real systems, e.g., in very hot gases where the attractive energies are much less than k_BT.

The equation of state was an early result of molecular dynamics. To present it, the virial equation is convenient and we digress to refer to that. In Hirschfelder *et al* (1954), equation (4.7) is derived from classical mechanics and its first term is two thirds of the equipartition value of the mean kinetic energy. For the class of pairwise additive potentials to which equation (4.10) belongs, $\Phi = \frac{1}{2}\Sigma_{i,j}\, \varphi(r_{ij})$ and the second term of equation (4.7) can be developed with the help of §2.3. In classical mechanics the averages are time averages but will be replaced by ensemble averages when necessary.

Now

$$\frac{1}{3}\left\langle \sum_{i=1}^{N} \boldsymbol{r}_i \cdot \boldsymbol{F}_i \right\rangle = -\frac{1}{3}\left\langle \sum_{i=1}^{N} \boldsymbol{r}_i \cdot \frac{\partial \Phi}{\partial \boldsymbol{r}_i} \right\rangle$$

$$= -\frac{1}{6}\int\int n^{(2)}(\boldsymbol{r}_1, \boldsymbol{r}_2) r_{12} \frac{\partial \varphi_{12}}{\partial r_{12}} \,\mathrm{d}\boldsymbol{r}_1 \,\mathrm{d}\boldsymbol{r}_2.$$

Using equation (2.6*b*), putting $\boldsymbol{r}_1 + \boldsymbol{r} = \boldsymbol{r}_2$, and keeping $\boldsymbol{r}_1$ fixed for the integration over $\boldsymbol{r}_2$, we find

$$\frac{1}{3}\left\langle \sum_{i=1}^{N} \boldsymbol{r}_i \cdot \boldsymbol{F}_i \right\rangle = -\tfrac{1}{6} n_0 N \int g(r)\, r \frac{\partial \varphi}{\partial r} \,\mathrm{d}\boldsymbol{r}.$$

Finally, equation (4.7) gives

$$pV = Nk_{\mathrm{B}}T - \tfrac{2}{3}\pi N n_0 \int r^3 g(r) \frac{\partial \varphi}{\partial r} \,\mathrm{d}r \tag{4.11}$$

which is the same as equation (3.13).

The second term, representing the departure from ideal gas behaviour, is a function of state and can be expanded in powers of V^{-1}. Thus, for the molar volume, V_{M}:

$$z \equiv \frac{pV_{\mathrm{M}}}{RT} = 1 + \frac{B(T)}{V_{\mathrm{M}}} + \frac{C(T)}{V_{\mathrm{M}}^2} + \frac{D(T)}{V_{\mathrm{M}}^3} + \cdots \tag{4.12a}$$

which can be alternatively expressed as:

$$z \equiv \frac{pV}{Nk_{\mathrm{B}}T} = 1 + B_2(T) n_0 + B_3(T) n_0^2 + \cdots \tag{4.12b}$$

where N is the number of particles and $n_0 \equiv N/V$.

For real fluids, the virial coefficients, B_2, B_3, etc, can be found by experiment. For van der Waals' equation of state, elementary arguments show $B(T) = (b - a/RT)$ where b is the covolume of hard spherical cores, namely $b = 2\pi N\sigma^3/3$. There are general statistical mechanical formulae for B_l but for $l > 3$ they require the evaluation of complicated integrals even for simple potentials (Hirschfelder *et al* 1954). However, the numerical evaluation of complicated integrals is what the MC and MD methods are for and this is exemplified in Ree and Hoover (1967), Erpenbeck and Wood (1984) where B_7 to B_{10} are given.

We therefore have MC results up to B_7 for equation (4.12) and MD results by Alder and Wainwright for $pV/Nk_{\mathrm{B}}T$. These are both for hard spheres and in this special case the absence of attractive forces and the abrupt cut-off of the repulsive potential at $r = \sigma$ imply that the virial coefficients are independent of temperature.

It was later noticed by Carnahan and Starling (1969) that the accurate MD values of z could be represented by the following simple formula:

$$z = \frac{1 + \eta + \eta^2 - \eta^3}{(1 - \eta)^3} \tag{4.13}$$

where

$$\eta \equiv \frac{4}{3} \frac{N\pi}{V} \left(\frac{\sigma}{2}\right)^3 = \tfrac{1}{6} n_0 \pi \sigma^3. \tag{4.14}$$

Table 4.1 summarises some numerical results (Carnahan and Starling 1969).

Table 4.1

V/V_0	z(MD)	$z(B_6)$	$z(B_7)$	z(equation (4.13))
1.5	12.5	10.46	11.27	12.43
1.6	10.17	8.95	9.50	10.16
1.7	8.59	7.79	8.18	8.56
2.0	5.89	5.59	5.73	5.83
3.0	3.05	3.01	3.03	3.03
10.0	1.36	1.36	1.36	1.36

In table 4.1, V_0 is the hard-sphere volume for crystalline closest packing, namely $N\sigma^3/\sqrt{2}$ and $z(B_n)$ means z calculated with terms up to B_n. The good agreement of Carnahan and Starling's formula with the MD results is clear; so is the accuracy of the virial series to six or seven terms. As might be expected the latter is better at lower densities.

Carnahan and Starling's formula has been much used. The procedure referred to in §4.1 for calculating $F_1 - F_2$ can be applied to it and the entropy follows from $S = -(\partial F/\partial T)_V$. F_1 can be chosen to be the same as that of an ideal gas. The result is

$$\Delta S_{\mathrm{E}} \equiv S_{\mathrm{HS}} - S_{\mathrm{ideal}} = \frac{Nk_{\mathrm{B}}\eta(3\eta - 4)}{(1 - \eta)^4}. \tag{4.15a}$$

Other thermodynamic properties follow by using thermodynamic relations. One example is the compressibility:

$$\kappa_{\mathrm{T}} = \frac{(1 - \eta)^4}{n_0 k_{\mathrm{B}} T[2\eta(4 - \eta) + (1 - \eta)^4]}. \tag{4.15b}$$

Similar but more complicated formulae exist for binary mixtures of different sized spheres and are useful in the theory of liquid mixtures (§6.9). The pair distribution function for hard-sphere liquids has been calculated by MC methods and is shown in figure 4.2.

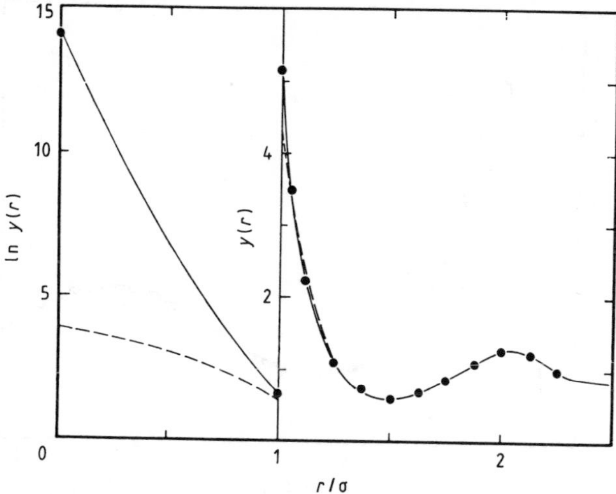

Figure 4.2 Pair distribution for hard spheres with $n_0\sigma^3 = 0.9$. The full curve is a continuous function parametrised to represent the MC results which are the full circles. The broken curve is the PY approximation (§5.6). Note that $y(r) \equiv g(r) \exp(\beta\varphi(r))$ and $y(r) = g(r)$ for $r \geqslant \sigma$. (See Barker and Henderson 1972.)

Since hard spheres can pack with LRO, a phase transition between liquid and crystalline hard-sphere systems is a possibility and an early triumph of MC was to detect it. To do so the p versus n_0 curve was computed for both fluid and solid isotherms and the corresponding entropy also. This enabled the points of equal chemical potential in the two phases to be located; this gave the values of p and n_0 for equilibrium at melting. The result was to identify a first-order transition at the following densities relative to that of closest LRO packing: 0.736 ± 0.003 and 0.667 ± 0.003 for solid and fluid respectively. Details of this investigation are given in Hoover and Ree (1968).

4.4 Some results for the Lennard-Jones model

The Lennard-Jones, or '6–12' potential is:

$$\varphi(r) = 4\varepsilon\left[\left(\frac{\sigma}{r}\right)^{12} - \left(\frac{\sigma}{r}\right)^{6}\right] \tag{4.16}$$

where ε and σ are respectively the energy and length parameters which characterise the chemical species. The attractive term represents the induced dipole pair interaction of closed-shell atoms and the r^{-12} gives the strong repulsion at small separations. This form of $\varphi(r)$ is too simple to be quantitatively correct but it is qualitatively satisfactory for closed-shell atoms and molecules.

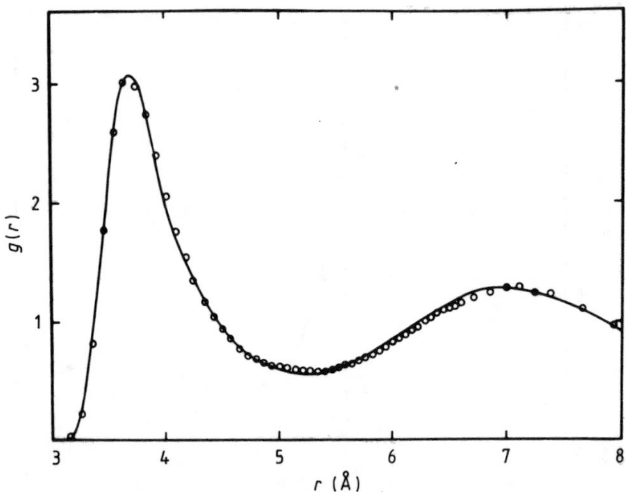

Figure 4.3 $g(r)$ for Ar at 85 K, $n_0 = 0.02125$ atoms Å^{-3}. Full curve, experimental results; circles, MC values. The MD values are indistinguishable on this scale from the MC values (see Ree and Hoover 1967, Erpenbeck and Wood 1984). (From Yarnell *et al* 1973.)

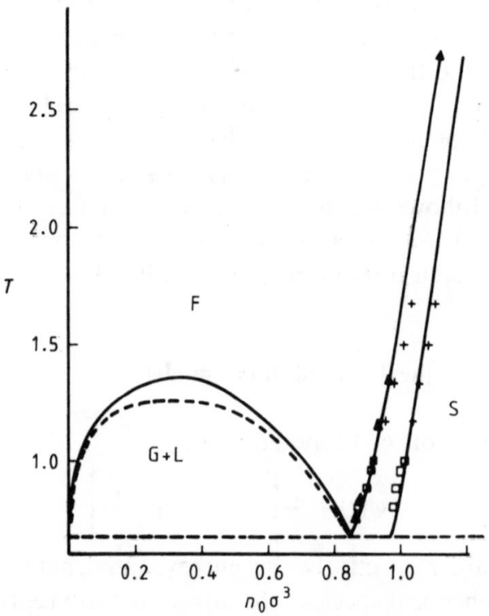

Figure 4.4 Phase diagram for Ar (T is in units of ε/k_B). Full curve, results of MC using the LJ potential; broken curve, squares and crosses, experimental data; triangles, points where $S(Q) = 2.85$—a suggested crystallisation criterion. The separation of the rising lines on the right corresponds to ΔV on melting. (From Hansen and Verlet 1969.)

The existence of both attractive and repulsive forces means that the equation of state shows critical phenomena. These, as well as the thermodynamic properties, $g(r)$ and melting, *inter alia*, have been calculated by the MD and MC methods. Figure 4.3 is a typical result taken from Yarnell *et al* (1973). Hansen and Verlet (1969) obtained the phase diagram in figure 4.4 with MC calculations. More collected data are in Barker and Henderson (1976). For Ar, which has been simulated frequently using equation (4.16), $\varepsilon = 0.0103$ eV and $\sigma = 3.405$ Å.

The MC and MD results for the 6–12 potential will be required in the next chapter.

4.5 Some results for a liquid metal potential

It may seem unplausible to refer to a pair interaction for a metallic liquid. The interparticle forces will clearly depend on the conduction electron gas and its density as well as on the ion–ion and ion–electron interactions. Nevertheless, for a given N and V an effective ion–ion pair potential can be devised and it is this that largely determines the structure. This matter will be discussed in a later chapter dealing with liquid metals and for the present we

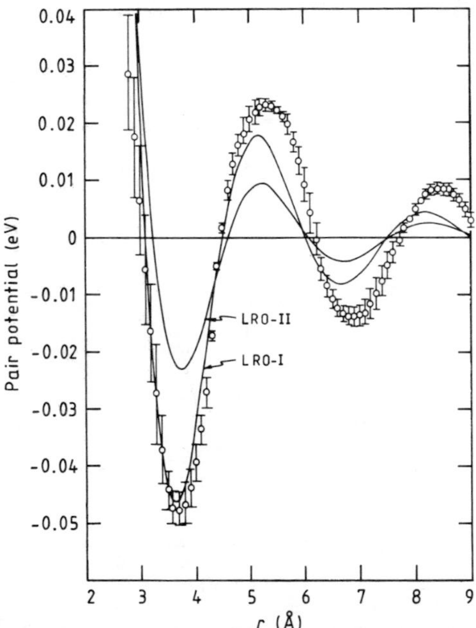

Figure 4.5 Oscillatory potentials for liquid Na. Φ, points derived by Johnson *et al* from structural data (see §5.9); LRO-I, an approximation to Φ; LRO-II, chosen for better fit to experimental structural data. (From Paskin and Rahman 1966.)

note that suitable liquid metal pair potentials have a somewhat softer repulsive part than the Lennard-Jones potential and a long-range damped oscillatory tail. The latter is a characteristically metallic feature: the oscillations derive ultimately from the sharp Fermi cut-off of conduction electron momentum. The sharper this is the more pronounced the oscillations.

A semiempirical potential suitable for liquid Na was given by Paskin and Rahman (1966). It is referred to as the LRO-II potential and is

$$\varphi(r) = \frac{A\cos(k_F r + B)}{r^3} + E\exp\left(F - G\frac{r}{r_0}\right). \tag{4.17}$$

This potential is shown in figure 4.5 and its first term gives the long-range oscillatory behaviour. k_F is the Fermi wavenumber calculated from the conduction electron density. The parameter values which make $\varphi(r)$ applicable to liquid Na at 377 K and density 0.927 g cm^{-3} are given in Paskin and Rahman (1966), Schiff (1969) and Tanaka and Fukui (1975). The latter report extensive MD calculations and use a 864-particle computation with a time step of 2.5×10^{-14} s and with the methods described in §4.2 generates tables of g and $g^{(3)}$. Figure 4.6 presents a MD and an experimental $g(r)$. $g^{(3)}$ will be used in the next chapter. A more recent analytical pair potential for simple metals, valuable for MD calculations, was given by Pettifor and Ward (1984).

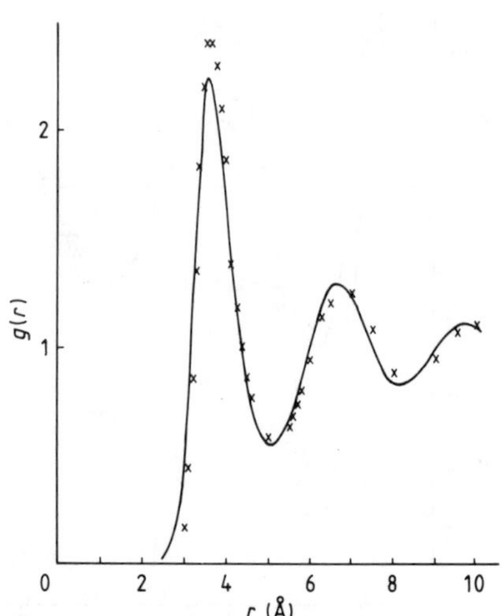

Figure 4.6 $g(r)$ in liquid Na. Full curve, MD with LRO-II at $T = 104$ °C (from Tanaka and Fukui 1975). Crosses, experimental points from Waseda (1980) (in Chapter 3) for $T = 105$ °C.

4.6 Quenching and computer glasses

In the course of an MD calculation energy can suddenly be withdrawn by reducing the particle speeds according to some prescription. This simulates the laboratory process of quenching which is used to produce glasses—in fact quenching is the routine procedure for making metallic glass samples. Suppose that all $\boldsymbol{r}$ and $\boldsymbol{v}$ are known in an MD simulation of a liquid in equilibrium at temperature $T_0 = (m/3k_BN)\Sigma_i v_i^2$. One quenching method is to reduce every velocity suddenly by a factor $1 + \alpha\{(T_f/T_0)^{1/2} - 1\}$ where $0 < \alpha < 1$ and T_f is the desired final temperature. This can be done several times in succession allowing a relaxation or equilibration time after each step or according to any other desired cooling sequence. Very high rates of cooling can be achieved—10^{12} K s^{-1} or more.

Such quenchings can be applied to constant volume or constant pressure MD computations and may be aimed at studying supercooling in liquids, nucleation, crystallisation or transition to glassy phases. The nature of glass transitions will be raised in a later chapter; here we note that MD quenching can simulate laboratory produced glasses or can create computer glasses that are unlikely to be made in the real world because the rates of cooling required exceed what is experimentally realisable. For example, argon glass can be made on a computer.

Criteria are needed for recognition that a glass has been formed and they include a shape of $g(r)$ typical of amorphous solids and without features associated with periodicity, and vanishingly small values of the diffusion coefficient computed from the velocity autocorrelation function (Angell *et al* 1981). Once a glass has been identified the MD will then give its thermodynamic and structural properties and also quantities important for the discussion of thermal motion—see, e.g., $\psi(t)$ and $S(Q, \omega)$ in Chapter 8.

In the course of MD quenchings, quench echoes were discovered. According to the elegant exposition by Nagel *et al* (1983) this phenomenon can be described and explained as follows. To a good approximation the motion of a glass system is a superposition of normal harmonic modes approximately obeying the equipartition principle. If all the kinetic energy is suddenly removed at time $t = 0$ there will still be potential energy and when the system is unclamped it will return to equilibrium at about half the initial temperature. Before that, however, suppose a second quench is performed at time t_1. Those modes with period τ close to $2t_1$ have little or no kinetic energy to lose at t_1 and will be unaffected by the second quench; all others will have their energy reduced. Subsequently the $\tau = 2t_1$ modes regain kinetic energy and significantly weight the computed temperature. However, after a further t_1 they have lost it again and this causes a temporary dip in $T \propto \Sigma v_i^2$. The result is shown in figure 4.7 and the dip is called the quench echo.

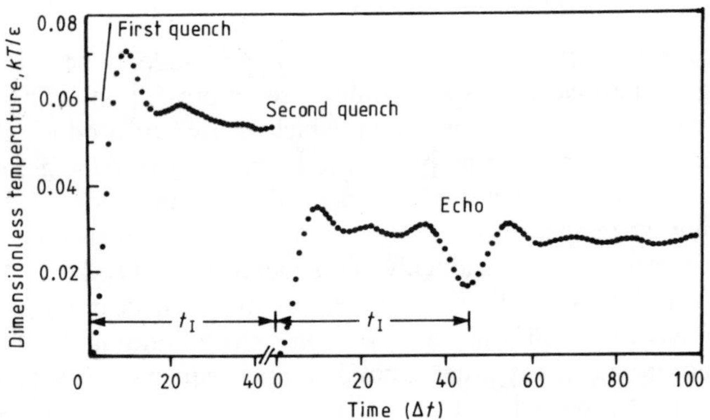

Figure 4.7 The quench echo—from Nagel *et al* (1983).

There are more subtleties to the quench echo phenomenon but here we simply note its value for exploring thermal motions. For example, by repeated quenching after successive periods t_I, the predominance of the mode or modes with $\omega = \pi/t_I$ can be made very great which allows the characteristics of a particular mode to be studied separately (Nagel *et al* 1981). One such characteristic is the degree of localisation in space. Localisation in more general terms is taken up in Chapter 9. Here we note that one quantity which helps to distinguish localised from extended modes is the

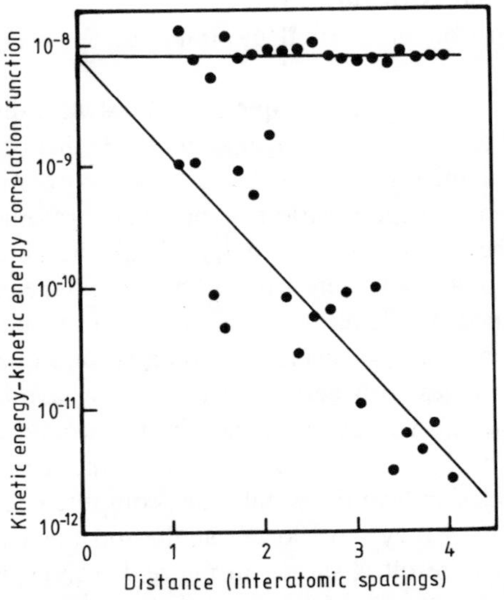

Figure 4.8 $l(r)$ for a localised and an extended mode—from Nagel *et al* (1983).

kinetic energy correlation function

$$l(r) \equiv \left(\sum_{i,j} \delta(\boldsymbol{r} - \boldsymbol{r}_{ij})\right)^{-1} \sum_{i,j} v_i^2 v_j^2 \delta(\boldsymbol{r} - \boldsymbol{r}_{ij}).$$

$l(r)$ remains roughly constant for extended modes because two particles r_{ij} apart can both have large v's; in localised modes the amplitude is confined to a spatial domain by definition so $v_i^2 v_j^2$ will fall as r_{ij} increases. Figure 4.8 shows the effect and identifies two different modes in a LJ glass. This is a powerful method for investigating thermal motion in amorphous solids (see also §8.5).

References

Alder B and Wainwright T E 1960 *J. Chem. Phys.* **33** 1439

Andersen H C 1980 *J. Chem. Phys.* **72** 2384

Angell C A, Clarke J H R and Woodcock L V 1981 *Adv. Chem. Phys.* **48** 397

Barnett R N, Cleveland C L and Landman U 1985 *Phys. Rev. Lett.* **54** 1679

Barker J A and Henderson D 1972 Theories of liquids *Ann. Rev. Phys. Chem.* **23** 439

—— 1976 *Rev. Mod. Phys.* **48** 587

Car R and Parinello M 1985 *Phys. Rev. Lett.* **55** 2471

Carnahan N F and Starling K E 1969 *J. Chem. Phys.* **51** 635

Erpenbeck J J and Wood W W 1977 *Modern Theoretical Chemistry* vol 6, ed B J Berne (New York: Plenum)

—— 1984 *J. Stat. Phys.* **35** 321

Hansen J P and Verlet L 1969 *Phys. Rev.* **184** 151

Hirschfelder J O, Curtiss C F and Bird R B 1954 *Molecular Theory of Gases and Liquids* (New York: Wiley)

Hockney R W and Eastwood J W 1981 *Computer Simulations Using Particles* (New York: McGraw-Hill)

Hoover W G and Ree R M 1968 *J. Chem. Phys.* **49** 3609

Kushick J and Berne B J 1977 *Modern Theoretical Chemistry* vol 6, ed B J Berne (New York: Plenum)

Metropolis N, Metropolis A W, Rosenblath M N, Teller A H and Teller E 1953 *J. Chem. Phys.* **21** 1087

Nagel S R, Grest G S and Rahman A 1983 *Phys. Today* **36** 24

Nagel S R, Rahman A and Grest G S 1981 *Phys. Rev. Lett.* **47** 1665

Parinello M and Rahman A 1981 *J. Appl. Phys.* **52** 7182

Paskin A and Rahman A 1966 *Phys. Rev. Lett.* **16** 300

Pettifor D G and Ward M A 1984 *Solid State Commun.* **49** 291

Rahman A 1964 *Phys. Rev.* **136** A405

Ree F H and Hoover W G 1967 *J. Chem. Phys.* **46** 4181

Schiff D 1969 *Phys. Rev.* **186** 151

Tanaka M and Fukui Y 1975 *Prog. Theor. Phys.* **53** 1547

Valleau J P and Torrie G M 1977 *Modern Theoretical Chemistry* vol 5, ed B J Berne (New York: Plenum)

Valleau J P and Whittington S G 1977 *Modern Theoretical Chemistry* vol 5, ed B J Berne (New York: Plenum)

Verlet L 1967 *Phys. Rev.* **159** 98

Wood W W 1968 *Physics of Simple Liquids* ed H N V Temperley, J S Rowlinson and G S Rushbrooke (Amsterdam: North-Holland)

Yarnell J L, Katz M J, Wentzel R G and Koenig S H 1973 *Phys. Rev.* A **7** 2130

5

THE STATIC PROPERTIES OF SIMPLE LIQUIDS

An attempt to introduce the theory of liquids must be based on a drastic selection of subject matter. In the following sections the theory of the structure and the thermodynamic properties of simple liquids will be sampled, mostly for non-metallic liquids. The ideal liquid in this context is liquid argon and the characteristic properties of metallic liquids will be treated in later chapters. It cannot be said that the theory of simple liquids is a solved problem because it remains very difficult to start from first principles and derive, for example, the correct equation of state. However, the nature of the difficulties is well understood and in many cases the failure to reach a conclusive result is not due to the inherent mystery of the subject, but rather to formidable technical problems. Barker and Henderson (1976) write:

> For ordinary liquid phases we now have excellent qualitative understanding of these questions and in simple cases this can lead to fairly rigorously quantitative predictions.

Many of the most common and important liquids are not as simple as liquid argon but it will not be attempted in this chapter to extend the theory to more complicated systems. Important classes of fluid thereby omitted include ionic liquids in which the long-range interparticle forces bring special difficulties; quantum liquids, notably He; liquids with non-spherical molecules in which rotatory degrees of freedom ought to be incorporated into the statistical mechanics; and liquid crystals. Liquid mixtures will also not be included but to exemplify something of the physics of liquid mixtures binary alloys will be treated in a later chapter.

5.1 Intermolecular potentials

The main aim of the theory of liquids is to use statistical mechanics to derive properties of the liquid state in agreement with observations and a necessary starting point is an adequate interparticle energy function. As in §4.1 we shall assume that classical mechanics is appropriate and that $E = T + \Phi$. More generally than in §4.3 however, we write

$$\Phi(\boldsymbol{r}_1, \boldsymbol{r}_2, \ldots, \boldsymbol{r}_n) = \frac{1}{2}\sum_{i,j} \varphi(r_{ij}) + \sum\sum\sum_{1 \leqslant i \leqslant j \leqslant k \leqslant N} \varphi_3(r_{ij}, r_{jk}, r_{ki}) + \cdots . \quad (5.1)$$

The object of equation (5.1) is to express the fact that Φ is not necessarily pairwise additive; if it were the φ-term alone would suffice. The triplet term, φ_3, is defined to vanish if any one of the three atoms in the argument is at an infinite distance from the other two. φ_3 therefore represents the interference with the interparticle force of a pair of atoms by a third atom in the vicinity. In the case of rare gas atoms one physical mechanism for φ_3 is the induced dipole–dipole interaction responsible for van der Waals' forces. If the quantum mechanical perturbation theory of this is taken to third order, a triplet interaction appears, a result obtained in 1943 by Axilrod and Teller. This and many other aspects of Φ for rare gas atoms in solid, liquid and gaseous systems are discussed in Barker (1976) where the evidence is presented that the φ_3-term in equation (5.1) contributes about 7 % of the cohesive energy of solid argon. For close agreement with sensitive experiments—e.g., to reproduce the third virial coefficient of argon as a function of temperature—incorporation of three-body forces into the theory are necessary even for these so called simple spherical atoms. Nevertheless, it is also true that approximate, sometimes only qualitative, agreement with experiments can be obtained without three-body forces and even with the model LJ potential of equation (4.16). It is therefore not unreasonable to introduce some liquid state theory in the pairwise additive approximation and this will be done as a simplification. Figure 5.1 shows some interatomic potentials which could be used in numerical exploitation of the formulae in this chapter. If the

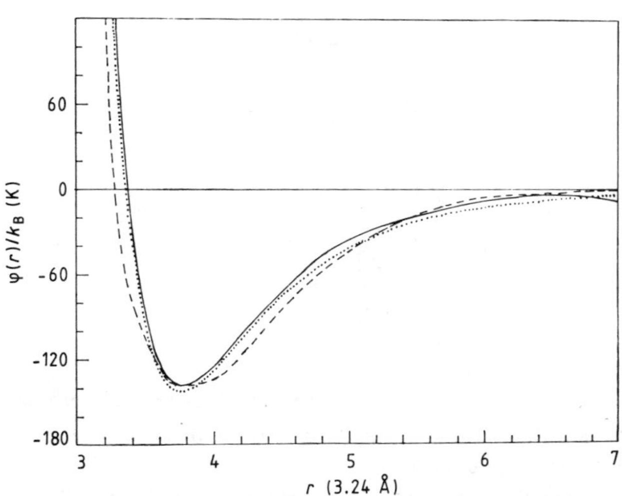

Figure 5.1 Typical rare gas potentials; these are for Ar. Full curve, inferred from neutron diffraction; dotted and broken curves, multiparameter equations from a variety of experimental input data from, respectively, Barker, Fisher and Watts and Dymond and Alder (see Barker 1976).

parameters of two-body potentials are determined by fitting data from liquids some many-body effects will be therby incorporated in any case.

5.2 Connections between thermodynamic quantities and $g(r)$

In an *ad hoc* way in §4.3 we have already connected $g(r)$ with U and with p. To recapitulate,

$$U = \tfrac{3}{2}Nk_{\mathrm{B}}T + \tfrac{1}{2}Nn_0 \int_V \varphi(r)g(r)\,\mathrm{d}\mathbf{r} \tag{5.2}$$

$$p = n_0 k_{\mathrm{B}}T - \tfrac{1}{6}n_0^2 \int_V rg(r)\frac{\partial\varphi}{\partial r}\,\mathrm{d}\mathbf{r}. \tag{5.3}$$

A formal way of deriving such relations starts with the grand canonical partition function, $\mathscr{Z}$, and its associated formulae. From statistical mechanical texts these are (see also §3.12):

$$\mathscr{Z} \equiv \sum_N \sum_j \exp\frac{\mu N - E_j}{k_{\mathrm{B}}T} \tag{5.4}$$

$$P(N, E_j) = \mathscr{Z}^{-1}\exp\frac{\mu N - E_j}{k_{\mathrm{B}}T} \tag{5.5}$$

$$\langle X\rangle = \sum_{N,j} X_{N,j}P(N, E_j) \tag{5.6}$$

$$U \equiv \langle E\rangle = k_{\mathrm{B}}T^2\left(\frac{\partial \ln \mathscr{Z}}{\partial T}\right)_{V,\mu} + \mu\langle N\rangle \tag{5.7}$$

$$\Omega \equiv -k_{\mathrm{B}}T\ln\mathscr{Z} = -pV \tag{5.8}$$

$$p = -\left(\frac{\partial\Omega}{\partial V}\right)_{T,\mu} \tag{5.9}$$

$$S = -\left(\frac{\partial\Omega}{\partial T}\right)_{\mu,V} \tag{5.10}$$

$$\langle N\rangle = -\left(\frac{\partial\Omega}{\partial\mu}\right)_{T,V}. \tag{5.11}$$

All these formulae refer to a system with fixed V, T and μ; both particles and energy can be interchanged with the environment so E and N have to be averaged over their fluctuations.

To derive equations (5.2) and (5.3) formally from equation (5.4) is somewhat tedious and will be omitted especially as equation (5.2) is intuitively obvious for pairwise interactions and equation (5.3) has been discussed in §4.3. A full treatment can be found in Rice and Gray (1965). However, there is an example of a more formal deduction in §3.12 where it was shown from equation (5.5) that

$$S(0) = n_0 k_B T \kappa_T. \tag{5.12a}$$

Equation (3.11*b*) leads to

$$S(0) = 1 + n_0 \int_0^\infty (g(r) - 1) 4\pi r^2 \, \mathrm{d}r \tag{5.12b}$$

consequently

$$k_B T \left(\frac{\partial n_0}{\partial p} \right)_T = 1 + n_0 \int_0^\infty h(r) 4\pi r^2 \, \mathrm{d}r \tag{5.12c}$$

where

$$h(r) \equiv g(r) - 1. \tag{5.13}$$

$h(r)$ is called the pair correlation function and tends to zero as $r \to \infty$. Equation (5.12*c*) is the *compressibility equation of state* since it relates n_0, p and T. Unlike equation (5.3), which does the same, the compressibility equation does not rest on the assumption of pairwise additivity of $\varphi(r)$ and is quite general for isotropic fluids.

It has previously been pointed out (§§4.1 and 4.3) that F and consequently S, G and μ can be obtained by integrating p/n_0^2 with respect to n_0. To do this starting with equations (5.3) or (5.12*c*) would require knowledge of the density dependence of $g(r)$. Similarly to derive a heat capacity, $(\partial U/\partial T)_V$, from equation (5.2) requires the temperature derivative of $g(r)$. It has already been remarked (§3.13) that the density derivative of $g(r)$ is connected theoretically with $g^{(3)}$ and the same is true of the temperature derivatives. Therefore some of the thermodynamic quantities require for their calculation knowledge that is equivalent to a knowledge of $g^{(3)}$ even if the pairwise additive assumption prevails and $g^{(3)}$ does not appear in U or p.

This section makes clear that if $g(r)$—more completely expressed as $g(\mathbf{r}_1, \mathbf{r}_2; T; n_0)$—could be calculated from $\varphi(r)$, a theory of the thermodynamic properties of liquids would exist. The theory of the pair distribution will now be introduced. This is an important and unfinished chapter in the theory of disordered matter. It is dealt with extensively in many treatments of liquid state physics such as Rice and Gray (1965), Croxton (1974, 1978), Hansen and McDonald (1976), Barker and Henderson (1976), Temperley *et al* (1968), Haymet *et al* (1981a,b), Meeron (1957) and Salpeter (1958), and many other works quoted therein.

5.3 Attempts to derive $g(r)$ from $\varphi(r)$

The first move is to write down a formula for $n^{(2)}$ which was defined in §2.3. This can be done with either the canonical or grand canonical formulae.

Using the classical version of the former (see §4.1) and equation (4.3c) in particular:

$$n^{(2)}(1, 2) = n_0^2 g(1, 2) = \frac{N(N-1)}{Q} \int \exp(-\beta\Phi)\,\mathrm{d}\boldsymbol{r}_3 \cdots \mathrm{d}\boldsymbol{r}_N. \quad (5.14)$$

This—as the definition of $n^{(2)}$ requires—is the probability of *any* configuration of $(N-2)$ particles at $\boldsymbol{r}_3, \ldots, \boldsymbol{r}_N$ together with two more particles definitely at $\boldsymbol{r}_1$ and $\boldsymbol{r}_2$. The factor $N(N-1)$ is the number of ways of choosing the latter two out of N particles.

We now differentiate $n^{(2)}(1, 2)$ with respect to the position coordinate $\boldsymbol{r}_1$. This gives

$$\nabla_1 n^{(2)}(1, 2) = \frac{N(N-1)}{Q} \int (-\beta) \exp(-\beta\Phi)(\nabla_1\Phi)\,\mathrm{d}\boldsymbol{r}_3 \cdots \mathrm{d}\boldsymbol{r}_N.$$

Using the definition of $g^{(3)}$ and thinking of the interactions of particles 1, 2 and 3 in figure 5.2 we may write

$$\nabla_1\Phi = \nabla_1\varphi(r) + \int n_0 g^{(3)}(r, s, t)\nabla_1\varphi(s)\,\mathrm{d}\boldsymbol{r}_3.$$

Substituting this into the previous expression, and recalling the equation (5.14) for $n^{(2)}$ and g, we find

$$-k_{\mathrm{B}}T\nabla_1 g(r) = g(r)\nabla_1\varphi(r) + g(r)\int n_0 g^3(r, s, t)\nabla_1\varphi(s)\,\mathrm{d}\boldsymbol{r}_3. \quad (5.15)$$

This is an equation for g in terms of $g^{(3)}$. Similar equations exist for $g^{(k)}$ in terms of $g^{(k+1)}$, $k \geqslant 3$.

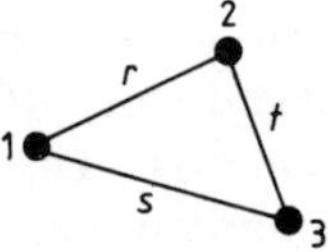

Figure 5.2 Diagram for particle interaction.

Before considering this equation let us rederive it by a physically more intuitive method. A quantity called the potential of the mean force can be defined by

$$\psi(r) \equiv -k_{\mathrm{B}}T \ln g(r) \quad (5.16)$$

and it can be interpreted by turning the definition into a Boltzmann factor $g(r) = \exp(-\psi(r)/k_{\mathrm{B}}T)$ which expresses the probability that two particles interacting with potential $\psi(r)$ should be found in thermal equilibrium separated by r. This accords with the meaning of $g(r)$. However, $\psi(r)$ is not the two-body interaction $\varphi(r)$ because all the other particles are present. The average force on particle 1 in the liquid can be written $-\nabla_1\psi(r)$ and must be the direct effect, $-\nabla_1\varphi(r)$, of particle 2 plus the contributions of all the others, namely

$$-\nabla_1\psi(r) = k_B T\nabla_1 \ln g(r) = -\nabla_1\varphi(r) - \int n_0 g^{(3)}(r, s, t)\nabla_1\varphi(s)\,\mathrm{d}\boldsymbol{r}_3.$$

This is equation (5.15) again; it is called the Born–Green–Yvon or BGY equation.

$g(r)$ cannot be calculated from this unless $g^{(3)}$ is already known. In general, $g^{(k+1)}$ is required to derive $g^{(k)}$. This poses the 'closure problem'. The most well known attempt to circumvent it is Kirkwood's *superposition approximation* of 1935, namely

$$g^{(3)}(r, s, t) \simeq g(r)g(s)g(t). \qquad (5.17)$$

This equation means that the pair distribution of any two particles is unaffected by the presence of a third—a proposition that is guaranteed to be true only if the third is very distant. The question is whether this approximation is good enough to reduce equation (5.15) to a useful equation for g. Leaving comment on this point until later we make a few general remarks preparatory to a reference to the work of Haymet *et al* (1981a,b) on the solution of the BGY equation.

It is a feature of theories of liquids, whether of their structural, thermodynamic or electronic properties, that desired quantities such as partition functions, configuration integrals and electronic densities of states cannot be evaluated directly from the formally exact expressions for them because of their inherent complexity. This complexity results from the disorder of the structure and the large number of interacting particles. A common expedient is to find for the quantity of interest a series expansion of which it is hoped a few terms will be sufficiently accurate even if only within a restricted range of parameters. In the theory of the thermodynamics of fluids, expansions in ascending powers of the density are common (see e.g., equation (4.12)) and these will be the more satisfactory the lower the density of the fluid. In the electronic theory of disordered metals expansions in powers of the potential will occur and low-order terms will suffice when the electron–ion potential is weak (Chapter 7). The coefficients in these expansions embody the structural information and, typically, the higher the order of the term the more complex the cluster of particles whose structure has to be known or approximated. For this reason the notation of general terms in the expansion can be complex and the methods of evaluation and partial summation can involve specialised techniques which combine mathematical and numerical work.

An early and well known example of this is the use of cluster integrals in expressions for the virial coefficients in the theory of dense imperfect gases. Starting from the canonical partition function a considerable train of algebra leads to the following expression for the fourth virial coefficient in equation (4.12*a*):

$$D(T) = -\frac{N^3}{8V}\iiiint (3f_{12}f_{23}f_{34}f_{14} + 6f_{12}f_{13}f_{14}f_{23}f_{34} + f_{12}f_{13}f_{14}f_{23}f_{24}f_{34})\, \mathrm{d}\boldsymbol{r}_1\, \mathrm{d}\boldsymbol{r}_2\, \mathrm{d}\boldsymbol{r}_3\, \mathrm{d}\boldsymbol{r}_4. \quad (5.18)$$

This result is demonstrated in detail (see Hirschfelder *et al* 1954 in Chapter 4) and depends on the pairwise additivity of $\varphi(r_{ij})$. f_{ij} is called a Mayer function and its definition is

$$f_{ij} \equiv f(r_{ij}) \equiv \exp(-\beta\varphi(r_{ij})) - 1. \quad (5.19)$$

A cluster integral is an integral of which the integrand is a sum of products of Mayer functions. The latter connect the various pairs of atoms in an extended group or cluster—in this case a group of four. The expression for $D(T)$ illustrates the remarks in the preceding paragraph. $C(T)$ involves cluster integrals concerning only three atoms; higher coefficients than $D(T)$ involve bigger clusters. As mentioned in §4.3, MC methods might have to be invoked to calculate the complicated cluster integrals in the sixth or seventh virial coefficients.

To return now to the problem of the closure of equation (5.15): this can be achieved with formal exactness by inserting

$$g^{(3)}(r, s, t) = g(r)g(s)g(t)\exp[\tau(r, s, t; n_0)] \quad (5.20a)$$

where

$$\tau(r, s, t; n_0) = \sum_{n=1}^{\infty} n_0^n \delta_{n+3}(r, s, t). \quad (5.20b)$$

In equation (5.20*b*) δ is a coefficient in the expansion of τ in powers of density and it is a cluster integral of Mayer functions. An example, for $n = 1$, is

$$\delta_4(r, s, t) \equiv \int f_{14}f_{24}f_{34}\, \mathrm{d}\boldsymbol{r}_4. \quad (5.20c)$$

Equations (5.20) and a general expression for δ_{n+3} were obtained by Meeron (1957) and Salpeter (1958). (The latter reference is a detailed and illuminating discussion of Mayer cluster integrals and their use.) τ in equation (5.20*a*) is a measure of the exactness of the superposition approximation and its expression in equation (5.20*b*) shows that the essence of the approach is another density expansion of which it is to be hoped the first term or two would make a significant improvement over the uncorrected superposition hypothesis.

It is in the latter spirit that Haymet *et al* (1981a,b) attempted to evaluate τ for a specified n_0 and $\varphi(r)$. As with much else in this chapter, the technical details of the calculation must be left for the interested reader to study in the original papers. For example, although δ_4 and δ_5 were, and δ_6 might be, obtained, Haymet *et al* give reasons for preferring to calculate the early terms of an alternative density expansion using the $h(r_{ij})$ of equation (5.13) instead of f_{ij}. The technical work involved in this leads ultimately to numerical values of τ and therefore of $g^{(3)}$. The φ used were those of the Lennard-Jones potential and the LRO-II liquid sodium potential defined in Chapter 4. These were chosen because MC or MD results are available for comparison.

One comparison could be of $g^{(3)}$ itself but in addition $g(r)$ and the thermodynamic properties are obtainable by solving equation (5.15) with (5.20*a*) inserted into it. On substitution for $g^{(3)}$, equation (5.15) becomes

$$-\frac{\partial}{\partial r}[k_B T \ln g(r) + \varphi(r)] = \pi n_0 \int_0^\infty \mathrm{d}s \, \frac{\mathrm{d}\varphi(s)}{\mathrm{d}s} g(s) \times \int_{|r-s|}^{r+s} \mathrm{d}t \, t g(t) \, \mathrm{e}^\tau \left(\frac{s^2 + r^2 - t^2}{r^2}\right). \tag{5.21}$$

This equation is exact, but approximations enter when only part of the series expansion for τ is inserted. At that point numerical methods for the solution of integrodifferential equations have to be invoked to obtain $g(r)$. Finally the equations of §5.2 enable thermodynamic properties to follow from $g(r)$.

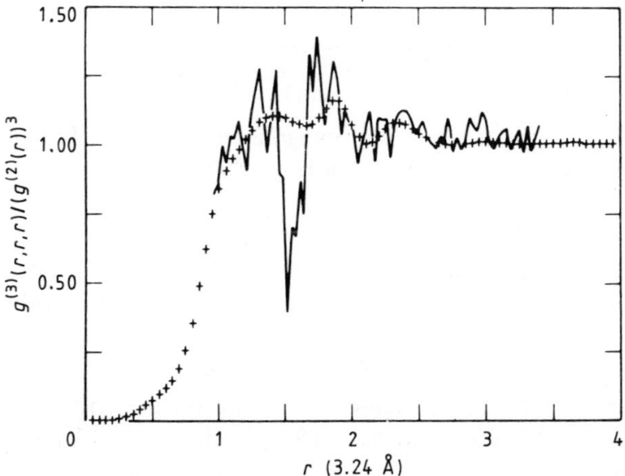

Figure 5.3 A test of the superposition approximation with the LRO-II potential. Full line, MD calculations; +, from equation (5.20). (From Haymet *et al* 1981a.)

This formidable programme of work illustrates the intractability of first-principles theories of liquids. This is a field in which the value of numerical simulation methods (Chapter 4) is very apparent—they enable the results of the theory to be checked. Some specimen results will now be presented from Haymet *et al* (1981a,b).

Dealing first with the liquid sodium potential, figure 5.3 tests the superposition approximation (equation (5.17)) by plotting exp $\tau(r, r, r)$ (equation (5.20*a*)) for equilateral triplet configurations. The MD results are 'noisy' but agree pretty well with the theory except for the large dip which may be an artefact (Haymet *et al* 1981a,b). The superposition approximation begins to fail for $r \leqslant 3.2$ Å which is approximately the first zero of φ (figure 4.5), but is quite good elsewhere. Figure 5.4 shows $g^{(3)}$ for an isosceles configuration; the agreement is excellent and typical of other triangular correlations that were tested.

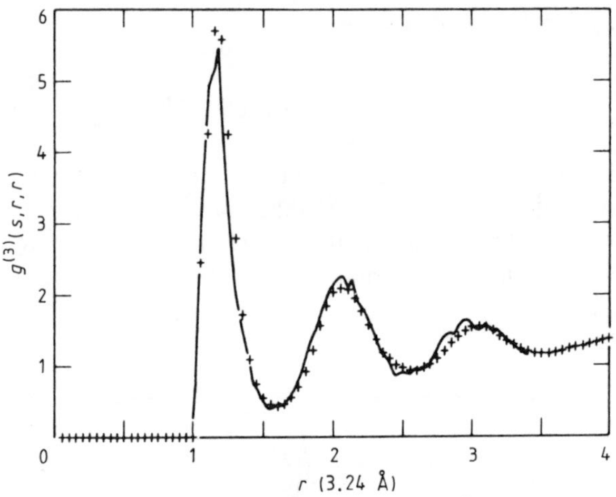

Figure 5.4 $g(s, r, r)$ for the isosceles configuration with $s = 1.10$. Full curve, MD calculations; +, equation (5.20). (From Haymet *et al* 1981a.)

Turning to the Lennard-Jones fluid, figure 5.5 shows the effect of corrections to the superposition approximation—which are considerable. Table 5.1 shows certain thermodynamic quantities for two states of the fluid. With reference to the parameters ε, σ of the LJ potential, state A has $T = 1.35\varepsilon/k_B$ and $n_0 = 0.65/\sigma^3$ which is a high-density state on the critical isotherm and state B has $T = 2.74\varepsilon/k_B$ and $n_0 = 0.80/\sigma^3$ which has higher density and temperature.

Assuming the Monte Carlo results are the right answer, the superiority of the corrected over the uncorrected superposition approximation is convincing and the rival first-principles theories (PY and HNC, §5.6) are less good on

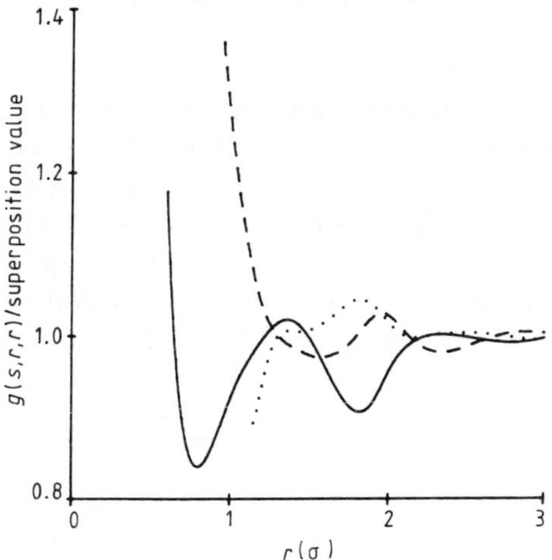

Figure 5.5 A test of the superposition approximation with the LJ potential. Full curve, $s = 1.10$; broken curve, $s = 1.65$; dotted curve, $s = 2.15$. $g^{(3)}(s, r, r)$ is from equation (5.20). (From Haymet *et al* 1981b.)

the whole. BH and WCA are not first-principles theories; by the present criterion they do rather well and deserve discussion later (§5.8).

At the time of writing the theory touched on in this section is probably better than any other first-principles theory and has served to illustrate the problems liquid theory presents and the nature of the approach needed to overcome them.

Table 5.1. This table has been extracted from Haymet *et al* (1981b).

Name of method	State A $\frac{p}{n_0 k_B T}$	State A $\frac{U}{N}$	State B $\frac{p}{n_0 k_B T}$	State B $\frac{U}{N}$
Monte Carlo[a]	0.85	−4.34	3.60	−4.28
SA[b]	0.18	−4.79	2.30	−4.83
Th 1[c]	0.64	−4.41	3.34	−4.47
Th 2 (PY)[d]	1.26	−4.32	3.60	
Th 3 (HNC)[d]	1.49	−4.26	4.53	−3.85
Th 4 (BH)[d]	0.74		3.70	
Th 5 (WCA)[d]	0.71		3.61	

[a]Monte Carlo results are taken from Hansen and Verlet (1969) given in Chapter 4.
[b]SA, superposition approximation assumed.
[c]Th 1, Haymet *et al* (1981a,b) described in text.
[d]Th 2 to Th 5, other theories—see §§5.6 and 5.8.

5.4 The superposition approximation and the BGY equation

It is difficult to answer in a word the question: how good is the superposition approximation? $g^{(3)}$ is not readily exhibited by graphs. It may be accurate for some ranges of its arguments, and not others. It may be better with some potentials, or at some densities, than others. It will certainly be more accurate at low densities. At high densities, figures 5.3 to 5.5 illustrate the partial success of the superposition approximation in LRO-II and LJ systems. Figure 5.6 tests it for hard spheres with promising results. Authorities sum it up by some such phrase as 'reasonably reliable at high densities'.

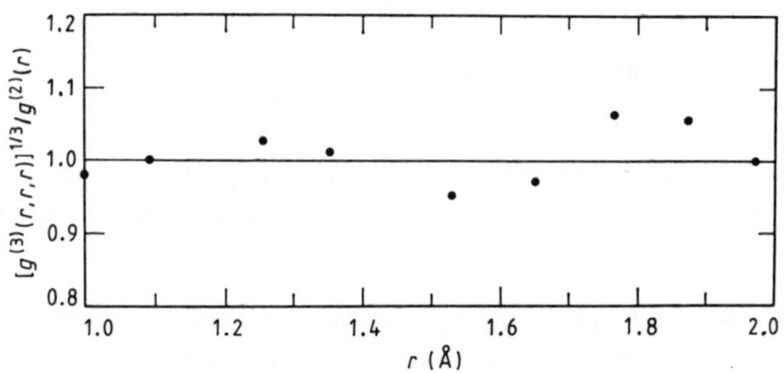

Figure 5.6 The superposition approximation tested by MD calculations for equilateral arrangements of hard spheres, $n_0 = 0.625$ of closest packing. (From Rice and Gray 1965.)

Authorities also concur with the view that the use of the superposition approximation in the equation for g, i.e. in equation (5.21) with $\tau = 0$, may lead to serious errors in both $g(r)$ and the thermodynamic properties because the equation magnifies the errors inherent in the approximate $g^{(3)}$. The equation can be manipulated into the form

$$k_B T \ln g(r) = -\varphi(r) + n_0 \int_V E(s)(g(t) - 1)\, d\boldsymbol{r}_3 \tag{5.22a}$$

where

$$E(r) \equiv \int_r^\infty \frac{d\varphi(t)}{dt} g(t)\, dt. \tag{5.22b}$$

This can be solved numerically for $g(r)$. In figure 5.7 can be seen the equation of state for hard spheres given by equations (5.21), (5.3) and (5.12c), and a comparison with computer simulations. The BGY equation is less satisfactory in this respect than certain other theories to which we now turn.

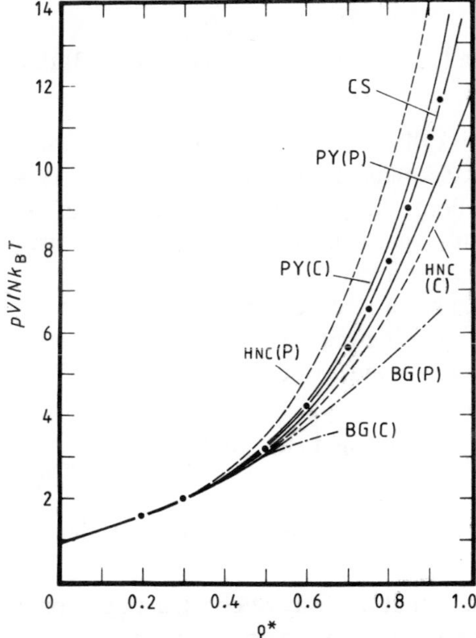

Figure 5.7 The equation of state of hard spheres. The points are MD calculations. P and C refer to the virial and the compressibility approximations—see, e.g., equations (5.32*f*) and (5.32*g*). The abbreviations are as in the text and $\rho^* \equiv N\sigma^3/V$. (From Barker and Henderson 1976.)

5.5 Density expansions. Direct correlation function

We have already referred to the use in fluid theory of density expansions and cluster integrals. In varying degrees of detail, Rice and Gray (1965), Croxton (1974, 1978), Hansen and McDonald (1976), Barker and Henderson (1976), Temperley *et al* (1968) and Thiele (1963) explain the mathematical techniques for forming expansions, for expressing the coefficients of n_0^2, n_0^3, etc as integrals, for representing the integrals as diagrams and for summing the series, at least in part, by selecting certain sets of diagrams for summation and rejecting others. To give some flavour of the results a few will be quoted though readers wishing to understand how they arise will have to study the references in some detail.

The quantity $h(r)$ of equation (5.13) can be written in terms of another function, $c(r)$, as follows:

$$h(r) = c(r) + n_0 \int c(s)h(t)\, \mathrm{d}\mathbf{r}_3 \qquad (5.23a)$$

or equivalently,

$$h(1, 2) = c(1, 2) + n_0 \int c(1, 3)h(2, 3)\, d\mathbf{r}_3. \qquad (5.23b)$$

This new quantity is called the direct correlation function and is often named after Ornstein and Zernike who first introduced it in 1914. Equation (5.23) defines $c(r)$ and in doing so splits up the correlation of atoms 1 and 2 into a part depending only on their positions and an indirect part depending through the integrand on the correlated positions of all the other particles. As we shall see, $c(r)$ has a range comparable with that of $\varphi(r)$ and therefore much shorter than the extent of $h(r)$. Substituting the left-hand side of equation (5.23) into its integrand we see

$$h(1, 2) = c(1, 2) + n_0 \int c(1, 3)c(3, 2)\, d\mathbf{r}_3 + n_0^2 \iint c(1, 3)c(3, 4)c(4, 2)\, d\mathbf{r}_3\, d\mathbf{r}_4 + \cdots. \qquad (5.23c)$$

This makes it evident that one way of constructing approximate theories for $h(r)$ is to adopt some expression for $c(r)$ in terms of $\varphi(r)$ and substitute it in equation (5.23). Further, a density expansion for $c(r)$ in terms of cluster integrals would lead to one for $h(r)$. The expansion for $c(r)$ was first given by Rushbrooke and Scoins and is (see, e.g., Rice and Gray (1965), Hansen and McDonald (1976), Barker and Henderson (1976))

$$\begin{aligned} c(1, 2) = {} & f_{12} + n_0 \int f_{12} f_{31} f_{23}\, d\mathbf{r}_3 \\ & + \frac{n_0^2}{2!} \iint d\mathbf{r}_3\, d\mathbf{r}_4 (f_{12}f_{23}f_{34}f_{14}f_{24}f_{13} + 2f_{12}f_{23}f_{34}f_{14} + 4f_{12}f_{23}f_{34}f_{14}f_{13} \\ & + f_{13}f_{23}f_{24}f_{14} + f_{12}f_{23}f_{14}f_{24}f_{13} + f_{23}f_{34}f_{14}f_{24}f_{13}) + \cdots \end{aligned} \qquad (5.24a)$$

where f_{ij} is defined in equation (5.19). The n_0^3 term has 238 clusters. The complexity of this is evident and the reader will no doubt be able to decode, and to appreciate the compactness of, the following diagrammatic representation by comparing it with equation (5.24a):

$$c(1, 2) = \; + n_0 \left(\; \right) + \frac{1}{2!} n_0^2 \left(2 \; + 4 \; + \; + \; + \; \right) + \cdots \qquad (5.24b)$$

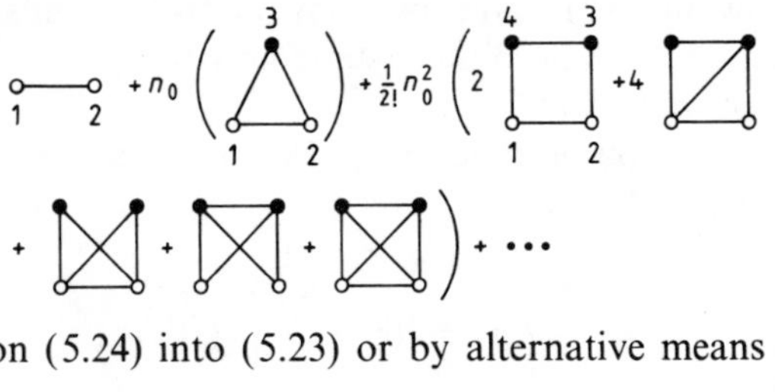

By inserting equation (5.24) into (5.23) or by alternative means a density expansion for $h(r)$ can be found

$$h(1,2) \equiv g(1,2) - 1 = f_{12} + n_0 \left(\int f_{13} f_{23} \, d\boldsymbol{r}_3 + \int f_{13} f_{23} f_{12} \, d\boldsymbol{r}_3 \right)$$

$$+ \frac{n_0^2}{2!} \iint d\boldsymbol{r}_3 \, d\boldsymbol{r}_4 (2 f_{12} f_{23} f_{34} f_{14} + 2 f_{23} f_{34} f_{14} + 4 f_{12} f_{23} f_{34} f_{14} f_{13}$$

$$+ 4 f_{23} f_{34} f_{14} f_{13} + f_{12} f_{23} f_{14} f_{13} f_{24} + f_{23} f_{14} f_{13} f_{24}$$

$$+ f_{12} f_{23} f_{34} f_{14} f_{24} f_{13} + f_{23} f_{34} f_{14} f_{34} f_{13}) + \cdots \qquad (5.25a)$$

or

$$h(1,2) = \text{o—o} + n_0 (\triangle + \wedge) + \frac{1}{2!} n_0^2 \left[2 (\sqcap + \square) + 4 (\cdots + \cdots) + \cdots + \cdots + \cdots + \cdots \right] + \cdots \qquad (5.25b)$$

A further useful connection between $c(r)$ and $h(r)$ is found by taking the Fourier transform of equation (5.23), namely

$$h(Q) = c(Q) + n_0 c(Q) h(Q) \qquad (5.26a)$$

or, using equation (3.11),

$$c(Q) = \frac{S(Q) - 1}{n_0 S(Q)}. \qquad (5.26b)$$

Insofar as $S(Q)$ is measured (see Chapter 3), $c(Q)$ is also determined and by

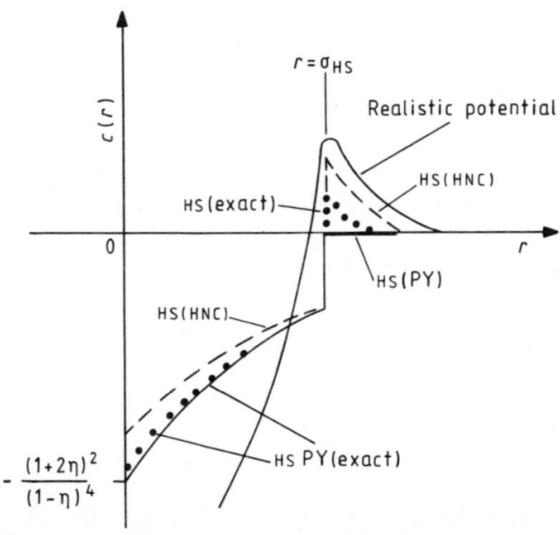

Figure 5.8 Schematic diagram to illustrate the Ornstein–Zernike correlation function for simple liquids far from the critical point. $c(r)$ depends on both T and n_0.

Fourier transforming back $c(r)$ would follow as well. Some curves of $c(r)$ are shown in figure 5.8 and these show its range is short. Inspection of equation (5.24) shows that most terms contain a factor f_{12} and this is also true of higher terms. The short range of $c(1, 2)$ follows from that of f_{12}; the contributions not containing f_{12} partially cancel one another.

5.6 The HNC and PY equations

If expansions like (5.24) and (5.25) could be summed completely the problem would be fully solved. This is not possible. Various approximations consist of summing certain terms only or certain classes of diagrams. Probably the simplest approach is to approximate $c(1, 2)$ by its first term, f_{12}, which leads from equation (5.23*c*) to

$$h(1, 2) = f_{12} + n_0 \quad + n_0^2 \quad + \cdots . \qquad (5.27)$$

However this is too simple to be adequate partly because an expression for $c(r)$ which ignores n_0, n_0^2 etc is unlikely to be reasonable for high densities. The two most discussed approximations are obtained by discarding from the expansion for $c(r)$, and therefore also from $h(r)$, not whole terms, but certain classes of diagrams.

The hypernetted chain or HNC version omits from the n_0^2-terms in equations (5.24*b*) and (5.25*b*), the last three and the last two diagrams respectively, with corresponding omissions in higher terms. This enables a compact expression to be obtained for $c(r)$, namely

$$c(r) \simeq h(r) - \ln g(r) - \varphi(r)/k_B T. \qquad (5.28)$$

This still implies an integral equation for $h(r)$; in fact, inserting equation (5.28) into (5.23) there follows

$$\ln g(r) + \frac{\varphi(r)}{k_B T} = n_0 \int \left(h(s) - \ln g(s) - \frac{\varphi(s)}{k_B T} \right) h(t) \, d\mathbf{r}_3. \qquad (5.29)$$

This corresponds to the integral equation, (5.22), of the BGY theory.

The Percus–Yevick or PY version omits from $c(r)$ the last four diagrams of the n_0^2-term and corresponding diagrams from higher terms. The effect is equivalent to

$$c(r) \simeq g(r) \left(1 - \exp \frac{\varphi(r)}{k_B T} \right) \qquad (5.30)$$

which, with equation (5.23), leads to the PY equation

$$g(r) \exp(\varphi(r)/k_B T) = 1 + n_0 \int g(s)[1 - \exp(\varphi(s)/k_B T)] h(t) \, d\mathbf{r}_3. \quad (5.31)$$

It is not obvious *a priori* which of the three integral equations, BGY, HNC and PY, lead to the best values of $h(r)$. It is not even obvious that omitting more classes of diagrams—which is done in PY as compared with HNC—will lead to worse results because the effects of two omitted classes may partially cancel out. Indeed, authorities regard the PY version as numerically superior for the HS and LJ potentials though not for longer range ionic interactions where the HNC does better. Some comparisons with computer simulations have been given in table 5.1 and figure 5.7. The PY approximation is particularly good at high temperatures even up to high densities. It also gives quite good virial coefficients for imperfect gases—better than those of the HNC.

A valuable and much exploited feature of the PY approximation is that it has an exact solution for the HS potential. With φ for hard spheres inserted, equation (5.31) can be solved in a way first given by Wertheim (1963) and by Thiele (1963). There results a closed expression for $c(r)$ from which $g(r)$ follows. The thermodynamic properties are then derived from equations (5.2), (5.3), (5.12c), etc. It is typical of the use of approximate theories for $g(r)$ that the virial and compressibility equations of state ((5.3, (5.12c)) give different results. It is then for investigation which one works better. The following equations summarise the PY hard-spheres solution; it is convenient to express r in terms of σ by $r/\sigma \equiv s$, $\eta \equiv \pi n_0 \sigma^3/6$ as in §4.3:

$$c(s) = -(\alpha + \beta s + \gamma s^3) \qquad s < 1 \tag{5.32a}$$

and

$$c(s) = 0 \qquad s > 1$$

where

$$\alpha \equiv \frac{(1+2\eta)^2}{(1-\eta)^4} \qquad \beta \equiv -\frac{6\eta(1+\frac{1}{2}\eta)^2}{(1-\eta)^4} \qquad \gamma \equiv \tfrac{1}{2}\eta\alpha \tag{5.32b}$$

$$c(\sigma, Q) = -4\pi\sigma^3 \int_0^1 \mathrm{d}s\, s^2 \frac{\sin sQ\sigma}{sQ\sigma}(\alpha + \beta s + \gamma s^3). \tag{5.32c}$$

We then have from equation (5.26b),

$$S(Q, \sigma) = \frac{1}{1 - nc(Q, \sigma)} \tag{5.32d}$$

$$S(0) = n_0 \kappa_T k_\mathrm{B} T = \frac{(1-\eta)^4}{(1+2\eta)^2} \tag{5.32e}$$

from equation (5.3)

$$z \equiv \frac{pV}{Nk_\mathrm{B}T} = \frac{1 + 2\eta + 3\eta^2}{(1-\eta)^2} \qquad \text{(virial equation)} \tag{5.32f}$$

from equation (5.12c)

$$z \equiv \frac{pV}{Nk_\mathrm{B}T} = \frac{1 + \eta + \eta^2}{(1-\eta)^3} \qquad \text{(compressibility equation)} \tag{5.32g}$$

from (5.32*f*)

$$\frac{F}{Nk_BT} = 3 \ln \frac{hn_0}{(2\pi mk_BT)^{1/2}} - 1 + 2 \ln(1-\eta) + \frac{6\eta}{1-\eta} \tag{5.32h}$$

from (5.32*g*)

$$\frac{F}{Nk_BT} = 3 \ln \frac{hn_0}{(2\pi mk_BT)^{1/2}} - 1 - \ln(1-\eta) + \frac{3\eta \;\; 2-\eta}{2(1-\eta)^2} \tag{5.32i}$$

from (5.32*f*)

$$S = k_B \ln\left[\frac{e}{n_0}\left(\frac{emk_BT}{2\pi\hbar^2}\right)^{3/2}\right] + k_B\left[-2\ln(1-\eta) + 6\left(1 - \frac{1}{1-\eta}\right)\right] \tag{5.32j}$$

and from (5.32*g*)

$$S = k_B \ln\left[\frac{e}{n_0}\left(\frac{emk_BT}{2\pi\hbar^2}\right)^{3/2}\right] + k_B\left[\ln(1-\eta) + \tfrac{3}{2}\left(1 - \frac{1}{(1-\eta)^2}\right)\right]. \tag{5.32k}$$

The leading terms in z, F and S are perfect gas terms in the limit $\eta \to 0$. $g(r)$ can be calculated from equations (5.32*a*) and (5.23) but is only piecewise analytic (Smith and Henderson 1970). It fits MC calculations well except that it is too low at $r = \sigma$ and the oscillations at large r are somewhat out of phase. Figure 5.9, from an influential paper by Ashcroft and Lekner (1966), exemplifies the quite good agreement between measured structure factors for simple liquid metals and equation (5.32*d*) for $\eta \simeq 0.45$.

The difference between equations (5.32*d*) and (5.32*g*) increases with the density and figure 5.7 compares them.

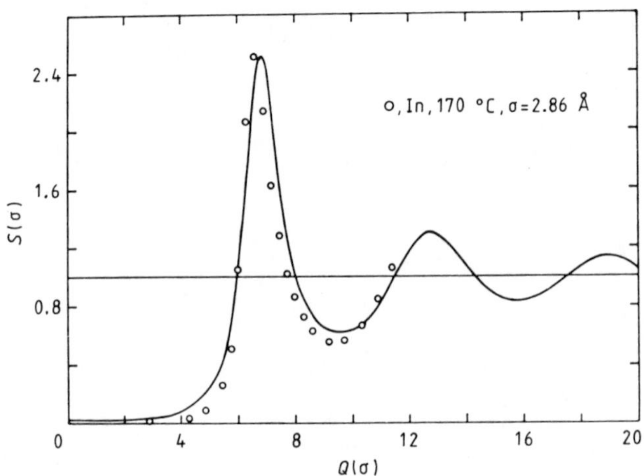

Figure 5.9 $S(Q)$ for liquid In (from Ashcroft and Lekner 1966). Full curve, PY theory; circles, experimental points.

The PY theory with hard-sphere potentials also gives an exact solution for binary mixtures. This has proved particularly useful in developing a theory of binary liquid alloys and will be referred to in that context (see §6.9).

For less simple potentials than that for hard spheres the PY and HNC theories have to be worked out numerically. At supercritical temperatures the agreement with computer simulations is quite good especially for PY but the lower the temperature the worse the approximations. Critical constants computed from the LJ potential by the BGY, PY and HNC equations differ very seriously with PY giving the best results for argon, e.g., the ratios of the PY theoretical to the experimental results are 0.99, 0.92, 1.01 for T_c, d_c and z_c as given by the virial equation; the results from the compressibility equation are inferior. At subcritical liquid densities the thermodynamic values are rather bad. These comparisons can be pursued in detail in Hansen and McDonald (1976) and Barker and Henderson (1976).

The inconsistency of the virial and compressibility equations of state for approximate theories has just been mentioned and illustrated in equation (5.32). Suppose an approximate theory, or possibly a combination of two such theories, were modified with the help of an adjustable parameter and suppose the latter could be chosen so that the two equations of state became consistent: would not this lead to an overall improvement? This ingenious suggestion has been successfully implemented in several ways in recent years by modifying or combining the PY and HNC theories. Manipulation of equations (5.28) and (5.30) leads to

$$g(r) \simeq \exp(-\varphi(r)/k_B T)\exp(\gamma(r)) \qquad \text{(HNC)}$$

$$g(r) \simeq \exp(-\varphi(r)/k_B T)(\gamma(r)+1) \qquad \text{(PY)}$$

where $\gamma(r) \equiv h(r) - c(r)$.

Suppose a mixture of these two equations was formed by

$$g(r) \simeq \exp(-\varphi(r)/k_B T)\left(1 + \frac{\exp(\gamma(r)f(r)) - 1}{f(r)}\right)$$

where $f(r) \equiv 1 - e^{-\alpha r}$, $\alpha \geqslant 0$. We note that $r \to 0$, $r \to \infty$ lead, respectively, to PY and HNC. It is now possible to compute some equation of state property, say the bulk modulus, from the virial and compressibility equations and to choose α to make the results the same. The hope is that $g(r)$ will now agree better than before with MD or MC and so therefore will all the thermodynamic properties. The success of this was demonstrated by Rogers and Young in 1984 following a series of important papers by Rosenfeld and Ashcroft (1979), Foiles *et al* (1984) and others, in which thermodynamic consistency was achieved by various methods and the nature of the approximations analysed. In so far as the parameter adjusted for thermodynamic consistency is regarded as correcting equation (5.28) by adding in the sum of omitted diagrams (called the bridge function), the result is called the modified hypernetted chain theory (MHNC). The general success of this approach has been excellent and has been extended to binary systems by Hansen and Zerah (1985) in whose paper most of the important references including those mentioned above can be found.

5.7 More recent theories, MSA and others

Although the approximate integral equations already referred to occupy a large proportion of the literature of liquid state theory there is a considerable and increasing number of other options, some of them representing improvements. In this section a few of these will be touched on sufficiently to indicate their defining characteristics, some of their applications and references in which they can be pursued. Theories may work better with some potentials than others or may answer some questions better than others. Parameters, e.g., core diameters and well depths, have to be selected for numerical calculations and a particular selection may happen to cause accentuation or cancellation of errors. Assessing the performance of theories is therefore a somewhat delicate matter which will be left to the specialist literature except for a few remarks. An illuminating example of such an assessment can be found in Kahl and Hafner (1982) where theories shortly to be referred to as MSA, ORPA and OCT are compared in detail for their treatment of the square-well potential which is a primitive combination of attraction and repulsion, namely,

$$\begin{aligned} \varphi(r) &= \infty && r < \sigma \\ &= -\varepsilon && \sigma < r < \gamma\sigma \\ &= 0 && r > \gamma\sigma. \end{aligned} \tag{5.33}$$

It appears that for this potential the PY and HNC theories are clearly inferior to the MSA and the latter is inferior to the OCT and ORPA for both $g(r)$ and the thermodynamic properties.

The mean spherical approximation (MSA) consists in writing

$$c(r) = -\beta\varphi(r) \tag{5.34a}$$

when φ is attractive. Suppose φ has a hard repulsive core, φ_{HS}, and is attractive outside, namely,

$$\varphi(r) = \varphi_{\mathrm{HS}} + \varphi_1(r). \tag{5.34b}$$

Then equation (5.34*a*) applies for $r > \sigma$ and

$$g(r) = 0 \qquad \text{for } r < \sigma. \tag{5.34c}$$

If $\varphi_1 = 0$, the MSA reduces to the PY (see equation (5.30)). The MSA allows exact solutions in closed form for several φ's including square wells, hard spheres with tails of the form $r^{-1}\exp(-\alpha r)$, charged hard spheres, dipolar and quadrupolar hard spheres and various mixtures. With non-spherical φ, the theory must lead not just to $h(r)$ but to $h(r, \boldsymbol{\theta})$ where $\boldsymbol{\theta}$ symbolises the relative orientation of particles separated by r. The MSA is capable of this but the results are only moderately satisfactory (Barker and Henderson 1976).

Remembering that the Ornstein–Zernike relation (equation (5.23)) always gives a route from an assumed $c(r)$ to $h(r)$ or $g(r)$, let us write

$$c(r) = c_{\mathrm{HS}}(r) + c_1(r) \tag{5.35a}$$

where

$$c_1(r) = -\beta\varphi_1(r) \qquad \text{for } r > \sigma \tag{5.35b}$$

and $c_{\mathrm{HS}}(r)$ is the exact HS result. The approximation (5.35*b*) would in general give a contribution to $g(r)$ for $r < \sigma$ and to prevent this outcome—inconceivable with hard cores—Andersen and Chandler (1972) 'optimised' c, which means that $c_1(r)$ *inside* the core was chosen to make equation (5.34*c*) true. Numerical calculation is required for this and also for the subsequent step of calculating $g(r)$ by inserting (5.35*a*) into the Ornstein–Zernike relation. Since $g(r)$ may be written

$$g(r) = g_{\mathrm{HS}}(r) + g_1(r) \tag{5.36}$$

the optimising procedure leads to a series expression and thence to a numerical result for $g_1(r)$. To use this method is to adopt the optimised random-phase approximation or ORPA.

This by no means exhausts the useful possibilities for approximation. Andersen and Chandler (1972) also considered writing

$$g(r) = g_{\mathrm{HS}} \exp g_1(r). \tag{5.37}$$

This ensures that $g(r) = g_{\mathrm{HS}}(r) = 0$ inside the core and is known as the exponential or EXP approximation. Its linearised version, namely

$$g(r) = g_{\mathrm{HS}}(1 + g_1(r)) \tag{5.38}$$

has the initials LEXP (Andersen and Chandler 1972). In Hansen and McDonald (1976), Barker and Henderson (1976) and Kahl and Hafner (1982) examples can be found in which the ORPA, EXP and LEXP give somewhat similar results for $g(r)$, better agreement with computer simulation values coming sometimes from one, sometimes from another, approximation.

Each version of the theory will lead to its own value of F and the other thermodynamic properties. Let us rewrite the pair potential as

$$\varphi(r, \lambda) = \varphi_{\mathrm{HS}} + \lambda\varphi_1(r) \tag{5.39}$$

where λ is called a coupling parameter and $0 \leqslant \lambda \leqslant 1$. $\lambda\varphi_1$ will cause a contribution to F over and above that due to the hard spheres. From equation (4.3) this contribution will be

$$\Delta F(\lambda) \equiv F(\lambda) - F_{\mathrm{HS}} = -k_{\mathrm{B}}T \ln Q(\lambda)$$

where

$$Q(\lambda) = \int \exp\left(-\tfrac{1}{2}\beta \sum_{i,j} \lambda\varphi_1(r_{ij})\right) \mathrm{d}\boldsymbol{r}^{(N)} \equiv \int \exp(-\beta\Phi_1(\lambda))\, \mathrm{d}\boldsymbol{r}^{(N)}.$$

Therefore $\partial\Delta F(\lambda)/\partial\lambda = \langle\partial\Phi_1(\lambda)/\partial\lambda\rangle_\lambda$, where an argument or subscript λ indicates the system with the potential (5.39). It then follows that

$$F(\lambda) = F_{\mathrm{HS}} + \int_0^\lambda \left\langle \frac{\partial}{\partial\lambda}\,\Phi_1(\lambda)\right\rangle \mathrm{d}\lambda. \tag{5.40a}$$

Substituting for $\Phi_1(\lambda)$ and introducing $g(r, \lambda)$ by arguments like those leading to equation (4.11), the canonically averaged integrand becomes

$$\tfrac{1}{2}Nn_0 \int \varphi_1(r)g(r, \lambda)\, \mathrm{d}\boldsymbol{r}. \tag{5.40b}$$

Consequently for the potential (5.34*b*), i.e. $\lambda = 1$,

$$F = F_{\mathrm{HS}} + \tfrac{1}{2}Nn_0 \int_0^1 \mathrm{d}\lambda \int \varphi_1(r)g(r, \lambda)\, \mathrm{d}\boldsymbol{r}. \tag{5.41}$$

Evaluating F for the various approximations to g can be done but is too complicated to do here. The EXP version of $g(r)$ and its corresponding value of F from equation (5.41) constitute the optimised cluster theory or OCT (see e.g. Kahl and Hafner (1982) and others therein).

As the last few equations illustrate, these methods have been applied to potentials which have hard cores. They amount to ways of calculating the effects on $g(r)$, F etc of the attractive tail of the potential. The hard-sphere contribution is well known from MC or MD calculations (see Chapter 4) or from the PY solution (equation (5.32)) if that is sufficient. The emphasis in all this work on the hard-sphere system is not justified solely by its relative simplicity and consequent extensive study. It is also that the strong and abrupt repulsive forces are the major influence on the structure of a dense liquid (see e.g. §2.1 or figure 5.9). Any slight softening of the repulsion or any attractive forces can then be expected to cause relatively minor modification to hard-sphere structure. The equation for φ could however be generalised to

$$\varphi(r, \lambda) = \varphi_0(r) + \lambda\varphi_1(r) \tag{5.42}$$

where φ_0 now stands for the potential of any well understood system to be known as the *reference system*. There are not many systems sufficiently well investigated to be reference systems and φ_0 is often, but not necessarily, put equal to φ_{HS}. φ_1 is called the perturbing potential and equation (5.41) is the starting point for perturbation and variational theories.

5.8 Perturbation and variation methods

The perturbation method requires a decision as to the way the potential is divided between φ_0 and φ_1, and a means of expanding any required quantity, e.g. the free energy F, in a convergent series of powers of a small parameter. The hope is that a small number of such terms will suffice and that they can be calculated from φ and the known properties of the reference system. Perturbation methods achieved important successes in the late 1960s and the 1970s (see Hansen and McDonald 1976, Barker and Henderson 1976, Henderson and Barker 1971).

Suppose equation (5.39) applies. Let Φ_1 denote the full perturbation energy for $\lambda = 1$, namely, $\frac{1}{2}\Sigma_{ij}\,\varphi_1(r_{ij})$. $\Phi_1(\lambda)$ denotes the corresponding quantity for $\lambda < 1$. It is possible to obtain a series for F by writing a Taylor expansion for $\langle \partial\Phi_1(\lambda)/\partial\lambda \rangle_\lambda$ about its value at $\lambda = 0$. This leads without difficulty to

$$\langle \partial\Phi_1(\lambda)/\partial\lambda \rangle_\lambda = \Phi_1 + \lambda\beta(\langle \Phi_1 \rangle^2_{\lambda=0} - \langle \Phi_1^2 \rangle_{\lambda=0}) + \mathrm{O}(\lambda^2).$$

From equation (5.40*a*), for $\lambda = 1$, we find

$$F = F_{\mathrm{HS}} + \Phi_1 + \tfrac{1}{2}\beta(\langle \Phi_1 \rangle^2 - \langle \Phi_1^2 \rangle) + \mathrm{O}(\beta^2). \tag{5.43}$$

This is called the high-temperature expansion and is the more readily applicable the smaller the value of β. Φ_1 is given by equation (5.40*b*) evaluated for $\lambda = 0$, i.e., the perturbation is averaged over the reference system. The next term is much more troublesome to evaluate and turns out to involve $g^{(3)}$ and $g^{(4)}$ for the reference system. This term is fully discussed in Barker and Henderson (1976) and Henderson and Barker (1971), and if it is to be evaluated numerically the superposition approximation for $g^{(3)}$ and $g^{(4)}$ would be needed. There is a corresponding expansion for the pair distribution function.

This approximation was tried out on the square-well potential with considerable success by Henderson and Barker (1971). The equation of state at several temperatures and densities obtained from second-order perturbation theory, i.e. from equation (5.43), agreed very well with MD and MC calculations. The ratio of the second-order term in F to the first order was about 0.1. Even the critical constants and coexistence curve could be well derived.

This success of perturbation theory was extended to potentials without hard cores. A single example, that of the Lennard-Jones potential (§4.4), will illustrate this. Figure 5.10 shows two ways (BH and WCA) in which φ could be subdivided into φ_0 and φ_1. Both have been successfully implemented but for illustration only one will be briefly outlined here. It is that of Weeks,

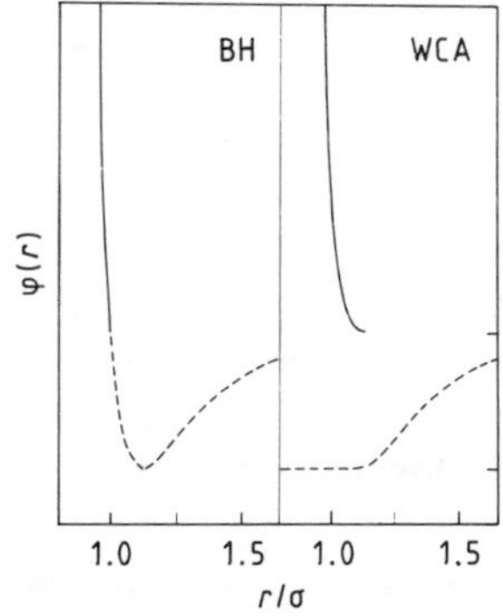

Figure 5.10 Two ways of subdividing the Lennard-Jones φ into φ_0 and φ_1 for perturbation theory. BH, Barker and Henderson; WCA, Weeks, Chandler and Andersen. (From Hansen and McDonald 1976.)

Chandler and Andersen (WCA, Weeks *et al* 1971). Explicitly,

$$\begin{aligned} \varphi(r) &= 4\varepsilon\left[\left(\frac{\sigma}{r}\right)^{12} - \left(\frac{\sigma}{r}\right)^{6}\right] = \varphi_0(r) + \varphi_1(r) \\ \varphi_0(r) &= \varphi(r) + \varepsilon \qquad r < 2^{1/6}\sigma \\ &= 0 \qquad r \geqslant 2^{1/6}\sigma \\ \varphi_1(r) &= -\varepsilon \qquad r < 2^{1/6}\sigma \\ &= \varphi(r) \qquad r \geqslant 2^{1/6}\sigma. \end{aligned} \tag{5.44}$$

In the WCA method, φ_0 is the repulsive and φ_1 the attractive part. In equation (5.41) therefore, with F_0 instead of F_{HS}, the second term represents the effect of bringing in the attractive part of the LJ potential. This required $g_0(r)$ and F_0 which could have been calculated using MC or MD. However an approximate method was simpler and adequate, namely the reference system was chosen to be a hard-sphere system with its diameter selected to make its $g(r)$ outside the range of the repulsive interaction agree with $g_0(r)$ or

$$g_0(r) \simeq y_d(r)\exp(-\beta\varphi_0(r)) \tag{5.45a}$$

where the suffix d denotes a HS system with core diameter d and $y_d(r) = g_d(r)$ outside the core and is a continuous extrapolation inside (see figure 4.2). d is derived numerically from equation (5.45*a*) by putting

$$\int \mathrm{d}\mathbf{r}\,[y_d(r)\exp(-\beta\varphi_0) - 1] = \int \mathrm{d}\mathbf{r}\,[y_d(r)\exp(-\beta\varphi_d) - 1] \tag{5.45b}$$

where $\varphi_d = \varphi_{\text{HS}}$ with diameter d. These are integrals of the correlation functions, $h(r)$, and through the compressibility equation (5.12c) determine the equation of state. WCA therefore identify the thermodynamic properties of the reference system as those of the hard-sphere system selected by equation (5.45). *Faute de mieux*, y_d was taken from the PY solution (5.32). d depends on the density and temperature, being smaller for higher T or density.

The reference system having been defined, the first-order correction to F could be calculated and other thermodynamic properties follow. There was excellent agreement with the MC and MD results (see e.g. Verlet (1967), Hansen and Verlet (1969) in Chapter 4) for a range of T and n_0, especially at high densities. Indeed the treatment demonstrates a quantitatively successful first-order perturbation treatment of fluids with LJ potentials. The BH and WCA entries in table 5.1 and also figure 5.11 illustrate this. These calculations confirm numerically the remark in the previous section that the structure of fluids of moderate density is determined largely by the repulsive forces; the higher the density the less effect the attractive part has on the structure.

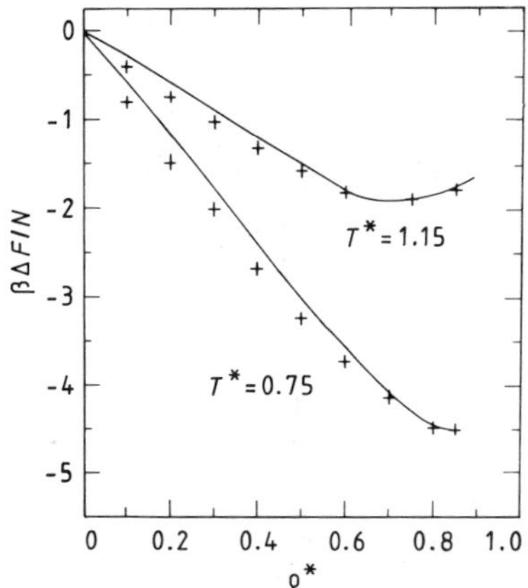

Figure 5.11 Excess Helmholtz free energy (i.e. $F - F_{\text{ideal}}$) per particle at two temperatures. Full curves, WCA theory; +, MC calculations. $\rho^* \equiv n_0\sigma^3$; $T^* \equiv k_B T/\varepsilon$. (From Weeks *et al* 1971.)

The variational methods are in a similar spirit and start from equation (5.42) with $\lambda = 1$. There is a remarkable theorem rigorously derivable within either quantum or classical statistical mechanics which states (Girardeau and Mazo 1973)

$$F \leqslant F_0 + \langle H - H_0 \rangle_0 \tag{5.46}$$

where F_0 is the free energy of the reference system and H, H_0 are the Hamiltonians of the total and reference systems, i.e., $(H - H_0)$ is the perturbation which will usually be the difference, Φ_1, in the potential energies. The brackets $\langle\rangle_0$ signify that $(H - H_0)$ is averaged over the ensemble of the reference system. (In quantum mechanics the *contents* of the brackets would be the quantum mechanical average for a member of the ensemble usually also written with angular brackets as $\langle\psi^*(H - H_0)\psi\rangle$.) The result is known as the Gibbs–Bogoliubov inequality.

The inequality can be used in the following way. The reference system is described by one or more parameters. For example, σ would be the parameter for the hard-sphere system. $\langle H - H_0\rangle_0$, or $\langle\Phi_1\rangle_0$, is then calculated, added to F_0, and the sum minimised by varying the parameter or parameters. The minimum of the RHS of equation (5.46) should be a good approximation to the LHS if the reference system is well chosen. An obvious way of testing this would be to calculate the known value of F for a LJ fluid from that of an HS fluid by using $(\varphi_{\mathrm{LJ}} - \varphi_{\mathrm{HS}})$ as the perturbation. Mansoori and Canfield (1969) did this and obtained thermodynamic properties in very good agreement with the exact (computer-simulated) values. Subsequently the Gibbs–Bogoliubov method has been very influential in the theory of liquid metals and an extended example will be given in Chapter 6.

5.9 Extracting $\varphi(r)$ from $g(r)$

The main problem of the preceding sections has been to derive $g(r)$ and the thermodynamic properties from a specified interparticle pair potential. To establish the latter is of course a problem in its own right and the reader might well have wondered whether the wealth of liquid state theory could not be used to extract the true $\varphi(r)$ from the observed $g(r)$ or $S(Q)$. The classic paper in this field is by Johnson *et al* (1964).

These authors pointed to the relation between $g(r)$, $\varphi(r)$ and the potential of the mean force (see equations (5.16)) and obtained $\varphi(r)$ numerically from $g(r)$. To do this, the unknown triplet distribution has to be evaded in some way and the BGY, HNC and PY integral equations were all tried. The main result was that for several liquid metals $\varphi(r)$ appeared with a long-range oscillatory tail whereas for Ar it did not. This is a significant difference and the Na potential LRO-II in figure 4.5 is derived in part from this analysis. However there are difficulties with this type of calculation, namely, the inaccuracy of the experimental input data, the numerical accuracy of the data processing and the approximations in the theories. The first two of these were ameliorated in later work and by 1983, using the MHNC, it became possible to relate observed $S(Q)$ to corresponding $\varphi(r)$ with considerable precision (Howells and Enderby 1972, Ballentine and Jones 1973, Dharma-Wardana and Aers 1983, Levesque *et al* 1985).

$\varphi(r)$ is of course alternatively available from atomic or molecular theory and experiment for non-metals and, for metals, as part of the general theory of dense metallic systems (see §6.7).

5.10 Cell and allied theories

Most of this chapter has been devoted to a formal first-principles theory which is embodied in distribution and partition functions. It has been pointed out that although the difficulties of this are formidable considerable success has rewarded some decades of effort. For a comparable period a much more intuitive approach has also been followed though it would probably be fair to say that it is now on the whole overshadowed by the formal methods. The names of some of the intuitive theories—'lattice', 'hole', 'free volume', 'tunnel' or 'cell'— indicate that their basis is a particular view of the nature of the liquid state which helps to postulate a partition function, perhaps semiempirical in nature. Only two of these theories will be touched on here; they and others can be found in Barker and Henderson (1976), Barker (1963) and Eyring and Jhon (1969).

A well known early theory was the Lennard-Jones and Devonshire cell theory. The idea was that every molecule is hemmed into a cell or cage by its neighbours. All cells are supposed to be occupied and have volume V/N. An expression could be derived for the potential energy, ε, of a molecule in its cell due to the field of its nearest neighbours as a result of its being displaced from the centre of its cell. The integral

$$\int \exp(-\varepsilon/k_B T)\, d\boldsymbol{r}$$

over the cell defines a quantity called the free volume, v_f, which is a weighted average considerably smaller than V/N. The quantity Q in the partition function of equation (4.3) becomes $(v_f)^N \exp(-\Phi_0/k_B T)$ where Φ_0 is the potential energy of the N atoms all situated at their cell centres. v_f has to be worked out from the interparticle forces and the assumed geometry, e.g. the LJ potential and a face-centered cubic cage. Clearly this approach relies heavily on crystal lattice theory and indeed looks on the liquid as a disordered crystal. It works better for solids than liquids but does have the formal virtue of relating the interparticle forces, via v_f and Q, to the thermodynamic properties. Though not very successful itself as a theory of liquids, it stimulated valuable developments, many of them concerned with introducing unoccupied cells or holes, or space of some kind, and thus allowing for more ease of motion.

In Eyring's theory of significant structures (Eyring and Jhon 1969), developed over a long period with many collaborators, the observed volume difference $(V_{liq} - V_{sol})$ is regarded as the source of free space in the liquid. This is not to be thought of as localised vacancies but more as space potentially available to such molecules as are energetic enough to knock others out of the way and create room. Eyring's descriptive phrase for this space is 'fluidised vacancies'. Molecules that avail themselves of this space have, temporarily, translatory or gas-like degrees of freedom; the rest have oscillatory motion which is solid-like and treated by the Einstein model. The partition function therefore contains gas-like and solid-like factors and a number of quantities, such as the Einstein temperature, θ_E, obtained by experiment from the solid. Eyring proposes the following partition function for a simple liquid like argon:

$$Z = \left\{ \frac{\exp(E_s/RT)}{[1-\exp(-\theta_E/T)]^3} \left[1 + n\left(\frac{V_{liq}-V_{sol}}{V_{sol}} \right) \right. \right.$$
$$\left. \left. \times \exp\left(-\frac{aE_s V_{sol}}{(V_{liq}-V_{sol})RT} \right) \right] \right\}^{N(V_{sol}/V_{liq})}$$
$$\times \left(\frac{(2\pi m k_B T)^{3/2}}{h^3} (V_{liq}-V_{sol}) \right)^{N(V_{liq}-V_{sol})/V_{liq}} \left[\left(\frac{N(V_{liq}-V_{sol})}{V_{liq}} \right)! \right]^{-1}. \tag{5.47}$$

In the first, solid-like, factor, n is a constant of proportionality connecting the number of possible positions a sufficiently energetic molecule can 'appropriate' with the fractional excess volume $(V_{sol} - V_{liq})/V_{sol}$. E_s is the energy of sublimation which enters because Eyring proposes that the energy required by a molecule to 'pre-empt' space from competing neighbours is $aE_s V_{sol}/(V_{liq} - V_{sol})$. In the second, gas-like, factor the factorial is to disallow interchange of identical gas-like particles as a mechanism for changing the state of the system. For a simple liquid theoretical estimates can be given for n and a, leaving no adjustable parameters.

The power of this partition function to reproduce liquid behaviour, i.e. to correlate liquid properties with certain solid properties, is very impressive. The theory is also wide ranging because partition functions applicable to metals, ionic melts, water and other systems are available and extensions of the approach to embrace surface tension and transport properties is also successful; in fact, Jhon and Eyring claim that 'there seems to be no liquid which cannot be usefully examined using this model'.

5.11 Specimen comparisons of theory and experiment

It is not wholly satisfying, perhaps especially for experimental physicists, to compare a theory only with computer simulation studies although there are excellent reasons for doing so. Ultimately the experimental results have to be explained and in this section a few more will be quoted to add to those in figures 4.3, 4.4, 4.6 and 5.9.

Figure 5.12 shows $g(r)$ derived from both x-ray and neutron diffraction experiments and from the HNC theory with two different potentials. Figure 5.13 is similar for the PY theory (Khan 1964). It is clear that there is quite good agreement in general but in detail differences between two potentials, between two theories and between two experiments are all significant. The

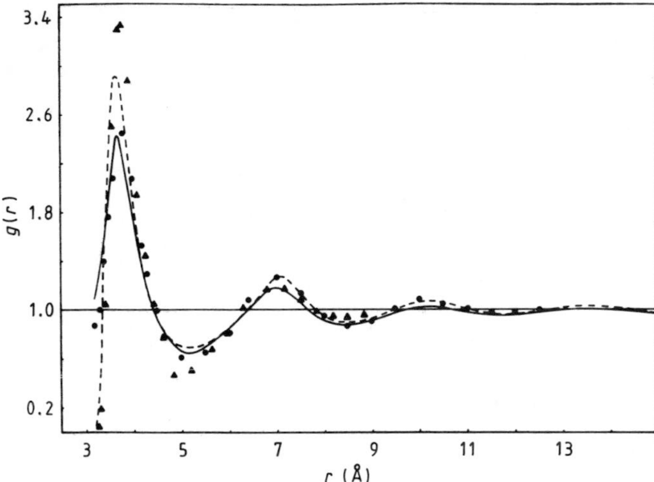

Figure 5.12 $g(r)$ for Ar from two experiments and two model potentials in the HNC theory. Later theories and experiments converge better. Broken curve, Lennard-Jones; full curve, Guggenheim–McGlashan; ●, neutron diffraction; ▲, x-ray diffraction. $T = 84.4$ K and $n = 2.113 \times 10^{-2}$ Å^{-3}. (From Khan 1964.)

next pair of graphs, figures 5.14 and 5.15, compare measured $S(Q)$'s for Na and Ar with MC results for potentials proportional to r^{-4} and r^{-12} respectively (Hansen and Schiff 1973). The good agreement emphasises the point often made that the structure is largely determined by the repulsive force. The latter is clearly much steeper for the non-metal than the metal. Even better agreement for Ar is found for the full LJ (see, e.g., Yarnell *et al*

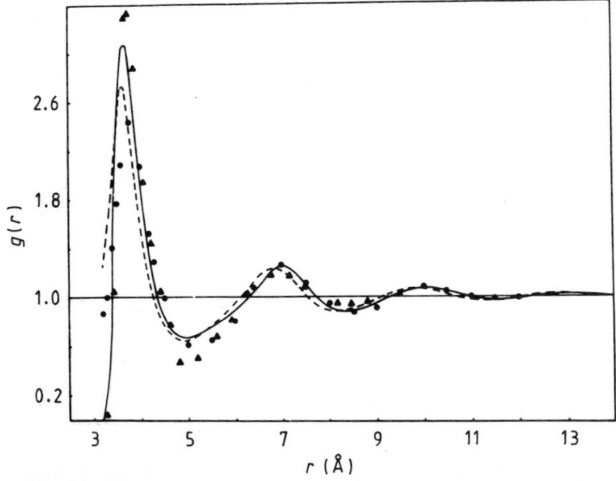

Figure 5.13 As in figure 5.12 but with the PY theory. Full curve, Lennard-Jones; broken curve, Guggenheim–McGlashan. (From Khan 1964.)

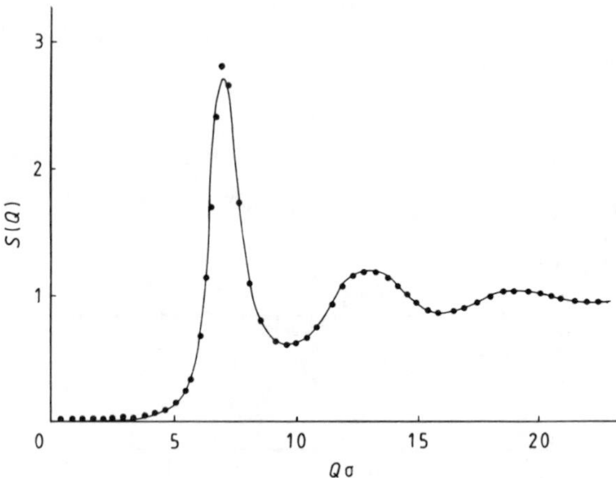

Figure 5.14 $S(Q)$ for liquid Na. Full curve, MC for an r^{-4} fluid. Points, experimental data from Greenfield *et al* (1971) (in Chapter 3). $\sigma = n_0^{-1/3}$. (From Hansen and Schiff 1973.)

1984 in Chapter 4) because the longer-range attractive interaction modifies $S(Q)$ at low Q.

As for the equation of state, figure 5.16 from Klein and Green (1963) shows that exact, i.e. MC, calculations for the LJ potential get nearest to the experimental result. The remaining discrepancy is presumably due to the inability of the LJ pair potential to represent the many-body interaction of

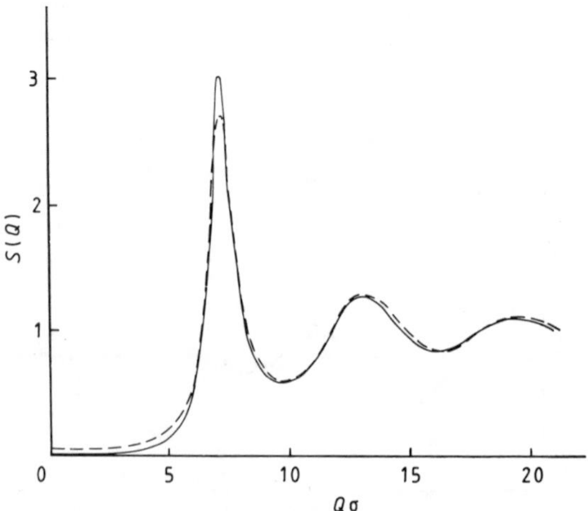

Figure 5.15 $S(Q)$ for Ar. Full line, MC for an r^{-12} fluid. Broken curve, experimental data from Hansen and Schiff (1973). $\sigma = n_0^{-1/3}$.

real atoms. The further discrepancies of the HNC are due to the approximations in that theory. It seems probable that better theories (see §5.7) with more realistic potentials could reproduce the experimental curve more or less exactly. The few examples in this section therefore support the view expressed in the quotation from Barker and Henderson (1976) at the beginning of the chapter.

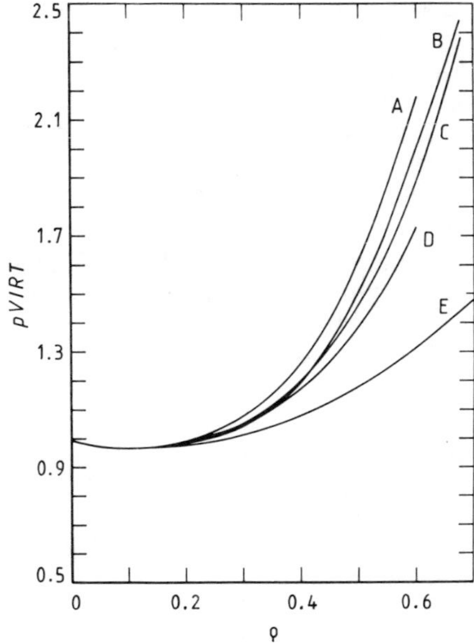

Figure 5.16 z versus density for Ar and a LJ fluid at 328.15 K. The two HNC curves (A, D) correspond to the PY equations (5.32f) and (5.32g). The MC curve (B) is from Wood and Parker. The virial curve (E) is for a three-term virial equation (§4.3). The Ar curve (C) is experimental—from Michels *et al* (1958). (From Klein and Green 1963.)

References

Andersen H C and Chandler D 1972 *J. Chem. Phys.* **57** 1918

Ashcroft N W and Lekner J 1966 *Phys. Rev.* **145** 83

Ballentine L E and Jones J C 1973 *Can. J. Phys.* **51** 1831

Barker J A 1963 *Lattice Theories of the Liquid State* (Oxford: Pergamon)

—— 1976 *Rare Gas Solids* vol 1, ed M L Klein and J A Venables (London: Academic) ch 4

Barker J A and Henderson D 1976 *Rev. Mod. Phys.* **48** 587

Croxton C A 1974 *Liquid State Physics* (Cambridge: Cambridge University Press)

—— (ed) 1978 *Progress in Liquid Physics* (Chichester: Wiley)

Dharma-Wardana M W C and Aers G C 1983 *Phys. Rev.* B **28** 1701
Eyring H and Jhon M S 1969 *Significant Liquid Structures* (New York: Wiley)
Foiles S M, Ashcroft N W and Reatto L 1984 *J. Chem. Phys.* **80** 4441
Girardeau M D and Mazo R M 1973 *Adv. Chem. Phys.* **24** 187
Hansen J-P and McDonald I R 1976 *Theory of Simple Liquids* (London: Academic)
Hansen J-P and Schiff D 1973 *Mol. Phys.* **25** 1281
Hansen J-P and Zerah G 1985 *Phys. Lett.* **108A** 277
Haymet A D J, Rice S A and Madden W G 1981a *J. Chem. Phys.* **74** 3033
—— 1981b *J. Chem. Phys.* **75** 4696
Henderson D and Barker J A 1971 *Physical Chemistry* vol 8A, ed H Eyring, D Henderson and W Jost (New York: Academic)
Howells W S and Enderby J E 1972 *J. Phys. C: Solid State Phys.* **5** 1277
Johnson M D, Hutchinson P and March N H 1964 *Proc. R. Soc.* A **282** 283
Kahl G and Hafner J 1982 *Phys. Chem. Liq.* **12** 109
Khan A A 1964 *Phys. Rev.* **134** A367
Klein M and Green M S 1963 *J. Chem. Phys.* **39** 1367
Levesque D, Weiss J J and Reatto L 1985 *Phys. Rev. Lett.* **54** 451
Mansoori G A and Canfield F B 1969 *J. Chem. Phys.* **51** 4958
Meeron E 1957 *J. Chem. Phys.* **27** 1238
Rice S A and Gray P 1965 *The Statistical Mechanics of Simple Liquids* (New York: Interscience)
Rogers F J and Young D A 1984 *Phys. Rev.* A **30** 999
Rosenfeld Y and Ashcroft N W 1979 *Phys. Rev.* A **20** 1208
Salpeter E E 1958 *Ann. Phys., NY* **5** 183
Smith W R and Henderson D 1970 *Mol. Phys.* **19** 411
Temperley H N V, Rowlinson J S and Rushbrooke G S 1968 *The Physics of Simple Liquids* (Amsterdam: North-Holland)
Thiele E 1963 *J. Chem. Phys.* **39** 474
Weeks J D, Chandler D and Andersen H C 1971 *J. Chem. Phys.* **54** 5237
Wertheim M S 1963 *Phys. Rev. Lett.* **10** 321

6

SIMPLE LIQUID METALS

To attempt a theory of the structure and thermodynamic properties of liquid metals on the lines used for Ar in the preceding chapter would presuppose that atomic interactions in metals can be treated as pairwise. Within certain limitations this can be done—a circumstance which helped to stimulate the extensive development of liquid metal theory in recent years. However, the obvious fact that melting a metal crystal does not turn it into a non-metal implies that much of the interest in molten metals is in the behaviour of their conduction or valence electrons. This interest has been both technological and scientific. The chemical, thermal and rheological properties interest metallurgists in general and engineers, and the scientific concern has much to do with the motion of unbound electrons in a disordered array of ions. This motion presents a problem not fully solved yet and determines both the electron states and the electron transport properties and ultimately the other properties as well.

After some brief remarks about experimentation with liquid metals this chapter will introduce a few concepts which enable many properties of many metallic liquids to be explained while deferring the general problem of electrons in disordered matter. This is done by using the nearly-free-electron (NFE) model. The latter, and the underlying concepts, do not relate solely to liquids of course but have been taken over from solid state physics. With the help of these ideas it will be shown how the liquid theory of Chapter 5 can be applied to structural and thermodynamic properties of liquid metals and alloys. Subsequently electron transport will be discussed.

The whole subject can be pursued in Beer (1972), Faber (1972), Shimoji (1977), Adams *et al* (1967), Takenchi (1973), Evans and Greenwood (1977), Cyrot-Lackmann and Desré (1980) and Wagner and Johnson 1984), and in all the other references therein.

6.1 Experimentation with liquid metals

In general, metallic cohesion is strong and results from valence electrons which in other contexts promote chemical reactivity. Melting points are therefore usually high, sometimes very high, and typically a sample is a hot aggressive liquid. Measurements which on solid metals would be relatively easy are therefore made difficult, the problem often being that of finding inert solid containers and insoluble electrodes. Usually refractory insulators like alumina and silica, and refractory conductors like W, Mo, graphite and stainless steel have to used. In measurements such as those of the Hall effect, where sample geometry is critical, the fluidity of liquid metals and the presence of gravitational and electrodynamic forces make for difficulties in the design of sample holders. An additional barrier to accuracy is diffusion under concentration and thermal gradients which are most likely to be troublesome in hot fluid systems. There are also magnetohydrodynamically induced motions which can interfere with galvanomagnetic measurements. Reasonable solutions have been found for most of these problems and many reliable measurements now exist of conductivity (σ), thermoelectric power (α) and the Hall coefficient (R_H) as functions of temperature and composition (in alloys) and occasionally of pressure too.

Thermodynamic properties including those of alloy mixing such as the entropy of mixing, can be measured by traditional methods of calorimetry dilatometry, vapour pressure measurement, etc, subject of course to the difficulties already referred to. The comparatively recent advent of refractory solid electrolytes, such as Na β-alumina, has been something of a boon to experimental alloy thermodynamics because they can be used as membranes permeable to ions but not electrons in concentration cells. It may be shown (Neale and Cusack 1982, Hoshino and Endo 1982) that the electromotive force, $E(c, T, p)$, of such a cell measured accurately as a function of concentration, temperature and pressure, can be processed to yield *all* the thermodynamic properties of a binary alloy as functions of c.

The structures of liquid metals and alloys are investigated by the methods of Chapter 3. Recent years have seen the development of transient methods, such as exploding wires, to investigate the liquid phases of high-melting-point metals, such as Ta, W, Nb (Gathers). Techniques common in solid state physics such as nuclear magnetic resonance (NMR), positron annihilation, photoemission and optical reflectivity are used where appropriate.

6.2 The general description of a metallic liquid

A liquid metal may be thought of as a disordered array of ions immersed in a gas of electrons. As in solid state physics, when the electron gas is derived from the outer s and p electrons, as in Na or Al, and when the mean free path for electron–ion collisions is several nearest-neighbour distances or more, the metal is regarded as simple (a relative term) and the electrons as nearly free. A characteristic mode of approach in metal physics is then to regard the effect of the ions on the electrons as a perturbation on the motion of a free-electron gas or, in other words, to calculate ‘corrections’ to the motion of free electrons caused by the presence of ions. Whether this is a plausible treatment depends partly on whether the electrons can be clearly divided between the ion cores and the gas and also on how strong the forces are between the electrons and ions. The prevailing view is that in many metals and alloys—the ones called simple—the situation indeed justifies this perturbation approach. Exceptions needing special care are transition metals (see §7.4, 7.7), expanded metals (see Chapter 11), and liquid alloys of special compositions where chemical bonding or negative-ion formation is suggested by the properties (see Chapter 11).

The unperturbed free-electron gas is therefore important to the discussion and its properties will be familiar to the reader. An electron passing through an ion will be deflected by the electron–ion potential. The ions are positive and the electron gas mobile, therefore the ions exert a long-range influence on the gas by attracting electrons to themselves. Meanwhile the electrons are also repelling each other and a complex balance of forces sets in whose ultimate effect is to envelop each ion in an electron cloud. This cloud, being negative, neutralises the electrostatic effect of the ion rather completely at large distances (say several atomic diameters) and less completely at small distances.

The cloud of electrons is the screening charge and the total effect on an approaching electron of the ion and its screen is the *screened potential.* In summary we may say that from the point of view of one electron passing through a liquid metal (or a crystalline one for that matter) that it encounters the screened potentials of one ion after another and is scattered by them. The scattering alters the motion from what would otherwise have been free-electron motion and perturbs the free-electron gas, e.g., it causes corrections to the density of states. The summed effects of all the electrons undergoing this scattering is to produce the very electron clouds which screen the ions, so the process and its descriptive mathematics should be self-consistent. A formalism is needed to express these ideas and the full discussion contains many subtleties some of which will now be touched upon.

6.3 Pseudopotentials and model potentials

Suppose ψ is the eigenfunction representing a valence electron in the single-electron model. Then

$$H\psi \equiv (T + U)\psi = \varepsilon\psi \tag{6.1a}$$

where U is the potential in a metal. Let φ_t be a core function treated as a linear sum of core states of the separate atoms, then

$$H\varphi_t = \varepsilon_t\varphi_t \qquad \varepsilon_t < \varepsilon. \tag{6.1b}$$

A convenient approach to pseudopotentials is through orthogonalised plane waves, $\chi_{\boldsymbol{k}}$, where

$$\chi_{\boldsymbol{k}} \equiv \exp(\mathrm{i}\boldsymbol{k}\cdot\boldsymbol{r}) - \sum_t a_t\varphi_t. \tag{6.2a}$$

The a_t's are chosen to make the $\chi_{\boldsymbol{k}}$ orthogonal to all the core states φ_t and therefore have to be

$$a_t = \int \exp(\mathrm{i}\boldsymbol{k} \cdot \boldsymbol{r})\varphi_t^* \, \mathrm{d}\boldsymbol{r}. \tag{6.2b}$$

If now ψ is expanded as $\Sigma_{\boldsymbol{k}} \, c(\boldsymbol{k})\chi_{\boldsymbol{k}}$, it follows that

$$\psi = \varphi - \sum_t \left(\int \varphi\varphi_t^* \, \mathrm{d}\boldsymbol{r} \right) \varphi_t \tag{6.3a}$$

where

$$\varphi \equiv \sum_{\boldsymbol{k}} c(\boldsymbol{k}) \exp(\mathrm{i}\boldsymbol{k} \cdot \boldsymbol{r}). \tag{6.3b}$$

The point of this has been firstly to recognise that ψ must be orthogonal to φ_t and must therefore contain such oscillations and nodes as are dictated by those of the core function φ_t and, secondly, to divide ψ into a part that contains all these oscillations and a smooth remainder, φ, which is the superposition of some plane waves. $\varphi = \psi$ where the φ_t are negligible—usually referred to as 'outside the cores'.

On substituting (6.3a) into (6.1a) we find

$$(T + U + U_{\mathrm{R}})\varphi = \varepsilon\varphi \tag{6.4a}$$

where

$$U_{\mathrm{R}}\varphi \equiv \sum_t (\varepsilon - \varepsilon_t)\left(\int \varphi\varphi_t^* \mathrm{d}\boldsymbol{r} \right)\varphi_t. \tag{6.4b}$$

A number of features important for our present purpose follow from these results. Formally, Schrödinger's equation (6.1a) for ψ has been turned into another Schrödinger equation for the smoother function φ with the same energy eigenvalue but a different potential. The latter is the *pseudopotential*, $U_{\mathrm{ps}} \equiv U + U_{\mathrm{R}}$. This is in principle an integral operator, as equation (6.4b) shows, though in many applications it is approximated by a simpler ordinary potential. It is significant that $|U_{\mathrm{ps}}| < |U|$ or, in other words, U_{R} is repulsive and partly cancels out the attractive potential U. One way of putting this is to relate U_{R} to the kinetic energy of the electron inside the core because it derives from the oscillatory part of ψ where the curvature $\partial^2\psi/\partial r^2$ is large. The partial cancellation of U_{R} and U is then an expression of the opposite signs of the classical kinetic and potential energies. This fact is used to justify treating U_{ps} as small in the sense that equation (6.4a) could be solved for φ and ε by low-order perturbation theory. Finally we note that U_{ps} and the corresponding φ in equation (6.4a) are not a unique pair. Any linear combination of φ_t could be added to φ making a new φ' which would also be a solution of equation (6.4a) with the same ε. This circumstance means that a particular φ' could be selected for some desired property, e.g., the smoothest φ' or the one that makes U_{ps} smallest. φ is called a pseudowavefunction.

The ideas so far introduced have been exceedingly important in the theory of metals and reference to Phillips and Kleinman (1959), Cohen and Heine (1961), Cohen *et al* (1970), Harrison (1966) and Ziman (1964) is recommended. Once it is recognised that ψ and φ are identical at points outside the cores it is clear that a solution ψ representing the scattering of an electron of energy ε by U will imply the same scattering cross section, or phaseshifts, for φ of energy ε scattered by U_{ps}; the scattering properties of U and U_{ps} are identical. In relation to scattering by spherical potentials, the condition that electrons of prescribed energy should have identical scattering cross sections for two different potentials is that the logarithmic derivative of ψ, $\mathrm{d}\ln\psi/\mathrm{d}\ln r$, at some radial distance, R_0, should be identical. This is satisfied for ψ or φ at points outside the core for the two potentials U and U_{ps} for a single ion. This allows a further extension of the pseudopotential idea: a purely fictitious or model potential, U_m, could be devised which satisfied the requirement for scattering identical with that of U but which, inside R_0, was otherwise arbitrary. U_m, unlike U_{ps}, does not have to be calculated by first principles from the core states; it can be invented and specified in terms of one or more parameters which are adjusted to give the required scattering properties. An example will be given shortly. It is clear that there are infinitely many U_{ps} and U_m in principle; certain of which have been much used and the terminological distinction between 'pseudo-' and 'model' is rarely insisted on.

From now on we will assume that the real potential of an ion in a liquid metal has been replaced by a pseudo- or model potential weak enough to justify the subsequent use of perturbation theory for the non-core electrons. For one ion situated at $\boldsymbol{R}_i$ we call this potential $u_\mathrm{I}(\boldsymbol{r}-\boldsymbol{R}_i)$ where I denotes ion. The total pseudopotential in a monatomic liquid is then

$$U(\boldsymbol{r}) = \sum_i u_\mathrm{I}(\boldsymbol{r}-\boldsymbol{R}_i) \tag{6.5}$$

where we have assumed pseudopotentials are additive like ordinary ones. Indeed let us assume further, for simplicity, that the non-local operator character of u_I is inessential so that $U(\boldsymbol{r})$ is an ordinary function of $\boldsymbol{r}$. This restriction could be removed in more complete treatments but considerable complexity results.

In perturbation theory, scattering matrix elements, between initial and final plane-wave states normalised in the volume V and with wavevectors $\boldsymbol{k}_i$ and $\boldsymbol{k}_f$, are required. The square modulus of such a matrix element is

$$\left|\frac{1}{V}\int \exp(-\mathrm{i}\boldsymbol{k}_f\cdot\boldsymbol{r})U(\boldsymbol{r})\exp(\mathrm{i}\boldsymbol{k}_i\cdot\boldsymbol{r})\,\mathrm{d}\boldsymbol{r}\right|^2 = V^{-2}NS(\boldsymbol{Q})|u_\mathrm{I}(\boldsymbol{Q})|^2 \tag{6.6a}$$

where

$$u_\mathrm{I}(\boldsymbol{Q}) \equiv \int \exp(\mathrm{i}\boldsymbol{Q}\cdot\boldsymbol{r})u_\mathrm{I}(\boldsymbol{r})\,\mathrm{d}\boldsymbol{r} \tag{6.6b}$$

and $S(\boldsymbol{Q})$ is the structure factor discussed in Chapter 3. This shows that any perturbation discussion of the electron states up to second order will make use of such structural information as is contained in the pair distribution. $u_{\mathrm{I}}(\boldsymbol{Q})$ is a Fourier transform of $u_{\mathrm{I}}(\boldsymbol{r})$ and is often called the bare-ion form factor. For isotropic liquids $S(\boldsymbol{Q}) \rightarrow S(Q)$ as before. Employment of pseudopotentials, perturbation theory and $S(Q)$ characterises the NFE theory of liquid metals.

We now illustrate u_{I} by one of its simplest examples called the Ashcroft empty-core potential (figure 6.1(a)). This is defined by

$$\begin{aligned} u_{\mathrm{I}}(r) &= 0 \qquad\qquad r < R_{\mathrm{c}} \\ &= \frac{-ze^2}{4\pi\varepsilon_0 r} \qquad r \geqslant R_{\mathrm{c}} \end{aligned} \tag{6.7a}$$

where ze is the ionic charge. It follows that

$$u_{\mathrm{I}}(\boldsymbol{Q}) = u_{\mathrm{I}}(Q) = -\frac{ze^2}{\varepsilon_0 Q^2}\cos QR_{\mathrm{c}}. \tag{6.7b}$$

$u_{\mathrm{I}}(Q)$ is in figure 6.1(b) which is similar in shape to many pseudopotential form factors in spite of the simplicity of the present one. Having $u_{\mathrm{I}} = 0$ inside the core represents in an extreme or overprecise form the cancellation of U and U_{R} referred to above. To fix R_{c} it is usual to calculate some measurable quantity with u_{I} and to choose R_{c} to give the correct answer; u_{I} is then used to calculate other things. The following values of R_{c} in angstroms were derived from the observed resistivity of liquid metals (see §6.12):

Na 0.88; Ag 0.55; Zn 0.67; Al 0.59; Pb 0.76.

They are not far from the ionic radii. Other model potentials put u_{I} = a constant, A_l, inside R_{c} with a value of A_l which varies with the angular

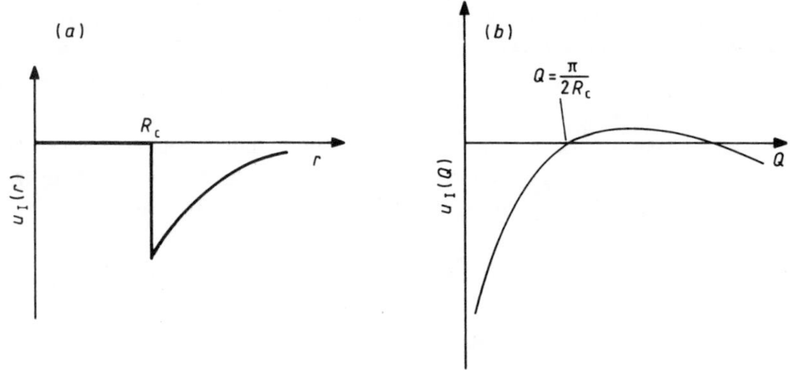

Figure 6.1

momentum (l) of the eigenstate. This gives more parameters to be settled by semiempirical means—see Phillips and Kleinman (1959), Cohen and Heine (1961), Cohen *et al* (1970), Harrison (1966) and Ziman (1964), for examples and details.

6.4 Dielectric screening functions

$u_{\rm I}$ expresses the influence of an ion on the electrons. The influence of the electrons on each other must now be incorporated. This is usually expressed by dividing $u_{\rm I}(Q)$ by a factor which quantifies the effect on a valence electron of the screening cloud of all other valence electrons. This factor is $\varepsilon(Q)$, the dielectric screening function, and the modified potential is

$$u_{\rm A}(Q) \equiv u_{\rm I}(Q)/\varepsilon(Q). \tag{6.8}$$

For the whole metal, equation (6.5) now becomes

$$U(\boldsymbol{r}) = \sum_i u_{\rm A}(\boldsymbol{r} - \boldsymbol{R}_i). \tag{6.9}$$

A in $u_{\rm A}$ denotes 'atom' and the system of ion plus screening cloud is sometimes called a pseudoatom though of course the screening electrons are not bound to the ion (see Ziman 1964). Determining $\varepsilon(Q)$ is another major problem in metal physics. The most elementary example of screening is the exponentially screened Coulomb potential, namely,

$$\begin{aligned} u_{\rm I}(r) &= \frac{-ze^2}{4\pi\varepsilon_0 r} \qquad & u_{\rm A}(r) &= \frac{-ze^2\exp(-qr)}{4\pi\varepsilon_0 r} \\ u_{\rm A}(Q) &= \frac{-ze^2}{\varepsilon_0(Q^2+q^2)} \qquad & u_{\rm I}(Q) &= \frac{-ze^2}{\varepsilon_0 Q^2} \end{aligned} \tag{6.10}$$

where q^{-1} is a screening length of the order of an atomic radius. Since this is usually inadequate for real problems the following outlines the calculation of a self-consistent screening function and leads to a much used formula. $u_{\rm A}(r)$ can be written $(u_{\rm I}(r) + u_{\rm e})$ where $u_{\rm e}$ is the extra potential of the screening cloud. Suppose the electron pseudowavefunctions φ are computed to first order by perturbation theory using $u_{\rm A}(r)$ as the potential. The corresponding charge densities are $-|e|\,|\varphi|^2$ which can be inserted in Poisson's equation to obtain $u_{\rm e}$ and thus achieve self-consistency. The calculation is carried out using Fourier components of the quantities and results in equation (6.8) with

$$\varepsilon_{\rm H}(Q) = 1 + \frac{me^2}{8\pi^2\varepsilon_0\hbar^2 k_{\rm F}\eta^2}\left(\frac{1-\eta^2}{2\eta}\ln\left|\frac{1+\eta}{1-\eta}\right| + 1\right) \tag{6.11a}$$

where

$$\eta \equiv Q/2k_{\rm F}. \tag{6.11b}$$

This $\varepsilon_H(Q)$, first found by Lindhard, is derived from the first-order perturbation and corresponds to those approximations in solid state physics, namely, the Hartree model, which ignore exchange interactions and electron correlation. It is therefore called the Hartree dielectric function. It is not surprising that k_F is a parameter since integration over all the electrons with $k \leqslant k_F$ must enter the calculation. At $\eta = 1$ there is a discontinuity of slope (see figure 6.2) arising from the assumed sharp cut-off of k at k_F. In Fourier transforms of $u_A(Q)$ back into r-space this leads to an oscillating tail to the potential (see §4.5). In a disordered metal the sharpness of the cut-off is questionable and the oscillations are probably less important (see §7.1).

There have been numerous attempts to improve on the Hartree–Lindhard $\varepsilon_H(Q)$ by incorporating electron–electron interactions more rigorously. Suppose the electron gas is disturbed by an external influence such as an ion, which increases the electrostatic potential energy of an electron by δU_1. The electron gas density distribution will alter by δn_e. Let us express these quantities through their Fourier components $\delta U_1(Q)$ and $\delta n_e(Q)$ and define a linear-response function $\chi_1(Q)$ by

$$\chi_1(Q) \equiv \delta n_e(Q)/\delta U_1(Q). \tag{6.12}$$

However δn_e is not produced wholly by δU_1 because any shift of electron density creates a further increment of potential δU_2. By Fourier transforming Poisson's equation $\nabla^2 U = e^2 n_e/\varepsilon_0$, we see that $\delta U_2(Q) = e^2 \delta n_e(Q)/\varepsilon_0 Q^2$. In the Hartree approximation we can define another response function, $\chi_H(Q)$, by

$$\chi_H(Q) \equiv \delta n_e(Q)/(\delta U_1(Q) + \delta U_2(Q)) \tag{6.13}$$

and substituting for δU_1 and δU_2 we obtain

$$\frac{1}{\chi_H(Q)} = \frac{1}{\chi_1(Q)} + \frac{e^2}{\varepsilon_0 Q^2}. \tag{6.14}$$

To relate this to $\varepsilon_H(Q)$, use equation (6.8) from which

$$\frac{1}{\varepsilon_H(Q)} = \frac{\delta U_1 + \delta U_2}{\delta U_1} = 1 + \frac{e^2 \chi_1(Q)}{\varepsilon_0 Q^2} \tag{6.15a}$$

or

$$\varepsilon_H(Q) = 1 - e^2 \chi_H/\varepsilon_0 Q^2. \tag{6.15b}$$

From equations (6.11), (6.14) and (6.15) we now find

$$\chi_H(Q) = -\frac{mk_F}{2\pi^2 \hbar^2}\left(1 + \frac{1-\eta^2}{2\eta} \ln\left|\frac{1+\eta}{1-\eta}\right|\right). \tag{6.16}$$

Now δU_1 and δU_2 apply to any test charge $-|e|$ but if the latter is an electron it will have exchange and correlation interactions with other electrons in the screening cloud. This influence could be called δU_3 and treated as a correction to δU_2 by changing the latter to $(1 - G(Q))\delta U_2$. $G(Q)$ formally expresses the extra effect. We now assume that the corrected response, $\delta n'_e$, is obtained from the corrected disturbance, $\delta U_1 + (1 - G(Q))\delta U_2$, by the same response function as before, namely, $\chi_H(Q)$. Thus from equation (6.13)

$$\delta n'_e(Q) = \chi_H(Q)\left(\delta U_1 + \frac{(1 - G(Q))e^2\delta n'_e(Q)}{\varepsilon_0 Q_2}\right) \tag{6.17}$$

where we have entered for $\delta U_2(Q)$ its value from Poisson's equation.

Substituting $\delta n'_e$ into equation (6.12) we find the revised χ_1 from

$$\frac{1}{\chi_H} = \frac{1}{\chi_1} + \frac{(1 - G(Q))e^2}{\varepsilon_0 Q^2}. \tag{6.18}$$

The dielectric function is always defined as δU_1 divided by the total change in potential felt by a test charge so it is still given by equation (6.15*a*) but with the revised δU_2 implying the new χ_1. From equations (6.18) and (6.15*a*) we now have, instead of (6.15*b*),

$$\varepsilon(Q) = 1 - \frac{e^2\chi_H(Q)}{\varepsilon_0 Q^2 + e^2\chi_H(Q)G(Q)}. \tag{6.19}$$

This reverts to $\varepsilon_H(Q)$ when $G(Q) = 0$.

The function $G(Q)$ is positive and of order unity; its shape varies considerably according to the theory used for estimating it. One function in common use at the time of writing (1985) is shown in figure 6.2 which is taken from Vashishta and Singwi (1972) who discuss other versions in their paper. Another is by Ichimaru and Utsumi (1981).

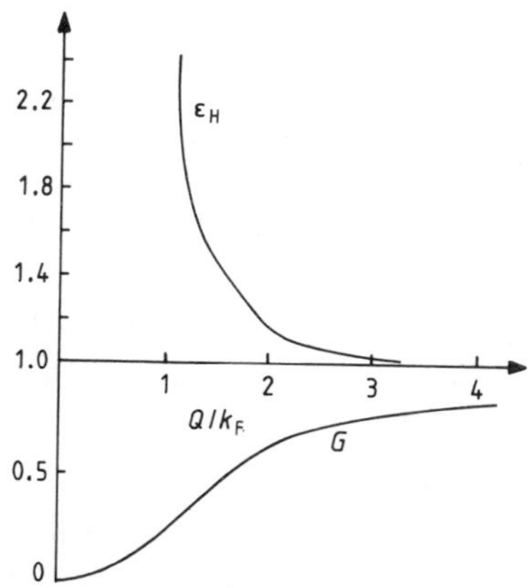

Figure 6.2 $\varepsilon_H(Q)$ for aluminium and $G(Q)$ according to Vashishta and Singwi (1972).

6.5 Input data for NFE calculations

The preceding two sections should have made it clear that any theory of a liquid metal property will probably call upon a pseudopotential, a structure factor and a dielectric screening function. So far as $S(Q)$ is concerned there will often be alternative experimental determinations to select from, as well as the option of using a theoretical $S(Q)$ such as the PY one (equation (5.32)). Both $u_1(Q)$ and $\varepsilon(Q)$ have various versions in the literature. Consequently, the theoretical result may well depend on the choice of the input data. How serious this is must depend on the sensitivity of the calculated object to the sizes and shapes of the assumed functions, but in general it must be accepted that variations of at least 20 % may result from changing the inputs. This is in addition to any underlying uncertainty as to whether the NFE model is applicable to the case at all.

No doubt the ideal approach is to use the soundest available of all the theoretical and experimental inputs regardless of the ensuing complications. In practice experienced theoreticians make whatever simplifications they think justified to achieve the end in view and there is a considerable variety of both ends and means.

It is therefore unreasonable to expect the NFE model to give unambiguous numerically accurate estimates of measurable quantities. What can be hoped for is qualitative descriptions and helpful insights about what microscopic aspects of the liquid metal are significant determinants of the properties. Since model potential parameters are often fixed by making calculated results agree with an observation, the subsequent theory relates one property with another or allows an inference about the temperature or pressure dependence of a property.

The applications below are in this spirit and it cannot be denied that the NFE model has great value in liquid metal physics within its undoubted limitations.

6.6 Energy of a liquid metal

To second order the perturbing potential U changes the electron energy $\hbar^2k^2/2m$ to

$$\frac{\hbar^2k^2}{2m} + \langle \boldsymbol{k}|U|\boldsymbol{k}\rangle + \sum_{\boldsymbol{Q}}{}' \frac{\langle \boldsymbol{k}+\boldsymbol{Q}|U|\boldsymbol{k}\rangle\langle \boldsymbol{k}|U|\boldsymbol{k}+\boldsymbol{Q}\rangle}{(\hbar^2/2m)(k^2-|\boldsymbol{k}+\boldsymbol{Q}|^2)} \tag{6.20}$$

where $\langle \boldsymbol{k}_1|U|\boldsymbol{k}_2\rangle$ can be written as

$$\int V^{-1/2}\exp(-\mathrm{i}\boldsymbol{k}_1\cdot\boldsymbol{r})UV^{-1/2}\exp(\mathrm{i}\boldsymbol{k}_2\cdot\boldsymbol{r})\,\mathrm{d}\boldsymbol{r}$$

and the prime on Σ' means that the term with $Q=0$ is omitted. The total energy of the electrons, ignoring exchange and correlation, is the sum of this expression over all occupied k's. This is conveniently obtained by writing Σ_k as $(4\pi^3)^{-1}V\int \mathrm{d}\boldsymbol{k}$ where $(4\pi^3)^{-1}V$ is the density of states in k-space allowing for two spin orientations and the integral is within the Fermi surface. In a disordered metal it is reasonable to expect the latter to be spherical though it will not represent a sharp discontinuity between occupied and unoccupied states. However, we shall assume that in the NFE approximation this hardly affects the total electron energy. Calling the latter E_e *per ion* we find

$$E_e = E_{eke} + z\overline{\langle \boldsymbol{k}|U|\boldsymbol{k}\rangle} + \frac{V}{4\pi^3 N}\int \mathrm{d}\boldsymbol{k}\sum_{\boldsymbol{Q}}{}' \frac{\langle \boldsymbol{k}+\boldsymbol{Q}|U|\boldsymbol{k}\rangle\langle \boldsymbol{k}|U|\boldsymbol{k}+\boldsymbol{Q}\rangle}{(\hbar^2/2m)(k^2-|\boldsymbol{k}+\boldsymbol{Q}|^2)} \tag{6.21}$$

where

$$E_{eke} = z\left(\frac{3}{5}\frac{\hbar^2 k_F^2}{2m}\right).$$

The first matrix element in the third term can be written

$$V^{-1}\sum_i \exp(-\mathrm{i}\boldsymbol{Q}\cdot\boldsymbol{R}_i)\int \exp[-\mathrm{i}(\boldsymbol{k}+\boldsymbol{Q})(\boldsymbol{r}-\boldsymbol{R}_i)] \times u_A(\boldsymbol{r}-\boldsymbol{R}_i)\exp[\mathrm{i}\boldsymbol{k}\cdot(\boldsymbol{r}-\boldsymbol{R}_i)]\,\mathrm{d}\boldsymbol{r}$$

or

$$\sum_i \exp(-\mathrm{i}\boldsymbol{Q}\cdot\boldsymbol{R}_i)\langle \boldsymbol{k}+\boldsymbol{Q}|u_A|\boldsymbol{k}\rangle. \tag{6.22}$$

This process, known as factorising the matrix element, enables the third term in E_e to be turned into a product of two factors, one depending on the structure through $\{\boldsymbol{R}_i\}$ and the other on u_A. The third term is

$$E_3 = \frac{1}{4\pi^3 N}\sum_{\boldsymbol{Q}}{}'\left|\sum_i \exp(\mathrm{i}\boldsymbol{Q}\cdot\boldsymbol{R}_i)\right|^2 F_1(\boldsymbol{Q}) = \frac{1}{4\pi^3}\sum_{\boldsymbol{Q}}{}' S(\boldsymbol{Q})F_1(\boldsymbol{Q}) \tag{6.23a}$$

where

$$F_1(\boldsymbol{Q}) \equiv V\int \mathrm{d}\boldsymbol{k}\frac{\langle \boldsymbol{k}+\boldsymbol{Q}|u_A|\boldsymbol{k}\rangle\langle \boldsymbol{k}|u_A|\boldsymbol{k}+\boldsymbol{Q}\rangle}{(\hbar^2/2m)(k^2-|\boldsymbol{k}+\boldsymbol{Q}|^2)}. \tag{6.23b}$$

The analogy with equation (3.6) is obvious.

Let us make again the assumption that u_A is an ordinary potential, not an operator. Then

$$F_1(\boldsymbol{Q}) = \frac{|u_A(\boldsymbol{Q})|^2}{V} I \tag{6.23c}$$

where the integral I can be evaluated and is in fact

$$I = 2\pi^3\chi_H(Q) = \frac{2\pi^3\varepsilon_0 Q^2(1-\varepsilon_H)}{e^2}. \tag{6.23d}$$

Because U contains the screening potentials the interpretation of equation (6.21) requires some careful energy accounting. The second term is $zV^{-1}\int U\,\mathrm{d}\boldsymbol{r}$. This is the potential energy of all the screened ions with respect to a uniformly smeared out electron cloud of number density z/V; thus it is $n_e \int u_A(\boldsymbol{r})\,\mathrm{d}\boldsymbol{r}$ where $n_e \equiv Nz/V$. But u_A may be written

$$\frac{-ze^2}{4\pi\varepsilon_0 r} + u_{\mathrm{rem}} + u_e$$

where the three terms are the potential energies of an electron with respect to a point ion, the remainder of the bare ion and the screening charge. The second term of equation (6.21) will accordingly be written ($E^{\mathrm{unif}}_{\mathrm{pt\,ion}} + E^{\mathrm{unif}}_{\mathrm{rem}} + 2E^{\mathrm{unif}}_{\mathrm{self}}$). In the last of these, $E^{\mathrm{unif}}_{\mathrm{self}}$ means the self-energy of the smeared out valence electron cloud obtained by integrating the potential energy of every element with respect to every other element and halving to avoid doubly counting the interactions. In $n_e \int u_e\,\mathrm{d}\boldsymbol{r}$ double counting actually occurs so $2E^{\mathrm{unif}}_{\mathrm{self}}$ is required. The superscripts make explicit the uniformity of the electron cloud involved in evaluating the first-order term.

The total energy per ion of the whole metal may now be written

$$E_{\mathrm{total}} = E_{\mathrm{eke}} + E_{\mathrm{ike}} + (E^{\mathrm{unif}}_{\mathrm{pt\,ion}} + E^{\mathrm{unif}}_{\mathrm{self}} + E^{\mathrm{pt\,ion}}_{\mathrm{pt\,ion}}) + E^{\mathrm{unif}}_{\mathrm{rem}} + E_3 - (E^{\mathrm{e}}_{\mathrm{e}} - E^{\mathrm{unif}}_{\mathrm{self}}). \tag{6.24a}$$

In this total, $E^{\mathrm{pt\,ion}}_{\mathrm{pt\,ion}}$ accounts for the mutually repulsive interaction of the ions at the positions $\{\boldsymbol{R}_i\}$, because treating them as point ions is reasonable at typical interionic separations in metals. E_{ike} is the ionic kinetic energy. $E^{\mathrm{e}}_{\mathrm{e}}$ is the electron–electron mutual potential energy; it refers to the screening clouds and has to be subtracted to compensate for the double counting of the e–e interaction implied by the use of screened potentials in equation (6.20). We now rewrite E_{total} using more conventional and compact but less explicit terms as follows:

$$E_{\mathrm{total}} = E_{\mathrm{eke}} + E_{\mathrm{ike}} + E_{\mathrm{M}} + E_0 + E_{\mathrm{bs}} \tag{6.24b}$$

where E_{M} is the electrostatic potential energy of a system consisting of point ions immersed in a uniform neutralising background; $E_0 \equiv E^{\mathrm{unif.}}_{\mathrm{rem}}$; $E_{\mathrm{bs}} \equiv E_3 - (E^{\mathrm{e}}_{\mathrm{e}} - E^{\mathrm{unif}}_{\mathrm{self}})$. The suffix bs stands for band structure because the second-order term modifies the free-electron density of states.

One purpose of all this is to isolate terms which are calculable. For example, techniques exist for solving the classical electrostatic problem presented by E_{M} (Harrison 1966). In fact for a liquid

$$E_{\mathrm{M}} = \frac{z^2e^2}{4\pi^2\varepsilon_0}\int (S(Q) - 1)\,\mathrm{d}Q. \tag{6.25}$$

It now remains to find expressions for E_0 and E_{bs}. As for E_0 it is

$$E_0 = n_e \int \left(u_{\mathrm{I}} + \frac{ze^2}{4\pi\varepsilon_0 r} \right) \mathrm{d}\boldsymbol{r} \tag{6.26}$$

and can be computed for any chosen bare-ion pseudopotential $u_{\mathrm{I}}(\boldsymbol{r})$. $E_{\mathrm{e}}^{\mathrm{e}}$ per ion is

$$(2N)^{-1}\int n_{\mathrm{e}}(\boldsymbol{r})U_{\mathrm{e}}(\boldsymbol{r})\,\mathrm{d}\boldsymbol{r}$$

and it is convenient to express n_{e} and U_{e} in Fourier components thus:

$$E_{\mathrm{e}}^{\mathrm{e}} = (2N)^{-1}\int \mathrm{d}\boldsymbol{r} \sum_{\boldsymbol{Q}'} V^{-1} n_{\mathrm{e}}(\boldsymbol{Q}')\exp(-\mathrm{i}\boldsymbol{Q}'\cdot\boldsymbol{r})V^{-1}\sum_{\boldsymbol{Q}} U_{\mathrm{e}}(\boldsymbol{Q})\exp(-\mathrm{i}\boldsymbol{Q}\cdot\boldsymbol{r}) \tag{6.27}$$

where

$$U_{\mathrm{e}}(\boldsymbol{r}) = \sum_i u_{\mathrm{e}}(\boldsymbol{r}-\boldsymbol{R}_i) \qquad \text{and} \qquad U_{\mathrm{e}}(\boldsymbol{Q}) = \int \exp(\mathrm{i}\boldsymbol{Q}\cdot\boldsymbol{r})U_{\mathrm{e}}(\boldsymbol{r})\,\mathrm{d}\boldsymbol{r}.$$

Now

$$E_{\mathrm{e}}^{\mathrm{e}} = (2NV^2)^{-1}\int \mathrm{d}\boldsymbol{r}\sum_{\boldsymbol{Q}} n_{\mathrm{e}}(-\boldsymbol{Q})U_{\mathrm{e}}(\boldsymbol{Q})$$

since only terms with $\boldsymbol{Q}' = -\boldsymbol{Q}$ survive. $n_{\mathrm{e}}(-\boldsymbol{Q}) = n_{\mathrm{e}}^*(\boldsymbol{Q})$, consequently

$$E_{\mathrm{e}}^{\mathrm{e}} = \frac{1}{2NV}\sum_{\boldsymbol{Q}}{}' \, n_{\mathrm{e}}^*(\boldsymbol{Q})U_{\mathrm{e}}(\boldsymbol{Q}) + \frac{1}{2NV}n_{\mathrm{e}}^*(0)U_{\mathrm{e}}(0).$$

The second term, with $Q = 0$, is just $E_{\mathrm{self}}^{\mathrm{unif}}$ since it refers to the non-oscillating parts of $n_{\mathrm{e}}(\boldsymbol{r})$ and $U_{\mathrm{e}}(\boldsymbol{r})$, therefore

$$E_{\mathrm{bs}} = E_3 - \frac{1}{2NV}\sum_{\boldsymbol{Q}}{}' \, n_{\mathrm{e}}^*(\boldsymbol{Q})U_{\mathrm{e}}(\boldsymbol{Q}). \tag{6.28}$$

As in §6.4, Poisson's equation is used to replace $n_{\mathrm{e}}^*(\boldsymbol{Q})$ by $\varepsilon_0 Q^2 U_{\mathrm{e}}^*(\boldsymbol{Q})/e^2$ and from equation (6.23*a*) we have

$$E_{\mathrm{bs}} = \frac{1}{4\pi^3}\sum_{\boldsymbol{Q}}{}' \, S(\boldsymbol{Q})F_1(\boldsymbol{Q}) - \frac{1}{2NV}\frac{\varepsilon_0}{e^2}\sum_{\boldsymbol{Q}}{}' \, Q^2|U_{\mathrm{e}}(\boldsymbol{Q})|^2.$$

Further simplification follows from factorising the matrix elements $U_{\mathrm{e}}(\boldsymbol{Q})$ in the same way as before, namely $|U_{\mathrm{e}}(\boldsymbol{Q})|^2 = NS(\boldsymbol{Q})|u_{\mathrm{e}}(\boldsymbol{Q})|^2$. Finally, from equations (6.8), (6.23*c*) and (6.23*d*)

$$E_{\mathrm{bs}} = \sum_{\boldsymbol{Q}}{}' \, N^{-1} S(\boldsymbol{Q})F(\boldsymbol{Q}) \tag{6.29a}$$

where†

$$F(\boldsymbol{Q}) \equiv \sum_{\boldsymbol{Q}}{}' \, \frac{N}{V}\frac{\varepsilon_0 Q^2}{2e^2}\frac{1-\varepsilon_{\mathrm{H}}}{\varepsilon_{\mathrm{H}}}|u_{\mathrm{I}}(\boldsymbol{Q})|^2. \tag{6.29b}$$

†Comparable formulae in the literature may appear slightly different because of alternative definitions of $u_{\mathrm{I}}(\boldsymbol{Q})$ and $S(\boldsymbol{Q})$ and the use of different systems of units.

In §6.8 a numerical illustration of the various contributions to E will be given. A more general comment is that exchange and correlation energies for the valence electrons are not yet included. To improve on this, ε_H in equation (6.29) could be replaced by $\varepsilon(\boldsymbol{Q})$ from equation (6.19) and E_{eke} in equation (6.24) could be supplemented by exchange and correlation energies taken from the theory of the unperturbed electron gas. This changes E_{eke} to

$$E_g \equiv E_{eke} + E_{exch} + E_{corr} = \frac{zA}{r_s^2} + \frac{zB}{r_s} + z(c + D \ln r_s) \qquad (6.30)$$

where E_g is the electron gas energy per ion, $(4\pi/3)\, r_s^3 \equiv n_e^{-1}$ and A, B, C and D are constants depending on the system of units.†

Further refinements are also possible. For example, the true ψ has nodes making the average $|\psi|^2$ inside the cores less than the average over the whole volume: there is said to be an orthogonalisation or depletion hole in the electron charge density inside the ion. The smooth pseudowavefunction φ misses this point and overestimates the electron density. This can be compensated by replacing z by an effective valency z^* a few per cent greater than z (Cohen *et al* 1970).

Collecting all the terms we now have

$$E_{total} = E_{ike} + E_g + E_M + E_0 + E_{bs} \qquad (6.31)$$

in which E_g and E_0 depend on V or n_e but not on the structure, while E_{bs} and E_M depend on both. The ionic motion may be treated classically so $E_{ike} = 3k_BT/2$. $S(\boldsymbol{Q})$ will be involved in E_{bs} and E_M and by using its average value and replacing $\boldsymbol{Q}$ by Q we make it refer to a configuration-average isotropic liquid. The summation over $\boldsymbol{Q}$ becomes an integral over Q.

6.7 Effective interionic potential

Since E_{bs} depends on $S(Q)$ it should be possible to re-express it as an effective interionic potential although one ion really acts on another with the electron gas as intermediary. To do this equation (6.29*a*) is rewritten:

$$E_{bs} = N^{-2} \sum_{\boldsymbol{Q}}{}' \sum_{i,j} \exp[\mathrm{i}\boldsymbol{Q} \cdot (\boldsymbol{R}_i - \boldsymbol{R}_j)] F(\boldsymbol{Q})$$

$$= \frac{1}{2N} \sum_{i,j}{}' \varphi(\boldsymbol{R}_i - \boldsymbol{R}_j) + \frac{1}{N} \sum_{\boldsymbol{Q}}{}' F(\boldsymbol{Q}) \qquad (6.32a)$$

where

$$\varphi(\boldsymbol{r}) \equiv \frac{2}{N} \sum_{\boldsymbol{Q}}{}' F(\boldsymbol{Q}) \exp(\mathrm{i}\boldsymbol{Q} \cdot \boldsymbol{r}). \qquad (6.32b)$$

†E_g is often expressed in atomic units per electron as follows: kinetic, $1.105/r_s^2$; exchange, $-0.458/r_s$; correlation, $0.0313 \ln r_s - 0.115$.

$\varphi(\mathbf{r})$ clearly gives rise to a pairwise energy just as if it were one of the φ in §5.1 but, unlike the LJ potential for example, it will alter if V changes because $F(\mathbf{Q})$ depends indirectly on n_e. The sum of $\varphi(\mathbf{r})$ and the direct Coulomb interaction of the ions is the effective ion–ion potential in metals. It controls structural rearrangements at constant volume. The second term in equation (6.32*a*) does not depend on structure.

For an isotropic liquid, equations (6.29) and (6.32) lead to

$$\varphi_{\text{eff}}(r) = \frac{\varepsilon_0}{2\pi^2 e^2} \int_0^\infty Q^4 \left(\frac{1-\varepsilon(Q)}{\varepsilon(Q)}\right) |u_1(Q)|^2 \frac{\sin Qr}{Qr}\, \mathrm{d}Q + \frac{z^2 e^2}{4\pi\varepsilon_0 r}. \quad (6.33)$$

As an example, substitute equation (6.7*b*) in this expression to give the Ashcroft empty-core interionic potential

$$\varphi_{\text{eff}}\,(\text{empty core}) = \frac{z^2 e^2}{4\pi\varepsilon_0 r}\left(1 + \frac{2}{\pi}\int_0^\infty \frac{1-\varepsilon(Q)}{\varepsilon(Q)} \frac{\cos^2 QR_c \sin Qr}{Q}\, \mathrm{d}Q\right). \quad (6.34)$$

φ_{eff} is exemplified by LRO-II in §4.5 and can be extracted from observed $S(Q)$ as explained in §5.9. It is a moot point whether even the best observed structural data uniquely defines φ_{eff} with sufficient precision to compare with a theoretical calculation based on equation (6.33). In any case, $\varphi_{\text{eff}}(r)$ depends on the input to the calculation as explained in §6.5 and illustrated in figure 6.3. An analytical form of $\varphi_{\text{eff}}(r)$, valuable for MD calculations, was given by Pettifor and Ward (1984).

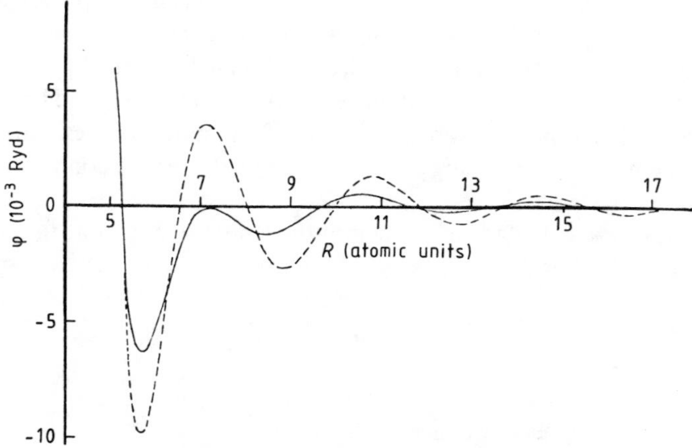

Figure 6.3 Effective interionic potentials in liquid Sn. Broken curve, based on the empty-core potential (equation (6.7)); full curve, based on a more complex non-local potential by Shaw. (From Kumaravadivel and Evans 1976.)

6.8 Thermodynamic properties

The main intention of this section is to employ the thermodynamic variational method, particularly equation (5.46), to find the thermodynamic properties of pure metallic liquids. This was a significant development of the 1970s first attempted by H D Jones in 1971. Young (in Evans and Greenwood (1977), p 1), Hafner (1977) and Ashcroft and Stroud (1978) provide numerous points of entry into the subject.

It is assumed that the nature of the material and the ion number density $n_0 \equiv N/V$ are specified and that each ion contributes z electrons to the free-electron gas of which the number density $n_e \equiv zn_0$. The underlying idea is that the real system can be treated as a reference system with Hamiltonian H_0, potential energy Φ_0, and Helmholtz free energy F_0 plus a perturbation $H_1 \equiv H - H_0$. The prescription in §5.8 then requires H_1 to be averaged over the configurations of the reference system followed by the minimisation of $(F_0 + F_1)$. There are two reasons for attempting this with a hard-sphere reference system. First, its properties are well known (§§4.3, 5.6) and secondly its $S(Q)$ fits observed $S(Q)$ for simple liquid metals moderately well (see, e.g., figure 5.9). H_1 leads to the non-hard-sphere part of the energy, i.e., to everything except the kinetic energy of the ions. The idea is just the same as that in figure 5.10 but now the repulsive part is being approximated by a hard-sphere potential and H_1 includes the attractive and oscillatory parts of the effective ion–ion potential and all the other energy terms. Since $F = U - TS$ we may write the Helmholtz free energy per ion as

$$F \leqslant F_0 + F_1 = (\tfrac{3}{2}k_BT - TS_{HS}) + E_g + E_M + E_0 + E_{bs} - TS_g$$

where S_{HS} and S_g are the entropies per ion of the hard spheres and the electron gas. S_{HS} may be taken from §4.3 (MD) or §5.6 (PY) with σ as the variation parameter. TS_g for a free-electron gas is $(zT^2/\sigma)(\pi k_B^2/3)g(E_F)$ where $g(E_F)$ is the electronic density of states at the Fermi energy. The $S(Q)$ required for calculating E_M and E_{bs} is that of the hard spheres (MD or PY) because the perturbation energy must be averaged over the configurations of the reference system.

The rest of the calculation is a computing exercise aimed at finding what value of σ minimises F for the specified temperature and the corresponding density which is taken from experiment. Table 6.1 gives some specimen energies for Na at 373 K (Umar *et al* 1974, Yokoyama *et al* 1977).

Table 6.1 Energies in atomic units per ion

E_{ckc}	E_{exch}	E_{corr}	$E_g - TS_g$	E_0	E_M	$\frac{3}{2}k_BT$	$-TS_{HS}$	E_{bs}
0.06701	−0.11281	−0.03578	−0.08159	0.06246	−0.21017	0.00177	−0.00911	−0.01369

Once the optimum σ and corresponding F have been found the former leads to theoretical values of $S(Q)$ and the latter to thermodynamic properties such as entropy $S = -(\partial F/\partial T)_V$. Most of S is contributed by the ideal gas term, namely, the leading term in equation (5.32 j or k), so a stricter test is to subtract this and calculate $\Delta S_E \equiv S - S_{ideal}$ as in equation (4.15). The theoretical and experimental values of $\Delta S_E/k_B$ for Na are -3.52 and -3.49 respectively. The theoretical internal energy is also correct to a few per cent. For more extensive comparisons of experiment and theory for pure metals see Hafner (1977). A general comment would be that the theory is moderately successful overall for ΔS_E, U (the internal energy), C_v, κ_T and $S(Q)$. The results are of course subject in detail to the choices of $\varepsilon(Q)$ and the pseudopotential. Another comment might be that the calculation of the thermodynamic properties of a liquid metal from first principles is a considerable achievement.

Other versions of thermodynamic perturbation theory could be tried and the WCA theory of §5.8 has been applied successfully to liquid metals by Kumaravadivel and Evans (1976). The ORPA (§5.7) and pseudopotentials have also been combined to calculate $S(Q)$ and other quantities (Brettonet and Regnaut 1985, Hafner and Kahl 1984, Regnaut 1986).

6.9 Thermodynamic properties of liquid alloys

The first part of this section summarises some binary alloy thermodynamics in phenomenological terms. There is an enormous wealth of different behaviours and many models with a variety of parameters.

Let $c_1, c_2 = 1 - c_1$ be the concentrations of two components of a binary alloy. If there are N_1, N_2 moles of each then

$$c_1 = \frac{N_1}{N_1 + N_2} \qquad c_2 = \frac{N_2}{N_1 + N_2}.$$

The Gibbs free energy of mixing is usually expressed per mole of mixture, denoted by ΔG, and defined by

$$\Delta G(T, p) \equiv G(T, p) - N_1 G_1^{(0)}(T, p) - N_2 G_2^{(0)}(T, p) \tag{6.35}$$

where G, $G_1^{(0)}$ and $G_2^{(0)}$ are the free energies per mole of the alloy and the pure components and $N_i = c_i$. There are corresponding definitions for the entropy, energies, enthalpy and volume of mixing ΔS, ΔU, ΔF, ΔH and ΔV.

The partial molar Gibbs energy is defined by

$$g_i = \left(\frac{\partial G}{\partial N_i}\right)_{T,p,N_j} \tag{6.36}$$

and is usually called the chemical potential μ_i. In binary alloys, i, j are 1 or 2. Partial molar S, V etc are similarly defined; they are intensive properties and functions of p, T and c. If Y is any extensive quantity and y_i is the partial molar Y, an important result is

$$Y = \sum_i N_i y_i. \tag{6.37}$$

For example, an alloy of Na with concentration $c_1 = c$ with Cs of concentration $(1 - c)$ has

$$\Delta G \text{ (per mole)} = (c\mu_{\text{Na}}(c) + (1 - c)\mu_{\text{Cs}}(c)) - c\mu_{\text{Na}}^{(0)} - (1 - c)\mu_{\text{Cs}}^{(0)}. \tag{6.38}$$

Activity, $a(c)$, is defined by equations of the form

$$\mu_{\text{Na}}(c) = \mu_{\text{Na}}^{(0)} + RT \ln a_{\text{Na}}(c) \tag{6.39}$$

and the activity of the other member of the alloy, say Cs, can be shown to be

$$a_{\text{Cs}}(c) = \exp\left(- \int_0^c \frac{c}{1 - c} \frac{\mathrm{d} \ln a_{\text{Na}}(c)}{\mathrm{d}c} \, \mathrm{d}c \right). \tag{6.40}$$

A final example of a thermodynamic quantity will be taken from table 3.2, namely, $S_{cc}(0)$ or S_{cc} for short:

$$S_{cc} = \frac{RT}{(\partial^2 G/\partial c^2)_{T,\,p,\,N}} \tag{6.41}$$

where N is the total number of moles.

All these quantities are functions of T and p as well as c. They are measurable and potentially obtainable from the theories discussed above, at least for simple metals. As we have seen the calculation is somewhat elaborate and involves many approximations and it has proved helpful to go some way towards interpreting measured thermodynamic properties with semiempirical simple models which avoid the pseudopotential theory.

Some simple models which fit some alloy data but which function largely to stimulate discussion of departures from them are as follows. Detailed discussions are given in Gaskell (1973) and data are given in Hultgren *et al* (1973).

The *ideal solution* is defined by

$$\Delta S = \Delta H = \Delta V = 0 \tag{6.42a}$$

$$\Delta G = \Delta F = -T\Delta S^{\text{ID}} \equiv -RT(c_1 \ln c_1 + c_2 \ln c_2) \tag{6.42b}$$

where the superscript 'ID' means 'ideal' and the ideal entropy of mixing comes from the statistical mechanics of ideal gas mixtures. It follows that

$$S_{cc}^{\text{ID}} = c_1 c_2 = c(1 - c). \tag{6.42c}$$

The *regular solution* has

$$\Delta S = \Delta S^{\mathrm{ID}} \qquad \Delta V = 0 \qquad \Delta U = \Delta H = \chi c_1 c_2 \tag{6.43a}$$

$$\Delta G = \Delta F = -T\Delta S^{\mathrm{ID}} + \chi c_1 c_2 \tag{6.43b}$$

where χ is an energy per mole independent of p, T and c. χ represents some energy of interaction between the two components. For this model,

$$S_{\mathrm{cc}} = \frac{c(1-c)}{1 - 2\chi c(1-c)/RT}. \tag{6.43c}$$

In spite of its apparent simplicity, the regular solution has a considerable richness of properties. For example, it is clear that for particular values of c and $\chi > 0$, S_{cc} will become infinite. Since S_{cc} is a measure of concentration fluctuations, $S_{\mathrm{cc}} = \infty$ is a sign of complete phase separation or immiscibility.

Flory models are a group intended to allow for large differences in molar volume of the pure components by using volume fraction instead of concentration in the entropy expression, i.e.,

$$-R(c_1 \ln c_1 + c_2 \ln c_2) \rightarrow -R(c_1 \ln \varphi_1 + c_2 \ln \varphi_2)$$

where

$$\varphi_i \equiv c_i v_i / V.$$

Here v_i is the partial molar volume which as a further simplification could be replaced by the actual molar volume of the pure component. V is the molar volume of alloy which simplifies to the ideal molar volume $(c_1 V_1^{(0)} + c_2 V_2^{(0)})$ on the assumption that $\Delta V = 0$. Originally Flory derived the expression from a lattice model of a polymer liquid and its extension to liquid alloys lacks rigorous justification. It can be supplemented by interaction terms as in the regular solution by writing

$$\Delta G = -RT(c_1 \ln \varphi_1 + c_2 \ln \varphi_2) + \tfrac{1}{2}\chi(c_1\varphi_2 + c_2\varphi_1) + p\Delta V \tag{6.44}$$

where the last term allows for a volume change on mixing. Endless evolution of such models is possible. For example, the first term could be replaced by $-T\Delta S_{\mathrm{HS}}$ from equation (A6.2) to make use of hard-sphere formulae. Since $S = -(\partial G/\partial T)_p$, an additional entropy contribution would arise from any assumed temperature dependence of χ.

In figure 6.4 some properties of liquid Na–Cs are shown fitted to a model using ΔS_{HS} and $\chi = \chi(T)$ (Neale and Cusack 1984); simpler models like the regular solution and equation (6.44) were inadequate. It remains to be seen whether the required parameters σ_1, σ_2 and $\chi(T)$ in such a model can be derived from liquid state theories and pseudopotentials.

Liquid alloys in which the interaction is so strong that compounds, immiscibility or ionicity intervene are outside the scope of NFE theory. Thermodynamic models which are extensions of equation (6.44) have been devised for them and are referred to in §11.11.

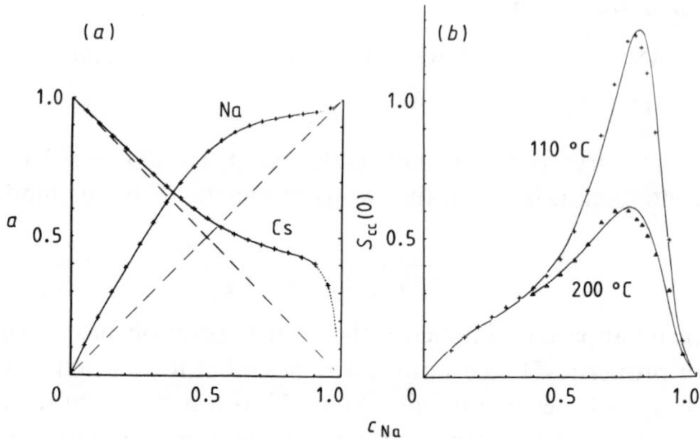

Figure 6.4 Measured thermodynamic properties of Na–Cs fitted to a thermodynamic model. (*a*) $T = 110$ °C; $p = 1$ bar; +, observations; full curves, model. (*b*) $p = 1$ bar; +, ▲, observations; full curves, model. (From Neale and Cusack 1984.)

We now apply the theory of §6.8 to binary alloys. This is straightforward in principle but distinctly more complicated because two pseudopotentials and three partial structure factors are involved. Furthermore, the electron density and therefore most other parameters become functions of concentration, c. For a given concentration, $F = F(T, V, \sigma_1, \sigma_2)$ where σ_i is a hard-sphere diameter. To minimise F in the Gibbs–Bogoliubov method $(\partial F/\partial \sigma_i) = 0$. The latter condition implies that $-(\partial F/\partial T)_V$, which is the entropy, is given by $-(\partial F/\partial T)_{V, \sigma_1, \sigma_2}$. The vapour pressure is given by $p = n^2(\partial F/\partial n)_{T, \sigma_1, \sigma_2}$.

A choice that has to be made in such calculations is the extent to which observed quantities will be fed in. For example, the variation of molar volume with concentration may well be taken from experiment in order to calculate something else.

An Appendix to this chapter collects some binary hard-sphere formulae needed for such calculations. Apart from that the reader may like to see some of the theoretical achievements in Hafner (1977), Umar *et al* (1974) and Yokoyama *et al* (1977). It has proved possible to calculate reasonably well the entropies, enthalpies and volumes of mixing for a number of binary alloys of simple metals. The σ_i are found to change on alloying partly because z^* alters and partly because the effective interionic potential is a function of electron gas density and therefore of concentration. Some results are shown in figure 6.5. Table 6.2 is extracted from the many results in Hafner (1977) for equiatomic compositions and shows that qualitative agreement is possible—though not necessarily in particularly difficult cases such as Cs alloys in which the Cs d states may not be properly accounted for in the theory.

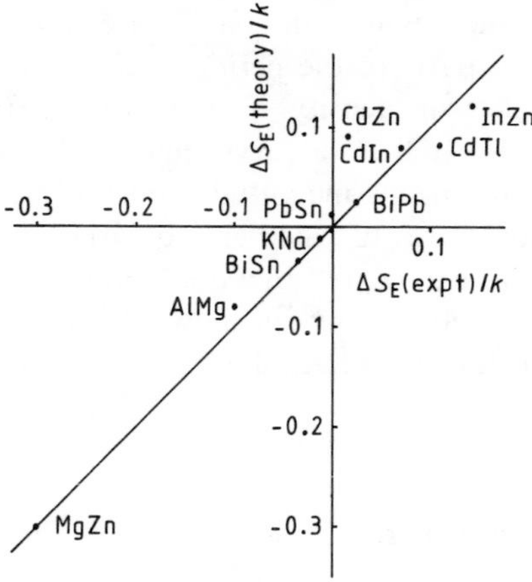

Figure 6.5 $\Delta S_{\mathrm{E}} \equiv \Delta S - \Delta S^{\mathrm{id}}$. The alloys have $c = 0.5$ and the procedure was to select a unique set of the parameters in the equations for F so that the Gibbs–Bogoliubov method gave the entropies of all the metals and alloys satisfactorily. (For details see Young in Evans and Greenwood 1977.)

All in all the thermodynamic perturbation theory of liquid metals and alloys built on the Gibbs–Bogoliubov theorem and the HS reference system gives much insight and some reasonable numerical results. It could hardly be expected that a hard-sphere mixture will have S_{ij} or S_{NN} and S_{cc}, capable of expressing the chemical attractions which exist even in some seemingly simple alloys. Consequently, strongly attracting metals, forming alloys with $S_{\mathrm{cc}} \ll c_1 c_2$, for certain concentrations, certainly need a deeper approach. The WCA approximation for the repulsive part and the ORPA for the attractive part of the potential may lead to better results and although it meets some technical difficulties when applied to binary systems it is being tried (Hafner and Kahl 1984, Young 1985). Compound-forming alloys will be treated later in §11.11.

Table 6.2

		ΔH (cal mol^{-1})		ΔS (cal mol^{-1} K^{-1})	
Alloy	T(°C)	Th.	Exp.	Th.	Exp.
NaK	100	380	174	0.98	1.35
NaCs	100	1640	218	0.30	1.20
KRb	100	160	30	1.22	—
LiMg	700	− 3870	− 2477	− 0.54	1.201 ± 0.6
AlMg	665	− 2170	− 806	1.20	1.18 ± 0.2

The effective ion–ion potentials (§6.7) are not of course identical for metals in different parts of the periodic table because of the trends in core structure and electron density. The shapes of $S(Q)$ are not identical either; for example, unlike the Na curve (figure 3.5), those for Hg, Sn and Ga have shoulders of various magnitudes on the high-Q side of the first peak. So the question arises of whether the variations of $S(Q)$ can be explained in terms of those in $\varphi_{\text{eff}}(r)$. In a qualitative way the pseudopotential theory does this quite well as has been shown by Hafner and by Young and their collaborators (Hafner and Kahl 1984, Young 1985).

6.10 The one-component plasma

It should not be thought that spherical ions whether 'hard' or 'soft' form the only useful reference systems. At the time of writing interest is increasing in the one-component plasma (OCP) as a model system for variational calculations. It consists of n ions per unit volume, of charge ze each, in a uniform unresponding neutralising background. It is a simple system in that a single parameter will determine its state. This parameter, Γ, is the ratio of the Coulomb energy, $(ze)^2/r_s$, to $k_B T$ where $r_s \equiv (3/4\pi n)^{1/3}$, thus

$$\Gamma \equiv \left(\frac{4\pi n}{3}\right)^{1/3} \frac{(ze)^2}{k_B T}. \tag{6.45}$$

Γ increases with density at constant temperature and the properties can be followed as functions of Γ by a series of MC calculations such as those made by Hansen (1973) who used the methods of §4.1 to obtain U, F, C_v, $g(r)$, $S(Q)$, $c(r)$ etc for values of Γ up to 300. At $\Gamma \gtrsim 160$ the lowest free energy of the OCP represents a crystal but for $\Gamma \lesssim 160$ $g(r)$ and $S(Q)$ have shapes typical of fluids. The freezing criterion corresponds to a first peak height in $S(Q)$ of about 2.71.

Sufficient of its properties are known for the OCP to be used as a reference system for liquid metals and the question arises: is it superior to the HS system for this purpose? Not surprisingly the answer is not straightforward and may depend on the details of the potential and the choice of the dielectric function. Mon *et al* (1981) found that the Gibbs–Bogoliubov method (§5.8) gave better results for liquid Na with the OCP than with the HS reference system but found the reverse for Al. An example of the exploitation of the OCP can be found in Young and Ross (1986) where it is used to calculate the equation of state of the alkali metals at high pressures and is found to be in good agreement with experiment.

6.11 Electron transport properties; introduction

In this section a few expressions are collected which are useful for discussing electron transport in disordered metallic conductors. Complete treatments from first principles are given in Smith *et al* (1967), Ziman (1960) and Dresden (1961).

In the one-electron approximation the Fermi function, $f^{(0)}(E, T)$, expresses the equilibrium distribution of electrons amongst the allowed energies. Over a wide regime in crystal physics and in liquid metals where the NFE model is adequate, unperturbed electron states can be thought of as propagating wavepackets labelled with a vector $\boldsymbol{k}$ and the problems of transport tackled with simple perturbation theory. The latter provides the probability of scattering per unit time from $\boldsymbol{k}$ to $\boldsymbol{k}'$ and the Boltzmann equation relates this probability to the effect of applied electric and magnetic fields. The resulting semiclassical solution gives the increase, $f^{(1)}$, in $f^{(0)}$ due to the combined effects of the scattering and applied fields in the steady state. In a much used approximate solution, $f^{(1)}$ is shown to be derivable from the following expression in which $\tau_{\boldsymbol{k}}$ is a relaxation time which may depend on the electron state but not on the applied fields:

$$f^{(1)} = -\tau_{\boldsymbol{k}}\boldsymbol{v} \cdot \boldsymbol{F}\frac{\partial f^{(0)}}{\partial E} \qquad \text{for } \boldsymbol{B} = 0 \qquad (6.46a)$$

where

$$\boldsymbol{F} \equiv T(E(\boldsymbol{k}) - \zeta)\nabla(T^{-1}) - \nabla\zeta \qquad (6.46b)$$

and

$$\zeta \equiv \mu + e\varphi. \qquad (6.46c)$$

In these equations, μ and ζ are the chemical and electrochemical potentials of the electron gas and φ which causes them to differ is the potential of any macroscopic electric field. $\boldsymbol{v}$ is the electron velocity. The magnetic field $\boldsymbol{B}$ is assumed to be zero but the expression is general enough to cover thermal and thermoelectric effects when $\nabla(T^{-1}) \neq 0$. Electric and thermal currents can be written down fairly easily in terms of $f^{(1)}$ and $\boldsymbol{v}$ and in fact for isotropic matter with forces and currents along the x-axis, the electric and thermal current densities j_x, Q_x are:

$$j_x = \frac{-e}{4\pi^3}\int v_x^2\tau_{\boldsymbol{k}}\left[-\frac{\partial\zeta}{\partial x} + T(E-\zeta)\frac{\partial}{\partial x}\left(\frac{1}{T}\right)\right]\frac{\partial f^{(0)}}{\partial E}\,\mathrm{d}\boldsymbol{k} \qquad (6.47a)$$

$$Q_x = -\frac{1}{4\pi^3}\int v_x^2\tau_{\boldsymbol{k}}(E-\zeta)\left[-\frac{\partial\zeta}{\partial x} + T(E-\zeta)\frac{\partial}{\partial x}\left(\frac{1}{T}\right)\right]\frac{\partial f^{(0)}}{\partial E}\,\mathrm{d}\boldsymbol{k}. \qquad (6.47b)$$

$\tau_{\boldsymbol{k}}$ has ultimately to be derived from the scattering probability and therefore depends on the mechanism supposed to cause the scattering, e.g., collisions with random impurities or, in a crystal, with phonons.

When the temperature gradient is zero, j_x is proportional to $\partial\varphi/\partial x$ so Ohm's law and the expression for conductivity follow at once. The detailed calculation of this formalism is given in the references. For metals $\partial f^{(0)}/\partial E$ is sharply peaked at $E = E_F$ which confines our present interest to electrons of this energy making $\sigma = \sigma(E_F)$. The important results are:

$$\sigma = \frac{e^2}{12\pi^3\hbar}\int_{FS}\int \tau v \, dS \tag{6.48}$$

$$\alpha = -\frac{\pi^2 k_B^2 T}{3|e|}\left(\frac{\partial \ln \sigma(E)}{\partial E}\right)_{E=E_F} \tag{6.49}$$

$$\lambda = \frac{1}{3}\left(\frac{\pi k_B}{|e|}\right) T\sigma. \tag{6.50}$$

In equation (6.49), known as Mott's formula, α is the thermoelectric power defined macroscopically in figure 6.6. Equation (6.50) is the Wiedmann–Franz law for thermal conductivity, λ.

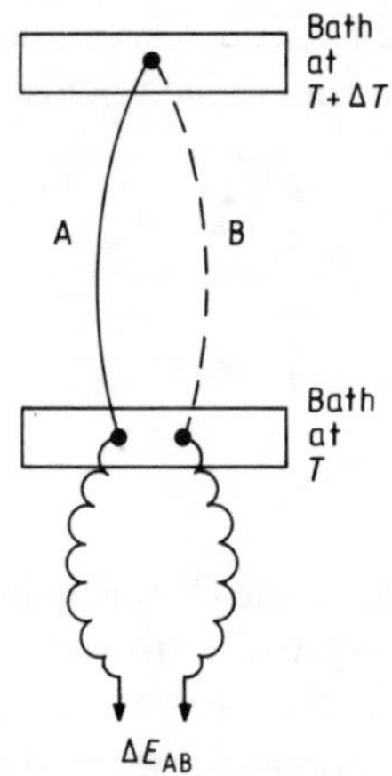

Figure 6.6 The definition of α, thermoelectric power. If ΔE_{AB} is the open-circuit Seebeck voltage, $\lim_{\Delta T\to 0}(\Delta E_{AB}/\Delta T) = \alpha_{AB}(T)$ which is conventionally positive if ΔE_{AB} would act to send current from A to B across the junction at $T + \Delta T$. $\alpha_{AB} = \alpha_B - \alpha_A$ and $\alpha = \int_0^T \sigma_T \, dT/T$ where σ_T is the Thomson coefficient.

In equation (6.48) the integral is over the Fermi surface. Defining a mean free path by

$$\Lambda = \tau v \tag{6.51}$$

we may obtain other versions of σ, e.g., if Λ depends only on E, not on $\boldsymbol{k}$,

$$\sigma = \frac{e^2 \Lambda_F S_F}{12\pi^3\hbar} \tag{6.52a}$$

where S_F is the area of the Fermi surface. If, further, $E = \hbar^2 k^2/2m$, then

$$\sigma = \rho^{-1} = \frac{n_e e^2 \tau}{m} \tag{6.52b}$$

where n_e is the free-electron density.

We now take a special case which gives a simple expression for τ. The assumptions are that the electron collisions are elastic, that E depends only on $|\boldsymbol{k}|$, and that the scattering probability per unit time depends only on the scattering angle and the magnitude $|\boldsymbol{k}|$ of the initial wavevector. Given these simplifying assumptions,

$$\frac{1}{\tau} = 2\pi \int_0^{\pi} P(\theta)(1 - \cos\theta) \sin\theta \, d\theta \tag{6.53}$$

where $P(\theta)$ means the probability per unit time that an electron is scattered through θ from an initial wavevector $\boldsymbol{k}$ into $\boldsymbol{k}'$ without change of energy.

If a magnetic field, $\boldsymbol{B}$, is present the formulae become considerably more complicated. However with the assumptions just made, the expression for the Hall coefficient is

$$R_H = R_H^{FE} \equiv -\frac{1}{n|e|}. \tag{6.54}$$

This formula is well known to be obtainable by elementary means from the free-electron model and indeed it remains a matter of considerable theoretical difficulty in liquid metals to find models which lead to any other expression (see, e.g., Ballentine in Evans and Greenwood (1977)). In crystals $R_H \neq R_H^{FE}$ in general because either or both τ and E depend on $\boldsymbol{k}$ not $|\boldsymbol{k}|$ but isotropy in liquid metals always seems a reasonable assumption.

The mean free path can be estimated from the measured conductivity by using equations (6.51) and (6.52). Typical values for liquid metals near their melting points are given in the table 6.3.

Table 6.3

Metal	Li	Na	Cu	Zn	Hg	Ga	Sn	Pb	Sb	Bi
Λ(Å)	45	157	34	13	7	17	10	6	4	4

An intuitive criterion for the applicability of the NFE model and the semi-classical use of the Boltzmann equation to transport is that Λ should be much greater than both the interatomic spacing and the Fermi wavelength; wavepackets, propagating between relatively rare collisions, then appear reasonable. The nearest-neighbour distance and λ_F are typically 3 and 4 Å, respectively, so it is clear that the validity of the NFE model is at least open to question in some polyvalent metals. The systems excluded in §6.2 from the NFE treatment will certainly need other transport equations.

6.12 Electron transport properties and NFE theory

The easiest way to obtain an expression for σ or ρ is to calculate τ from equation (6.53) and insert it in (6.52*b*). We may replace $P(\theta)$ by $n_s v_F \sigma(\theta)$ where n_s is the number density of scattering centres and $\sigma(\theta)$ the scattering cross section for a single scattering centre.

In the event that each centre scatters independently we find

$$\rho = \frac{m}{n_e e^2 \tau} = \frac{2\pi m n_s v_F}{n_e e^2} \int_0^\pi \sigma(\theta)(1 - \cos\theta)\sin\theta\, d\theta. \tag{6.55}$$

This would serve as an expression for the resistivity of a sparse random array of impurity centres in a metal at such low temperature that scattering by phonons in the host was negligible or at least separable. Equation (6.55) is therefore the basis of a theory of residual resistivity of dilute alloys first given by Mott and subject to decades of subsequent development. A theory of $\sigma(\theta)$ would be required. For liquid pure metals, however, coherent scattering is more plausible than independent scattering, so $n_s \to N/V$ and $\sigma(\theta)$ stands for the cross section *per ion* of the whole assembly of which the scattering potential—i.e. the screened pseudopotential—is given by $U(r)$ in equation (6.9). In accordance with the idea that the electrons are nearly free we assume that $\sigma(\theta)$ can be calculated by the Born approximation, namely

$$N\sigma(\theta) = \frac{4\pi^2 m^2}{h^4} \left| \int \exp(\mathrm{i}\boldsymbol{Q}\cdot\boldsymbol{r}) \sum_i u(\boldsymbol{r} - \boldsymbol{R}_i)\, d\boldsymbol{r} \right|^2 \tag{6.56a}$$

whence

$$\sigma(\theta) = \frac{4\pi^2 m^2}{h^4} S(Q)|u(Q)|^2. \tag{6.56b}$$

The u is the screened pseudopotential called u_A in §6.4 and elsewhere. We may replace $\sin\theta/2$ by $Q/2k_F$ since $\boldsymbol{k}_F$ is the incident wavevector. Substituting (6.56) into (6.53) and thence into (6.52*b*) and using the various relations between E_F, k_F, v_F and n_e we find

$$\rho = \frac{3\pi}{e^2 \hbar v_F^2} \left(\frac{N}{V}\right) \frac{1}{4k_F^4} \int_0^{2k_F} Q^3 S(Q)|u(Q)|^2\, dQ. \tag{6.57}$$

This equation expresses the idea that resistivity is the result of coherent but not very strong scattering from all the ions. The multiple scattering is omitted just as it is from the analogous x-ray scattering formulae such as (3.11*c*). In ρ the integration is over all scattering angles weighted with the factor $(1 - \cos\theta)$ or $2(Q/2k_F)^2$ to emphasise large-angle scattering. A resistivity formula on this basis was first proposed in 1948 by Krishnan and Bhatia but its great current importance stems from its rediscovery in 1961 by Ziman who realised the importance of using the pseudopotential not just for numerical evaluation but for justifying an attempt to treat the scattering as weak. Ziman's formula—as it is now frequently called—was very influential in establishing that something could be said from first principles about metallic liquids.

For compactness in subsequent discussions let us define

$$\langle\langle F(Q) \rangle\rangle \equiv \frac{1}{4k_F^4} \int_0^{2k_F} Q^3 F(Q)\, \mathrm{d}Q \tag{6.58}$$

then

$$\rho = \frac{3\pi}{e^2 \hbar v_F^2} \left(\frac{N}{V}\right) \langle\langle F(Q) \rangle\rangle \tag{6.59a}$$

where for a pure metal

$$F(Q) = S(Q)|u(Q)|^2. \tag{6.59b}$$

Substituting this into equation (6.49) we find

$$\alpha = \frac{-\pi^2 k_B^2 T}{3|e|E_F}(3 - 2q - \tfrac{1}{2}r) \tag{6.60a}$$

where

$$q \equiv \frac{F(2k_F)}{\langle\langle F(Q) \rangle\rangle} \tag{6.60b}$$

and

$$r \equiv \frac{k_F \langle\langle (\partial F(Q)/\partial k)_{k_F} \rangle\rangle}{\langle\langle F(Q) \rangle\rangle}. \tag{6.60c}$$

The r-term arises only if $u(Q)$ is supposed to have a k-dependence, i.e., we have an energy-dependent pseudopotential. Often, as in §6.6, u is supposed to be independent of k and r is then zero.

It was not long before Ziman's formula was extended to binary alloys. To do this $F(Q)$ in equation (6.59*b*) has to be replaced by $F(Q)$ as defined in equation (3.15*b*) with, of course, u_A and u_B for electron scattering replacing f_A and f_B for x-rays. The necessity for the partial structure factors in the resistivity formula follows from the same physical reasoning as in §3.5 and can be obtained directly from equation (6.56*a*) by allowing for both types of atom in the sum. With the proper $F(Q)$, formulae (6.58) to (6.60) apply to binary alloys of any concentration. Since R_H does not contain τ in equation (6.54) the expectation is that even in alloys the free-electron value of the Hall coefficient should be found.

6.13 Successes and limitations of the theory

All the provisos of §6.5 apply to transport. We do not therefore expect equation (6.59*a*) to get exactly the right answer. To see what it can do consider figure 6.7. The upper limit of integration, corresponding to 180° scattering, involves chiefly the first peak of $S(Q)$. It appears that polyvalent metals should have a higher resistivity than monovalent metals and this is true. Since k_F and $u(Q)$ are insensitive to temperature, $(T/\rho)(\partial\rho/\partial T)$, which is typically 0.1 to 0.8 in liquid metals at the melting point, results from $\partial S(Q)/\partial T$ and in many cases the numerical predictions are in reasonable agreement with experiment. Divalent metals are known to have small or negative temperature coefficients of resistance. This is accommodated by the theory because of the unique proximity of $2k_F$ to the peak position in divalent cases: as T rises the peak is both lowered and broadened and the integral can increase or decrease according to the exact form of the curves. Calculations of ρ using $S(Q)$ from various experiments, or different versions of bare-ion potential and dielectric function, give correct orders of magnitude but differ among themselves. In aluminium, for instance, theoretical results bracket the observed 25 $\mu\Omega$ cm with values from 20 to 37. This is typical and shows why it is probably fruitless to compare theory and experiment with too detailed an attention to numerical precision. In fact sometimes the theory is trusted sufficiently to use measured values of ρ to determine parameters of the pseudopotential (see §6.3). The same applies, only more so, to the thermoelectric power. Examination of equation (6.60) shows that its numerical value depends quite sensitively on the ordinates in figure 6.7 at $Q = 2k_F$ and these are subject to the experimental error in $S(Q)$ and the theoretical uncertainty in $u(Q)$. Nevertheless, although the theory cannot predict the observed values reliably, the latter do not disprove the theory—rather they direct attention to the difficulties of testing it.

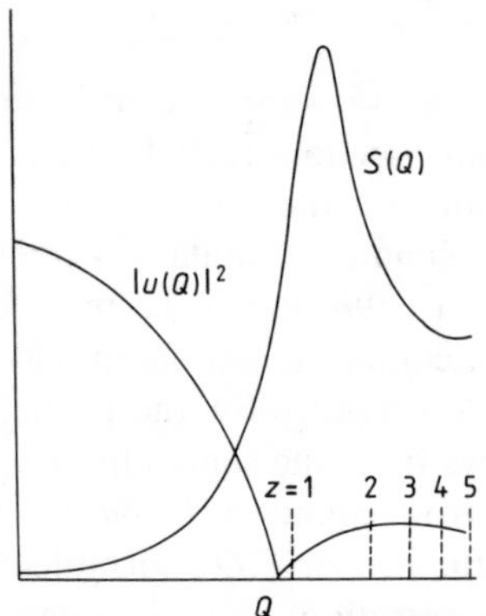

Figure 6.7 Terms in the integrand of equation (6.57). $2k_F$ for different valencies is shown schematically.

It will be appreciated that this is all the more true for alloys because information is so scarce about partial structure factors and about the concentration dependences of u_A and u_B. To illustrate the ability of the theory to explain the general behaviour in alloys of simple metals, suppose equation (6.59) were evaluated with the following assumptions:

(i) S_{ij} is calculated from the PY solutions for a hard-sphere mixture;

(ii) the electron density is given by $n_e = (z_A c_A + z_B c_B)N/V$ where z is the valence;

(iii) equation (6.8) (the empty-core potential) is used with a value of R_c for each ion;

(iv) equation (6.19) gives $\varepsilon(Q)$. The resistivity then follows by computation but not before choosing the hard-sphere diameters and their dependence on concentration, the R_c and the function $G(Q)$ in $\varepsilon(Q)$ and then taking the variation in molar volume with concentration from experiment.

Naturally these multiple choices are guided by other knowledge, e.g., R_c could be chosen to give the correct resistivity in the pure components, the hard-sphere diameters could be those which work best in some other application—to thermodynamic properties perhaps or some aspect of the solid state. Thus there is considerable scope for informed judgement. Figure 6.8 shows how quite satisfactory fits can result from plausible choices of input data and also exemplifies the considerable variety of $\rho(c)$ functions there are to be accounted for. There are many other examples in the literature and similar remarks apply to the thermoelectric power.

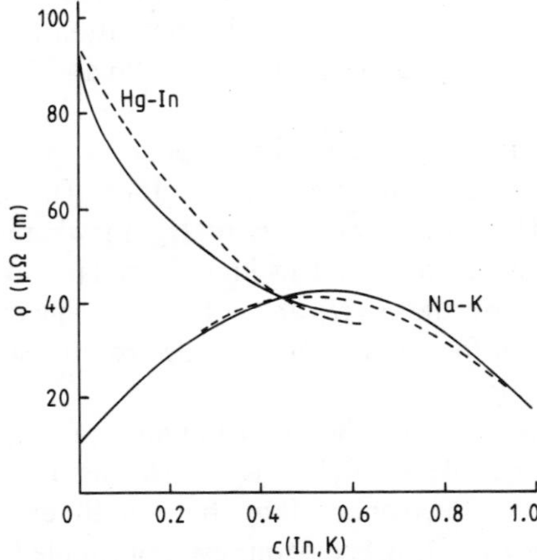

Figure 6.8 Calculated and experimental resistivities of two binary alloy systems selected from Ashcroft and Langreth (1967). The curves are fitted at the end points. Full curves, theory using equation (6.59*a*) and $S_{ij}(Q)$ from PY; broken curves, experimental values.

6.14 The Hall effect and the thermal conductivity

In about twenty or so simple liquid metals (including molten Ge) and a similar number of binary alloys the Hall coefficient has the free-electron value $R_H^{FE} = -(n_e|e|)^{-1}$ to within the experimental error of a few per cent assuming that n_e is calculated from the valencies. Sometimes R_H^{FE} is reached on melting by a striking jump from R_H (crystal) > 0, as in Zn. R_H is found to be independent of T. These observations gave much of the original credence to the NFE theory of simple liquid metals in the 1960s.

There are however a few puzzling exceptions such as the observation that although R_H^{FE} occurs in Hg and In it does not do so in Hg–In alloys. Furthermore, $(R_H/R_H^{FE} - 1)$ has the values $+0.14$, -0.24, -0.31 in Sb, Tl and Bi respectively. The situation with Pb is unclear.

An analysis of the consequences of the Boltzmann equation with the magnetic field $B \neq 0$ shows that the condition for $R_H \neq R_H^{FE}$ is the existence of skew scattering. This is scattering such that $P(\theta, \boldsymbol{k} \to \boldsymbol{k}') \neq P(\theta, \boldsymbol{k}' \to \boldsymbol{k})$ and one mechanism for bringing this about is the spin–orbit (SO) interaction which is the stronger the higher the atomic number. The SO interaction adds a term to the relaxation time which resembles an increment of magnetic field. There is another effect proportional to the spin susceptibility because the SO effect, being linear in the spin, acts oppositely on spin-up and spin-down electrons with a net result proportional to the polarisation. The outcome of the theory (Ballentine and Huberman 1977) is that $(R_H/R_H^{FE} - 1)$ depends on the product of $S(Q)$ and another function $U_s(Q)$ which embodies the skew-scattering. Because $S(Q)$ is a peaked function and $U_s(Q)$ varies smoothly from positive to negative values, the result depends for its size and sign on where the first peak of $S(Q)$ happens to lie in relation to the zero of $U_s(Q)$. Apparently the product is significant in Sb, Tl, Pb and Bi and about zero in Au, Hg.

In alloys like Hg–In the corresponding product involves the partial structure factor whose form can vary with concentration. The fortuitous cancellation of effects in Hg presumably fails in Hg–In whereas in pure In the spin–orbit effect is much smaller. This kind of explanation, though difficult to demonstrate because the $S_{ij}(Q)$ are not known, means that R_H cannot always be relied upon for determining the electron density through the assumption that $R_H = R_H^{FE}$.

According to the NFE theory the Lorenz number, $L \equiv \lambda/T\sigma$, has the value $L^{FE} = \pi^2(k_B/e)^2/3$ or $2.443 \times 10^{-8}\ \mathrm{V^2\,K^{-2}}$. This assumes that the thermal conductivity is entirely electronic. Many, but not all, experiments find considerable departures from this behaviour even in simple liquid metals. Conceivably departures from L^{FE} would occur if the ions contributed significantly or if electron–electron collisions—which always reduce L—were important. Theoretical pursuit of such possibilities leads only to small corrections nowhere near the 20 to 30 % departures sometimes found.

Suspicion has therefore fallen on some of the experimental results especially because of the technical difficulty of removing the contribution of convection from the observed heat flow. At the time of writing the position is not entirely satisfactory though recent careful experiments do suggest that $L(\text{obs})$ is indeed close to L^{FE} in Ga, Hg, Hg–In, Sn (Haller *et al* 1977) and that in liquid alkali metals the values of $L(\text{obs})/L^{\text{FE}} \sim 0.9$ are reasonably accounted for by electron–electron scattering (Cook 1982).

6.15 Optical properties

Before commenting on the optical properties of liquid metals the opportunity will be taken to collect some phenomenological formulae which will be useful in other contexts too. Under sinusoidal fields all ions and electrons whether free or bound contribute by their response to both the polarisation and the conduction. There will be components of motion in and out of phase. The response may therefore be represented by either a complex conductivity, $\sigma(\omega) = \sigma_{\text{R}}(\omega) - \mathrm{i}\sigma_{\text{I}}(\omega)$, or a complex permittivity $\varepsilon(\omega) = \varepsilon_{\text{R}}(\omega) + \mathrm{i}\varepsilon_{\text{I}}(\omega)$ and relations must exist between the real and imaginary parts. From Maxwell's equations the same must apply to the complex refractive index, $n_c(\omega) = n(\omega) + ik(\omega)$. In SI units some of the relations are as follows, where it has been assumed for simplicity that the matter is isotropic and the magnetic permeability is real. Omitting the argument ω,

$$\varepsilon_{\text{R}} - \varepsilon_0 = \sigma_{\text{I}}/\omega \qquad \varepsilon_{\text{I}} = \sigma_{\text{R}}/\omega \qquad \varepsilon - \varepsilon_0 = \sigma/\omega \tag{6.61a}$$

$$n^2 - k^2 = \varepsilon_{\text{R}}/\varepsilon_0 \qquad 2nk = \varepsilon_{\text{I}}/\varepsilon_0. \tag{6.61b}$$

The quantities σ_{R}, ε_{I} and k represent power absorption; for example, the optical absorption coefficient is

$$\alpha(\omega) = 2\omega(\varepsilon_0\mu_0)^{1/2}k. \tag{6.61c}$$

All three complex quantities are widely used as alternatives to represent the results of optical or lossy dielectric measurements and in all cases the real part could, in principle, be calculated from the measured frequency dependence of the imaginary part, or vice versa, from the Kramers–Kronig relations, e.g.,

$$\varepsilon_{\text{R}}(\omega) - \varepsilon(\infty) = \frac{2}{\pi}\int_0^{\infty} \frac{y\varepsilon_{\text{I}}(y)\,\mathrm{d}y}{y^2 - \omega^2} \tag{6.62a}$$

$$\varepsilon_{\text{I}}(\omega) = -\frac{2\omega}{\pi}\int_0^{\infty} \frac{(\varepsilon_{\text{R}}(y) - \varepsilon(\infty))\,\mathrm{d}y}{y^2 - \omega^2}. \tag{6.62b}$$

According to the NFE theory of liquid metals, the frequency dependence of n and k should be that of a free-electron gas provided $h\nu$ is too small to excite plasmons or core electrons from below the conduction band. It is a question whether excitations to energy bands above E_F occur in liquid metals and cause absorption peaks as they do in crystals. Drude's formulae for $\sigma(\omega)$ can be derived by elementary means from classical free-electron theory and recur with the same form in quantum theory, namely,

$$\sigma_R(\omega) = \omega\sigma_I(\omega) = \frac{n_e e^2 \tau}{m}(1 + \omega^2\tau^2)^{-1} = \frac{\sigma(0)}{1 + \omega^2\tau^2} \tag{6.63a}$$

$$\sigma_I(\omega) = -\omega\tau\sigma_R(\omega). \tag{6.63b}$$

Faber (1972) contains a detailed discussion of the derivation which, *inter alia*, makes the point that it is probably necessary to replace m by an effective mass m^* to take account of several complexities of real metals. However, the extensive measurements of $\sigma_R(\omega)$ in infrared and visible light using either ellipsometry or reflectivity measurements enable four generalisations to be made, namely,

(i) peaks identifiable as transitions to higher-energy bands disappear on melting but excitations from lower levels, e.g., from full d bands in noble metals, persist in the liquid;
(ii) low-energy (infrared) transitions within the conduction band are rather well represented by equations (6.63);
(iii) taking $m^*/m = 1$ is not always good enough and may be a few per cent out either way while in particular cases, e.g., Al, the discrepancy may be over 10 %;
(iv) the fitted parameter $\sigma(0)$ which ought to equal $\sigma(\text{DC})$ is usually somewhat too low.

The rather good overall agreement of the results with the Drude equations undoubtedly helped to establish the NFE model as applicable to simple liquid metals. (See the relevant chapters in Beer (1972), Faber (1972), Shimoji (1977).) Nevertheless the question of the proper value of m^* raises serious matters which the NFE is really formulated to evade. Among the reasons why $m^* \neq m$ are the possible departures of the real density of states, $g(E)$, from the free-electron form especially at E_F, the influence of exchange and correlation on $g(E_F)$ and the effect on the formula for $\sigma(0)$ of the use of pseudowavefunctions instead of real wavefunctions (Brown and Jarzynski 1974). These points have wider than optical significance and will be referred to, if not fully clarified, in the next section and Chapter 7. Meanwhile it should be mentioned that the Drude value of $\varepsilon_I(\omega)$ should really be augmented by a term derived from whatever interband transitions there are. Photoemission studies (see §7.6) are now providing some knowledge of $g(E)$ which enables the contribution of interband transitions to be estimated and added to ε_I (Drude) as a small correction. By this means, equations (6.63) with the expected values of n_e and $\sigma(0)$ can be made to agree excellently with the observations on liquid Pb and Bi (Wotherspoon *et al* 1979).

6.16 Modifications to the resistivity formula

The simplicity of concepts and derivation leading to equation (6.57) has stimulated many attempts to improve on it. We defer any reference to the Kubo–Greenwood or similar approaches to electron transport which supplant the NFE method by a radically different and more general approach. Here we touch on a few attempts to improve or stretch the Ziman formula without changing its appearance very much.

In perfect crystals the electron–ion interaction creates the Bloch waves without electrical resistance. The latter results from vibratory motion at finite temperature and to calculate it the phonon spectrum is required. In §6.12 there was no reference to thermal motion and $S(Q)$ derives from the average static distribution of positions; the disorder of the latter causes the resistance. In principle the thermal motion ought to enter the theory and this can be arranged formally by introducing $S(Q, \omega)$. This is a quantity describing inelastic scatterings which transfer energy $\hbar\omega$ as well as momentum $\hbar \boldsymbol{Q}$ to the ion system. $S(Q, \omega)$ will be more fully discussed in Chapter 8. Baym (1964) proved that in equation (6.57) $S(Q)$ should be replaced by

$$S^{\rho}(Q) \equiv \int_{-\infty}^{+\infty} \mathrm{d}\omega \, \frac{\hbar\omega}{k_{\mathrm{B}}T} \frac{S(Q, \omega)}{[\exp(\hbar\omega/k_{\mathrm{B}}T) - 1]}. \tag{6.64}$$

Since $\hbar\omega$ is an excitation energy of the ion system it is typically $k_{\mathrm{B}}\theta_{\mathrm{D}}$ or less (θ_{D} is the Debye temperature). Since $T > \theta_{\mathrm{D}}$ in liquid metals, $\hbar\omega/k_{\mathrm{B}}T \ll 1$ and $S^{\rho}(Q)$ is approximately

$$\int_{-\infty}^{+\infty} S(Q, \omega) \, \mathrm{d}\omega = S(Q).$$

Ziman's formula is thus the perfectly appropriate high-temperature limit of a more general approach, sometimes called the 'generalised Faber–Ziman theory'. If we were considering a metallic glass or supercooled liquid with $T \ll \theta_{\mathrm{D}}$ the generalised theory would be necessary and it is typical of the physics of disordered matter that $S(Q, \omega)$, though in principle measurable by inelastic neutron scattering, is rarely known (see also §12.4).

An entirely different correction to equation (6.57) might be urged on the grounds that $g(E)$ is not exactly equal to $g_{\mathrm{FE}}(E)$ and consequently we may not use the free-electron relations between n_{e}, v_{F}, k_{F}, etc. In particular $|\boldsymbol{v}|$ should be $\hbar^{-1}|\nabla_{\boldsymbol{k}}E|$ or $\hbar k/m^*$ not simply $\hbar k/m$. Since

$$g(E) = (4\pi^3)^{-1} \int \mathrm{d}S(E)/\nabla_{\boldsymbol{k}}(E)$$

we should expect $\rho \propto (m^*)^2 \propto [g(E_F)]^2$. If, however, the electron–ion interaction is supposed to cause a significant departure from $g_{FE}(E)$ would not τ have to be revised too? When this is done to second order in perturbation theory the resulting correction cancels out the previous one so any dependence of ρ on $g(E)$ disappears. This reinstates the NFE formula with a wider validity than might first have been expected and also removes any temptation to replace m by m^* in equation (6.52*b*) on density of states grounds—at least for simple liquid metals. This cancellation is discussed in an illuminating article by Ziman (see Adams *et al* 1967) and in the references to Mott, Edwards and others therein. However, it is a somewhat subtle problem and recently the concellation has been shown not to be exact (Itoh 1984). Partly for that reason we shall not pursue in detail any other of the various suggestions for applying a correction factor of the form (m^*/m) to the Ziman formula. These suggestions include corrections for electron–electron interactions, for non-locality of the pseudopotential, and for the fact that, if the true wavefunction is properly normalised, a plane pseudowavefunction used to represent it should have a corrected normalising constant. Estimates of all these correction factors to equation (6.57) have not significantly altered its ability to agree with observations (Brown and Jarzynski 1974).

The Ziman formula is not expected to apply to 3d transition metals because the unfilled d states cause strong resonant scattering which seems inappropriate for description by a pseudopotential. Nevertheless, Evans *et al* (1971) put forward a version of equation (6.57) in which $|u(Q)|^2$ was replaced by the exact value of the squared matrix element for the scattering of a plane wave by a transition metal ion. The potential of the latter was taken to be a muffin-tin potential as derived in solid state physics for band structure calculations and the exact scattering can be calculated by the phaseshift method. For resonant scattering by 3d metals the d phaseshift ($l = 2$) dominates the process. This approach—which is the 'extended Ziman theory'—has been applied to transition and noble metal alloys of high resistivity with the same kind of qualitative success that the ordinary formula has for simple metals. Nevertheless there are some conceptual problems, namely, when scattering is so strong, why is it reasonable to ignore multiple scattering? In any case, what rational prescription is there for the number of conduction electrons per atom—the quantity which determines k_F—when hybridisation exists between s and d electrons? The reader can pursue an attempt to answer these questions in Esposito *et al* (1978), but it may be that high-resistance liquid metals are really beyond the scope of equation (6.57), however ingeniously it is modified.

Finally, we return to the fact that the NFE theory assumes that k_F is a well defined wavevector and that the Fermi surface is sharp. This cannot be so and intuitively it seems clear that the shorter the mean-free path the greater the uncertainty in k_F. In other words, the scattering makes the wavefunctions irregular and their Fourier transforms will contain a superposition of free-electron waves with a distribution of amplitude $a(k)$. In the event that the $a(k)$ distribution has a single well defined peak for electrons of energy E_F, this peak could define k_F and the width of the peak, Δk_F, measures the blurring of the Fermi surface (see also §7.1). Taking these ideas into account would surely change the predicted values of τ, Λ or ρ; conversely a revised Λ would redefine Δk_F. In short a self-consistency problem arises and can be solved approximately by an additional factor, depending on Λ, in the integrand of the Ziman formula. The latter can then be solved by iteration until the Λ given by the integral equals that in the integrand. This method raises the expected resistivities by a few per cent in several simple metals but by about 20 % in Li, Cs, though these figures depend on the inputs (see §6.5). These considerations are really taking the discussion out of the realm of nearly-free-electron theory as well as making the derivations considerably more complicated. The reader may therefore like to pursue them in McCaskill and March (1982) and Khajil and Tomak (1986).

Appendix

Hard-sphere formulae for binary systems

σ_i, m_i, n_i, c_i are the diameter, mass, number density and concentration, respectively, of spheres of type i.

$$\eta = \text{packing fraction} = (\text{volume of spheres})/(\text{volume of system})$$

$$= \tfrac{1}{6}\pi n_0(c_1\sigma_1^3 + c_2\sigma_2^3) \qquad n_0 \equiv N/V \tag{A6.1}$$

PY = Percus–Yevick (§5.6) CS = Carnahan–Starling (§4.3)

$$S_{\text{HS}} = S_{\text{gas}} + S_{\text{c}} + S_\eta + S_\sigma \tag{A6.2}$$

per atom, where

$$S_{\text{gas}} = \text{ideal gas term} = k_B \ln\left[\frac{e}{n_0}\left(\frac{em_1^{c_1}\, m_2^{c_2} k_B T}{2\pi\hbar^2}\right)^{3/2}\right]$$

$$S_{\text{c}} \text{ or } S^{\text{id}} = \text{concentration term} = -k_B(c_1 \ln c_1 + c_2 \ln c_2) \tag{A6.3}$$

$$S_\eta = \text{packing term} = k_B \ln(1-\eta) + \tfrac{3}{2}k_B\left[1 - \frac{1}{(1-\eta)^2}\right] \qquad \text{PY}$$

$$= k_B\left(1 - \frac{1}{1-\eta}\right)\left(3 + \frac{1}{1-\eta}\right) \qquad \text{CS}$$

$$S_\sigma = \text{misfit term}$$

$$= \frac{k_B\pi c_1 c_2 n_0(\sigma_1-\sigma_2)^2[12(\sigma_1+\sigma_2) - \pi n_0(c_1\sigma_1^4 + c_2\sigma_2^4)]}{24(1-\eta)^2} \qquad \text{PY}$$

$$= k_B c_1 c_2(\sigma_1-\sigma_2)^2\{(x_1+x_2)[\zeta(\zeta-1) - \ln\zeta] + 3(\zeta-1)x_1\} \qquad \text{CS}$$

where

$$\zeta = \frac{1}{1-\eta} \qquad x_1 = \frac{\sigma_1 + \sigma_2}{c_1\sigma_1^3 + c_2\sigma_2^3} \qquad x_2 = \frac{\sigma_1\sigma_2(c_1\sigma_1^2 + c_2\sigma_2^2)}{(c_1\sigma_1^3 + c_2\sigma_2^3)^2}$$

NB $S_\sigma = 0$ when $\sigma_1 = \sigma_2$

$$F_{\text{HS}} \text{ per atom} = \tfrac{3}{2}k_{\text{B}}T - TS_{\text{HS}} \tag{A6.4}$$

$$E_{\text{M}} \text{ per atom} = \sum_{i,j} c_i c_j z_i z_j I_{ij} \tag{A6.5}$$

where

$$I_{ij} \equiv \frac{e^2}{4\pi^2\varepsilon_0}\int_0^\infty (S_{ij}(Q) - 1)\,\mathrm{d}Q$$

$$p_{\text{HS}}/k_{\text{B}}T = \frac{n_0(1+\eta+\eta^2) - \frac{1}{2}\pi n_1 n_2(\sigma_1-\sigma_2)^2(\sigma_1+\sigma_2+\sigma_1\sigma_2 X)}{(1-\eta)^3} \qquad \text{PY} \tag{A6.6}$$

where

$$X \equiv \tfrac{1}{6}\pi(n_1\sigma_1^2 + n_2\sigma_2^2)$$

$$p_{\text{HS}}/k_{\text{B}}T = \frac{n_0(1+\eta+\eta^2-\eta^3)}{(1-\eta)^3} \qquad \text{CS.} \tag{A6.7}$$

References

Adams P O, Davies H A and Epstein S G (ed) 1967 *Proc. Conf. on The Properties of Liquid Metals (Brookhaven National Laboratory, NY) 1966* (London: Taylor and Francis)

Ashcroft N W and Langreth D C 1967 *Phys. Rev.* **159** 500

Ashcroft N W and Stroud D 1978 *Solid State Phys.* **33** 2 (New York: Academic)

Ballentine L E and Huberman M 1977 *J. Phys. C: Solid State Phys.* **10** 4991

Baym G 1964 *Phys. Rev.* **135** A1691

Beer S Z (ed) 1972 *Liquid Metals* (New York: Dekker)

Brettonet J and Regnaut C 1985 *Phys. Rev.* B **31** 5071

Brown C and Jarzynski J 1974 *Phil. Mag.* **30** 21

Cohen M H and Heine V 1961 *Phys. Rev.* **122** 1821

Cohen M L, Heine V and Weaire D 1970 *Solid State Phys.* **24** (New York: Academic)

Cook J G 1982 *Can. J. Phys.* **60** 1759

Cyrot-Lackmann F and Desré P (ed) 1980 *Proc. 4th Int. Conf. on Liquid and Amorphous Metals (J. Physique* **41** C8 Suppl. No 8)

Dresden M 1961 *Rev. Mod. Phys.* **33** 265

Esposito E, Ehrenreich H and Gelatt C D 1978 *Phys. Rev.* B **18** 3913

Evans R and Greenwood D A (ed) 1977 *Liquid Metals (Bristol) 1976* (Inst. Phys. Conf. Ser. 30)

Evans R, Greenwood D A and Lloyd P 1971 *Phys. Lett.* **35A** 57

Faber T E 1972 *An Introduction to the Theory of Liquid Metals* (Cambridge: Cambridge University Press)

Gaskell D R 1973 *Introduction to Metallurgical Thermodynamics* (Tokyo: McGraw-Hill Kogaku-Sha)

Gathers D *Rep. Prog. Phys.*

Hafner J 1977 *Phys. Rev.* A **16** 351

Hafner J and Kahl G 1984 *J. Phys. F: Met. Phys.* **14** 2259

Haller W, Güntherodt H-J and Busch G 1977 *Liquid Metals (Bristol) 1976* (Inst. Phys. Conf. Ser. 30) p 207

Hansen J P 1973 *Phys. Rev.* A **8** 3096

Harrison W A 1966 *Pseudo-potentials in the Theory of Metals* (New York: Benjamin)

Hoshino H and Endo H 1982 *Phys. Chem. Liq.* **11** 327

Hultgren R, Desai P D, Hawkins D T, Gleiser M and Kelley K K 1973 *Selected Values of the Thermodynamic Properties of Binary Alloys* (Ohio: Am. Soc. Metals)

Ichimaru S and Utsumi K 1981 *Phys. Rev.* B **24** 7385

Itoh M 1984 *J. Phys. F: Met. Phys.* **14** L179

Kahl G and Hafner J 1985 *J. Phys. F: Met. Phys.* **15** 1627

Khajil T M A and Tomak M 1986 *Phys. Status Solidi* b **134** 321

Kumaravadivel R and Evans R 1976 *J. Phys. C: Solid State Phys.* **9** 3877

McCaskill J S and March N H 1982 *Phys. Chem. Liq.* **12** 1

Mon K K, Gann R and Stroud D 1981 *Phys. Rev.* A **24** 2145

Neale F E and Cusack N E 1982 *J. Phys. F: Met. Phys.* **12** 2839

—— 1984 *J. Non-Cryst. Solids* **61/62** 169

Pettifor D G and Ward M A 1984 *Solid State Commun.* **49** 291

Phillips J C and Kleinman L 1959 *Phys. Rev.* **116** 287

Regnaut C 1986 *J. Phys. F: Met. Phys.* **16** 295

Shimoji M 1977 *Liquid Metals* (New York: Academic)

Smith A C, Janek J F and Adler R B 1967 *Electronic Conduction in Solids* (New York: McGraw-Hill)

Takeuchi S (ed) 1973 *Proc. Conf. on The Properties of Liquid Metals (Tokyo) 1972* (London: Taylor and Francis)

Umar I H, Meyer A, Watabe M and Young W H 1974 *J. Phys. F: Met. Phys.* **4** 1691

Vashishta P and Singwi K S 1972 *Phys. Rev.* B **6** 875

Wagner C N J and Johnson W L (ed) 1984 *Proc. 5th Int. Conf. on Liquid and Amorphous Metals 1984* (*J. Non-Cryst. Solids* **61/62**)

Wotherspoon J T M, Rodway D C and Norris C 1979 *Phil. Mag.* B **40** 51

Yokoyama I, Meyer A, Stott M J and Young W H 1977 *Phil. Mag.* **35** 1021 (Erratum 1979 *Phil. Mag.* A **40** 729)

Young W H 1985 *J. Physique Coll.* **46** C8 427

Young W H and Ross 1986 in press

Ziman J M 1960 *Electrons and Phonons* (London: Oxford University Press)

—— 1964 *Adv. Phys.* **13** 89

7

ELECTRON STATES IN DISORDERED MATTER

A central problem which has been avoided in the preceding chapter by the use of the NFE model is that of the electron dynamics in materials like disordered alloys and liquid metals. The problem has several aspects. First we need theoretical quantities suitable for describing the dynamics—for it is not obvious that $E(\boldsymbol{k})$ relations or Fermi surfaces are meaningful though densities of states in energy may well be. Then we need methods for calculating the required quantities and of forming ensemble averages of them. It would be desirable to embrace transport as well as equilibrium properties and to include both weak and strong electron–ion potentials. Experimental tests of the resulting theories would be welcome also.

This chapter will touch on all these points but it will soon become clear that the central problem is both difficult and incompletely solved and apt to lead quite rapidly into specialised theoretical methods. The aim will therefore be to indicate the ideas behind a number of the theories and to give examples of valuable applications. The diagrammatic summations necessary to establish some of the results will not be set out but instead references will be given in which detailed explanations can be found.

A subject ignored in this chapter is localisation. If the disorder is great enough, electron states become localised and this phenomenon will be described in Chapter 9.

7.1 The Green and spectral functions

Suppose our problem is specified by a Hamiltonian H with $H|\psi_n\rangle = E_n|\psi_n\rangle$ where the $|\psi_n\rangle$ are a complete orthonormal set of one-electron states. Let us define operators

$$G^{\pm}(E \pm \mathrm{i}\varepsilon) \equiv \frac{1}{E - H \pm \mathrm{i}\varepsilon} \tag{7.1}$$

in which ε is a small positive quantity which is allowed to go to zero after mathematical operations—a procedure necessary because $(E - H)^{-1}$ is not a well defined operator. Now let

$$\lim_{\varepsilon \to 0} G^{\pm}(E \pm \mathrm{i}\varepsilon) \equiv G_{\mathrm{p}}(E) \mp \mathrm{i}\pi\rho(E) \tag{7.2a}$$

whence

$$\lim_{\varepsilon \to 0} (G^{+} - G^{-}) = -2\mathrm{i}\pi\rho(E) = 2\mathrm{i}\,\mathrm{Im}\,G^{+}. \tag{7.2b}$$

It is not difficult to show that $(E - E_n \pm \mathrm{i}\varepsilon)^{-1}$ is an eigenvalue of $(E - H \pm \mathrm{i}\varepsilon)^{-1}$ and since the completeness of $\{\psi_n\}$ can be expressed by

$$1 = \sum_n |\psi_n\rangle\langle\psi_n|$$

we may write

$$G^{\pm} = \sum_n \frac{|\psi_n\rangle\langle\psi_n|}{E - E_n \pm \mathrm{i}\varepsilon}. \tag{7.3}$$

It follows that

$$\rho(E) = \frac{1}{\pi} \lim_{\varepsilon \to 0} \sum_n \frac{\varepsilon|\psi_n\rangle\langle\psi_n|}{(E - E_n)^2 + \varepsilon^2} = \sum_n |\psi_n\rangle\langle\psi_n|\delta(E - E_n) \tag{7.4}$$

where the last step invokes the δ-function property

$$\pi^{-1} \lim_{\varepsilon \to 0} \varepsilon/(x^2 + \varepsilon^2) = \delta(x).$$

The vanishingly small imaginary part of E enables the singularities when $E = E_n$ to be handled with some rigour. Normally integrations involving G are performed before proceeding to the limit $\varepsilon \to 0$ from above or below the real axis. Mathematical points concerning the existence and analyticity of $(z - H)^{-1}$ where z is a general complex energy are discussed in Rickayzen (1980).

The point of all this can be seen by considering and interpreting representations of the operators G and ρ which are called, respectively, Green and spectral operators.

As with any operator, matrix elements of G and ρ could be formed with sets of states of which $|\boldsymbol{r}\rangle$ and $|\boldsymbol{k}\rangle$ are two examples. These are position and momentum eigenstates and $\boldsymbol{r}$ can be assumed to comprise both spatial and spin coordinates if necessary. In this way sets of matrix elements such as $\langle\boldsymbol{r}|G|\boldsymbol{r}'\rangle$ or $\langle\boldsymbol{k}|\rho|\boldsymbol{k}'\rangle$ arise and any set completely characterises its operator. The matrix elements are functions which can alternatively be written $G(\boldsymbol{r}, \boldsymbol{r}', E)$ or $\rho(\boldsymbol{k}, \boldsymbol{k}', E)$. To take the latter, from equation (7.4)

$$\rho(\boldsymbol{k}, \boldsymbol{k}', E) = \sum_n \langle\boldsymbol{k}|\psi_n\rangle\langle\psi_n|\boldsymbol{k}'\rangle\delta(E - E_n). \tag{7.5a}$$

This is 'a matrix element of ρ in the momentum representation'. Similarly, $\langle\boldsymbol{r}|G|\boldsymbol{r}'\rangle$, or $G(\boldsymbol{r}, \boldsymbol{r}', E)$, is a spatial representation of $G(E)$ and is called a Green function. $\langle\boldsymbol{r}|\psi_n\rangle$ and $\langle\boldsymbol{k}|\psi_n\rangle$ stand respectively for the probability amplitudes that the electron in a state $|\psi_n\rangle$ with energy E_n has coordinate $\boldsymbol{r}$ or momentum $\hbar\boldsymbol{k}$. Thus $\langle\boldsymbol{r}|\psi_n\rangle$ is the Schrödinger wavefunction, ψ_n, in terms of which $\langle\boldsymbol{k}|\psi_n\rangle$ is $\Omega^{-1/2} \int \psi_n(\boldsymbol{r}) \exp(-\mathrm{i}\boldsymbol{k} \cdot \boldsymbol{r})\, \mathrm{d}\boldsymbol{r}$ where Ω is the total volume of the system. An alternative to equation (7.5a) would be, from equation (7.4),

$$\rho(\boldsymbol{r}, \boldsymbol{r}', E) = \sum_n \psi_n(\boldsymbol{r})\psi_n^*(\boldsymbol{r}')\delta(E - E_n). \tag{7.5b}$$

From equations (7.5*a*) and (7.2*b*) the element $\langle \boldsymbol{k}|\rho|\boldsymbol{k}\rangle$ is

$$\rho(\boldsymbol{k}, E) = \sum_n |\langle \boldsymbol{k}|\psi_n\rangle|^2 \delta(E - E_n) = -\frac{1}{\pi} \operatorname{Im} G^+(E, \boldsymbol{k}) \tag{7.6}$$

and has an important interpretation. It is the probability that an electron of energy E would have momentum $\hbar\boldsymbol{k}$. $\rho(\boldsymbol{k}, E)$ is called the spectral function and obviously contains significant information about the dynamics. If H describes a free electron, $E = \hbar^2k^2/2m$ and $\rho(\boldsymbol{k}, E)$ is a delta function in k-space at the appropriate $\boldsymbol{k}$. If H refers to a crystal, $\rho(\boldsymbol{k}, E)$ is a set of delta functions separated by reciprocal lattice vectors in k-space because ψ_n, being a Bloch function, is a superposition of harmonic waves with just those $\boldsymbol{k}$. In disordered matter $\psi_n(\boldsymbol{r})$ is an irregular function expressible as a Fourier integral of harmonic waves and $\rho(\boldsymbol{k}, E)$ is in general a continuous function of $\boldsymbol{k}$ for a given E. It immediately follows that an $E(\boldsymbol{k})$ relation can be defined, if at all, only in an approximate sense in disordered matter.

The density of states in energy, per unit volume of matter, can be written for a system of volume Ω as

$$g(E) = \Omega^{-1} \sum_n \delta(E - E_n) \tag{7.7}$$

because this ensures that the number of states with energies in the range ΔE is given by $\Omega \int_{\Delta E} g(E)\, \mathrm{d}E$. Summing $\rho(\boldsymbol{k}, E)$ over the allowed values of $\boldsymbol{k}$ by letting $\Sigma_{\boldsymbol{k}} \to \Omega(2\pi)^{-3} \int \mathrm{d}\boldsymbol{k}$ we then find

$$\frac{1}{(2\pi)^3} \int \rho(\boldsymbol{k}, E)\, \mathrm{d}\boldsymbol{k} = g(E). \tag{7.8}$$

Thus the integral of the spectral function over $\boldsymbol{k}$ leads to the density of states. A further factor of two is required to include both spin orientations for electrons. Similar arguments show that the local density of states per unit volume of system can be written as a function of position $\boldsymbol{r}$ for one spin orientation

$$g(E, \boldsymbol{r}) = \Omega^{-1}\rho(\boldsymbol{r}, E) = -\Omega^{-1}\pi^{-1} \operatorname{Im} G(\boldsymbol{r}, \boldsymbol{r}, E) \tag{7.9}$$

and, furthermore

$$g(E) = \Omega^{-1} \int \rho(\boldsymbol{r}, E)\, \mathrm{d}\boldsymbol{r}. \tag{7.10}$$

Among many other relevant and useful formulae, most of which are discussed in Rickayzen (1980) or Rodberg and Thaler (1967) are the following in which the argument $(E \pm \mathrm{i}\varepsilon)$ of G is omitted:

$$G_0^{\pm}(\boldsymbol{r}, \boldsymbol{r}') = \frac{1}{(2\pi)^3} \int \frac{\mathrm{d}\boldsymbol{k}' \exp[\mathrm{i}\boldsymbol{k}' \cdot (\boldsymbol{r} - \boldsymbol{r}')]}{E - \hbar^2k'^2/2m \pm \mathrm{i}\varepsilon} \tag{7.11}$$

$$G_0^{\pm}(\boldsymbol{k}, \boldsymbol{k}') = \frac{(2\pi)^3\Omega^{-1}\delta(\boldsymbol{k} - \boldsymbol{k}')}{E - \hbar^2k'^2/2m \pm \mathrm{i}\varepsilon} \tag{7.12}$$

where G_0 refers to free-particle motion and the relation

$$\langle \boldsymbol{k}|\boldsymbol{k}'\rangle = (2\pi)^3\Omega^{-1}\delta(\boldsymbol{k}-\boldsymbol{k}')$$

for momentum eigenfunctions normalised in Ω has been used.

If electronic properties derived from G or ρ are to be compared with observations they must be averaged over an ensemble. The latter would usually be the ensemble of all the configurations allowed to the atoms under specified thermodynamic conditions; for example, all the conceivable arrangements of atoms in a disordered binary alloy of fixed concentration and temperature. $\langle X\rangle$ denotes such an average of the quantity X. $\langle G\rangle$ and $\langle \rho\rangle$ are therefore the quantities of major importance.

Let $H = H_0 + V$ where H_0 may be, but is not necessarily, for a free particle and $G_0^{\pm} = (E - H_0 \pm \mathrm{i}\varepsilon)^{-1}$. From simple operator identities we find

$$G = G_0 + G_0VG \tag{7.13a}$$

$$G = G_0 + G_0VG_0 + G_0VG_0VG_0 + \cdots . \tag{7.13b}$$

Writing down the corresponding Green function $G(\boldsymbol{r}, \boldsymbol{r}')$ and using the completeness relation in the form $\int|\boldsymbol{r}\rangle\langle\boldsymbol{r}|\,\mathrm{d}\boldsymbol{r} = 1$ we have, omitting the argument $(E \pm \mathrm{i}\varepsilon)$ again,

$$G(\boldsymbol{r}, \boldsymbol{r}') = G_0(\boldsymbol{r}, \boldsymbol{r}') + \iint G_0(\boldsymbol{r}, \boldsymbol{r}_1)V(\boldsymbol{r}_1, \boldsymbol{r}_2)G_0(\boldsymbol{r}_2, \boldsymbol{r}')\,\mathrm{d}\boldsymbol{r}_1\,\mathrm{d}\boldsymbol{r}_2 + \cdots . \tag{7.13c}$$

If V is an ordinary rather than a non-local potential, $V(\boldsymbol{r}_1, \boldsymbol{r}_2)$ can be replaced by $V(\boldsymbol{r}_1)\delta(\boldsymbol{r}_1 - \boldsymbol{r}_2)$ but in any case $\langle G(\boldsymbol{r}, \boldsymbol{r}')\rangle$ is found by averaging the right-hand side of the last equation term by term and then resumming.

Whether the application postulates small perturbations of free electrons or starts from a tight-binding model, series expansions for the Green function, of which equation (7.13) is an example, usually emerge. Total summation is rarely possible and the various approximations therefore adopted are concerned with adding up some of the terms to infinite order and with estimating the effect of leaving the others out. A few examples will follow and a general account of Green functions for condensed matter can be found in Rickayzen (1980).

The Green function method was first applied to disordered matter in the early sixties and many attempts to derive the properties of electrons are aimed at finding the spectral or Green functions for use in the formulae of this section. The spectral function facilitates answers to general questions such as: 'Does a liquid metal have a Fermi surface?'. If the disordered potential is weak we expect the electron wavefunctions to be only slightly distorted from plane harmonic waves and, for a fixed E, $\rho(\boldsymbol{k}, E)$ will be strongly peaked at $|\boldsymbol{k}| = (2mE/\hbar^2)^{1/2}$. After averaging, $\rho(\boldsymbol{k}, E)$ will depend only on $|\boldsymbol{k}|$ so the locus of the peak in k-space is a sphere. If $E = E_\mathrm{F}$ the sphere could be called a 'Fermi surface' and the width of the peak could be referred to as the 'thickness of the Fermi surface'. The thickness will be less for weaker scattering and least for materials with the longest mean-free path. The latter is quite long in liquid Na (see §6.11) and it is not unreasonable to think of a thickened spherical Fermi surface in such a liquid metal. However it is a matter of degree, and the purist's answer to the question posed above is 'No'. In figure 7.1 $\rho(k, E)$ calculated for liquid Bi is illustrated.

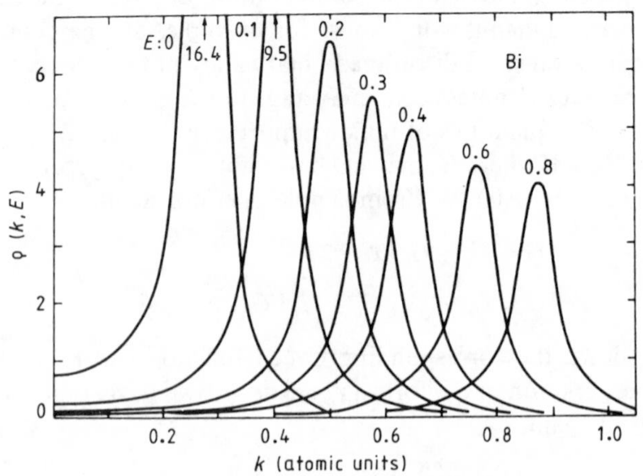

Figure 7.1 $\rho(k, E)$ for liquid Bi calculated by the method of §7.3 (from Ballentine 1975). The numbers on the curves are the (fixed) values of energy for that curve.

7.2 The coherent potential approximation

In 1967 Soven, Taylor, and Onodera and Toyozawa all independently introduced a method that has become recognised as an excellent way of finding $g(E)$ in disordered substitutional alloys—a case of cellular disorder (see §1.7). The problem is to determine $\rho(\boldsymbol{k}, E)$ and $g(E)$ for a binary alloy in which lattice sites $\{i\}$ can be occupied by A or B atoms with independent probabilities c and $1-c$ where c is the number concentration of A.

A first approximation would be to replace the actual potential, v_A or v_B, around each site, by an average potential $v_{Av} \equiv cv_A + (1-c)v_B$. This creates a hypothetical crystal with properties intermediate in some sense between A and B and the electron Bloch states would follow from conventional calculations. For $c \sim 0$ or $c \sim 1$ the results would be good but clearly the specific effects of disorder on the wavefunctions and on $\rho(\boldsymbol{k}, E)$ have been suppressed by averaging v at the start rather than ρ at the end. The method is called the virtual-crystal approximation (VCA) and works reasonably well if v_A and v_B do not differ very much.

The coherent potential approximation (CPA) also uses the idea of a hypothetical crystal which is formed by allocating an effective potential v_C, to each site. Let us imagine this material—called an effective medium—and isolate for consideration a single site in it where an atom with potential v_A or v_B is now placed. Electrons propagating in the effective medium will be scattered at this site becaue $v_A, v_B \neq v_C$. The CPA requirement is that v_C should be such that, if t_A, t_B are the scattering amplitude matrices for an A or a B atom embedded in the effective medium, then

$$ct_A + (1 - c)t_B = 0. \tag{7.14}$$

The remarkable fact is that this makes the Green function for the effective medium very close to the configuration average $\langle G \rangle$ for the original alloy.

To proceed a little further, let us define the T-matrix by

$$G = G_0 + G_0TG_0 \tag{7.15a}$$

which with equation (7.13) shows that

$$T = V + VG_0V + VG_0VG_0V + \cdots . \tag{7.15b}$$

T defines the scattering by V of waves propagating in the medium denoted by suffix zero. The latter might be free space, or a perfect crystal, and V the potential of impurity atoms. Now consider a fictitious crystalline medium, M, with the same lattice as the binary alloy and ordered potential V_C. The proposal is to choose V_C—which is an as-yet-unknown and potentially complex energy—so that $G_M = \langle G \rangle$. Now from equation (7.13), $G_0^{-1}G = 1 + VG$, so that

$$(G_0^{-1} - V_C)G = 1 + (V - V_C)G.$$

Choose V_C so that $G_0^{-1} - V_C \equiv G_M^{-1} = \langle G \rangle^{-1}$. Then

$$\begin{aligned} G &= G_M + G_M(V - V_C)G \\ &= G_M + G_M T_C G_M \end{aligned}$$

where T_C represents the scattering of waves propagating in M by the disordered potential $(V - V_C)$. Averaging G, which is the ultimate object of the exercise, now gives

$$\langle G \rangle = G_M + G_M \langle T_C \rangle G_M$$

or, from equation (7.15b),

$$0 = \langle T_C \rangle = \langle (V - V_C) \rangle + \langle (V - V_C)G_M(V - V_C) \rangle + \cdots . \tag{7.16}$$

In principle the last equation defines V_C, gives the properties of M and solves the problem. But the evaluation of the averaged infinite series is not practicable and approximation is necessary. If the matrix T_C were approximated by its first term, $V - V_C$, then $\langle T_C \rangle = 0$ implies $V_C = \langle V \rangle$ and we regain the virtual-crystal approximation which shows the relation of this approximation to the more general formulation.

The CPA applies the condition of no scattering on the average, i.e. $\langle T_C \rangle = 0$, to a single site i, i.e. $\langle t_i \rangle = 0$ and this may be shown to be the equivalent of summing to infinity many, but not all, kinds of terms in equation (7.16).

The proof of this last assertion and the derivation of formulae apt for numerical computation are given in specialised expositions: Yonezawa and Morigaki (1973) and Gyorffy and Stocks (1977) are particularly illuminating. The input to a calculation will depend on the model used for the alloy. If the TBA is used, the Hamiltonian will include the energies of one or more atomic electron states at A and B sites and the transfer integrals between sites. The CPA is often formulated from this starting point. However, a powerful method for band structure calculations in ordered crystals starts with non-overlapping or muffin-tin atomic potentials and considers the multiple scattering of electrons by this array. Since the scattering properties of a potential are expressible by the phaseshifts it produces, the inputs to such calculations are the phaseshifts. This method, the KKR, is described in books on band structure calculations and it was a substantial advance when it was shown that the CPA idea could be extended to KKR-type calculations thus opening these up for use with disordered alloys (Gyorffy and Stocks 1977). In all cases, however, the use of the single-site condition, equation (7.14), means that any effect on the electron states of clustering or other SRO is ignored.

An extensive application of the CPA–KKR method to the Cu–Ni system (Szotek *et al* 1984) shows that the disorder smears the Fermi surface in those alloys but not so much that it ceases to be reasonable to refer to the detailed topology of the Fermi surface such as the necks touching the zone boundary in Cu-rich alloys. This statement follows from the sharpness of the peaks in $\rho(k, E_F)$ when it is plotted against k as exemplified in figure 7.2. At the time of writing neither extensive computations nor comparisons with experiment are very common but §7.6 will show that experimental tests are possible and there is good reason to hope that the CPA provides a satisfactory first-

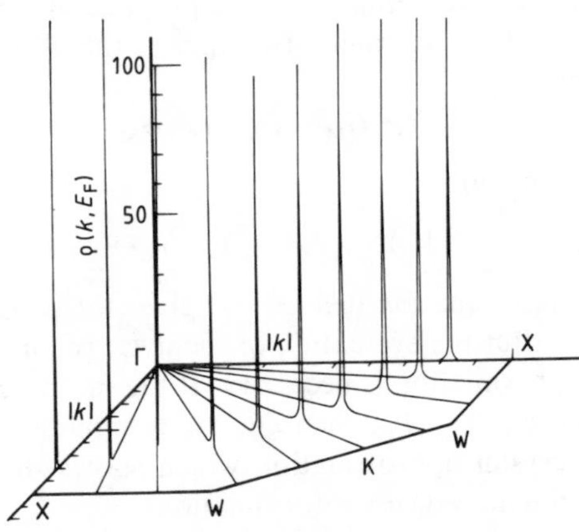

Figure 7.2 Spectral function in disordered solid $Ni_{0.13}Cu_{0.87}$ plotted against k for $E = E_F$. The peaks are sharp enough to define a Fermi surface with some precision. (See Szotek *et al* 1984, after Gordon *et al* 1981.)

principles explanation of electronic structures in disordered substitutional alloys.

7.3 Liquid metals

Even a simple liquid metal poses the substantial problem of structural or topological disorder (see §1.7). Equation (7.6) shows that if we want $g(E)$ via $\rho(\boldsymbol{k}, E)$ we must find $-\pi^{-1}\,\mathrm{Im}\langle G^+(\boldsymbol{k})\rangle$. So let G_0 refer to free particles and $V(\boldsymbol{r}) = \Sigma_m v_a(\boldsymbol{r} - \boldsymbol{R}_m)$ where $v_a(\boldsymbol{r} - \boldsymbol{R}_m)$ is the potential due to the atom at $\boldsymbol{R}_m$. Then if we form $\langle \boldsymbol{k}|G|\boldsymbol{k}\rangle$ from equation (7.13*b*) and insert $V(\boldsymbol{r})$ in the right-hand side we can imagine the resulting series being averaged term by term and summed to give $\langle G\rangle$.

To illustrate the problem consider the third term, namely, $\langle \boldsymbol{k}|G_0VG_0VG_0|\boldsymbol{k}\rangle$. The operators can be embedded in separate matrix elements by repeated use of the completeness relation

$$(2\pi)^{-3}\Omega \int \mathrm{d}\boldsymbol{k}\,|\boldsymbol{k}\rangle\langle \boldsymbol{k}| = 1.$$

Each G_0 matrix element is subject to equation (7.12) and turns into a factor of the form $(2\pi)^3(E - \hbar^2k_n^2/2m + \mathrm{i}\varepsilon)^{-1}$. The middle product VG_0V leads to

$$\frac{\Omega}{(2\pi)^3}\int \mathrm{d}\boldsymbol{k}_2 \frac{\langle \boldsymbol{k}|V|\boldsymbol{k}_2\rangle\langle \boldsymbol{k}_2|V|\boldsymbol{k}\rangle}{E - \hbar^2k_2^2/2m + \mathrm{i}\varepsilon}.$$

After factorising the matrix elements in the manner of §6.6, putting $\boldsymbol{k}_2 = \boldsymbol{k} + \boldsymbol{Q}$ and assuming v_a is an ordinary potential this integral becomes

$$\frac{n_0}{(2\pi)^3}\int \mathrm{d}\boldsymbol{Q} \frac{S(\boldsymbol{Q})|v_a(\boldsymbol{Q})|^2}{E - |\boldsymbol{k} + \boldsymbol{Q}|^2/2m + \mathrm{i}\varepsilon} \tag{7.17}$$

where n_0 is the number of atoms in the system—now taken to be unit volume—and $S(\boldsymbol{Q})$ is the configuration-averaged structure factor. In a liquid $S(\boldsymbol{Q}) \to S(Q)$.

This shows that the term with two V requires a knowledge of $S(Q)$. The term with one V can be made to vanish by a choice of the energy zero. Terms with three or more V need higher-order correlation functions of the liquid structure and as stated in Chapters 3 to 5 these are not available except through some hypothesis like the superposition approximation. This immediately illustrates the practical difficulty of evaluating the terms even if there were a method of summing the series. Actually the series cannot be summed but an approach can be made as follows.

Define an operator Σ by $\Sigma\, G = \langle VG\rangle$. This can be combined with the averaged equation (7.13*a*), namely $\langle G\rangle = G_0 + G_0\langle VG\rangle$ with the result:

$$\langle G\rangle = \frac{1}{E - H_0 - \Sigma}. \tag{7.18}$$

This is exact but no help without an expression for Σ. It is the main thrust of the papers by S F Edwards in 1961 and subsequent users of the method to provide a series expansion for Σ which is called the irreducible self-energy operator. The technique for doing this is very clearly set out by Ballentine (1975). The terms of the expression are successively more complex products of $G_0, v_a(Q)$ and liquid structure correlation functions.

After making a decision as to the nature—local or non-local, pseudo- or real—of $v_a(\boldsymbol{r})$, the easiest thing to do would be to approximate $\Sigma(\boldsymbol{k}, E)$ by its lowest non-trivial term which happens to be identical with the expression (7.17). The assumption is that the electrons are nearly free. $g(E)$ then comes from $-\pi^{-1}\,\mathrm{Im}\langle G^+\rangle$ in the limit $\varepsilon \to 0$. Ballentine showed that this approximation is too simple to give reliable results and an improvement is to replace the factor deriving from G_0 in the integrand of equation (7.17) by the corresponding one deriving from G itself in equation (7.18). Thus, instead of equation (7.17),

$$\sum(\boldsymbol{k}, E) = \frac{n_0}{(2\pi)^3}\int \mathrm{d}\boldsymbol{Q}\,\frac{S(\boldsymbol{Q})|v_a(\boldsymbol{Q})|^2}{E - \hbar^2|\boldsymbol{k}+\boldsymbol{Q}|^2/2m - \Sigma(\boldsymbol{k}+\boldsymbol{Q}, E)}. \tag{7.19}$$

This now has to be solved self-consistently for $\Sigma(\boldsymbol{k}, E)$ and is an approximation corresponding to the summation to infinity of a class of terms in the Σ-expansion. Figure 7.1 is the result of such a computation using a non-local pseudopotential for v_a; the details of this calculation are in Ballentine (1975).

Computations of this kind have been made for a number of simple metals and binary alloys using various pseudopotentials and model potentials. They correspond to weak electron–ion interactions. On the whole the $g(E)$ for simple liquid metals departs rather little from the free-electron shape except in the alkaline earths especially Sr and Ba. Calculations of $g(E)$ for simple liquid metals continue to evolve (Van Oosten and Geertsma 1985).

The preceding method is a developed and self-consistent perturbation theory in which electrons are only weakly affected by the potential. For stronger interactions it is natural to look for a theory based on the TBA. The latter is one of the starting points for the CPA for substitutional alloys so we might ask if there is something for liquid metals comparable with the CPA for solid alloys and based on the TBA.

The characteristic assumption of the TBA is that the one-electron wavefunction can be expressed as a sum of atomic orbitals, $\varphi_i(\boldsymbol{r}) \equiv \varphi(\boldsymbol{r} - \boldsymbol{R}_i)$, situated at the sites $\{i\}$. For simplicity assume one s-orbital per site but this is not essential. Expressing the TBA in Green functions terms, the assumption is

$$G(\boldsymbol{r}, \boldsymbol{r}') = \sum_{i,j} \varphi_i(\boldsymbol{r}) G_{ij} \varphi_j^*(\boldsymbol{r}') \tag{7.20}$$

where G_{ij} is a site representation of G for which an equation is needed. Since the coordinate representation of the unit operator is $\delta(\boldsymbol{r} - \boldsymbol{r}')$, equation (7.1) can be rewritten as

$$(z - H)G(\boldsymbol{r}, \boldsymbol{r}') = \delta(\boldsymbol{r} - \boldsymbol{r}') \tag{7.21}$$

where z is the complex energy $E \pm i\varepsilon$. The insertion of equation (7.20) into equation (7.21) and a little matrix algebra lead to an equation for G_{ij}, namely,

$$\sum_l (zS_{il} - H_{il})G_{lj} = \delta_{ij} \tag{7.22a}$$

where

$$S_{il} \equiv S(\boldsymbol{R}_{il}) = \int \varphi_i^*(\boldsymbol{r})\varphi_l(\boldsymbol{r})\,\mathrm{d}\boldsymbol{r} \tag{7.22b}$$

and

$$H_{il} \equiv \int \varphi_i^*(\boldsymbol{r})H\varphi_l(\boldsymbol{r})\,\mathrm{d}\boldsymbol{r}. \tag{7.22c}$$

S_{il} and H_{il} are, respectively, overlap and transfer integrals. If the φ's were orthonormal $S_{il} = \delta_{il}$ but this is not true in general. It is reasonable to assume that $H_{il} = H(\boldsymbol{R}_{il})$ and that $H_{ii} = \varepsilon_0 =$ constant but neither of these will be strictly valid because not all atoms in a liquid have the same environment.

In principle these equations enable $G(\boldsymbol{r}, \boldsymbol{r}')$ to be found, configuration averaged and its imaginary part used to derive $g(E)$. However the reader will not be surprised to find that G_{ij} has an expansion which needs to be averaged term by term and resummed and that this cannot be done, if only because the higher terms need high-order correlations of the liquid structure. Writing $\{H(\boldsymbol{R}_{ij}) - zS(\boldsymbol{R}_{ij})\} \equiv H'(\boldsymbol{R}_{ij})$, equation (7.22$a$) becomes

$$(z - \varepsilon_0)G_{ij} - \sum_{l \neq i} H'_{il}(\boldsymbol{R}_{il})G_{lj} = \delta_{ij}.$$

With $G_0 \equiv (z - \varepsilon_0)^{-1}$ iteration leads to

$$G_{ij} = G_0\delta_{ij} + G_0(1 - \delta_{ij})H'_{ij}G_0 + \sum_{\substack{l \neq i \\ l \neq j}} G_0H'_{il}G_0H'_{lj}G_0 + \cdots.$$

Since H' is $H'(\boldsymbol{R}_{ij})$, it is the configuration averaging of the successive terms that bring in quantities like $g^{(2)}(\boldsymbol{R}_1, \boldsymbol{R}_2)$, $g^{(3)}(\boldsymbol{R}_1, \boldsymbol{R}_2, \boldsymbol{R}_3)$ etc.

Although the preceding few equations are based on the TBA, it has been shown that formally analogous expansions can start with the KKR or muffin-tin concept of non-overlapping potentials among which free-electron waves are multiply scattered. Quantities expressing atomic Green functions and scattering phaseshifts in the muffin-tin method occupy similar positions in the mathematical structure to the G_0, G_{ij} and H' of the TBA. Similar approximations in the summations of the averaged series can therefore apply to both schemes. As it is probable that the TBA is most appropriate for covalently bonded systems and the muffin-tin method for metals, it has been argued that the schemes are complementary in this sense. But for either of them two kinds of approximation are needed:

(i) some selection of summable terms in the expansion of the Green function;

(ii) some way of expressing $g^{(N)}(\boldsymbol{R}_1, \ldots, \boldsymbol{R}_N)$ in terms of $g^{(2)}(\boldsymbol{R}_1, \boldsymbol{R}_2)$.

Authorities seem to be agreed that there is an approximate summation for $G(\boldsymbol{r}, \boldsymbol{r}')$ which corresponds to the CPA. It is tempting to try to express it descriptively by the considerations expressed in equation (7.14), i.e., by placing conditions on the scattering at a single site of waves propagating in an effective medium which, in a liquid, would be isotropic. There seems not to be a transparent way of doing this, nor indeed any substitute for an analysis of the Green function expansion term by term, accompanied by systematic decisions about which terms to leave in or take out. One decision concerns $g^{(N)}(\boldsymbol{R}_1, \ldots, \boldsymbol{R}_N)$ and this could be deemed to conform to the superposition approximation (§5.3). By making further decisions agree with those required for setting up the CPA on lattices—e.g., by forbidding multiple occupancy of sites and by excluding terms describing *repeated* scattering of an electron between two more sites—a single-site approximation for $G(\boldsymbol{r}, \boldsymbol{r}')$ results. It is called the effective-medium approximation (EMA), first introduced by Roth in 1974. For some detailed analysis of the options, of the correspondence between the CPA and EMA and of the diagrammatic methods for partial summation, Watabe (1977), Roth (1974), Yonezawa *et al* (1975) and many others cited therein can be consulted. Numerous other single-site approximations are known by their authors' initials (e.g. IY = Ishida–Yonezawa) or by acronyms (e.g. QCA = quasicrystalline approximation) and have been shown to involve less complete summations, or less satisfactory approximations for $g^{(N)}$, or non-analyticity. However, comparison of these theories is a very specialised matter and the reader will want to refer to the original papers.

In Chapter 5 the value of checking theories of liquids against exact results obtained by MC or MD methods was emphasised. Similarly one might hope to test the EMA against exact calculations. In the next section finite clusters will be referred to. Provisionally we note that, given sufficient computing power, $g(E)$ for a large but finite cluster of atoms can be calculated exactly. This was done for a DRP model of 315 hard spheres generated with Ichikawa's method (§2.5) by Fujiwara and Tanabe who imagined non-orthogonal hydrogen-like orbitals on each sphere and computed the density of states. With the same admittedly unrealistic model of a liquid metal, Aloisio *et al* (1981) used the TBA version of the EMA to calculate $g(E)$ and compared it with the exact $g(E)$—much to the credit of the EMA which gave rather good agreement and performed significantly better than other single-site theories. Together with pertinent theoretical results such as that the EMA reduces to the CPA for sites on a lattice (Roth 1974), this gives some confidence that the EMA could be used for real metals. Figure 7.3 gives an example for liquid Fe, though not many such calculations have been done so far.

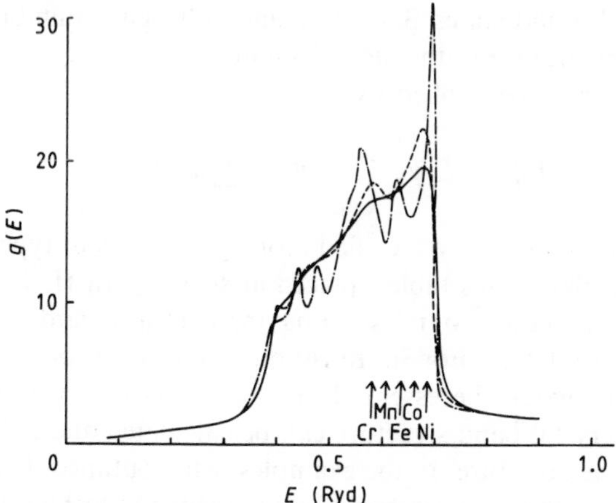

Figure 7.3 $g(E)$ calculated for liquid Fe by Asano and Yonezawa (1980) using the EMA. The arrows indicate E_F.

7.4 Clusters and the recursion method

Mention of liquid Fe raises the additional complication that in liquid or amorphous transition metals single bands from single orbitals in the TBA will be insufficient. Hybridisation between s- and d-like states will certainly cause complexity in the calculations and raise difficult questions such as: what kind of electron states contribute to the conductivity? To illustrate another method of dealing with disordered matter in general and transition metals in particular we refer to cluster calculations for liquid La and liquid and amorphous Fe (Ballentine 1982, Bose *et al* 1983).

Even clusters of a few atoms in crystalline order show electronic spectra with features recalling bulk crystals. With computing methods that can handle clusters of hundreds of atoms there is a good chance that bulk properties will be well approximated. Bulk liquids for example can be simulated by clusters which have densities and pair distributions equal to observed ones. To this extent the cluster approximates a configuration-averaged system though a true configuration average of the cluster is not taken.

A good cluster calculation of the 1980s contains a number of computing technicalities and subtle points, and only an outline will be given here. First, the cluster must be specified. A hard-sphere packing (§2.5) would be a possibility. But a more realistic approach would be to take an interatomic potential and use the MC method (§4.1), incorporating a Boltzmann factor, to realise the structure of lowest energy compatible with a chosen temperature. The potential itself might come from the physics of the solid, e.g., its elastic constants, or by choosing the one that reproduces the observed liquid $S(Q)$ through the HNC or other liquid theory (§5.9). To distinguish between the amorphous and liquid states Ballentine and colleagues suitably altered the density and temperature in the MC calculation.

The TBA may now be applied by

$$H\psi_n = E_n\psi_n \qquad \psi_n = \sum_{i,\alpha} a_{n,i}^{(\alpha)}|\varphi_i^{(\alpha)}\rangle \tag{7.23}$$

where $\{\varphi_i^{(\alpha)}\}$ are numerically specified atomic functions of type α, e.g. those of Herman and Skillman's tables, placed at sites $\{i\}$. In H, $V = \Sigma_i v_i$ where v_i is the atomic potential round site i having careful regard to the effect of exchange and correlation. For 3d transition metals the incorporation of both s and d orbitals would be essential. To settle the parameters of the TBA expansion the crystal band structure can be calculated and adjusted to that established in the literature. In the examples being outlined, band structures in the disordered states were derived by the recursion technique which is of general application. An introduction to it will therefore be given and the extended treatments in Heine (1980), Haydock (1980), Bullett (1980), Kelly (1980), Haydock and Nex (1985), Pettifor and Weaire (1984) and Ballentine and Kolář (1986) are strongly recommended.

From equations (7.4) and (7.9) we have

$$g(E, \boldsymbol{r}) = \sum_n |\psi_n(\boldsymbol{r})|^2\delta(E - E_n)$$

which, with equation (7.23), gives the following expression for the contribution to the density of states at energy E made by the orbital $\varphi_i^{(\alpha)}$:

$$g_i^{(\alpha)}(E) = \sum_n |a_{n,i}^{(\alpha)}|^2\delta(E - E_n). \tag{7.24a}$$

There is an implicit assumption here that the $\varphi_i^{(\alpha)}$ are orthonormal. While they can be normalised they may not be orthogonal and if this is not a negligible effect special correction procedures, ignored here, will be needed (see Heine (1980), Haydock (1980), Bullett (1980), Kelly (1980), Haydock and Nex (1985), Pettifor and Weaire (1984) and Ballentine and Kolář (1986)). Since $a_{n,i}^{(\alpha)} = \langle\varphi_i^{(\alpha)}|\psi_n\rangle$, we see that equation (7.24a) is a matrix element with $\varphi_i^{(\alpha)}$ of the operator $\rho(E)$ in equation (7.4). So we may also write, from equation (7.2b),

$$g_i^{(\alpha)}(E) = -\pi^{-1}\,\mathrm{Im}\langle\varphi_i^{(\alpha)}|G^+|\varphi_i^{(\alpha)}\rangle. \tag{7.24b}$$

So for obtaining the local orbital density of states we need the matrix element of the imaginary part of G^+ taken with $\varphi_i^{(\alpha)}$.

If $\{i\}$ is a Bravais lattice and only one orbital per site is involved, $g_i^{(\alpha)}(E)$ gives the density of states per atom. In a more complicated crystal $g(E)$ per unit cell would involve summing $g^{(\alpha)}(E)$ over all orbitals in the cell. In a pure disordered metal $\Sigma_\alpha g_i^{(\alpha)}(E)$ for one site could be calculated for several—not exactly equivalent—sites and averaged to find $g(E)$ per atom.

In the recursion method it turns out that the $\langle\varphi_i^{(\alpha)}|G^+|\varphi_i^{(\alpha)}\rangle$ can be expressed as a continued fraction and that each continuation corresponds to including the influence on $g_i^{(\alpha)}(E)$ of neighbours successively further away from site i. These influences progressively decrease and the error in the result comes from the two separate kinds of cut-off. One is the finite size of the cluster in which the site i is embedded; the other is the finite number of continuations to which the fraction is actually calculated. Experts in the field know how to optimise these choices (Heine 1980, Haydock 1980, Bullett 1980, Kelly 1980, Haydock and Nex 1985, Pettifor and Weaire 1984, Ballentine and Kolář 1986). This being done, the recursion method is efficient in computation and versatile in that it can be used for Bloch and disordered solids, surface states and for phonons and excitations other than electronic spectra.

To see how the continued fraction arises let us invent an orthonormal set of new basis states $|u_n\rangle$, $n = 0, 1, 2, \ldots$ where $|u_0\rangle$ is the orbital $|\varphi_i^{(\alpha)}\rangle$ sited near the centre of the cluster. $H|u_0\rangle$ is a function which could be expressed as a combination of $|\varphi_j^{(\alpha)}\rangle$ with the biggest contribution coming from sites j nearest to i. Let $a_0 \equiv \langle u_0|H|u_0\rangle$; then $(H - a_0)|u_0\rangle$, which is readily seen to be orthogonal to $|u_0\rangle$, can be normalised by a factor b_1^{-1} and then called $|u_1\rangle$. Thus $b_1|u_1\rangle = (H - a_0)|u_0\rangle$. Now define $a_1 \equiv \langle u_1|H|u_1\rangle$ and consider $b_2^{-1}(H|u_1\rangle - b_1|u_0\rangle - a_1|u_1\rangle) \equiv |u_2\rangle$. $|u_2\rangle$ is normalised by b_2^{-1} and is orthogonal to $|u_1\rangle$ and $|u_0\rangle$. Thus $b_2|u_2\rangle = (H - a_1)|u_1\rangle - b_1|u_0\rangle$. Continuing this process we see that

$$b_{n+1}|u_{n+1}\rangle = (H - a_n)|u_n\rangle - b_n|u_{n-1}\rangle \tag{7.25a}$$

and that

$$\begin{aligned}\langle u_m|H|u_n\rangle \equiv H_{mn} &= b_n && \text{if } n = m+1 \text{ or } m = n+1\\ &= a_n && \text{if } m = n\\ &= \text{zero} && \text{otherwise.}\end{aligned} \tag{7.25b}$$

Each time a new $|u_n\rangle$ is defined and made orthogonal to the preceding one a new layer of neighbours is involved.

Equation (7.25*b*) says that if H is represented by a matrix of elements, H_{mn}, it is tridiagonal, i.e. it consists of zeros except down the leading diagonal and one on either side of it. Now the object required is $\langle u_0|G^+|u_0\rangle$ where

G^+ may be written $(z-H)^{-1}$ as before, with $z = E + i\varepsilon$. The matrix $\mathbf{z} - \mathbf{H}$ is

$$\mathbf{z} - \mathbf{H} = \begin{pmatrix} z - a_0 & -b_1 & 0 & \\ -b_1 & z - a_1 & b_2 & \text{etc} \\ 0 & -b_2 & z - a_2 & \\ & \text{etc} & & \end{pmatrix}$$

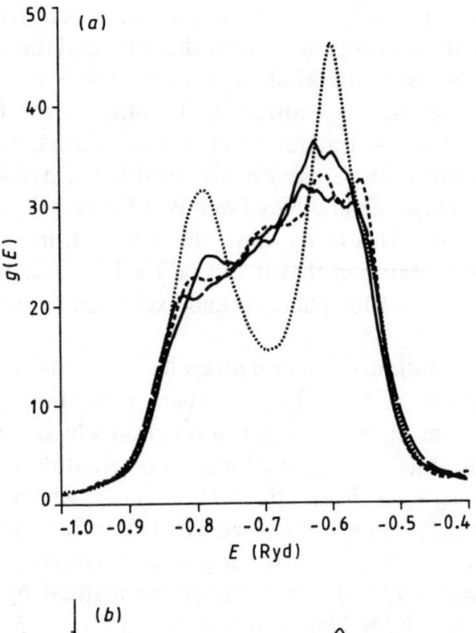

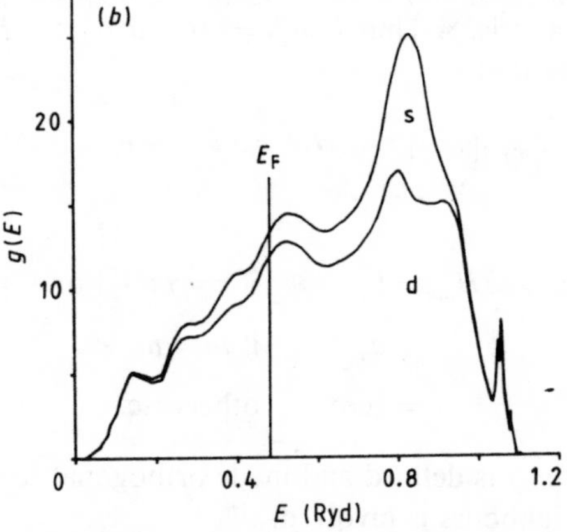

Figure 7.4 (*a*) $g(E)$ for crystalline (BCC, dotted curve; FCC, ---), amorphous (full curve) and liquid (— —) Fe (from Bose *et al* 1983). (*b*) $g(E)$ for liquid La (from Ballentine 1982).

and the first element, $\langle u_0|(z-H)^{-1}|u_0\rangle$, of the inverse follows from a short exercise in determinant algebra as the continued fraction

$$\langle u_0|G^+|u_0\rangle = \cfrac{1}{E-a_0-\cfrac{b_1^2}{E-a_1-\cfrac{b_2^2}{E-a_2\ldots}}} \tag{7.26}$$

The local density of states can therefore be computed from the a_n and b_n; the considerable number of technical points involved are extensively reviewed in Heine (1980), Haydock (1980), Bullett (1980), Kelly (1980), Haydock and Nex (1985), Pettifor and Weaire (1984) and Ballentine and Kolář (1986).

Pursuing this method for the Fe clusters, Bose *et al* (1983) obtained the results in figure 7.4(*a*). It is a feature of the method that $g(E)$ for s- and d-like orbitals can be studied separately as is shown for liquid La in figure 7.4(*b*). The similarity of $g(E)$ in figures 7.3 and 7.4(*a*) is interesting in view of their very different provenances.

An alternative to the recursion technique is the equation of motion method. This consists in integrating the time-dependent Schrödinger equation numerically, starting with an initial state, $\psi(0)$, which can be chosen to suit the problem to be solved. From $\psi(t)$ the Green function is computed and from its imaginary part the density of states follows. If the initial state was the plane wave with vector $\boldsymbol{k}$, then the result would be $-\pi^{-1}\,\mathrm{Im}\,G(E,\boldsymbol{k})$ which is the spectral function (equation (7.6)). How to implement this, and its relation to the recursion method, are compactly expounded by MacKinnon in Pettifor and Weaire (1984).

7.5 Other methods and miscellaneous points

It should be apparent that finding $g(E)$ in disordered metals is a problem which, if not entirely intractable, is still not fully solved. The directness and power of the Green function as compared with the Schrödinger wavefunction is shown by its appearance in different guises in the various methods. There are of course other theories not described here and, because approximations are always involved, there is not at the time of writing a consensus about all matters. Take for example the use of E–$\boldsymbol{k}$ relations. Since $\boldsymbol{k}$ is not a quantum number for the eigenfunctions of H, a k-value can be assigned to a state of energy E only through the peak of $\rho(k, E)$. For simplicity let us consider nearly free electrons subject to scattering by a disordered potential. From equation (7.18) we may write

$$\langle G(\boldsymbol{k}, E)\rangle = \left(E - \frac{\hbar^2 k^2}{2m} - \Sigma(k, E)\right)^{-1} \tag{7.27}$$

where Σ, called the self-energy, will normally require approximation from some suitably averaged expansion. Suppose

$$\Sigma(\boldsymbol{k}, E) = A(\boldsymbol{k}, E) - \mathrm{i}\,\Gamma(\boldsymbol{k}, E) \tag{7.28}$$

then from §7.1, a dispersion relation between E and $\boldsymbol{k}$ could be defined by solutions, $E(\boldsymbol{k})$, of

$$E - \frac{\hbar^2 k^2}{2m} + A(\boldsymbol{k}, E) = 0 \tag{7.29}$$

and also

$$\rho(\boldsymbol{k}, E) = \frac{1}{\pi} \frac{\Gamma}{(E - \hbar^2 k^2/2m + A)^2 + \Gamma^2}. \tag{7.30}$$

Σ therefore shifts the E–$\boldsymbol{k}$ relation from the free-electron parabola and the Fermi sphere, defined by $E(k_F) = E_F$ in isotropic matter, will not have the free-electron radius $(3\pi^2 n)^{1/3}$. It is therefore not obvious which Fermi radius is actually measured by certain techniques, e.g. the de Haas–van Alphen effect, in disordered alloys. For a given solution $E(k)$ of equation (7.29), $\rho(k, E(k))$ is smaller for greater values of Γ. One way of representing this situation is by electrons of effective mass m^* propagating with wavevector $\boldsymbol{k}$ but with a lifetime limited by scattering and inversely proportional to Γ. Defining m^* by $\hbar k/2m^* = \hbar^{-1}\, \partial E(k)/\partial k$ we find

$$\frac{m}{m^*} = \left(1 + \frac{m}{\hbar k}\frac{\partial A}{\partial k}\right)\left(1 - \frac{\partial A}{\partial E}\right)^{-1}. \tag{7.31}$$

For strong interactions, e.g. of d electrons in transition metals, this picture may have less immediate appeal. In figure 7.5 two E–k relations for transition metals are shown qualitatively. The curves in figure 7.5(*a*) are from Bose *et al* (1983) and show incidentally that the recursion method can be used with $|u_0\rangle$ defined as a running wave instead of an atomic orbital—a procedure necessary if the density of states for a particular $\boldsymbol{k}$ is required. The curve in figure 7.5(*b*) is from Morgan and Weir (1983) in which Morgan and collaborators use the equation of motion method for the Green function in the TBA with a hard-sphere distribution of atoms. The latter authors attach significance to the curve where $\partial E/\partial k$ is negative because this suggests a negative group velocity for electrons and a possible explanation of somewhat mysterious positive Hall coefficients. The former authors wonder whether, in this region of hybridisation between s and d states, an E–k relation is definable at all. This very interesting point requires resolution (see also §12.5).

From time to time attempts have been made to combine cluster calculations with more general theories. If the Green functions in a small cluster of muffin-tin potentials were found exactly they might be given boundary conditions which make them consistent with the surrounding matter treated as

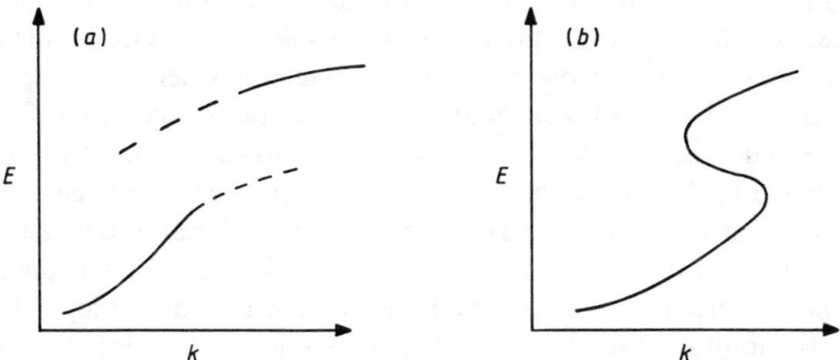

Figure 7.5 Possible E–k relations in disordered transition metals shown schematically for d-like states (see text) (from Bose *et al* 1983, Morgan and Weir 1983).

a homogeneous effective medium. This is an attempt to combine some of the virtues of the EMA with the ability of cluster calculations to deal with short-range order (John and Keller 1980). A formal expression for $\rho(k, E)$, which incorporates the idea that the very significant effect of local SRO in the averaged electronic structure can be expressed by a few local parameters, has been advanced by Beeby (1983) and Beeby and Hayes (1985).

The preceding sections have been concerned in the main with disordered metallic systems. The problem of amorphous semiconductors with CRN structures is not easier and will be outlined in Chapter 11.

7.6 Some relevant experiments

Experimental measurements of $g(E)$ or $\rho(k, E)$ are not very direct, common or precise. Consequently decisive confrontations of theory with observation are still much to be desired. Deferring semiconductors until Chapter 11 we will concentrate on metals.

Clues to the density of states come from experiments that reflect $g(E_F)$. For example, although electronic specific heat is not easy to measure in liquid metals, the Knight shift, also proportional to $g(E_F)$, is observed to change rather little on melting as a rule and this suggests that the values of $g(E_F)$ in hot metals and their melts are nearly equal. The infrared optical constants are known to be consistent with the view that $g(E)$ not far below E_F resembles the free-electron value in simple liquid metals. But what is really desirable is a spectroscopy that will explore the whole range of energy or momentum described by $\rho(k, E)$. The two possibilities we illustrate here are ultraviolet photoemission spectroscopy (UPS) and the angular correlation of positron annihilation radiation (ACAR). This is not to suggest that other spectroscopies—Auger, x-ray photoemission, soft x-ray, visible light ellipsometry, electron energy loss—cannot also contribute within their ranges of application.

Let E_i, E_f be the energies of the initial and final one-electron states in photoexcitation. E_f will be high enough to take the electron outside the metal, i.e. $E_f - E_F >$ the work function. $E_i + \hbar\omega = E_f$ where $\hbar\omega$ is the photon energy. $|P_{if}|^2$ is the transition probability per unit time which can be found by Fermi's Golden Rule in the way usual for optical excitations. Not all electrons excited to E_f will escape for geometrical and scattering reasons; conversely some electrons will emerge with less than E_f because they have been incident or target particles in inelastic collisions. Suppose $T(E)$ is the probability that an electron with E actually escapes with that energy. Then the energy distribution of electrons arriving at an energy-resolving detector can be written intuitively as

$$N(E, \hbar\omega) \propto \sum_{i,f} |P_{if}|^2 T(E)\delta(E_i + \hbar\omega - E_f)\delta(E - E_f). \qquad (7.32a)$$

Here the first δ-function conserves energy and the second selects E_f to be the energy of the detector setting. The Fermi energy is commonly selected as the energy zero. In crystals P_{if} vanishes unless $\boldsymbol{k}_i = \boldsymbol{k}_f$ for interband transitions in the reduced-zone scheme. In disordered matter the selection rule appears not to hold because $\boldsymbol{k}$ is not a quantum number for the states. If we now assume $|P_{if}|^2 T(E)$ is at most a slowly varying function of E, equation (7.32a) reduces to

$$N(E, \hbar\omega) \propto g(E - \hbar\omega)g(E). \qquad (7.32b)$$

A density of states inferred from such an equation is called an optical density of states (ODS) and contains an unknown distortion due to $|P_{if}|^2 T(E)$. Peaks in $N(E, \hbar\omega)$ due to those in $g(E - \hbar\omega)$ can be identified because they will move by $\Delta\hbar\omega$ if ω is altered while peaks due to $g(E)$ will not.

The assumptions in equation (7.32) imply that it is best used to interpet gross features of $N(E, \hbar\omega)$ in terms of $g(E)$ rather than as an accurate measurement of electronic structure. An example is taken from the work of Wotherspoon *et al* (1979). Their inference concerning the ODS of liquid Bi is shown in figure 7.6. While it shows 6s and 6p bands separated by a minimum, in qualitative agreement with independent x-ray photoemission, the correspondence with $g(E)$ calculated for liquid Bi (see §7.3) is not good. This comparison emphasises the somewhat tantalising distance between theory and observation. The former is based upon approximate calculation and inadequate structural information while the latter requires considerable data processing to remove the background of low-energy electrons. It also involves the assumptions mentioned above and a further one, probably very reasonable, that the light penetrates far enough to excite photoemission characteristic of the bulk.

It is well known that a positron–electron pair with their centre of mass at rest will undergo 2γ annihilation in which the photons leave in exactly opposite directions. If the photons make an angle $(\pi - \theta)$ then θ is a measure of

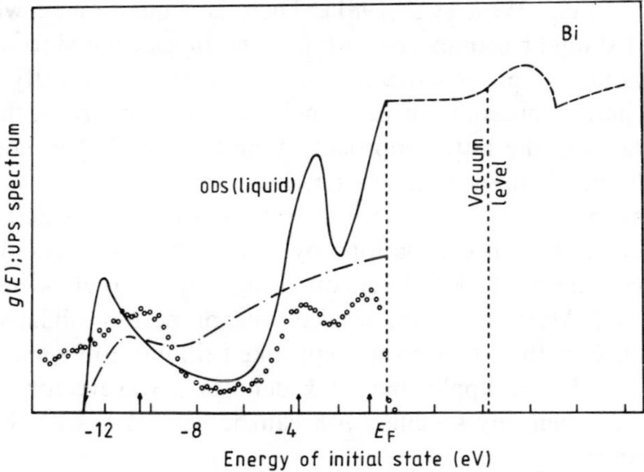

Figure 7.6 UPS results for liquid Bi. Full curve, $g(E)$ from UPS; chain curve, calculated $g(E)$ from Green function method of §7.3 (see figure 7.1). Circles, $g(E)$ from x-ray photoemission. The arrows are the peak positions of $g(E)$ in crystalline Bi. (From Wotherspoon *et al* 1979.)

the momentum of the annihilating pair. For annihilation after the thermalisation of positrons in condensed matter, θ is a few milliradians. If the positron is stationary at the moment of annihilation the distribution of θ measures the distribution of electron momentum and this is a reasonable first approximation because $p^- \gg p^+$ as a rule. But a thorough analysis of the annihilation process, at least in the one-electron approximation, gives a general expression which takes account of the wavefunctions of both particles. The following formula stems from the theory of charged particle interactions with the radiation field and gives the distribution of $\boldsymbol{p}$ of the photon pair coming from annihilation in condensed matter. The superscripts + and − distinguish between positron and electron properties, f is the Fermi function and either a bar or $\langle\ \rangle$ show that an appropriate configuration average has been taken; Ω is the sample volume:

$$\bar{\rho}_{2\gamma}(\boldsymbol{p}) = \Omega^{-1}\int \mathrm{d}\boldsymbol{r} \int \mathrm{d}\boldsymbol{r}' \exp[-\mathrm{i}\boldsymbol{p}\cdot(\boldsymbol{r}-\boldsymbol{r}')/\hbar] \int \frac{\mathrm{d}E^-}{\pi} f(E^-) \int \frac{\mathrm{d}E^+}{\pi} f(E^+)$$
$$\times \langle \operatorname{Im} G(\boldsymbol{r}, \boldsymbol{r}', E^-) \operatorname{Im} G(\boldsymbol{r}, \boldsymbol{r}', E^+)\rangle. \qquad (7.33)$$

The imaginary parts of the Green functions are $\rho(\boldsymbol{r}, \boldsymbol{r}', E)$ for the particles (see §7.1) and in a particular case of disordered matter would have to be calculated by methods such as those in the preceding sections. A positron will be repelled by ions and concentrated in the interstices, and the potential used for it will have to express this.

Measuring $\rho_{2\gamma}(\boldsymbol{p})$ for all three components of $\boldsymbol{p}$ is possible in principle but the best equipments in the 1980s integrate over one component and provide

$\rho_{2\gamma}(p_x, p_y)$ (see, e.g. West *et al* 1981). There is a question of whether the observations should be compared with $\int \rho_{2\gamma}(\boldsymbol{p})\, \mathrm{d}p_z$ calculated from equation (7.33) or should be processed into another more conveniently computed quantity. Where disordered substitutional alloys are concerned, the underlying lattice favours the latter approach. Equation (7.33) has not yet been applied to liquid or amorphous matter.

At the time of writing extensive preparations are being made to compare experimental ACAR with calculations by CPA on binary alloys. The Green functions derivable by the KKR version of the CPA (§7.2) can be inserted into equation (7.33). After averaging, the electronic properties will have periodicities determined by the lattice so concepts like Brillouin zones and reciprocal lattice vectors, $\boldsymbol{K}$, are applicable. If $\boldsymbol{k}$ denotes a wavevector in the first Brillouin zone, then any specified $\boldsymbol{p}/\hbar$ can be written $\boldsymbol{k} + \boldsymbol{K}$. Suppose we define a quantity

$$\bar{\rho}^{\mathrm{ep}}(\boldsymbol{k}) \equiv \sum_{K} \bar{\rho}_{2\gamma}(\boldsymbol{k} + \boldsymbol{K}).$$

This is within the first zone and it, or more correctly its integral over the z-component, could be computed from the observed two-dimensional momentum distribution. $\bar{\rho}^{\mathrm{ep}}(\boldsymbol{k})$ can also be computed theoretically, via equation (7.33), from the KKR–CPA. This is a somewhat formidable calculation which can be pursued in Szotek and Temmerman (1986) and other references cited therein. Theory and experiment can thus be confronted. For pure crystals the corresponding methods have been widely used and $\rho^{\mathrm{ep}}(\boldsymbol{k})$—with the configuration average now superfluous—indicates the shape of the Fermi surface because $\rho_{2\gamma}(\boldsymbol{p})$ contains information as to which $\boldsymbol{k}$'s are occupied (Kaiser *et al* 1986). Disorder in alloys distorts the Bloch functions and smears the Fermi surface but in principle, especially with the increasing resolving power in momentum of modern equipment, these effects could be tested through $\bar{\rho}^{\mathrm{ep}}(\boldsymbol{k})$. Positron annihilation is a particularly useful technique for such studies because the scattering resulting from disorder vitiates other powerful methods of studying Fermi surfaces such as the de Haas–van Alphen effect.

The UPS experiment mentioned above integrates over angles of photon incidence and electron emission. It is possible to define both angles precisely in an experiment thus performing angle-resolved UPS (ARUPS). This measures both the energy of the photoelectron and its component of momentum parallel to the surface. In experiments on well prepared surfaces of single crystals it becomes possible to explore E–$\boldsymbol{k}$ relations. In order to exploit this for disordered alloys it is necessary to perform for the process of photoemission a KKR–CPA calculation comparable with the one just referred to for ACAR. Matrix elements for both photoexcitation and escape can be included. An extensive application of this to Cu–Ni has been made by Allen *et al* (1983) which confirms from ARUPS observations that, in spite of disorder, a treatment of the alloy using wavevectors, Fermi surface and Brillouin zones is still valuable for describing the configuration-averaged system. The result amounts to quantitative agreement between experimental ARUPS and theoretical KKR–CPA. There are no comparably satisfying results for topologically disordered metals. How matters will turn out for crystalline alloys with stronger scattering than Cu–Ni remains to be seen, but the effects of order–disorder transformations in Cu_3Au have been clearly seen (Jordan *et al* 1985).

7.7 Electron transport

In this section are introduced some of the approaches to electron transport which transcend the elementary method in §6.12. They are necessary for treating alloys, amorphous semiconductors, transition metals, expanded liquid metals and any system where the electron–ion interaction cannot be treated as a mild perturbation of free electrons. Many of the formulae are elegant and general but difficult to apply numerically to real disordered matter.

Transport is limited by scattering which is fully expressed for the whole system by the T-matrix of equation (7.15*a*). It is helpful to expand T in terms of the exact scattering amplitude, or t-matrix, for a single scatterer. By analogy with equation (7.15*b*) this may be written in operator form

$$t_i = v_i + v_i G_0 v_i + v_i G_0 v_i G_0 v_i + \cdots . \tag{7.34}$$

It is shown in the Appendix, or more thoroughly in Rodberg and Thaler (1967), that

$$T = \sum_i t_i + \sum_{i \neq j} t_i G_0 t_j + \sum_{\substack{i \neq j \\ j \neq k}} t_i G_0 t_j G_0 t_k + \cdots \tag{7.35}$$

where the inequalities simply prevent *adjacent* indices from being equal. This series classifies the multiple scattering into events at a single site, then events involving one scattering at each of two sites, then three etc; what is ruled out is two successive scatterings at the same site. If T were approximated by its first term then no multiple scatterings occur and wavelets from single scatterings would interfere, as in diffraction theory. In other words, so far as electrical resistivity is concerned, we should have the 'extended Ziman theory' of §6.16. If, further, t_i is approximated by v_i then we have the Born approximation for single scattering and the original NFE model (equation (6.57)). This shows clearly that the problem in this section is to incorporate multiple scattering into transport theory.

Let us first recall some phenomenological relations. j_x is the x-component of electric current density and is set up by driving forces such as electric fields and temperature gradients (thermoelectric effect). The same is true of the thermal energy flux, Q_x. With suitably defined driving forces and with forces and fluxes in the same direction, we have the Onsager equations (Ziman 1960):

$$j_x = L^{11}\frac{E_x}{T} + L^{12}\frac{\mathrm{d}}{\mathrm{d}x}\left(\frac{1}{T}\right) \tag{7.36a}$$

$$Q_x = L^{21}\frac{E_x}{T} + L^{22}\frac{\mathrm{d}}{\mathrm{d}x}\left(\frac{1}{T}\right) \tag{7.36b}$$

$$L^{12} = L^{21}. \tag{7.36c}$$

Comparing these with the normal definitions of electrical and thermal conductivity and thermoelectric power we find

$$\sigma = L^{11}/T \qquad \alpha = L^{12}/TL^{11} \qquad \lambda = [L^{22} - (L^{12})^2/L^{11}]/T^2.$$

The complication of magnetic fields is assumed absent. In anisotropic matter

$$j_\mu = L^{11}_{\mu\nu}\frac{E_\nu}{T} + L^{12}_{\mu\nu}\nabla_\nu\left(\frac{1}{T}\right) \tag{7.37a}$$

$$Q_x = L^{21}_{\mu\nu}\frac{E_\nu}{T} + L^{22}_{\mu\nu}\nabla_\nu\left(\frac{1}{T}\right) \tag{7.37b}$$

$$L^{21}_{\mu\nu} = L^{12}_{\nu\mu} \tag{7.37c}$$

where the repetition of ν implies summation over the component ν and thus $\sigma_{\mu\nu} = L^{11}_{\mu\nu}/T$.

Another valuable equation of great generality is the Einstein relation

$$\sigma_{\mu\nu} = e^2 D_{\mu\nu}\left(\frac{\mathrm{d}n}{\mathrm{d}\bar{\mu}}\right)_T \tag{7.38}$$

where $D_{\mu\nu}$ is the diffusion coefficient of particles with charge e, chemical potential $\bar{\mu}$ and number density n (Butcher 1972).

The phenomenological relations make it clear that if a microscopic theory of the $L^{\alpha\beta}_{\mu\nu}$ were available the electron transport properties would follow. Very general expressions for the L do exist and are frequently called Kubo equations. Let us concentrate on $L^{11}_{\mu\nu}$ or $\sigma_{\mu\nu}$. To obtain Kubo's formula for $\sigma_{\mu\nu}$ is an exercise in linear-response theory (Kubo 1965). Here we shall simply pose the problem and quote the result.

A system with Hamiltonian H_0 is perturbed by an influence expressed by $H'(t) = -\hat{A}F(t)$ where $\hat{A}$ is an operator characterising the influence and $F(t)$ gives its magnitude and time variation. For example, if the influence were an electric field acting on charged particles at positions $\{\boldsymbol{r}_i\}$, then $H' = \Sigma_i e\boldsymbol{r}_i \cdot \boldsymbol{E}(t)$ with $-\hat{A} \to \hat{X} \equiv \Sigma_i e\boldsymbol{r}_i$ and $F(t) = \boldsymbol{E}(t)$. Suppose $\hat{B}$ is an operator characterising an observable quantity B and let $\langle \Delta B(t) \rangle$ be the average deviation caused by the perturbation from the equilibrium value of B, e.g., B could be an electric current density for which $\hat{\boldsymbol{j}} = \Sigma_i e\hat{\boldsymbol{v}}_i$ (summed over the charges in unit volume) and $\hat{\boldsymbol{v}}_i \equiv \hat{\boldsymbol{p}}_i/m$.

The problem in linear-response theory is to obtain an expression for $\langle \Delta B(t) \rangle$ on the assumption that the effect is weak enough to be linear in $F(t)$. It is not required that the interactions within the system itself—e.g. the electron–ion interaction contained in H_0—are weak. $F(t)$ may be supposed to increase in strength from zero at $t = -\infty$ when the system was in equilibrium until some time $t > 0$. The assumptions of causality and linear response lead to

$$\langle \Delta B(t) \rangle = \int_{-\infty}^{t} \varphi_{\mathrm{BA}}(t, t')F(t')\, \mathrm{d}t' \tag{7.39}$$

where φ_{BA} is independent of F. The hypothesis of weak linear response implies that the effect of a pulse of force τ seconds after its application depends only on τ not on what the system was like at the time of the pulse; therefore $\varphi(t, t') \to \varphi(t - t') = \varphi(\tau)$. Putting $F(t) = F_0 \exp(\mathrm{i}\omega t)$, and defining $\chi_{\mathrm{BA}}(\omega)$ by

$$\langle \Delta B(t) \rangle = \chi_{\mathrm{BA}}(\omega) F_0 \exp(\mathrm{i}\omega t)$$

we find

$$\chi_{\mathrm{BA}}(\omega) = \int_0^{\infty} \exp(-\mathrm{i}\omega\tau)\varphi_{\mathrm{BA}}(\tau)\, \mathrm{d}\tau. \tag{7.40}$$

φ is called a response function and χ_{BA} an admittance or susceptibility.

The general result for φ_{BA} can be expressed in various ways of which one is

$$\varphi_{\mathrm{BA}}(t) = (\mathrm{i}\hbar)^{-1}\langle [\hat{A}, \exp(\mathrm{i}H_0 t/\hbar)\hat{B}\exp(-\mathrm{i}H_0 t/\hbar)] \rangle. \tag{7.41}$$

The admittance could refer to electrical, magnetic or optical response according to the choices of $\hat{A}$ and $\hat{B}$. The statistical averaging brackets signify that the trace of $\hat{\rho}_0$ multiplied by the commutator has to be taken, where $\hat{\rho}_0$ is the density matrix of the unperturbed system.

In the special case that B is a current density and F an electric field, $\chi = \sigma$ and the linearity of response expresses Ohm's law. Kubo's result is

$$\sigma_{\mu\nu}(\omega) = L^{11}_{\mu\nu}/T = \frac{1 - \exp(-\beta\omega)}{2\omega} \int_{-\infty}^{+\infty} \langle \hat{j}_\mu(t)\hat{j}_\nu(0) \rangle \exp(-\mathrm{i}\omega t)\, \mathrm{d}t \tag{7.42}$$

where

$$\hat{j}_\mu(t) \equiv \exp(-\mathrm{i}H_0 t/\hbar)\hat{j}_\mu \exp(\mathrm{i}H_0 t/\hbar).$$

This very general result applies to all systems including disordered ones. Luttinger's discussion (Luttinger 1964) reestablishes both this formula and corresponding formulae for all the Onsager coefficients, $L^{\alpha\beta}_{\mu\nu}$. Such equations appear in numerous formally different, but equivalent, guises. They are not easy to apply to particular cases and it often helps to reduce the generality. For example, suppose we are concerned with an isotropic disordered material with the usual many-centred potential and with electrons described by one-electron wavefunctions. Then all the general expressions for $L_{\mu\nu}^{\alpha\beta}$ can be reduced to formulae of the type often designated Kubo–Greenwood (Greenwood 1958). This reduction is set out by Hindley (1970) and for conductivity results in

$$\mathrm{Re}\,\sigma(\omega) = \frac{2\pi e^2}{3\Omega\omega}\left\langle \sum_{\substack{m,n \\ m \neq n}} |v_{mn}|^2 (f(E_n) - f(E_n + \hbar\omega))\delta(E_m - E_n - \hbar\omega) \right\rangle \quad (7.43a)$$

where v_{mn} is a matrix element with eigenfunctions of H_0 and f is the Fermi function. Taking $\omega \to 0$

$$\sigma = -\frac{2\pi e^2\hbar}{3\Omega}\int \mathrm{d}E \frac{\mathrm{d}f}{\mathrm{d}E}\sum_{m,n}\langle |v_{mn}|^2 \delta(E - E_n)\delta(E - E_m)\rangle. \quad (7.43b)$$

This is still quite general but can be deduced directly by relatively elementary means. In fact if Fermi's Golden Rule is used to calculate P, the net rate of absorption of energy by all the electrons when a perturbation $exE \cos \omega t$ is applied, and if P is then equated to its macroscopic value $\Omega E_0^2 \sigma(\omega)/2$, equation (7.43) follows. From equation (7.43b) σ is often written in the form

$$\sigma = -\int \mathrm{d}E \frac{\mathrm{d}f}{\mathrm{d}E}\sigma(E). \quad (7.43c)$$

In equation (7.43), as in all the linear-response formulae, the transport coefficient is expressed in terms of properties of the unperturbed system, in this case matrix elements and electron energies. Since the latter are also described indirectly by the Green functions it would not be surprising to find expressions for σ in terms of Green functions. Rickayzen (1980) introduces Green functions in a general way into linear-response theory, but here we simply manipulate equation (7.43b) in its anisotropic form

$$\sigma_{\mu\nu} = -\frac{2\pi e^2\hbar}{\Omega}\left\langle \sum_{m,n} \frac{\partial f}{\partial E_m} \langle m|v_\mu|n\rangle\langle n|v_\nu|m\rangle \delta(E_n - E_m) \right\rangle. \quad (7.43d)$$

The expression to be averaged can be rewritten

$$\sum_{m,n}\left(\frac{\partial f}{\partial E}\right)_{E_m} \langle m|v_\mu \delta(E_n - E_m)|n\rangle\langle n|v_\nu|m\rangle$$

or

$$\int \mathrm{d}E \frac{\mathrm{d}f}{\mathrm{d}E}\sum_m \left\langle m \middle| v_\mu \sum_n \delta(E - E_n) \middle| n \right\rangle \langle n|v_\nu \delta(E - E_m)|m\rangle.$$

Making use of the completeness relation and the fact that $|m\rangle$, $|n\rangle$ and $|i\rangle$ are all orthonormal eigenstates of H_0 this becomes

$$\int \mathrm{d}E \frac{\mathrm{d}f}{\mathrm{d}E}\sum_{i,m,n} \langle m|i\rangle\langle i|v_\mu \delta(E - E_n)|n\rangle\langle n|v_\nu \delta(E - E_m)|m\rangle$$

and thus

$$\sigma_{\mu\nu} = -\frac{2\pi e^2\hbar}{\Omega}\int \mathrm{d}E\,\frac{\mathrm{d}f}{\mathrm{d}E}\,\mathrm{Tr}\langle v_\mu\delta(E-H_0)v_\nu\delta(E-H_0)\rangle$$

where we have rewritten Σ_m as the trace of the corresponding operator and restored the averaging brackets.

Now from §7.1,

$$\delta(E-H_0) = \rho(E) = -(2\pi\mathrm{i})^{-1}(G^+(E) - G^-(E))$$

whence finally

$$\sigma_{\mu\nu} = \frac{e^2\hbar}{\pi\Omega}\int \mathrm{d}E\,\frac{\mathrm{d}f}{\mathrm{d}E}\,\mathrm{Re}\,\mathrm{Tr}\langle v_\mu(G^+v_\nu G^+ - G^+v_\nu G^-)\rangle. \qquad (7.43e)$$

This formula in various versions has been the starting point of many discussions (Edwards 1958, Neal 1970) and shows that quantities of the form $\langle v_\mu G(E+\mathrm{i}\varepsilon)v_\nu G(E\pm\mathrm{i}\varepsilon)\rangle$ have to be studied. It follows that a formula for $\sigma_{\mu\nu}$ could in principle come from any theory such as the CPA or EMA that has an approximation for the Green functions. In practice this is difficult but it is being done in the EMA although some authorities (Itoh and Watabe 1984, Itoh *et al* 1981) remark that 'all the physics is hidden behind mathematical integral equations' and others (Roth and Singh 1982) more succinctly refer to their calculations as 'horrendous'. By simplifications such as putting the average of a product of Green functions equal to the product of their averages the theory can be made more tractable if less reliable. In this way σ for liquid transition metals was calculated (Asano and Yonezawa 1980) and, reasonable agreement was obtained with the observations.

A much simpler use of the Kubo–Greenwood formula was made by Hindley (1970) in the random-phase model (RPM). In its simplest form this is a primitive cubic lattice in which electron eigenfunctions have the form $\psi_m = \Sigma_i\, a_{mi}\varphi_i$ where φ_i are atomic wavefunctions at the sites $\{i\}$ and the a_{mi} are complex numbers with random phases and amplitudes that vary irregularly from site to site. This is similar to the Anderson model (§9.8) but no localisation of ψ is implied so $|a_{mi}|^2$ must be of the same order in all cells. ψ does not represent a freely propagating wave with a long mean-free path. On the contrary the irregularly varying occupation probability from site to site can be interpreted as a Brownian or diffusive motion of the electron governed by the transfer integrals (like equation (7.22*c*)) which can be taken to be zero except between neighbouring sites. For σ from equation (7.43*b*) we require

$$|\langle m|v|n\rangle|^2 = \sum_{i,i'} |a_{mi}|^2|a_{ni'}|^2|\langle\varphi_i|v|\varphi_{i'}\rangle|^2.$$

Now the average value of $|a_{mi}|^2$ gives the average occupation probability per site ($=a^3/\Omega$) where a is the lattice spacing. By summing over i we get a factor

Ω/a^3 back again because this is the total number of cells. Estimating $\Sigma_{i'}|\langle\varphi_i|v|\varphi_{i'}\rangle|^2$ from hydrogen 1s states a distance a apart Hindley found it to be $\lambda^2\hbar^2/m^2a^2$ where λ is of the order unity. So the average of $|v_{mn}|^2$ is approximately $(\lambda^2\hbar^2/m^2a^2)(a^3/\Omega)$. Using equation (7.7) to sum over m, n we find for a degenerate electron gas that

$$\sigma = \frac{2\pi}{3m^2} e^2\lambda^2\hbar^3 a(g(E_F))^2. \tag{7.44}$$

This result has been fairly widely used to interpret observations on strongly scattering non-crystalline metals and, when modified by a suitable factor, $\exp[-\beta(E_c - E_F)]$, for amorphous semiconductors also (Chapter 11). The proportionality $\sigma \propto (g(E_F))^2$ was originally introduced by Mott and makes σ a tool for exploring the density of states. However the RPM is a simple model and the proportionality cannot be regarded as a universal feature.

The Hall effect has also been treated in the RPM (Friedman 1971) with the result

$$R_H = \alpha R_{FE}/g \tag{7.45}$$

where α is of order unity and g, often called the 'Mott g-factor' is $g(E_F)/g_{FE}(E_F)$. More general treatments of R_H have led to considerable complexities. The phenomenological definition is

$$R_H = \frac{\sigma_{xy}(H)}{H\sigma_{xx}^2} \tag{7.46}$$

where $\sigma_{xy}(H)$ is the linear response in the presence of $H = (0, 0, H_z)$. Apart from the free-electron case the calculation of $\sigma_{xy}(H)$ runs into subtle difficulties such that a settled theory of the Hall effect cannot be said to exist (Itoh 1985, Morgan and Howson 1985). Attempts to remedy this situation will no doubt continue because a number of disordered materials have $R_H \neq R_H^{FE}$ and some like liquid La and some amorphous alloys even have $R_H > 0$.

It must not be supposed that the preceding paragraphs mention all the current approaches to $\sigma_{\mu\nu}$. Numerical methods based on clusters are possible for example (Ballentine and Hammerberg 1984). In fact for a degenerate electron gas equation (7.43*c*) gives $\sigma = \sigma(E_F)$ which by equation (7.38) is $e^2D(E_F)g(E_F)$ where $D(E)$ is an energy-dependent diffusivity. $D(E)$, like $\sigma(E)$, has a Kubo–Greenwood formulation which can be expressed in Green functions and computed by the recursion method. It is applied in Ballentine (1982) and Bose *et al* (1983) to liquid and amorphous Fe and gives results in excellent agreement with experiment. The possibility in these calculations of separating the contribution to σ of s- and d-like electron states showed that d states carry more current because their significantly lower $D(E_F)$ is more than compensated by their higher $g(E_F)$. But the analysis of transport in 3d transition metals will no doubt continue. At the time of writing transport in strong scattering disordered solids is an evolving subject. Attempts are being made to derive σ from a Boltzmann-like equation whose solutions are, however, much more general than the f_1 of §6.11 and which cover both weak and strong scattering and even localisation. Readers wishing to pursue this could consult Morgan *et al* (1985) and numerous other references cited therein.

One aspect of this recent work needs emphasis. In disordered materials at sufficiently low temperatures it is possible for electrons to undergo a number of elastic scatterings, changing the direction of momentum but not the energy, with the result that there is some probability of initial momentum $\boldsymbol{k}$ being reversed to $-\boldsymbol{k}$. In two dimensions this can happen by scattering sequences proceeding 'clockwise' and 'anticlockwise' leading to final amplitudes for both backscattered waves that are equal and in phase. The total backscattering intensity is thereby enhanced through constructive interference—an effect often called quantum interference. Although this is a possible phenomenon in any kind of wave propagation, its application to electrons in metals has led to the name '$2k_F$ scattering' (see Bergmann (1983) for the theory and many references). It can happen in 3-D also. If it is not mitigated, $2k_F$ scattering will reduce the conductivity and even lead to localisation of the electron states ($\sigma = 0$) as shown theoretically by, e.g., Morgan and Hickey (1985) in agreement with numerical calculations to be referred to later in §9.8.

$2k_F$ scattering can however be mitigated by influences that spoil the phase coherence and so diminish the backscattered intensity. Inelastic scattering by phonons will do this and a rise in temperature will then enhance the conductivity. Likewise $2k_F$ scattering can be reduced by electron–electron collisions, magnetic fields and spin–orbit interaction leading to rather complicated dependences of resistance on temperature and to magnetoresistance. This will be referred to again in §§9.11 and 12.5.

Appendix

If the potential $V = \Sigma_i v_i$ is substituted into a typical T-term from equation (7.15*b*) the term becomes a sum of extended products of the form $v_i G_0 v_j G_0 v_k G_0 \ldots$ in which i, j, k etc may or may not be equal to one another. Suppose that in such a product v_i occurs successively n_i times, followed by v_j successively n_j times etc, thus $j \neq i$. After the j-sequence the k-sequence starts with k necessarily different from j but possibly equal to i. So the product could be written $t_i^{(n_i)} G_0 t_j^{(n_j)} G_0 t_k^{(n_k)} \ldots$ with adjacent suffixes unequal and where $t_i^{(n_i)} = v_i G_0 v_i G_0 v_i \ldots$ for n_i repetitions of v_i. Now $t_i^{(n_i)}$ is the nth term of t_i as appears from inspection of equation (7.34) so $t_i = \Sigma_{n_i} t_i^{(n_i)}$. If this form of t_i is now substituted into a typical term of equation (7.35), we recapture the same extended products $t_i^{(n_i)}G_0t_j^{(n_j)}G_0t_k^{(n_k)}\ldots$ that came from the original expansion for T, showing that the two expansions for T are equivalent.

References

Allen N K, Durham P J, Gyorffy B L and Jordan R G 1983 *J. Phys. F: Met. Phys.* **13** 223

Aloisio M, Singh V A and Roth L M 1981 *J. Phys. F: Met. Phys.* **11** 1823

Asano S and Yonezawa F 1980 *J. Phys. F: Met. Phys.* **10** 75

Ballentine L E 1975 *Adv. Chem. Phys.* **31** 263

—— 1982 *Phys. Rev.* B **25** 6089

Ballentine L E and Hammerberg J E 1984 *Can. J. Phys.* **62** 692

Ballentine L E and Kolář M 1986 *J. Phys. C: Solid State Phys.* **19** 981

Beeby J L 1983 *Phil. Mag.* B **48** L23

Beeby J L and Hayes T M 1985 *Phys. Rev.* **32** 6464

Bergmann G 1983 *Phys. Rev.* B **28** 2914

Bose S K, Ballentine L E and Hammerberg J E 1983 *J. Phys. F: Met. Phys.* **13** 2089

Bullett D W 1980 *Solid State Phys.* **35** (New York: Academic)

Butcher P N 1972 *J. Phys. C: Solid State Phys.* **5** 3164

Edwards S F 1958 *Phil. Mag.* **3** 1020

Friedman L 1971 *J. Non-Cryst. Solids* **6** 329

Gordon B E A, Temmerman W E and Gyorffy B L 1981 *J. Phys. F: Met. Phys.* **11** 821

Greenwood D A 1958 *Proc. Phys. Soc.* **71** 585

Gyorffy B L and Stocks G M 1977 *Electrons in Finite and Infinite Structures* ed P Phariseau and L Scheire (New York: Plenum) p 144

Haydock R 1980 *Solid State Phys.* **35** (New York: Academic)

Haydock R and Nex C M M 1985 *J. Phys. C: Solid State Phys.* **18** 2235

Heine V 1980 *Solid State Phys.* **35** (New York: Academic)

Hindley N K 1970 *J. Non-Cryst. Solids* **5** 17

Itoh M 1985 *J. Phys. F: Met. Phys.* **15** 1715

Itoh M, Niizeki K and Watabe M 1981 *J. Phys. F: Met. Phys.* **11** 1605

Itoh M and Watabe M 1984 *J. Phys. F: Met. Phys.* **14** 1847

John W and Keller W 1980 *J. Physique Coll.* **41** C8 400

Jordan R G, Sohal G S, Gyorffy B L, Durham P J, Temmerman W M and Weinberger P 1985 *J. Phys. F: Met. Phys.* **15** L135

Kaiser J H, West R N and Shiotani N 1986 *J. Phys. F: Met. Phys.* **16** 1307

Kelly M J 1980 *Solid State Phys.* **35** (New York: Academic)

Kubo R 1965 *Statistical Mechanics of Equilibrium and Non-equilibrium* ed J Meixner (Amsterdam: North-Holland) p 81

Luttinger J M 1964 *Phys. Rev.* A **135** 1505

Morgan G J and Hickey B L 1985 *J. Phys. F: Met. Phys.* **15** 2473

Morgan G J and Howson M A 1985 *J. Phys. C: Solid State Phys.* **18** 4327

Morgan G J, Howson M A and Saub K 1985 *J. Phys. F: Met. Phys.* **15** 2157

Morgan G J and Weir G F 1983 *Phil. Mag.* B **47** 177

Neal T 1970 *Phys. Fluids* **13** 249

Pettifor D G and Weaire D (ed) 1984 *The Recursion Method and its Applications* (Berlin: Springer)
Rickayzen G 1980 *Green's Functions and Condensed Matter* (New York: Academic)
Rodberg L S and Thaler R M 1967 *Introduction to the Quantum Theory of Scattering* (New York: Academic)
Roth L M 1974 *Phys. Rev.* B **9** 2476
Roth L M and Singh V 1982 *Phys. Rev.* B **25** 2522
Szotek Z, Gyorffy B L, Stocks G M and Temmerman W M 1984 *J. Phys. F: Met. Phys.* **14** 1984 (and references therein)
Szotek Z and Temmerman W M 1986 *J. Phys. F: Met. Phys.* **16** 17
Van Oosten A B and Geertsma W 1985 *Physica* B/C **133** 55
Watabe M 1977 *Liquid Metals (Bristol) 1976* (Inst. Phys. Conf. Ser. 30) p 288
West R N, Mayers J and Walters P A 1981 *J. Phys. E: Sci. Instrum.* **14** 478
Wotherspoon J T M, Rodway D C and Norris C 1979 *Phil. Mag.* B **40** 51
Yonezawa F and Morigaki K 1973 *Prog. Theor. Phys. Suppl.* **53** 1
Yonezawa F, Roth L and Watabe M 1975 *J. Phys. F: Met. Phys.* **5** 435
Ziman J M 1960 *Electrons and Phonons* (London: Oxford University Press)

8

THERMAL MOTION

As with many phenomena in condensed matter, the thermal motion of atoms is relatively easier to understand in non-interacting gases and perfect crystals than in liquids or amorphous solids. A perfect gas atom moves rectilinearly between collisions and the mean-square displacement in time t, $\langle r^2(t) \rangle$, is $\overline{v^2}t^2$ where $\overline{v^2}$ is given by the appropriate statistics. The atoms in crystals vibrate about their equilibrium positions and, in the Einstein model, do so with a single frequency of about 10^{13} Hz. In more complicated situations, perhaps in a liquid, an atom might be thought to move like a diffusing Brownian particle so that its $\langle r^2(t) \rangle$ is $6Dt$ as given by the diffusion equation for large t. In a liquid we should expect an atom eventually to diffuse to infinity and intuitively we might guess that the motion is an irregular oscillation about a centre whose position wanders. In solids finite and localised vibrations are to be expected. This is, however, to think of the motion of one atom at a time. Cooperative or collective motions will surely occur. The propagation of phonons through periodic structures is a well known example and for a given wavevector, $\boldsymbol{k}$, there is a corresponding angular frequency ω. There is a well developed theory of the dispersion relation, i.e. the function $\omega(\boldsymbol{k})$, and the thermal motion and thermal properties are explicable by the excitation of phonons, longitudinal or transverse, in the acoustic and optical branches of the spectrum.

Long-wavelength acoustical waves propagate in liquids also, with velocities determined by density and compressibility. It is not obvious however that phonons of wavelength comparable with the interatomic spacing can be said to exist in liquids or, if they can, whether their $\boldsymbol{k}$ is well defined with a precise dispersion relation or, on the contrary, heavily damped. The problem is analogous to that of the electron E–k relation (see §7.5).

It follows that it would be desirable to have laboratory methods for investigating both single atom and collective or coherent motion and the largest single source of experimental information is neutron inelastic scattering. This is because neutron beams with wavelengths suitable for structure studies have energies comparable with thermal excitation energies in condensed matter. X-rays of similar wavelengths have energies higher by a factor of 10^5. Neutron and other experiments and fundamental theory have in recent years been much aided by MD calculations. Nevertheless thermal motion in disordered matter remains a somewhat obscure field.

8.1 $G(r, t)$, $S(Q, \omega)$ and density correlations

Suppose that in a sample of condensed matter there is an atom m at $\boldsymbol{R}_m(0)$ when $t = 0$. Let us consider the possibility of finding some atom, call it n, near $\boldsymbol{R}_n(t)$, t seconds later and let $\boldsymbol{r} = \boldsymbol{R}_n(t) - \boldsymbol{R}_m(0)$. The average number of atoms in a small volume v around $\boldsymbol{r}$ can be expressed in two ways. One is by $\int_v G_{\mathrm{cl}}(\boldsymbol{r}, t)\, \mathrm{d}\boldsymbol{r}$ where $G_{\mathrm{cl}}(\boldsymbol{r}, t)$ means the average number density at $(\boldsymbol{r}, t)$ given that there was an atom at $(0, 0)$. The other way is by

$$N^{-1} \int_v \mathrm{d}\boldsymbol{r} \left\langle \sum_{m,n}^{N} \delta\left(\boldsymbol{r} - \boldsymbol{R}_n(t) + \boldsymbol{R}_m(0)\right) \right\rangle$$

because the δ-function counts one every time the relative separation of the m-atom at $t = 0$ from the n-atom at $t = t$ equals $\boldsymbol{r}$, Σ_n sums over all possible atoms n and $N^{-1}\, \Sigma_m$ averages over all possible atoms m. Thus we may write

$$G_{\mathrm{cl}}(\boldsymbol{r}, t) \equiv N^{-1} \left\langle \sum_{m,n} \delta\left(\boldsymbol{r} - \boldsymbol{R}_n(t) + \boldsymbol{R}_m(0)\right) \right\rangle. \tag{8.1}$$

$G_{\mathrm{cl}}(\boldsymbol{r}, t)$ is called the classical space–time correlation function and there was nothing to stop the atom at $\boldsymbol{R}_n(t)$ from being the same one that was at $\boldsymbol{R}_m(0)$.

Recalling the classical density function $\Sigma_i\, \delta(\boldsymbol{r} - \boldsymbol{R}_i)$ (equation (2.2*a*)) we might guess that $G_{\mathrm{cl}}(\boldsymbol{r}, t)$ has something to do with correlations of density at points $(\boldsymbol{r}, t)$ apart. That this is so follows somewhat naturally from a theory of neutron scattering by the sample. For if the sample density were continuous and uniform there would be no scattering and indeed the scattering cross section turns out to be a measure of the correlated variations of density in space and time. However, the cross section has to be calculated quantum mechanically and the consequent definition of $G(\boldsymbol{r}, t)$ is more general than equation (8.1) suggests.

To illustrate this in the simplest case we generalise the treatments of x-ray or electron scattering in Chapters 3 and 6. The full theory is given in Lovesey (1984) and the seminal paper is by van Hove (1954). We assume nuclear scattering by identical spinless nuclei to avoid complications due to isotopic, magnetic and spin-dependent interactions.

Adopting the usual expression for a potential, namely, $V(\boldsymbol{r}) = \Sigma_m\, v(\boldsymbol{r} - \boldsymbol{R}_m)$, we use Fermi's Golden Rule to write down the probability per unit time of scattering from an initial neutron state with wavevector $\boldsymbol{k}_i$ to a final state with $\boldsymbol{k}_f$ as a result of which the target sample goes from state $|i\rangle$ to $|f\rangle$. This probability is

$$P_{if} = \frac{2\pi}{\hbar}\, |V_{fi}|^2 \delta(\hbar\omega + E_i - E_f)$$

where the sample is supposed to acquire energy $\hbar\omega$ from the neutron and where $V_{fi} = \langle \boldsymbol{k}_f f | V(\boldsymbol{r}) | \boldsymbol{k}_i i \rangle$. The symbol $|\boldsymbol{k}_\alpha \alpha\rangle$ is a product of the state functions for the neutron and target and the incident neutron beam will be supposed normalised to have a flux $\hbar k_i/m$.

Factorising the matrix element (cf §6.6),

$$P_{if} = \frac{2\pi}{\hbar} |v(\boldsymbol{Q})|^2 \sum_{m,n} \langle i| \exp(-\mathrm{i}\boldsymbol{Q}\cdot\boldsymbol{R}_m)|f\rangle\langle f| \exp(\mathrm{i}\boldsymbol{Q}\cdot\boldsymbol{R}_n)|i\rangle \delta(\hbar\omega + E_i - E_f)$$

where

$$v(\boldsymbol{Q}) \equiv \int v(r) \exp(\mathrm{i}\boldsymbol{Q}\cdot\boldsymbol{r})\, \mathrm{d}\boldsymbol{r}.$$

The δ-function can be written in the form:

$$(2\pi\hbar)^{-1} \int_{-\infty}^{+\infty} \mathrm{d}t \exp[-\mathrm{i}(E_i/\hbar - E_f/\hbar + \omega)t].$$

Moreover for any eigenvalue E_g of the unperturbed target Hamiltonian H_0, relations like

$$\exp(-\mathrm{i}E_g t/\hbar)\langle g| = \langle g| \exp(-\mathrm{i}H_0 t/\hbar)$$

can be used as required. Therefore

$$P_{if} = \frac{1}{\hbar^2} |v(\boldsymbol{Q})|^2 \int_{-\infty}^{+\infty} \mathrm{d}t \exp(-\mathrm{i}\omega t) \sum_{m,n} \langle i| \exp(-\mathrm{i}\boldsymbol{Q}\cdot\boldsymbol{R}_m)|f\rangle$$
$$\times \langle f| \exp(\mathrm{i}H_0 t/\hbar) \exp(\mathrm{i}\boldsymbol{Q}\cdot\boldsymbol{R}_n) \exp(-\mathrm{i}H_0 t/\hbar)|i\rangle.$$

We now define operators $\hat{\boldsymbol{R}}_l(t)$ by

$$\exp(\mathrm{i}\boldsymbol{Q}\cdot\hat{\boldsymbol{R}}_l(t)) = \exp(\mathrm{i}H_0 t/\hbar) \exp(\mathrm{i}\boldsymbol{Q}\cdot\boldsymbol{R}_l) \exp(-\mathrm{i}H_0 t/\hbar) \tag{8.2}$$

whence

$$P_{if} = \hbar^{-2}|v(\boldsymbol{Q})|^2 \int_{-\infty}^{+\infty} \mathrm{d}t \exp(-\mathrm{i}\omega t)$$
$$\times \sum_{m,n} \langle i| \exp(-\mathrm{i}\boldsymbol{Q}\cdot\hat{\boldsymbol{R}}_m(0))|f\rangle\langle f| \exp(\mathrm{i}\boldsymbol{Q}\cdot\hat{\boldsymbol{R}}_m(t))|i\rangle.$$

This transition rate should be summed over all possible final states $|f\rangle$ and the relation $\Sigma_f |f\rangle\langle f| = 1$ secures this.

The transition rate should also be thermally averaged over the initial states and if we multiply by $\exp(-\beta E_i)$ and sum over i, divide by $Z \equiv \Sigma_i \exp(-\beta E_i)$, and recall that $H_0|i\rangle = E_i|i\rangle$ we reach a statistical mechanical expression of the form $Z^{-1}\,\mathrm{Tr}(\exp(-\beta H_0)\hat{O})$ where $\hat{O}$ is an operator and Z the partition function. This is the formula for the thermal average of the variable O so we now have

$$\sum_f \langle P_{if} \rangle = \hbar^{-2} |v(\boldsymbol{Q})|^2 \int_{-\infty}^{+\infty} \mathrm{d}t \exp(-\mathrm{i}\omega t))$$

$$\times \sum_{m,n} \langle \exp(-\mathrm{i}\boldsymbol{Q} \cdot \hat{\boldsymbol{R}}_m(0)) \exp(\mathrm{i}\boldsymbol{Q} \cdot \hat{\boldsymbol{R}}_n(t)) \rangle \tag{8.3}$$

where $\langle\ \rangle$ indicates the thermal average.

Suppose the final neutron path is into a small solid angle $\mathrm{d}\Omega$; then a number $mk_f\,\mathrm{d}\Omega\,\mathrm{d}E_f/(2\pi)^3\hbar^2$ of wavevectors are available. The cross section for such an event is

$$\sum_f \langle P_{if} \rangle \frac{mk_f\,\mathrm{d}\Omega\,\mathrm{d}E_f}{(2\pi)^3\hbar^2} \left(\frac{\hbar k_i}{m}\right)^{-1}$$

So the cross section per unit solid angle and per unit energy range is

$$\frac{\mathrm{d}^2\sigma}{\mathrm{d}\Omega\,\mathrm{d}E_f} = \frac{k_f}{k_i} \frac{m^2}{(2\pi)^3\hbar^5} |v(\boldsymbol{Q})|^2 \int_{-\infty}^{+\infty} \mathrm{d}t \exp(-\mathrm{i}\omega t)$$

$$\times \sum_{m,n} \langle \exp(-\mathrm{i}\boldsymbol{Q} \cdot \hat{\boldsymbol{R}}_m(0)) \exp(\mathrm{i}\boldsymbol{Q} \cdot \hat{\boldsymbol{R}}_n(t)) \rangle. \tag{8.4}$$

At this point we define a quantum mechanical $G(\boldsymbol{r}, t)$ by

$$G_{\mathrm{qu}}(\boldsymbol{r}, t) \equiv \frac{1}{(2\pi)^3} \int \mathrm{d}\boldsymbol{Q} \exp(-\mathrm{i}\boldsymbol{Q} \cdot \boldsymbol{r}) N^{-1}$$

$$\times \sum_{m,n} \langle \exp(-\mathrm{i}\boldsymbol{Q} \cdot \hat{\boldsymbol{R}}_m(0)) \exp(\mathrm{i}\boldsymbol{Q} \cdot \hat{\boldsymbol{R}}_n(t)) \rangle \tag{8.5}$$

and shall now relate it to both the cross section and the density correlations. First define an important quantity $S(\boldsymbol{Q}, \omega)$ by

$$\frac{\mathrm{d}^2\sigma}{\mathrm{d}\Omega\,\mathrm{d}E_f} = \frac{Nm^2}{(2\pi)^2\hbar^4} \frac{k_f}{k_i} |v(\boldsymbol{Q})|^2 S(\boldsymbol{Q}, \omega) \tag{8.6}$$

whence from equation (8.4)

$$S(\boldsymbol{Q}, \omega) = \frac{1}{2\pi N\hbar} \int_{-\infty}^{+\infty} \mathrm{d}t \exp(-\mathrm{i}\omega t)$$

$$\times \sum_{m,n} \langle \exp(-\mathrm{i}\boldsymbol{Q} \cdot \hat{\boldsymbol{R}}_m(0)) \exp(\mathrm{i}\boldsymbol{Q} \cdot \hat{\boldsymbol{R}}_n(t)) \rangle. \tag{8.7}$$

From equations (8.5) and (8.7), by Fourier transformation,

$$S(\boldsymbol{Q}, \omega) = \frac{1}{2\pi\hbar} \int_{-\infty}^{+\infty} \mathrm{d}t \exp(-\mathrm{i}\omega t) \int \mathrm{d}\boldsymbol{r} \exp(\mathrm{i}\boldsymbol{Q} \cdot \boldsymbol{r}) G_{\mathrm{qu}}(\boldsymbol{r}, t) \tag{8.8a}$$

and

$$G_{\mathrm{qu}}(\boldsymbol{r}, t) = \frac{\hbar}{(2\pi)^3} \int \mathrm{d}\omega \exp(\mathrm{i}\omega t) \int \mathrm{d}\boldsymbol{Q} \exp(-\mathrm{i}\boldsymbol{Q} \cdot \boldsymbol{r}) S(\boldsymbol{Q}, \omega) \tag{8.8b}$$

which show that the cross section is essentially a spatial and temporal

Fourier transform of $G_{qu}(\boldsymbol{r}, t)$. Furthermore the averaged quantity in equation (8.5) for $G_{qu}(\boldsymbol{r}, t)$ can be written

$$\int \mathrm{d}\boldsymbol{r}' \langle \exp(-\mathrm{i}\boldsymbol{Q} \cdot \hat{\boldsymbol{R}}_m(0) + \mathrm{i}\boldsymbol{Q} \cdot \boldsymbol{r}')\delta(\boldsymbol{r}' - \hat{\boldsymbol{R}}_n(t)) \rangle \tag{8.9}$$

so integrating over $\boldsymbol{Q}$ we obtain a δ-function making

$$G_{qu}(\boldsymbol{r}, t) = \frac{1}{N} \sum_{m,n} \left\langle \int \mathrm{d}\boldsymbol{r}' \delta(\boldsymbol{r} - \boldsymbol{r}' + \hat{\boldsymbol{R}}_m(0))\delta(\boldsymbol{r}' - \hat{\boldsymbol{R}}_n(t)) \right\rangle. \tag{8.10}$$

We now return to the density operator which in its quantum mechanical form is

$$\hat{\rho}(\boldsymbol{r}, t) \equiv \sum_r \delta(\boldsymbol{r} - \hat{\boldsymbol{R}}_n(t)). \tag{8.11}$$

Inserting this into equation (8.10) gives

$$G_{qu}(\boldsymbol{r}, t) = N^{-1} \int \mathrm{d}\boldsymbol{r}' \langle \hat{\rho}(\boldsymbol{r}' - \boldsymbol{r}, 0)\hat{\rho}(\boldsymbol{r}', t) \rangle \tag{8.12}$$

and if the $\hat{\rho}$ were classical commuting quantities G_{qu} would be identical with G_{cl} in equation (8.1). $G(\boldsymbol{r}, t)$ clearly has to do with density correlations in space and time and therefore they determine the neutron scattering. Consequently, information about thermal motion must reside in $S(\boldsymbol{Q}, \omega)$ which is called the scattering law and is a response function in the context of linear-response theory (§7.7). $G(\boldsymbol{r}, t)$ and $S(\boldsymbol{Q}, \omega)$ are clearly generalisations of the $g(\boldsymbol{r})$ and $S(\boldsymbol{Q})$ of Chapters 2 and 3. In fact $G(\boldsymbol{r}, 0) = z(\boldsymbol{r})$ (equation (2.8a)) and $\int_{-\infty}^{+\infty} \mathrm{d}\omega\, S(\boldsymbol{Q}, \omega) = S(\boldsymbol{Q})$. Like their static counterparts $G(\boldsymbol{r}, t)$ and $S(\boldsymbol{Q}, \omega)$ relate only to pair correlations and are properties of the unperturbed target. Unless it is necessary to include them, suffixes cl and qu will be dropped from G henceforward.

8.2 Neutron inelastic scattering

The quantity $|v(\boldsymbol{Q})|^2$ in the cross section is the Born approximation for scattering by a single nucleus. For the neutron energies suitable for $S(\boldsymbol{Q}, \omega)$ measurements only isotropic s-wave scattering is important and, as mentioned in §3.7 and explained at length in Lovesey (1984), the amount of scattering is determined by a scattering length b characteristic of each isotope and independent of neutron energy. The fictitious $v(\boldsymbol{r})$ which, when inserted into the Born approximation, gives the correct nuclear scattering from a nucleus at the origin is $2\pi\hbar^2 b\delta(\boldsymbol{r})/m$. So $|v(\boldsymbol{Q})|^2$ is $4\pi^2\hbar^4 b^2/m^2$ (compare the use of the electron–ion pseudopotential in Chapter 6).

This applies only to a sample of identical spinless isotopes. If there is really a random distribution of spin orientations and isotopes, though only one chemical species, then in calculating $|v(\boldsymbol{Q})|^2$, factors $b_m^* b_n$ would arise and instead of $|b|^2$ we should have $\overline{b_m^* b_n}$ where the bar averages over spins and isotopes. If $m = n$, this product is $\overline{|b|^2}$; if $m \neq n$, it is $|\bar{b}|^2$ because there is no correlation of nuclear properties between sites. So in general

$$b^2 \rightarrow |\bar{b}|^2 + \delta_{mn}(\overline{|b|^2} - |\bar{b}|^2).$$

A consequence of this is that $\mathrm{d}^2\sigma/\mathrm{d}\Omega\,\mathrm{d}E_f$ can be rewritten in two parts by substituting for $|v(\boldsymbol{Q})|^2$ in equation (8.6):

$$\left(\frac{\mathrm{d}^2\sigma}{\mathrm{d}\Omega\,\mathrm{d}E_f}\right)_{\mathrm{coh}} = N\frac{k_f}{k_i}|\bar{b}|^2 S(\boldsymbol{Q}, \omega) \tag{8.13a}$$

$$\left(\frac{\mathrm{d}^2\sigma}{\mathrm{d}\Omega\,\mathrm{d}E_f}\right)_{\mathrm{inc}} = N\frac{k_f}{k_i}(\overline{|b|^2} - |\bar{b}|^2)\, S_{\mathrm{inc}}(\boldsymbol{Q}, \omega). \tag{8.13b}$$

The incoherent cross section can be seen to be correctly so called because the b-term is $-\overline{|b - \bar{b}|^2}$ and random deviations $(b - \bar{b})$ will not give constructive interference (cf §3.6). Furthermore, since equation (8.13b) comes from the case $m = n$, equation (8.7) gives

$$S_{\mathrm{inc}}(\boldsymbol{Q}, \omega) = \frac{1}{2\pi N\hbar}\int_{-\infty}^{+\infty} \mathrm{d}t\, \exp(-\mathrm{i}\omega t) \times \sum_n \langle \exp(-\mathrm{i}\boldsymbol{Q}\cdot\hat{\boldsymbol{R}}_n(0))\exp(\mathrm{i}\boldsymbol{Q}\cdot\hat{\boldsymbol{R}}_n(t))\rangle \tag{8.14}$$

from which a new quantity, $G_{\mathrm{s}}(\boldsymbol{r}, t)$, can be defined by analogy with equations (8.5) and (8.8)

$$\mathrm{S}_{\mathrm{inc}}(\boldsymbol{Q}, \omega) = \frac{1}{2\pi\hbar}\int_{-\infty}^{+\infty} \mathrm{d}t\, \exp(-\mathrm{i}\omega t)\int \mathrm{d}\boldsymbol{r}\, \exp(\mathrm{i}\boldsymbol{Q}\cdot\boldsymbol{r})G_{\mathrm{s}}(\boldsymbol{r}, t) \tag{8.15a}$$

with

$$G_{\mathrm{s}}(\boldsymbol{r}, t) = \frac{\hbar}{(2\pi)^3}\int \mathrm{d}\omega\, \exp(\mathrm{i}\omega t)\int \mathrm{d}\boldsymbol{Q}\, \exp(-\mathrm{i}\boldsymbol{Q}\cdot\boldsymbol{r})S_{\mathrm{inc}}(\boldsymbol{Q}, \omega). \tag{8.15b}$$

The $m = n$ case of equation (8.10), with the $\hat{R}$ thought of as ordinary position vectors, shows that classically $G_{\mathrm{s}}(\boldsymbol{r}, t)$ means the probability that the same particle is at $(0, 0)$ and $(\boldsymbol{r}, t)$; the suffix s is for 'self' and G is the self pair correlation function, $G_{\mathrm{s}}(\boldsymbol{r}, 0) = \delta(\boldsymbol{r})$. If necessary a 'distinct' pair correlation, $G_{\mathrm{d}}(\boldsymbol{r}, t)$, can be invoked such that

$$G(\boldsymbol{r}, t) = G_{\mathrm{s}}(\boldsymbol{r}, t) + G_{\mathrm{d}}(\boldsymbol{r}, t) \tag{8.16a}$$

and

$$G_{\mathrm{d}}(\boldsymbol{r}, 0) = n_0 g(\boldsymbol{r}). \tag{8.16b}$$

Some properties of $S(\boldsymbol{Q}, \omega)$ are worth noting. While both positive and negative values of $\hbar\omega$ are possible, $S(\boldsymbol{Q}, \omega) \neq S(-Q, -\omega)$. In fact

$S(\boldsymbol{Q}, \omega) = \exp(\hbar\omega/k_B T)S(-\boldsymbol{Q}, -\omega)$ which expresses the principle of detailed balance. For analysing experimental data containing both $\pm\hbar\omega$ it is convenient to construct an even function in ω, namely the symmetrised scattering law. For isotropic matter it is

$$\tilde{S}(Q, \omega) \equiv \exp(-\hbar\omega/2k_B T)S(Q, \omega) = \tilde{S}(Q, -\omega). \tag{8.17}$$

The quantity $\int_{-\infty}^{+\infty} \omega^n S(Q, \omega)\, d\omega$ is called the nth frequency moment. The zeroth one is $S(Q)$ and for classical fluids the second is $Q^2 k_B T/m$ for both S and S_{inc}. The latter are even functions of ω for classical fluids so their odd moments are zero. In general however the first moment is $\hbar Q^2/2m$. Both good measured and model theoretical functions $S(Q, \omega)$ ought to have the proper values of the moments and in the experiments quoted in the next section the check makes for confidence, though it is a sign of the technical difficulties that the agreement is not perfect.

As indicated in §3.11 and reviewed extensively in neutron scattering literature the measurement of $S(\boldsymbol{Q}, \omega)$ requires major central research facilities such as the reactor at the Institut Laue–Langevin or the pulsed neutron sources at the Argonne and Rutherford Laboratories. Considerable efforts of data reduction are involved but we omit such considerations and treat $S(\boldsymbol{Q}, \omega)$ as an experimentally available quantity. Whether a measurement gives $S(\boldsymbol{Q}, \omega)$, $S_{inc}(\boldsymbol{Q}, \omega)$ or a weighted average of them depends on the properties of the nuclei and resolving the two scattering laws may be difficult. It also depends on the state of polarisation of the beam and polarising spectrometers may help to separate S and S_{inc}. Some separate isotopes, e.g. ^{4}He, ^{36}Ar, have wholly coherent cross sections but natural Ar does not; scattering from V and H is almost wholly incoherent. As a result of integrating the isotropic scattering over Ω, the whole nuclear cross section is $4\pi\overline{|b|^2}$. $|\bar{b}|^2$ may therefore be substituted by $\sigma_{coh}/4\pi$ and $(\overline{|b|^2} - |\bar{b}|^2)$ by $\sigma_{inc}/4\pi$. These σ's are called coherent and incoherent cross sections. $\sigma \equiv 4\pi\overline{|b|^2}$ is called the total cross section. Values of σ are tabulated in Lovesey (1984).

Generally, from equation (8.13),

$$\frac{d^2\sigma}{d\Omega\, dE_f} = \frac{N}{4\pi}\frac{k_f}{k_i}(\sigma_{coh}\, S(\boldsymbol{Q}, \omega) + \sigma_{inc} S_{inc}(\boldsymbol{Q}, \omega)). \tag{8.18}$$

Formulae for polyatomic systems are given by Powles (1973).

8.3 Specimen experiments

We now outline examples of illuminating experiments, first on liquid Ar and Rb. In the former, $S(Q, \omega)$ and $S_{inc}(Q, \omega)$ were separated for the first time when Sköld *et al* (1972) performed two independent measurements, one on nearly pure ^{36}Ar, an almost wholly coherent scatterer, and the other on a mixture of ^{36}Ar and ^{40}Ar chosen to maximise $\sigma_{inc}/\sigma_{coh}$. Knowing the cross sections and having two measured functions satisfying equation (8.18), the two simultaneous equations could be solved for S and S_{inc}. In the Rb experiment (Copley and Rowe 1974a, b) S_{coh} is measured because $\sigma_{coh}/\sigma_{inc} \sim 1800$.

This type of experiment might have $5 < E_i < 50$ meV, $0.5 < Q < 6$ Å^{-1}, $-10 < \hbar\omega < 10$ meV; but much wider ranges of these variables are in common use at the advanced centres.

Samples of the large number of data are shown in figures 8.1 to 8.3. The central peaks in $S(Q, \omega)$ versus ω curves, where $\Delta E \sim$ few meV, are called quasi-elastic peaks and are usually interpreted as the result of energy losses or gains by the neutrons when they exchange small amounts of translational or rotational energy with diffusing nuclei. The separate, or inelastic, side peaks at larger ΔE (as in figures 8.2(*a*) and 8.7) represent collective progagating modes. These are excitations of energy $\hbar\omega$ and wavelength $2\pi/Q$ comparable in the Rb case to the interatomic spacing. The ω–Q curve given by these peaks (see figure 8.3(*a*)) has been called a dispersion relation but without the implication that phonons of long lifetime are propagating indefinitely long distances through the fluid. The breadth of the peak indicates strong damping and the contrast with phonon spectra in crystals is marked (compare the discussion of $\rho(k, E)$ for electrons in §7.5). If inelastic peaks are absent, or make a more or less perceptible shoulder on the central peak, it

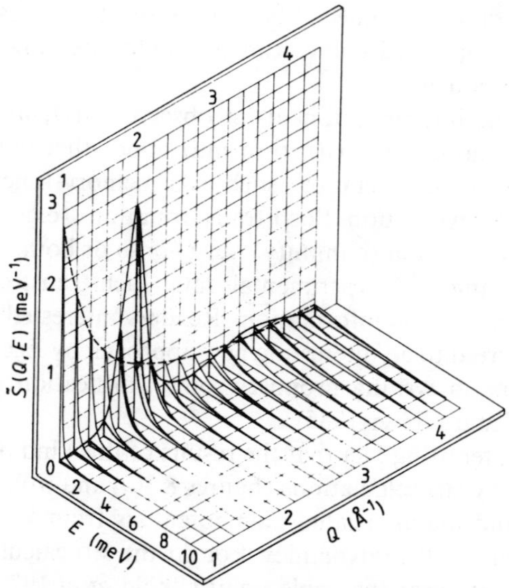

Figure 8.1 Symmetrised coherent and incoherent scattering laws for liquid Ar. The broken and full curves show the envelopes of the incoherent and coherent scattering functions, respectively. $E = \hbar\omega$. (From Sköld *et al* 1972.)

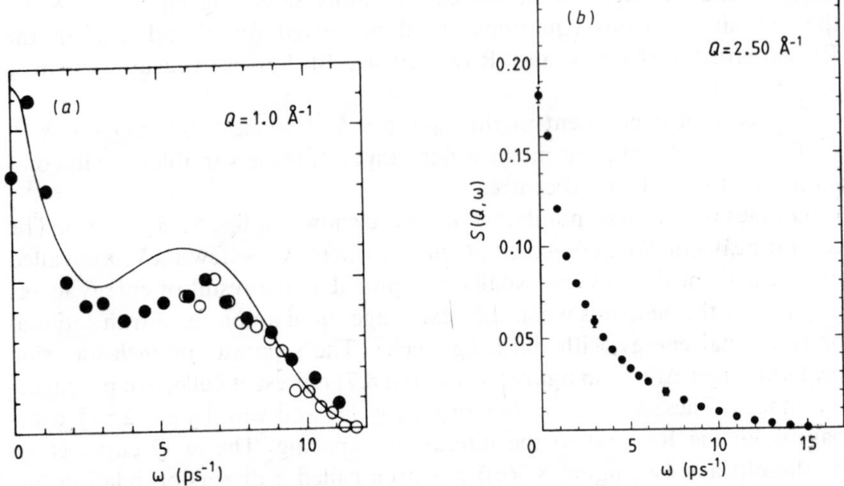

Figure 8.2 $\tilde{S}(Q, \omega)$ plotted against ω at $Q = 1.0$ and $2.5\ \text{Å}^{-1}$ for liquid Rb. Full circles, energy gain data; open circles, energy loss data. Note the peak in the $Q = 1.0$ curve (see Copley and Rowe 1974a).

does not follow that collective propagating modes do not occur. However, the criterion to be used to decide the presence or absence of such modes has been a matter of difficulty, indeed of controversy. It involves the question of whether and how observed curves can be validly decomposed into central and inelastic components.

In principle, the interpretation of an observed $S(Q, \omega)$ might be made easier by transforming it into $G(r, t)$ or into some other function defined in the hope of simpler interpretation, such as a memory function of the kind introduced in the next section. Such transformations (e.g., equation (8.8*b*)) usually involve integration to infinity over Q or ω or both. As this is impossible with finite ranges of experimental data, such transformations do not always give unambiguous information. Readers interested in this problem are therefore referred to an interesting new approach by Egelstaff and Gläser which at the time of writing promises some clarification of these difficult issues (Egelstaff and Gläser 1985).

It is a highly interesting fact that the results of this kind of experiment are well reproduced by MD calculations (figures 8.3(*a*) and (*b*)). This extends the success of MD and the associated pair potentials from static cases such as $S(Q)$ for Ar (figure 4.3) to dynamics. For Ar the MD calculations used a LJ potential for a state near the triple point (Sköld *et al* 1972, Ballucani *et al* 1983b) and for Rb a long-range oscillatory potential similar, but not identical, to those discussed for metals in §§4.5 and 6.7 (see Rahman 1974a, b and Ballucani *et al* 1983a). Although comparisons of the observed $S(Q)$ for the dense Kr gas with $S(Q)$ from MC calculations suggest there is room for

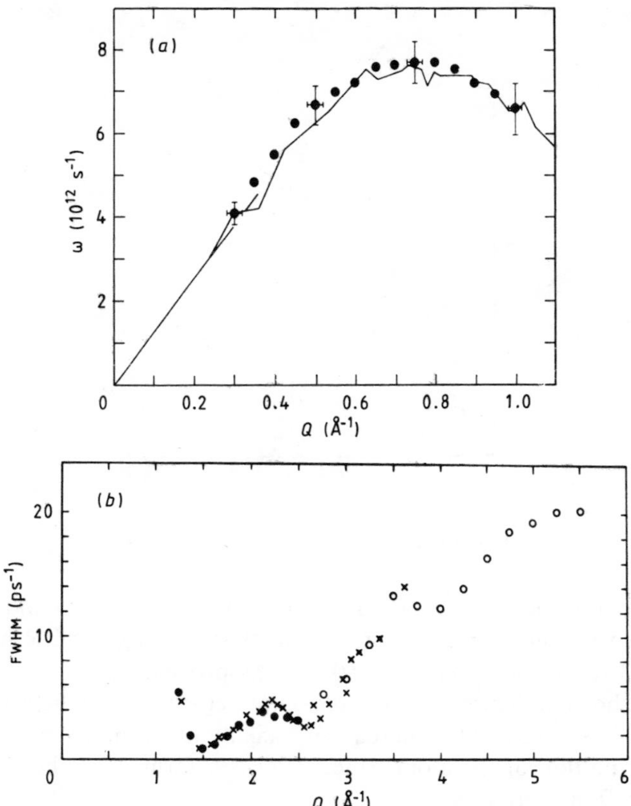

Figure 8.3 (*a*) Position of Rb peak versus Q. Circles, points from curves like that in figure 8.2 for $Q = 1.0$ Å^{-1}. Full line, MD calculation. (*b*) Full width at half maximum of $S(Q, \omega)$ in liquid Rb for $Q > 1.0$ Å^{-1}. Circles, experiment; crosses, MD calculation (from Rahman 1974a).

correction for n-body forces (Egelstaff *et al* 1983) we shall not pursue this but take the view that MD with pair potentials reproduces observations of real fluids sufficiently well that they can be relied on to obtain revealing features of the particle motion that cannot easily be derived from experiments. One such is $\langle r^2(t)\rangle$ and another is the classical normalised velocity autocorrelation function

$$\psi(t) \equiv \frac{\langle \boldsymbol{v}_i(0) \cdot \boldsymbol{v}_i(t)\rangle}{\langle v_i^2\rangle} = m\beta\langle v_x(0)v_x(t)\rangle. \tag{8.19}$$

$\langle r^2(t)\rangle$ is shown for Rb in figure 8.4. The timescale, in ps, indicates that the MD runs may represent about 10^{-10} to 10^{-9} s in the life of the sample or 10^4 to 10^5 steps of calculation. The linear dependence of $\langle r^2(t)\rangle$ on t which develops after about 3 ps is characteristic of diffusive motion and its slope gives $D = 2.66 \times 10^{-5}$ cm^2 s^{-1} at 332 K (Ballucani *et al* 1983a,b). An

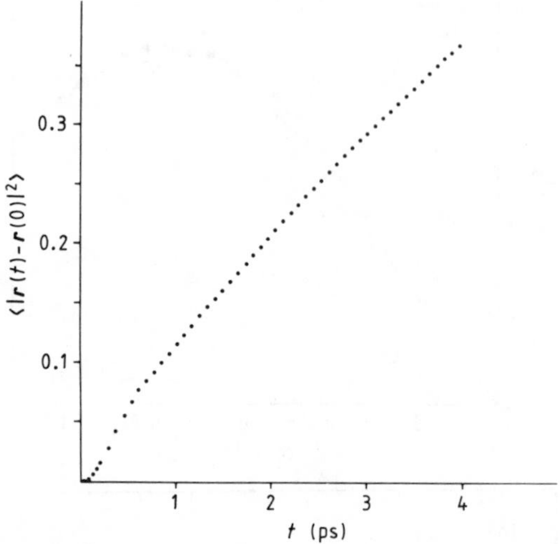

Figure 8.4 $\langle r^2(t) \rangle$ for liquid Rb from MD calculation (from Ballucani *et al* 1983a,b).

experimental value at 312 K is 2.62×10^{-5} cm ^{2} s $^{-1}$ (Norden and Lodding 1967). At very short times, before collisions set in, free-particle behaviour might be expected and indeed $\langle r^2(t) \rangle \propto t^2$ approximately.

$\psi(t)$ is shown in figure 8.5 and a region of negative correlation is conspicuous at 0.5 ps. This and the subsequent oscillations strongly suggest damped vibratory motion of the atom. Since there is a formula of the Kubo type (§7.7) for D, namely,

$$D = \frac{1}{3} \int_0^\infty \mathrm{d}t \langle \boldsymbol{v}(0) \cdot \boldsymbol{v}(t) \rangle \tag{8.20}$$

integration of $\psi(t)$ gives $D = 2.64 \times 10^{-5}$ cm s $^{-1}$—close to the value from $\langle r^2(t) \rangle$. Rb is notable for showing collective or coherent motion (figures 8.2 and 8.3) which is often described as 'phonon-like'. It is the same atoms that evince 'single particle' and 'coherent' motion so there ought to be some way of showing that characteristics of the one influence the other. An interesting way of doing this was used by Ballucani *et al* (1983b). Following their prescription, define a special correlation function

$$\Phi(a_0, b_0; t) \equiv \frac{\Sigma_i \langle \boldsymbol{v}_i(0) \cdot \boldsymbol{v}_i(t) + \Sigma'_{j \neq i} \boldsymbol{v}_i(0) \cdot \boldsymbol{v}_j(t) \rangle}{\Sigma_i \langle \boldsymbol{v}_i(0) \cdot \boldsymbol{v}_i(0) + \Sigma'_{j \neq i} \boldsymbol{v}_i(0) \cdot \boldsymbol{v}_j(0) \rangle}. \tag{8.21}$$

Without the sum over j this would be $\psi(t)$ but with Σ' it includes cross correlations between atom i at $t = 0$ and atoms j at t. The parameters a_0 and b_0 mean that $\Sigma'_{j \neq i}$ is to include only neighbours with $a_0 < r < b_0$. With a_0 equal to a little less than the nearest-neighbour distance, b_0 could be chosen to include the first shell of neighbours, or the first plus second, as defined by the troughs in $g(r)$. Figure 8.5 also shows $(\Phi - \psi)$ when b_0 includes the first shell and may be interpreted as follows. $(\Phi - \psi)$ is an i–j correlation which is clearly out of phase with ψ. Momentum is initially transferred from i to its neighbours and the maximum in $(\Phi - \psi)$ is close to where ψ changes sign. The picture is of an ion rebounding from its first shell. Meanwhile the centre of mass of the latter (not necessarily each individual j-atom) is itself entering on oscillatory motion but in opposite phase to that of atom i. ψ rises from its minimum and becomes slightly positive again, so part of the momentum it transferred is given back but the rest is transmitted to the second shell as the first shell's correlation with $v_i(0)$ falls to small values.

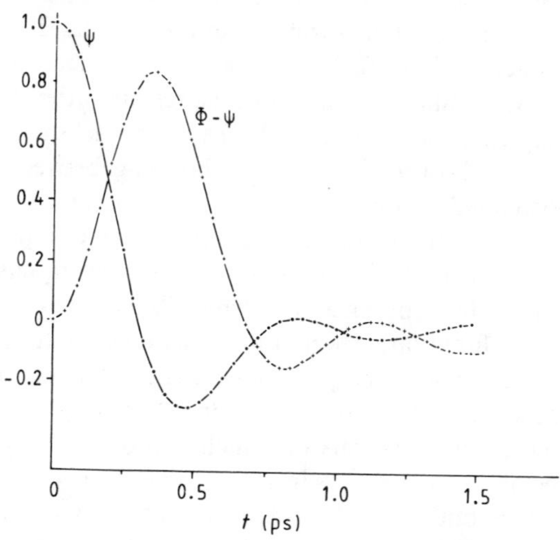

Figure 8.5 $\psi(t)$ (dots) and $\Phi(t) - \psi(t)$ (crosses) for liquid Rb from MD calculation (from Ballucani *et al* 1983a).

These arguments and the experiments indicate that the pair potential in Rb is responsible for damped collective motions, with wavelength comparable with the interatomic spacing, which are compatible with diffusive motion in the long term for the atoms separately. Collective motions and so-called dispersion relations have been observed in liquid Ne, Pb, K and Bi also and the conditions for their existence are discussed by Söderström *et al* (1980).

The quite different LJ potential does not lead to such phenomena. The behaviours in figures 8.2 and 8.3 do not occur with liquid Ar and the correlated oscillations of $(\Phi - \psi)$ and ψ do not appear in the MD results (Ballucani *et al* 1983b). However, at the time of writing, the short-wavelength collective motion in Ar, if any, is not without controversy among specialists (de Schepper *et al* 1984). The observed S_{inc} for Ar indicates simple diffusive motion setting in after the short time of about 2×10^{-12} s and suggests that vibrational modes must be strongly damped. This and also the shape of S_{coh} are in good agreement with MD computations with an LJ potential (Sköld *et al* 1972).

Among representative examples of work in disordered solids we could take the experiments of Suck *et al* (1983) on $Mg_{70}Zn_{30}$. By measuring $S(Q, \omega)$ and plotting it against $\hbar\omega$, peaks were revealed which, for the first time in a metallic glass, could be attributed to collective motions occurring at Q-values near Q_p, the first peak position in $S(Q)$. A dispersion curve of these peaks was roughly U-shaped with a minimum near Q_p. The peaks disappeared on crystallisation. The interpretation was in terms of the dispersion of transverse modes (see §8.5).

In $SrCl_2$, as in other solids of the fluorite structure, the anions acquire disordered translational motion above a transition temperature and become mobile carriers. Incoherent neutron scattering was used by Schabel *et al* (1983) to characterise the diffusive ion motion—in effect this is a measurement of $G_s(\mathbf{r}, t)$. It is difficult to interpret the results without invoking either theoretical or MD calculations and in this case a model theory existed which gave $S_{inc}(Q, \omega)$ for diffusion by random hopping between regular lattice sites. It was postulated that numerous vibrations occurred between random hops which were virtually instantaneous and that the hops and vibrations were entirely uncorrelated. This results in an S_{inc} which has a Lorentzian-shaped factor from the hopping and a Debye–Waller factor from vibrations. The measurements fitted rather well if it was assumed that the anion hopped to nearest- and next-nearest-neighbour vacancies. The MD calculations carried out by Jacucci and Rahman (1978) on a similar system give extensive information and detailed insights into such motions.

Another type of disorder occurs in so called plastic crystals which preserve translational order but allow disordered orientations of molecules (§1.7); examples are CCl_4, NH_4Cl, O_2, s-triazine. A combination of neutron inelastic scattering and MD enabled the disordered orientational motion of SF_6 to be understood as an adjustment to competitive ordering and disordering forces in the presence of acoustic phonon modes (Dove *et al* 1986).

A progressive understanding of thermal motion will continue to come from an intimate mixture of experiments and computer simulations with first-principle and model theories. The theoretical problem is that of the evolution in space and time of a fluctuation perhaps induced by an incident particle. In a perfect crystal the disturbance would propagate as phonons which would proceed to the sample boundaries in the absence of dissipative mechanisms. Disordered matter occurs in such variety that no theory will serve all systems; indeed in one system different ranges of ω and Q may require separate treatments. The motion is usually discussed in terms of the quantities already introduced, such as $G(\mathbf{r}, t)$, $S(Q, \omega)$ and $\psi(t)$ or various transforms or properties of them.

In a pioneering paper, Vineyard (1958) elucidated G_s and S_{inc} for ideal cases and figure 8.6 shows schematically the square of the width, W^2, of $G_s(\mathbf{r}, t)$ in a few cases in which it can be calculated exactly. In these curves the bound atoms show free-particle behaviour at very short times and we should expect this in liquids too. The rest of the curve for liquids and disordered

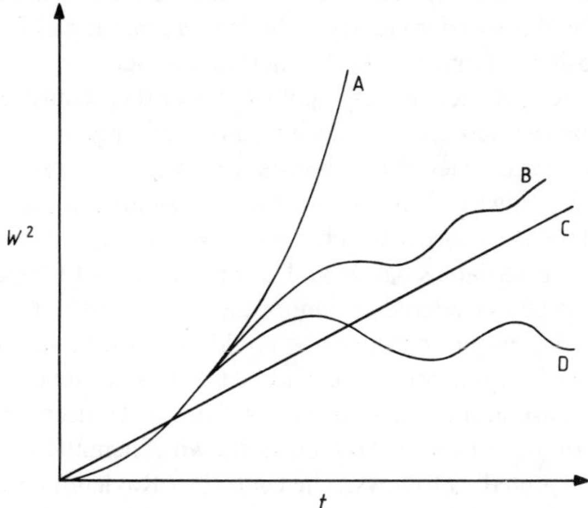

Figure 8.6 Schematic diagram showing the squared width $\langle W^2 \rangle$, obtained from the e^{-1} width of $G_s(r, t)$ in ideal cases. The particles are in: A, a perfect gas; B, a liquid; C, a system in which they obey the diffusion equation; D, a crystal.

solids is not obvious, although a linear dependence of W^2 on t—diffusive behaviour—might well be expected for a liquid at the limit $t \to \infty$. All the ideal systems have

$$G_s(r, t) = \left(\frac{1}{\pi W^2}\right)^{3/2} \exp(-r^2/W^2) \tag{8.22}$$

where $W^2 = 4Dt$ for diffusion and $2k_BTt^2/m$ for the perfect gas. Equation (8.22) has therefore been suggested as a good general purpose equation for liquids. Compared with MD calculations for a LJ potential it performs reasonably well near the triple point.

From equation (8.15*a*) it follows for a perfect gas that $S_{inc}(Q, \omega)$ is a Gaussian function of ω at fixed Q whereas for a diffusing particle it is a Lorentzian. In an interesting measurement of S_{inc} for liquid Na, Gläser and Morkel (1984) showed the transition from approximately Lorentzian towards approximately Gaussian form as Q increased from 1.4 to 2.45 Å^{-1}. This exemplifies the way in which single-particle motion in liquids falls in just the complex range between the ideal cases.

The major problem remains of deriving $G(\mathbf{r}, t)$ or $S(Q, \omega)$ from first principles or at least on some reasonable phenomenological basis.

8.4 Some theoretical approaches: fluids

The propagation of acoustic waves with speed $c = (mn_0\kappa_s)^{-1/2}$ is a well known phenomenon in fluids. The wavelength is much larger than the interatomic spacing and the dissipative mechanisms are viscous resistance and heat conduction. The evolution of density fluctuations in the limit of small Q and ω can be discussed similarly in hydrodynamic terms involving shear and bulk viscosities, η and ζ, and the thermal conductivity, λ. In fact the $\boldsymbol{r}$- and t- dependences of such density fluctuations or their transformed version, $S(Q, \omega)$, can be derived from basic hydrodynamic equations, namely, the conservation of mass, energy and momentum with the further assumption that the system is always in thermodynamic equilibrium locally. The local equilibrium state is accordingly characterised by any convenient pair of independent state variables say n and T or p and S. Fluctuations in one variable have to be considered assuming the other constant and vice versa. A fluctuation δn in number density could, for example, be written as $(\partial n/\partial p)_S \delta p + (\partial n/\partial S)_p \delta S$. $S(Q, \omega)$ calculated in this context is the hydrodynamic limiting case and its derivation is set out in Hansen and McDonald (1976) and Mountain (1966). $S(Q, \omega)$ is shown schematically in figure 8.7 and can be interpreted as follows. The central or Rayleigh peak has a width $2Q^2a$ where $a = \lambda/c_p$, the thermal diffusivity, c_p being the heat capacity per unit volume. It represents, in Fourier transform, entropy fluctuations at constant pressure which die away in time but do not propagate. The side or Brillouin peaks have a width $2Q^2\Gamma$ where Γ is the attenuation coefficient of acoustic waves deriving from viscosity and thermal conductivity. They represent, in Fourier transform, pressure fluctuations at constant entropy and propagate with the speed of sound.

Thinking of $S(Q, \omega)$ as a factor in a scattering cross section, we see that the Brillouin peaks indicate the ability of the target sample to be excited into propagating or 'phonon-like' modes with lifetimes inversely related to the width of the peaks. Figures 8.2 and 8.3 give neutron examples of this phenomenon though in fact they require a generalisation of the hydrodynamic

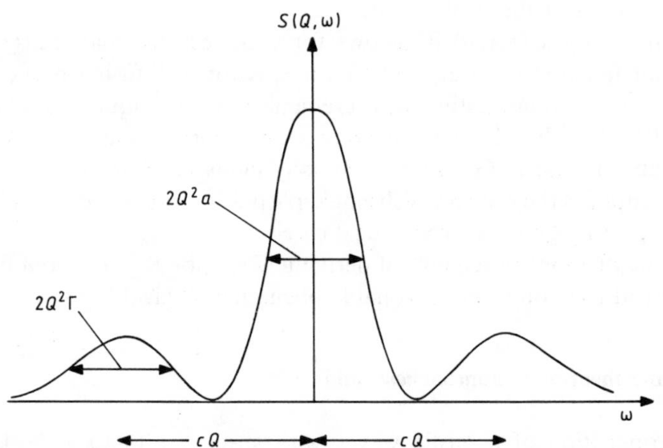

Figure 8.7 Schematic diagram of Rayleigh and Brillouin peaks in $S(Q, \omega)$.

theory for their full explanation because their wavelengths are not much greater than the interatomic distance (see below). True hydrodynamic Brillouin scattering of light is well known however, and can be detected with laser/interferometer systems. It is coherent with $\lambda \sim 6000$ Å and $E_i/\hbar\omega \sim 10^5$ (see §8.6 for an example).

It is not difficult to conceive a generalisation of hydrodynamic theory though the implementation of this requires rather extensive deductions. An atom which receives an impact of duration short compared with a characteristic time τ for the rearrangement of its neighbours reacts as if it were in a solid. A liquid therefore responds elastically to sufficiently high frequencies. τ is called the Maxwell relaxation time and the coefficient relating the shear stress to the time derivative of shear strain is η for $\omega\tau \ll 1$, $(-\mathrm{i}\omega)^{-1}$ times the shear modulus G for $\omega\tau \gg 1$ and $G[(G/\eta) - \mathrm{i}\omega]^{-1}$ in between, where $-\mathrm{i}\omega$ replaces $\partial/\partial t$ for a harmonic component of a general disturbance. Theories based on frequency-dependent responses of this kind are called viscoelastic approximations and offer an extension of hydrodynamic theory to cover fluctuations of larger Q and ω. $\tau = \eta/G$ is typically of the order of 10^{-13} s. Naturally it is necessary to have a formalism into which viscoelastic assumptions may be introduced and an important example stems from the Langevin equation.

In the theory of the Brownian motion an important starting point is the Langevin equation written here in one dimension

$$\dot{v}_x(t) = -Kv_x(t) + A_x(t) \tag{8.23}$$

where v_x is the speed of a particle and the K- and A-terms are m^{-1} times, respectively, a viscous resistance to the motion and a fluctuating force. The latter averages to zero and is uncorrelated with $v_x(0)$; it also has $\langle A(t)A(0)\rangle = 2A_0\delta(t)$ where A_0 is a measure of its strength. The Langevin equation gives a very good account of the Brownian motion of massive particles in a fluid bath at temperature T. The subtlety of it lies in the fact that both the friction and random forces are aspects of the same molecular bombardments. They are in fact connected, k_BTK/m being equal to A_0.

It is easy to derive the velocity autocorrelation function, $\psi(t)$, from equation (8.23) and if this is treated as a theory of ψ for molecules as distinct from macroscopic particles, it fails to agree with MD results because it is a positive exponentially decaying function whereas ψ is considerably more complicated (see, e.g. figure 8.5). The reason is that in relation to molecules the assumptions in equation (8.23) are oversimplified, especially the implication that the retarding force at t is proportional to the speed at t. We should rather expect the molecule and its neighbours to take time to adjust to each others' motion and the actual force on the molecule with speed $v_x(t)$ to depend on the recent history of the environment which has been adjusting itself to the molecule's motion at earlier moments $t' < t$. Another way of putting it is that the interaction of a specified or 'tagged' particle with its neighbours causes excitations in various modes which propagate through the fluid with a range of speeds and later feed back an influence on the tagged particle. This effect can be seen in MD calculations and it leads to a surprisingly long period of feedback causing $\psi(t)$ to decay slowly as $t^{-3/2}$ for large t (see, e.g., Sjögren and Sjölander 1979). A Brownian particle on the other hand is subjected simultaneously to large numbers of uncorrelated impacts and essentially responds to a slowly varying force. To embody the feedback effect for application to molecular motion the Langevin equation is generalised to

$$\dot{\boldsymbol{v}}(t) = -\int_0^t \mathrm{d}t'\, K(t-t')\boldsymbol{v}(t') + \boldsymbol{A}(t) \tag{8.24a}$$

where we have moved into three dimensions and assume that $\boldsymbol{A}(t)$ has the same properties as before. $K(t-t')$ is called a memory function and vanishes for $t < t'$. Multiplying equation (8.24a) by $v(0)/\langle v^2\rangle$ and averaging we find

$$\dot{\psi}(t) = -\int_0^t \mathrm{d}t'\, K(t-t')\psi(t') \tag{8.24b}$$

because $\boldsymbol{A}$ and $\boldsymbol{v}(0)$ are uncorrelated. Depending on the assumed form of $K(t-t')$ this provides for $\psi(t)$ a more complex behaviour than exponential decay. As with the original Langevin equation, K and A are related, in fact

$$K(t-t') = \langle A(t)A(t')\rangle/\langle v^2\rangle. \tag{8.24c}$$

In important papers, Mori (1965a,b) placed the generalised Langevin equation on a firm statistical mechanical basis and showed that equations of the same form, with appropriate memory functions, could be used for any other fluctuating dynamical variable which has the general properties that have been assumed for $v(t)$. For example, an equation of form (8.24a) is satisfied by the random variable $\boldsymbol{A}(t)$ itself, with another memory function, say $M(t)$, and another fluctuating quantity uncorrelated with $\boldsymbol{A}$ as second term on the right. K can be shown to obey an equation of form (8.24b) with K replacing ψ and M replacing K. Thus there is an equation for ψ in terms of K and for K in terms of M. An equation can be found for M in terms of yet another memory function and a hierarchy can be erected which can actually be solved for ψ in the form of a continued fraction. The latter can be used for successive approximations to ψ assuming that the continuous fraction can be closed off by adopting some plausible expressions for the memory functions. Indeed, if $\tilde{\psi}(s)$ is the Laplace transform of $\psi(t)$ it follows at once from equation (8.24b) that

$$\tilde{\psi}(s) = \frac{\psi(0)}{s+\tilde{K}(s)} = \frac{\psi(0)}{s+\dfrac{K(0)}{s+\tilde{M}(s)}} = \frac{\psi(0)}{s+\dfrac{K(0)}{s+\text{etc}}} \tag{8.25}$$

If the first memory function $K(t)$ is given the simple form $K_0 \exp(-\alpha|t|)$, and the others ignored, the resulting $\psi(t)$ is already more complex than that given by equation (8.23) and becomes oscillatory as well as exponentially decaying. Naturally the preceding statements need proof and an extended discussion of memory function theory is given in Sjögren and Sjölander (1979), Gaskell (1978) and Copley and Lovesey (1975).

Among the quantities which have been discussed in terms of memory functions are $\psi(t)$ and autocorrelation functions of the particle current, $\boldsymbol{j}(\boldsymbol{r}, t)$. The latter is an important quantity because the Fourier components, $\boldsymbol{j}(\boldsymbol{Q}, t)$, can be formed and correlations considered of the longitudinal and of the transverse components, i.e., those parallel and perpendicular to $\boldsymbol{Q}$. The particle current is a fluctuating quantity and its longitudinal and transverse autocorrelation functions enable longitudinal and shearing motions to be discussed separately; the former involve density fluctuations and bulk viscosity and the latter shear viscoelasticity. The necessary definitions are

$$\boldsymbol{j}(\boldsymbol{r}, t) \equiv \sum_i \boldsymbol{j}_i \equiv \sum_i \boldsymbol{v}_i \delta(\boldsymbol{r} - \boldsymbol{r}_i(t)) \tag{8.26a}$$

$$\boldsymbol{j}(\boldsymbol{Q}, t) = \sum_i \boldsymbol{v}_i(t) \exp(\mathrm{i}\boldsymbol{Q} \cdot \boldsymbol{r}_i(t)) \tag{8.26b}$$

$$C_\mathrm{l} \equiv \frac{1}{N} \left\langle \sum_j \boldsymbol{Q} \cdot \boldsymbol{v}_j(0) \exp(-\mathrm{i}\boldsymbol{Q} \cdot \boldsymbol{r}_j) \sum_i \boldsymbol{Q} \cdot \boldsymbol{v}_i(t) \exp(\mathrm{i}\boldsymbol{Q} \cdot \boldsymbol{r}_i) \right\rangle. \tag{8.27a}$$

If the z-axis is chosen in the $\boldsymbol{Q}$-direction the autocorrelation coefficient becomes

$$C_\mathrm{l}(Q, t) = \frac{Q^2}{N} \langle j_z(\boldsymbol{Q}, t) j_z(-\boldsymbol{Q}, 0) \rangle. \tag{8.27b}$$

Similarly

$$C_\mathrm{t}(Q, t) = \frac{Q^2}{N} \langle j_x(\boldsymbol{Q}, t) j_x(-\boldsymbol{Q}, 0) \rangle \tag{8.27c}$$

where l, t stand for longitudinal and transverse, respectively.

For single-particle or self-current autocorrelation

$$C_\mathrm{s}(Q, t) = \langle \boldsymbol{Q} \cdot \boldsymbol{j}_i(\boldsymbol{Q}, t) \boldsymbol{Q} \cdot \boldsymbol{j}_i(\boldsymbol{Q}, 0) \rangle. \tag{8.27d}$$

It can be shown (Hansen and McDonald 1976) that the Fourier transform of C_l, i.e. $C_\mathrm{l}(Q, \omega)$, is equal to $\omega^2 S(Q, \omega)$. C_t does not directly determine neutron scattering. Memory functions can be chosen for their physical plausibility and to satisfy overriding requirements such as that $S(Q, \omega)$ should have the right moments (see §8.2) and reduce to hydrodynamic results in the appropriate limit. In constructing memory functions characteristic viscoelastic hypotheses, such as making the Maxwell relaxation time a function of Q, are brought in. From a large and somewhat complex literature we take two typical results both of which follow from exponential memory functions for the currents (Hansen and McDonald 1976, Copley and Lovesey 1975):

$$C_\mathrm{t}(Q, \omega) = \frac{Q^2 k_\mathrm{B} T}{\pi m} \frac{\omega_\mathrm{t}^2 \tau_\mathrm{t}(Q)}{\omega^2 + \tau_\mathrm{t}^2(Q)(\omega_\mathrm{t}^2 - \omega^2)^2} \tag{8.28}$$

where $\omega_\mathrm{t}^2 = Q^2 G_\infty(Q)/n_0 m$ and G_∞ is the high-frequency limit of the shear modulus; $\tau_\mathrm{t}(Q)$ is a Maxwell relaxation time. The theory out of which equation (8.28) emerges gives a very good but not perfect fit to the MD data for a LJ fluid over a considerable range of Q and ω. It embraces the phenomena of shear wave dispersion predicted by MD for liquid Ar but apparently not yet accessible to experiment. The second example is

$$S(Q, \omega) = \frac{Q^2 k_B t}{\pi m} \frac{\tau_1(Q)(\omega_1^2 - Q^2 k_B T/mS(Q))}{[\omega\tau_1(Q)(\omega^2 - \omega_1^2)]^2 + (\omega^2 - Q^2 k_B T/mS(Q))^2} \tag{8.29}$$

where ω_1^2 is the ratio of the fourth moment to the second moment of $S(Q, \omega)$ and can be shown to depend on the radial distribution function and pair potential; $\tau_1(Q)$ is a Maxwell relaxation time. This equation deals reasonably satisfactorily with the shape of the Rb data in figures 8.2 and 8.3 showing that neutron scattering from this liquid can be accounted for by a viscoelastic theory of damped collective density fluctuations but there is rather poor agreement with MD for argon modelled with LJ potential.

The relation between single particle and collective motion, illustrated in §8.3, can be expressed by deriving $\psi(t)$ in terms of C_l and C_t. Using plausible physical assumptions, such as that momentum diffuses in dense liquids much faster than the particles themselves, Ballucani *et al* (1983a) achieved this successfully for Rb.

Such theories are intricate and to a large extent successful but they are not from first principles because memory functions have to be modelled and parameters decided upon for $\tau(Q)$. Achieving a first-principles theory is difficult but notable endeavours have been made of which only the briefest outline will be attemped here. Barker and Gaskell (1975) proceeded to an equation of type (8.24*b*) for the function $C_l(Q, t)$ defined by equation (8.27). This was done by differentiating C_l with respect to time and expressing the resulting velocity derivatives in terms of the interatomic forces—in other words a memory function equation for C_l was to be derived from the equation of motion of the particles. After a number of subsidiary assumptions C_l and $S(Q, \omega)$ could be calculated from interatomic potential data and $S(Q)$ with considerable success for Rb.

It is not, however, necessary to remain with the memory function formalism. Hubbard and Beeby (1969) used the linear-response theory approach. A small change in density, $\delta\rho(\boldsymbol{r}, t)$, was supposed to be produced by a weak perturbing potential, the two being connected by a response function like φ in §7.7 but depending on $\boldsymbol{r} - \boldsymbol{r}'$ as well as $t - t'$. This quantity is intimately related to $S(\boldsymbol{Q}, \omega)$. The response function is obtained—not without assumed approximations—by finding the small density changes caused by the displacements the particles suffer under the applied perturbing force. The calculation is done in two stages and the first amounts to an attack on the problem for solid amorphous materials because the atomic sites were fixed. In the second stage diffusive motion was allowed by making the position vectors time dependent. If exploited to the full the method is a complete solution, at least for linear responses, embracing a derivation of $S(Q)$ and $G_s(\boldsymbol{r}, t)$. In practice model functions and approximations were needed and the rule $\int S(\boldsymbol{Q}, \omega)\, d\omega = S(Q)$ was not well obeyed by the result. Apparently the model greatly underestimates the damping.

The basic quantity in another group of approaches is

$$f(\boldsymbol{r}, \boldsymbol{p}, t) \equiv \sum_i \delta(\boldsymbol{r} - \boldsymbol{r}_i(t))\delta(\boldsymbol{p} - \boldsymbol{p}_i(t))$$

which is the density distribution of particles in 6-D phase space. The autocorrelation function of f in phase space and time obeys an equation which is reminiscent of the Boltzmann equation and in fact reduces to it at low number densities. The autocorrelation function is rich in consequences and once obtained would yield $S(Q, \omega)$ and many other useful quantities by appropriate integrations. However the formal exact equation for it contains an unknown memory function so a first-principles theory requires a theory of the memory function. Jhon and Forster (1975) therefore proposed an approximation which was in the spirit of kinetic theory and was regarded by the authors as intermediate between viscoelastic or hydrodynamic model theories and a first-principles treatment starting with the particles and their interactions. It was conspicuously successful with liquid Ar and reduces to the hydrodynamic theory at the low-Q and -ω limit and has the proper moments of $S(Q, \omega)$.

Mode-coupling theory is the name given to an attempt to embody the delayed feedback of the liquid onto a moving particle into mathematical expressions for memory functions such as K in equation (8.24*b*). Especially in its self-consistent form it is technically too complicated to explain here and at the time of writing is still under development for liquids, supercooled liquids and glasses. It is quite capable of giving expressions for $\psi(t)$, $S_s(Q, \omega)$ and $S(Q, \omega)$ from a starting input of $\varphi(\mathbf{r})$ and $S(Q)$. The paper by Sjögren and Sjölander referred to above discusses mode coupling extensively in relation to $\psi(t)$ and obtains excellent agreement with MD results for liquid Rb.

Other theories will probably be developed since particle motion in fluids is not a closed subject. Because of relations like equation (8.20), the transport coefficients are also products of this work (see, e.g., Gerl and Bruson 1980).

8.5 Some theoretical approaches: solids

As usual a crystalline order of sites does not remove the problems of disorder if varieties of particles are randomly distributed on the sites. To take the simplest case: a linear chain of regularly spaced atoms consists of two types with different masses and different coupling constants sited according to some probability distribution. The aim is to find the frequency distribution of the allowed vibrations and to understand something of the nature of the modes. This immediately poses a far more difficult mathematical problem than that of the crystalline chain which is often used to introduce elementary phonon theory. The disorder problem was actually solved in 1953 by Dyson for a long chain of coupled harmonic oscillators with random masses or coupling constants (Dyson 1953). Although this gave an analytic solution for the frequency distribution it is not particularly revealing about the features nor readily generalised to two or three dimensions.

A clearer picture emerged from a computational attack on the problem by Dean and collaborators (Dean 1972). This was essentially a numerical solution of the classical equation for coupled oscillators. To taste the flavour of this type of work consider an atom of mass m_i at site i in a square lattice (see figure 8.8); it is shown surrounded by its nearest neighbours P, Q, R and S with site labels p, q, r and s. The equation of motion in the x-direction is

$$m\ddot{x}_i = \gamma_{iq}(x_q - x_i) + \gamma_{is}(x_s - x_i) + \gamma'_{ip}(x_p - x_i) + \gamma'_{ir}(x_r - x_i)$$

where γ_{ij}, γ'_{ij} are, respectively, central and non-central force constants between atoms at sites i and j. Assuming only nearest-neighbour harmonic interactions, $m\ddot{x}_i = -\omega^2 x_i$. This leads to a set of equations, one for each atom, which have to be satisfied simultaneously for each allowed value of ω. The mathematical problem is then to obtain the ω—i.e. the frequency spectrum—for given boundary conditions which could, for example, specify either clamped or free boundary atoms. The technique for this major computational problem was reviewed by Dean (1972) and this reference gives many examples for one, two and three dimensions. Decisions have to be made about the kind and amount of disorder, e.g., γ and γ' could be fixed but the distribution of two different masses over the lattice could be randomised to avoid LRO and SRO.

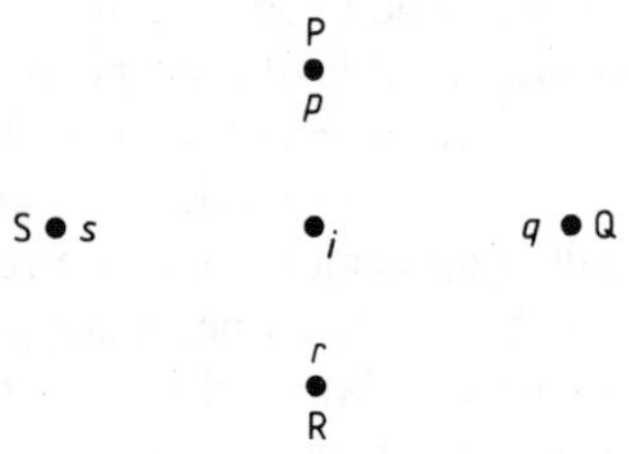

Figure 8.8 Atom in a square lattice.

In figure 8.9 Dean's results are shown for the comparable but simpler problem of the one-dimensional chain with equal force constants but random heavy (H) and light (L) atoms. The interesting interpretation is that the set of peaks at high frequencies is associated with localised modes of vibration each characteristic of a local structure. Peak A, for example, is associated with the vibration of a light atom embedded in heavy ones, namely, a structure like . . . HHHLHHH . . . . Other peaks, C, D, and E, come from . . . HHHLLHHH . . . structures and so on. The heavy atoms are hemming in the high-frequency vibrations of isolated or paired light atoms and localising the modes (see also §9.5). The assignment of peaks to structures fully explains the evolution of the spectrum as the proportion of light atoms increases. First, peak A intensifies as the concentration of L rises but later, peaks C, D and E increase at the expense of A as the probability of paired atoms rises relative to that of single ones. Exactly similar considerations apply to 2-D network spectra, the high-frequency peaks corresponding to isolated L atoms or pairings or triangular clusters. This type of work continues to be extended and its close relation with the recursion method of §7.4 clarified (Baer 1983).

Such calculations are based on hypothetical disordered 2-D lattices. To get

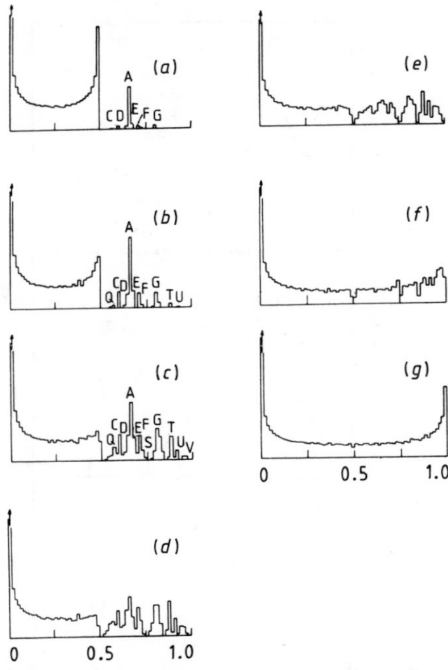

Figure 8.9 Distribution of ω^2 for a random chain of 8000 atoms with equal force constants and mass ratio 2 : 1. The fraction of light atoms, c_L, is: (*a*), 0.05; (*b*), 0.16, (*c*), 0.38; (*d*), 0.50; (*e*), 0.62; (*f*), 0.841; (*g*), 0.95. The abscissa is ω^2/ω_0^2 where ω_0 is the maximum frequency of a monatomic chain of light atoms. (From Dean 1972.)

nearer to real topologically disordered glasses Dean and his collaborators constructed the kind of 3-D ball and spoke model of SiO_2 referred to in §2.6. Using its coordinates and taking central and non-central forces from existing knowledge of SiO_2 bonds, Dean's computational methods gave spectra like figure 8.10 (Bell *et al* 1968). Much later, Laughlin and Joannopoulos (1977) calculated the SiO_2 spectra using the vibrational motion equivalent of the Green function methods used for electronic motion in Chapter 7. Some results are shown in figure 8.10 where the spectrum inferred from neutron inelastic scattering is given as well. The overall agreement is quite good. This still does not assign the different frequencies to identified atomic motions and Dean and collaborators emphasise that the modes are very complex being neither propagating waves like phonons in crystals nor highly localised vibrations centred on only a few atoms. Nevertheless tentative suggestions have been made such as that the peak at about 730 cm^{-1} is a bond-bending mode in which the O atoms move in the Si–O–Si planes perpendicularly to the Si–Si lines. Other identifications are bond stretching in which O vibrates in the Si–O–Si plane parallel to the Si–Si lines causing the peak at 1040 cm^{-1}, while bond rocking in which O moves at right angles to the plane accounts for the 410 cm^{-1} feature. Germania and beryllium fluorite glasses can be discussed in similar terms.

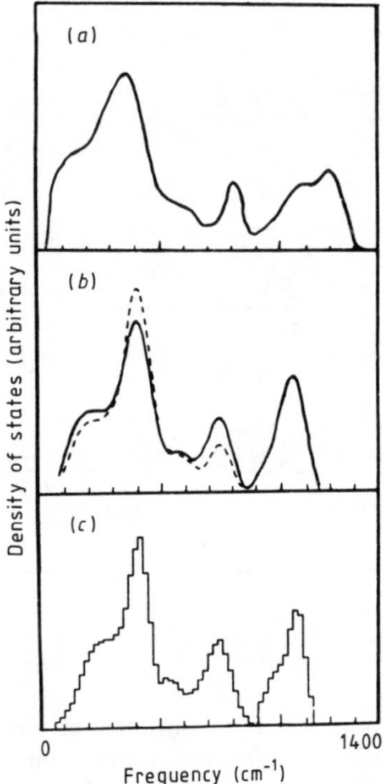

Figure 8.10 Frequency spectrum of vitreous SiO_2 inferred from: (*a*), inelastic neutron scattering; (*b*), Green function theory (broken curve, $S(Q, \omega)$; full curve, density of states); (*c*), density of states computed for the ball and spoke model. (From Laughlin and Joannopoulos 1977.)

For a theory or numerical calculation hypotheses are necessary for the structure and interparticle forces as exemplified in the last paragraph. For not too complicated metallic glass, say Mg_xCa_{1-x}, the three interparticle potentials emerge from the general pseudopotential theory of §§6.7 to 6.9. The structure is never completely known so it could be derived by a suitable version of the relaxation procedure in §2.5 using a dense random packing hard-sphere model (DRPHS) to start with and the interionic potentials to compute the relaxation. In a metal, however, the energy is both density- and structure-dependent and relaxation is carried out at some given starting density. This may not be the density for lowest energy so the relaxation has to be repeated after the first relaxed structure has been compressed slightly and the potentials reassessed for the new density. Pursuing this amounts to a way of seeking which density gives an energy minimum representing the metastability of the glass and the result should agree with the observed density. Assuming this is successful it is then a matter of numerical integration of the equations of motion of the atoms to obtain the frequency spectrum, $g(\omega)$, and $S(Q, \omega)$. Techniques for this have been given by von Heimendahl (1979). However, there are alternative possibilities. In §7.4 the recursion method for clusters was outlined in connection with $g(E)$ for electrons. $g(\omega)$ for thermal vibrations is a not dissimilar problem and the recursion method is certainly adaptable to it.

The first comprehensive examples of this programme, including successive relaxation for the density, are the calculations for glassy $Ca_{70}Mg_{30}$ made by Hafner (1983). From among the wealth of results for $g(\omega)$ and $S(Q, \omega)$, in general agreement with experiment, we refer to the dispersion relation for collective modes identified from the inelastic peaks when $S(Q, \omega)$ is plotted against ω for constant Q. Experiments made by Suck *et al*, comparable with those previously referred to for $Mg_{70}Zn_{30}$, are shown in figure 8.11. The agreement is qualitatively reasonable and the minima occur both at Q_p and the position of the second peak in $S(Q)$. These minima have an interesting explanation. In phonon and electron scattering in crystals 'Umklapp' scattering is a well known process; it involves wavevector changes of $\boldsymbol{Q} + \boldsymbol{K}$ instead of $\boldsymbol{Q}$ where $\boldsymbol{K}$ is a reciprocal lattice vector (see Ziman 1960 in Chapter 6). Since the peaks of $S(Q)$ represent the constructive interference of scattered wavelets—of which Bragg scattering in crystals is the extreme case leading to δ-functions instead of finite peaks—it seems possible that a scattering term especially associated with Q_p might be the residual vestige in disordered solids of Umklapp processes in crystals. That this is a reasonable hypothesis was shown by Hafner (1983, 1985) in connection with $Mg_{70}Zn_{30}$ with the result that the residual or diffuse Umklapp scattering term

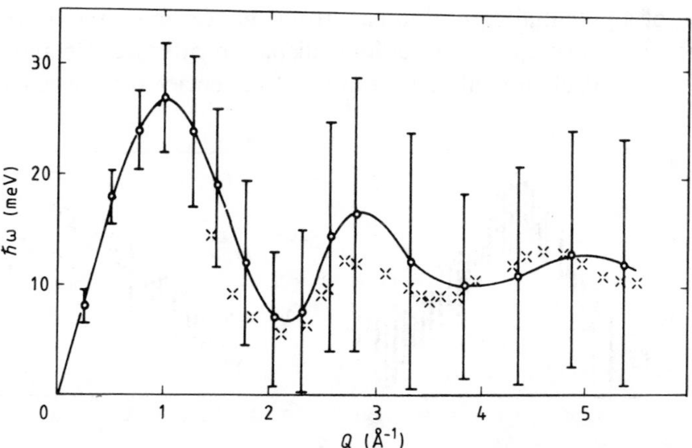

Figure 8.11 Dispersion of collective excitations in metallic glass $Ca_{70}Mg_{30}$. Circles and full line, theoretical results for which the bars show the width of the peak in $S(Q, \omega)$ at half maximum; crosses, observed peak positions. (From Hafner 1983.)

for long-wavelength acoustic modes could account well for the experimental results, with most of the effect coming from transverse modes. The cluster relaxation recursion calculation is also applicable to the problem of how disorder in masses, force constants, or local distortion, affects $S(Q, \omega)$ in a substitutional alloy based on a lattice (Hafner and Punz 1984).

The frequency spectrum, $S(Q, \omega)$, and the dispersion relation are various aspects of the thermal motion. Another aspect is the actual spatial distribution of a particular mode, i.e., the degree to which it is localised. All these properties can be investigated with computer simulations. §4.6 mentions how MD can create computer glasses and introduces the quench echo. An application of these techniques first revealed general properties of both longitudinal and transverse thermal modes in a Lennard-Jones glass (Grest *et al* 1982). A brief description is as follows. As the frequency rises to about 2.3×10^{12} Hz, the peak in $S(Q, \omega)$ representing longitudinal acoustic modes is present but decreasing in definition as ω rises indicating increasing damping of the phonons. At even higher frequencies, up to 5 or 6×10^{12} Hz, instead of vanishing completely, the peaks actually become more distinct and then apparently derive from highly localised modes. The dispersion relation of these peaks is shown in figure 8.12. In a crystal the minimum would occur at the first reciprocal lattice vector. Here, as in $Mg_{70}Zn_{30}$, it occurs at Q_p which is therefore sometimes called a pseudo-zone boundary. A quantity allied to C_t in equation (8.27*c*) can be computed to investigate transverse modes. They also exist in the same ω- and Q-ranges with similar though less well developed behaviour and a linear dispersion relation rising with Q. Again the higher-frequency modes are highly localised.

Later, Bhatia and Singh (1985) suggested a semiphenomenological model for dispersion relations. The force on an ion in a metal was expressed as the gradient of a potential derived in part from changes in local electronic density when an acoustic wave causes ionic displacements (see §6.6 and 6.7). To simplify the calculation of the vibration frequencies drastic assumptions

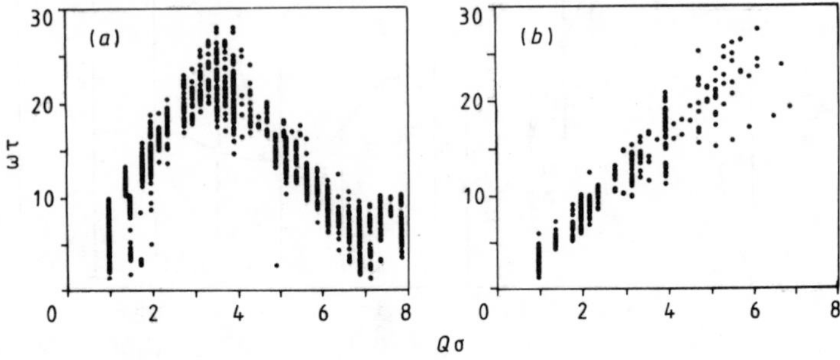

Figure 8.12 Dispersion relations for (*a*) longitudinal and (*b*) transverse excitations obtained by MD calculations for a LJ computer glass. σ, ε = LJ parameters; $\tau \equiv \sigma(m/\varepsilon)^{1/2}$. (From Grest *et al* 1982.)

were made including the restriction of the pair forces to nearest neighbours, the smearing out of the first coordination shell into a uniform spherical distribution of neighbours and the omission of any considerations leading to phonon damping. In spite of this, dispersion relations broadly comparable with both experiment and MD were derived for most of the systems discussed in previous sections including the LJ and metallic glasses and liquid Rb.

At the time of writing, it appears that a good deal remains to be learned about the nature of thermal excitations in even chemically simple liquids and amorphous solids, but it also appears that neutron inelastic scattering and molecular dynamics are a powerful combination of techniques for answering at least some of the questions.

8.6 Other systems and techniques

To introduce thermal motion emphasis has been laid on monatomic matter and neutron inelastic scattering. Before leaving the subject brief reference will be made to much wider ramifications. For not only its intrinsic physical interest, but also as a contribution to chemistry and biology, thermal motion is studied in water, aqueous solutions, organic liquids, liquid crystals, proteins and polymers. With larger molecules translation interacts to a greater or lesser extent with rotation, vibration, bending, twisting, folding, uncoiling and other complex motions. The dynamics of protein molecules in solution, for example, are a major determinant of their biochemical function, and flash photolysis, NMR, Mössbauer spectroscopy and neutron inelastic scattering are amongst the techniques being used to study them (McCammon 1984). Some polymers can be tailored to predetermined lengths which enables the thermal motion to be investigated as a function of chain length, in connection with recent proposals to treat polymers as fractals (Rosenberg 1985).

While neutron inelastic scattering can make great contributions to the study of thermal motion, important information also comes from various photon spectroscopies. Infrared spectra have peaks corresponding to those in figure 8.10 indicating that photons with energies of about 0.1 eV can explore vibratory motion in glasses. The Brillouin scattering of visible light has been used to study phase transitions (Dil 1982) and partially ordered biological matter (Cusack and Lees 1984). The scattering of Mössbauer γ-rays can in principle measure $S(Q, \omega)$ over certain ranges of the variables. The fractional change of energy when the γ-photon is scattered is very small but the very high energy resolution attainable in Mössbauer studies enables $\hbar\omega$ to be measured and inelastic events to be separated from elastic ones. Mössbauer studies of diffusive and vibrational motion in supercooled liquids and glassy solids are described by Champeney (1979).

Of the photon spectroscopies, infrared and Raman have been widely applied to amorphous solids but, as with UPS in §7.6, the observed spectra are transformed from the density of states, $g(\omega)$, by a transition probability factor. For infrared absorption the coupling between the photon and the vibration is through the dipole moment associated with charge displacement; for Raman spectra it is through the polarisability. To reproduce the observed spectra theoretically the following sequence would be required: adopt a model structure (e.g. the Polk model of §2.6); choose interatomic central and non-central forces; apply one of the numerical methods to calculate $g(\omega)$; make a model for the relation of particle displacement to dipole moment and polarisability and calculate the transition probabilities. A sequence of this kind, and much else of relevance, can be found for amorphous Si and Ge in the work of Alben *et al* (1975). The $g(\omega)$ for amorphous and crystalline forms are not identical, but not dissimilar either, from which it appears—since LRO is absent from the amorphous models—that the local structures and forces dominate the density of vibrational states.

It is possible to separate out from the total nuclear magnetic relaxation rate in NMR the contribution, R_Q, made to it by quadrupolar relaxation. The latter is the phenomenon which occurs when a nucleus has a finite electric quadrupole moment experiencing fluctuating electric field gradients from its moving neighbours. The fluctuations act on the nucleus through its quadrupole and induce transitions between the nuclear spin states which have been split by the applied magnetic field. The relaxation time of these states, measured by NMR techniques, is thereby affected. This suggests that observation of R_Q might contribute to understanding relative motion of atoms in liquids but this would only be so if R_Q could be related theoretically to some definite aspect of the motion. This is quite difficult because R_Q depends on the spatial and temporal correlation function of the number density of the probe nuclei undergoing NMR with that of all other nuclei and electrons which collectively produce the field gradients at the probe nuclei. Nevertheless there are theories of R_Q and the prospects for this type of investigation were examined by Bosse and Wetzel (1983), by Gaskell and Woolfson (1984) and by Dupree and Rogers (1985). This is an example of the use of NMR to explore thermal motion in liquids. A general survey of NMR in relation to diffusing or re-orientating molecules in solids has been made by Ailion (1983).

Finally to return to neutron inelastic scattering, it can be anticipated that the isotopic substitution method introduced for structure studies (§3.10) will prove valuable for dynamics also. The hydrogen–deuterium exchange has an obvious relevance to proton dynamics and an application of isotopic substitution to $S(Q, \omega)$ in molten salts can be seen in the work of McGreevey *et al* (1985).

References

Ailion D C 1983 *Methods Exp. Phys.* **21** 439

Alben R, Weaire D, Smith J E and Brodsky M H 1975 *Phys. Rev.* B **11** 2271

Baer S 1983 *J. Phys. C: Solid State Phys.* **16** 6939

Ballucani U, Vallauti R, Gaskell T and Gori M 1983a *Phys. Lett.* **102A** 109

Ballucani U, Vallauti R, Murthy C S, Gaskell T and Woolfson M 1983b *J. Phys. C: Solid State Phys.* **16** 5605

Barker M I and Gaskell T 1975 *J. Phys. C: Solid State Phys.* **8** 89

Bell R J, Bird N F and Dean P 1968 *J. Phys. C: Solid State Phys.* **1** 299

Bhatia A B and Singh R N 1985 *Phys. Rev.* B **31** 4751

Bosse J and Wetzel C 1983 *Phys. Rev.* A **28** 2459

Champeney D C 1979 *Rep. Prog. Phys.* **42** 1017

Copley J R D and Lovesey S W 1975 *Rep. Prog. Phys.* **38** 461

Copley J R D and Rowe J M 1974a *Phys. Rev.* A **9** 1656

—— 1974b *Phys. Rev. Lett.* **32** 49

Cusack S and Lees S 1984 *Biopolymers* **23** 337

Dean P 1972 *Rev. Mod. Phys.* **44** 127

Dil J G 1982 *Rep. Prog. Phys.* **45** 285

Dove M T, Pawley G S, Dolling G and Powell B M 1986 *Mol. Phys.* **57** 865

Dupree R and Rogers A T 1985 *J. Phys. C: Solid State Phys.* **18** L265

Dyson F J 1953 *Phys. Rev.* **92** 1331

Egelstaff P A and Gläser W 1985 *Phys. Rev.* A **31** 3803, 1986 *Ibid.* **34** 2121

Egelstaff P A, Gläser W, Litchinsky D, Schneider E and Suck J-B 1983 *Phys. Rev.* A **27** 1106

Gaskell T 1978 *Prog. Liq. Phys.* ed C A Croxton (New York: Wiley) p 309

Gaskell T and Woolfson M S 1984 *Phys. Lett.* **105A** 83

Gerl M and Bruson A 1980 *J. Physique Coll.* (*Suppl.*) **41** C8 335

Gläser W and Morkel Ch 1984 *J. Non-Cryst. Solids* **61/62** 309

Grest G S, Nagel S R and Rahman A 1982 *Phys. Rev. Lett.* **49** 1271

Hafner J 1983 *Phys. Rev.* B **27** 678

—— 1985 *J. Non-Cryst. Solids* **75** 253

Hafner J and Punz G 1984 *Phys. Rev.* B **26** 7336

Hansen J-P and McDonald I R 1976 *Theory of Simple Liquids* (New York: Academic)

von Heimendahl L 1979 *J. Phys. F: Met. Phys.* **9** 161

van Hove L 1954 *Phys. Rev.* **95** 249

Hubbard J and Beeby J L 1969 *J. Phys. C: Solid State Phys.* **2** 556

Jacucci G and Rahman A 1978 *J. Phys. Chem.* **69** 4117

Jhon M S and Forster D 1975 *Phys. Rev.* A **12** 254

Laughlin R B and Joannopoulos J D 1977 *Phys. Rev.* B **16** 2942

Lovesey S W 1984 *Theory of Neutron Scattering from Condensed Matter* (Oxford: Clarendon)

McCammon J A 1984 *Rep. Prog. Phys.* **47** 1

McGreevey R L, Mitchell E W J, Margaca F M A and Howe M A 1985 *J. Phys. C: Solid State Phys.* **18** 5235

Mori 1965a *Prog. Theor. Phys.* **33** 423

—— 1965b *Prog. Theor. Phys.* **34** 399
Mountain R D 1966 *Rev. Mod. Phys.* **38** 205
Norden A and Lodding D 1967 *Z. Naturf.* **22A** 215
Nose S and Yonezawa F 1985 *Solid State Commun.* **56** 1005
Powles J G 1973 *Adv. Phys.* **22** 1
Rahman A 1974a *Phys. Rev. Lett.* **32** 52
—— 1974b *Phys. Rev.* A **9** 1667
Rosenberg H M 1985 *Phys. Rev. Lett.* **54** 704
Schabel P, Hayes W, Hutchings M T, Lechner R E and Renker B 1983 *Radiat. Eff.* **75** 73
de Schepper I M, van Well A A, de Graaf L A and Cohen E D G 1984 *Phys. Rev. Lett.* **53** 402 (and references therein)
Sjögren L and Sjölander A 1979 *J. Phys. C: Solid State Phys.* **12** 4369
Sköld K, Rowe J M, Ostrowski G and Randolph P D 1972 *Phys. Rev.* A **6** 1107
Söderstrom O, Copley J R D, Suck J-B and Dorner B 1980 *J. Phys. Coll.* (*Suppl.*) **41** C8 230
Suck J-B, Rudin H, Guntherodt H-J and Beck H 1983 *Phys. Rev. Lett.* **50** 49
Vineyard G H 1958 *Phys. Rev.* **110** 999

9

PERCOLATION AND LOCALISATION

The two subjects of this chapter have some similarities and make some mutual connections. They are both of fairly recent origin and authorities in the two fields have remarked how after the initial launch of the subject rather little happened for a decade or more whereupon a sudden burst of activity transformed the scene. By 1982 Anderson could refer to localisation as 'a growth industry' and Domb has recounted how the literature on percolation grew in a remarkable manner during the 1970s.

The attempt in this introduction is chiefly to apply these subjects to electron motion in condensed matter but in both cases the potential value is far wider. Localisation theory could apply to any kind of wave propagation in disordered media and percolation is not even confined to physics.

The two subjects come together in a brief treatment of conductivity by hopping.

9.1 Percolation—an introduction

In abstract terms percolation is the passage of an influence through a medium which is irregularly structured in the sense that through some regions the influence can pass more easily than in others and in some places not at all. A fundamental question is: if the disorder of the medium is subject to some probability distribution, for what values of the probability can the influence move infinite distances through the medium and for what values is it confined to finite regions?

It is not difficult to reduce the abstract question to particular terms. The 'medium' could be a computer network and the 'influence' information; disorder in the medium would result from random line or switch failures. On the other hand, the medium might be an array of equally spaced plants and the influence a disease carried by an insect moving on the surface. Irregularity in the medium could be introduced by the random removal of plants or by random spraying with repellent that rendered some plant-to-plant paths impassable for the insect.

To proceed from such examples to a problem in physics let us consider the experiment of Last and Thouless (1971) which was simple in concept and illuminating in results. Let a sheet of electrically conducting paper be the medium and let the influence be electric current flowing between two opposite edges. Disorder is introduced gradually by punching holes at coordinates (x_i, y_i) chosen sequentially by random numbers. The whole set of coordinates forms a square lattice and the hole size is such that nearest- and next-nearest-neighbouring holes overlap; for example, if, for a given y, x_i and x_{i+1} are punched out, no current can pass from x_i to x_{i+1} but it could flow round the double hole. The conductance, G, fell as the number of holes increased and when about 40 % of the sites were punched no current path remained so G vanished. There are at least two difficult problems posed by this. Let p be the probability that a site remains intact, i.e. not punched out. Let $P(p)$ be the fraction of all intact sites which are in electrical contact with each other and which collectively form the conducting path across the paper. Then $P(p)$ and $G(p)$ vanish when $p \simeq 0.6$ and problems arise as to the shapes of these two functions including the question of how they approach zero. It was shown by this experiment that $G(p)$ and $P(p)$ are different and that G falls to zero less rapidly than P near $p \simeq 0.6$. A little reflection reveals that many of the sites in electrical contact will be largely irrelevant to the conduction since they lie in cul-de-sac side arms of a spidery cluster which conducts mostly along its main body or backbone. So the cluster can lose area (e.g. $P(p)$ falls by the disconnection of side arms) without much loss of conduction.

These examples illustrate two general points about the percolation problem. First, that it is a mathematical problem of abstract generality applicable to many phenomena by no means confined to physics. Historically it was

Table 9.1

Phenomenon or system	Transition
Flow of liquid in a porous medium	Local/extended wetting
Spread of disease in a population	Containment/epidemic
Communication or resistor networks	Disconnected/connected
Conductor–insulator composite materials	Insulator/metal
Composite superconductor–metal materials	Normal/superconducting
Discontinuous metal films	Insulator/metal
Stochastic star formation in spiral galaxies	Non-propagation/propagation
Quarks in nuclear matter	Confinement/non-confinement
Thin helium films on surfaces	Normal/superfluid
Metal–atom dispersions in insulators	Insulator/metal
Dilute magnets	Para/ferromagnetic
Polymer gelation, vulcanisation	Liquid/gel
The glass transition	Liquid/glass
Mobility edge in amorphous semiconductors	Localised/extended states
Variable-range hopping in amorphous semiconductors	Resistor–network analogue

first formulated when in 1954 Broadbent and Hammersley discussed hypothetical random mazes of fluid-bearing pipework at a time when Broadbent was interested in gas penetration into the porous carbon in gas masks for coal mines (see Hammersley 1983). Secondly, the examples suggest that when the probability of the relevant disorder reaches a critical value, the character of the medium changes radically, e.g. from conducting to non-conducting. The critical value is called a percolation threshold and it marks a transition or phase change. Table 9.1 taken from Zallen (1983) illustrates both of these points.

It appears that although the nature of the percolating influence may not be very important for the specification of the problem, the structure of the medium and the nature of the disorder are of the essence. Percolation theory is the mathematical science of setting up precisely and solving such problems and authorities appear agreed that it is a matter of some difficulty.

9.2 Site and bond percolation. Clusters

It is easiest to set up problems on 2-D lattices. There are analogous problems in 3-D but the solutions are not necessarily analogous because dimensionality makes significant differences. This is immediately obvious on considering the fraction of punched holes needed to prevent current flow in a 2-D sheet and a 1-D strip. Consider therefore a 2-D lattice. Figure 9.1 shows one with intersections for empty sites and stones for full sites with a probability p that a site is occupied. The section of lattice is finite with L^2 sites and to obtain properties independent of the existence of boundaries, $L \to \infty$ at suitable points in the theory.

A cluster is a group of sites such that each member is nearest neighbour to at least one other member of the group. The figure has finite clusters in black and also contains a cluster that spans the lattice in both directions; for $L \to \infty$ it would be an infinite cluster. If an influence could travel only between nearest neighbours the spanning or infinite cluster would enable the influence to percolate across the system and is therefore called a percolating cluster. Finite clusters will not contribute to percolation. $P(p)$, called the percolation probability, is the probability that an occupied site is in the percolating cluster. This probability will be zero if p is too small or, in other words, small p produces only finite clusters. In an infinite lattice, $P(p)$ rises from zero as p increases above a critical value p_c; in fact it is a theoretical result that as p approaches p_c from above

$$P(p) \sim (p - p_c)^\beta \tag{9.1}$$

where p_c depends on the lattice type and β on the dimensionality. Even if $p < p_c$, there may be a percolating cluster in a finite lattice but the low probability of this falls to zero as $L \to \infty$ and the probability that there is a percolating cluster is then a step function at p_c.

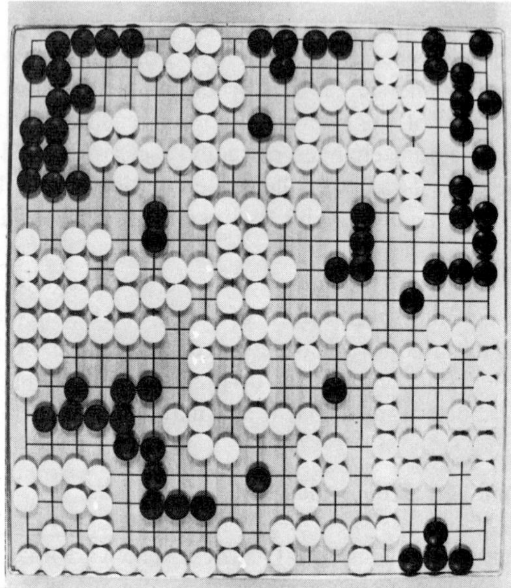

Figure 9.1 The board and stones of the game of 'go' afford excellent illustrations of 2-D lattice problems. Here the clusters in a site percolation distribution are shown with the percolating cluster picked out in white stones.

The preceding description and definitions of terms have analogues when we think of the presence or absence of bonds between sites instead of the occupation of sites. Figure 9.2 shows this case. Black bonds again indicate finite clusters and a percolating cluster of white bonds is shown which could act as a channel for electricity or fluid. p now means the probability that a bond exists or is unblocked; $q \equiv (1-p)$ is the probability of the absence or blockage of a path.

It was shown by Fisher in 1964 that all bond percolation processes could be reformulated as corresponding site processes on another lattice but not vice versa. In this sense site percolation is the more basic. It looks obvious, and was also proved rigorously by Fisher, that if there is an infinite cluster of occupied sites in 2-D it is the only one present and it cannot coexist with an infinite cluster of empty sites. This is not so in 3-D. Finite clusters are an interesting study in themselves and they are frequently called lattice animals apparently because of an analogous study in biology of the possible geometries of organisms with s cells. What is the probability, $n_2(p)$, of a cluster of two occupied sites? The probability of two adjacent occupations is p^2. On a square or rectangular lattice the probability of the pair being isolated is that of the six perimeter sites being empty, i.e. $(1-p)^6$ or q^6. The product

Figure 9.2 An illustration of bond percolation. The percolating cluster is in white.

p^2q^6 must be multiplied by two because the pair could point in either the x- or y-direction. So the required probability is $2p^2q^6$ which is the average number of such clusters per lattice site in a large lattice. It is easy to work out that $n_3(p)$ is $2p^3q^8 + 4p^3q^7$. In general,

$$n_s(p) = p^2 D_s(q) \qquad s = 1, 2, 3, \ldots \tag{9.2a}$$

where

$$D_s(q) \equiv \sum_t g(s, t) q^t. \tag{9.2b}$$

t is the number of perimeter sites and $g(s, t)$ is a weighting coefficient equal to the number of independent ways a cluster of size s and given shape can be laid down on the lattice. $g(2, 6)$ was two in the example above. The $D_s(q)$ are called perimeter polynomials and rapidly become formidable as s and t rise. The probability that a site belongs to a cluster of s sites will be $sn_s(p)$ and the mean size of finite clusters is defined as

$$S(p) \equiv \sum_s s^2 n_s(p) \left(\sum_s s n_s(p) \right)^{-1} = p^{-1} \sum_s s^2 n_s(p). \tag{9.3}$$

If the infinite cluster is present and the probability that a site belongs to it is

P_∞, then $p = P_\infty + \Sigma_s\, sn_s(p)$ since an occupied site must be either in a finite cluster or in the infinite one.

$S(p)$ would be expected to go to infinity as $p \to p_c$ from below and indeed $S(p)$ and a number of other quantities conform to power laws like equation (9.1) near the threshold. For example, the equations

$$S(p) \sim (p_c - p)^{-\gamma} \tag{9.4}$$

and

$$\sum_s D_s(q)p^s \sim (p_c - p)^{2-\alpha} \tag{9.5}$$

define the indices α and γ. We would also expect the RMS distance ξ between an occupied site and another site in the same cluster to diverge at $p = p_c$ and in fact

$$\xi \sim (p_c - p)^{-\nu}. \tag{9.6}$$

There are other indexed quantities as well and readers familiar with critical phenomena will note the analogy between equations (9.1) and (9.4) to (9.6) and the equations of vanishing or divergence at the liquid–vapour critical point or Curie point. This analogy is much explored in the literature, though percolation as so far described is a phenomenon of geometry and probability without physical forces (see also §2.9).

Apart from definitions, the statements made in this section, such as the index equations, the proposition that the indices depend only on the dimensionality, and also the values of p_c and the indices, all need proof or at least discussion. Many of the papers in the literature are concerned with such

Table 9.2. Results for p_c and the indices. Quite large uncertainties still exist in the 3-D indices. (e) indicates values which are known or believed to be exact.

	α	β	γ	ν
2-D	$-2/3$ (e)	5/36 (e)	43/18 (e)	4/3 (e)
3-D	-0.64	0.45	1.73	0.88

Lattice	Site or bond	p_c
2-D square	b	0.5 (e)
2-D square	s	0.59
3-D SC	b	0.249
3-D SC	s	0.312
2-D Δ'r	s	0.5 (e)
3-D diamond	s	0.43
3-D FCC	s	0.20

questions. For example, power series expansions, Monte Carlo calculations and the more recent techniques of scaling and the renormalisation group have been used to discuss critical percolation and to evaluate indices. Entry to these fields can be made by referring to Deutscher *et al* (1983), Balian *et al* (1979), Essam (1980), Stauffer (1979) and Castellani *et al* (1981) but Jug (1986) sounds a warning note. Some approximate results for p_c and the indices, mostly taken from Gaunt and Sykes (1983) are in table 9.2.

Figure 9.3 shows $P(p)$ for bond percolation on a simple cubic lattice.

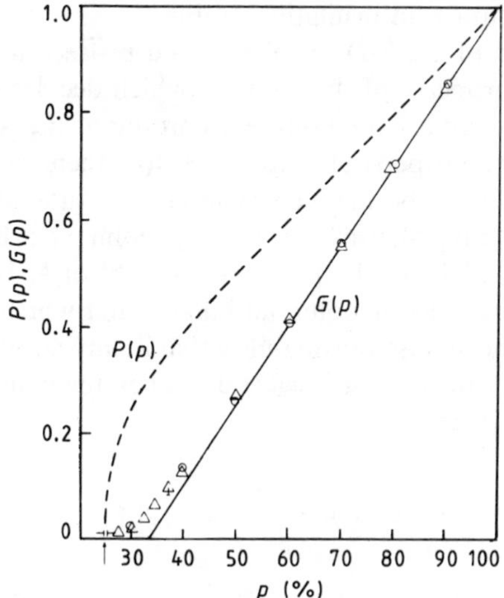

Figure 9.3 $P(p)$ (broken curve) and $G(p)$ (plotted points, obtained numerically, normalised to unity at $p = 1.0$) for bond percolation in a 3-D SC lattice. The straight line comes from the EMT (§9.4). (From Kirkpatrick 1973.)

9.3 A few more general problems

Once the basic ideas and problems of bond and site percolation on lattices have been introduced it is not difficult to envisage more complex problems. Many of these have been discussed in the literature and may well have relevance to physical problems of disordered matter (Deutscher *et al* 1983). Among them are: mixed site and bond problems; probabilities that are correlated so that the presence of a site or bond influences the likelihood of neighbouring ones; connections which stretch beyond nearest neighbours.

Lattices do not exhaust the possibilities however. Percolation may occur in networks with topological disorder. In this connection it was pointed out by Ziman that if n is the coordination number of a 3-D lattice and p_c the bond percolation threshold, $np_c \sim 1.5$. Thus percolation will occur if the influence can find, on the average, 1.5 open bonds to flow through from any site (Ziman 1968). Since this applies to several lattices it looks likely to be approximately true for random arrays such as DRP—but see §9.13 later.

Going even further from lattices, percolation can be envisaged through continuous media. It is not hard to imagine a map of towns and countryside in which all towns (distributed by some probability function) were bounded but which would connect into one infinite conurbation with finite rural patches if the town-to-country ratio increased to a threshold value. In 3-D an analogous problem is the percolation of electricity through a composite medium of conducting and insulating grains.

The specification of the 3-D problem could proceed as follows. Suppose $A(\boldsymbol{r})$ is a random property of the medium which decides whether the influence can or cannot occupy the point $\boldsymbol{r}$ according to the following criterion: if the influence has a property A' with $A' > A(\boldsymbol{r})$, then $\boldsymbol{r}$ is open for passage, otherwise not. $A(\boldsymbol{r})$ could be an inverse temperature or a potential energy, the corresponding A' being the inverse melting point of a liquid or the total energy of a particle. The medium can be thought of as having inaccessible volumes (freezing regions or potential hills) round which the influence can flow. Suppose the normalised probability that at any point $\boldsymbol{r}$ the medium has a value of A from A to $A + \mathrm{d}A$ is $\psi(A)\,\mathrm{d}A$. Then the volume fraction of the medium with $A < A'$ is

$$\varphi(A') = \int_{-\infty}^{A'} \psi(A)\,\mathrm{d}A.$$

This is the fraction of the medium accessible to the influence. Percolation should occur if $\varphi(A')$ reaches some critical value φ_c. It is difficult to calculate φ_c in 3-D and impossible without sufficient knowldege of the distribution of A. For the case that A is a random electrical potential $V(\boldsymbol{r})$, the problem is that of a percolating electric charge carried by particles of total energy $E \equiv A'$. $\varphi(A') = \varphi(E)$ and will be below φ_c for some energies but at some high enough energy $E = E_c$, $\varphi(E_c) = \varphi_c$. This was studied by Zallen and Scher (1971) using $\varphi_c = 0.15$. This value of φ_c is semiempirical and came from the observation that if touching metal spheres occupy sites in a lattice with probability p, then the volume of spheres is about 0.15 times the volume of the lattice when $p = p_c$. This follows easily from p_c like those given in table 9.2 irrespective of which 3-D lattice is considered and it is therefore a reasonable hypothesis for a continuum. For 2-D the corresponding estimate is 0.44 and for 1-D it is obviously 1.

In referring to the passage of electricity we are coming to problems in physics and the remaining references to percolation will be in the context of physical problems.

9.4 Electron transport: effective-medium theory

It is not difficult to think of electrical conduction problems which parallel the geometrical ones of the previous two sections. To take an obvious example: let the bonds of a simple cubic (SC) lattice be equal resistors and consider the conductance, G, measured between opposite faces of a large cube. Let the resistors be severed, i.e. replaced by insulators, at random. What is the dependence of G on the fraction, p, of resistors left in place? Monte Carlo and other computations show that there is a threshold and that G vanishes according to

$$G \sim (p - p_c)^t \qquad p > p_c. \tag{9.7}$$

If the resistive bonds were replaced by superconductors with probability p, then G increases with p, diverging near p_c according to

$$G \sim (p_c - p)^{-s} \qquad p < p_c. \tag{9.8}$$

Though easy to set up, problems like this and the analogous site problems are difficult to solve (Kirkpatrick 1973, Nakamura 1984, Straley 1983). The demonstration experiment in §9.1 shows that $G(p)$ and $P(p)$ are different. This is because bonds in the percolating cluster do not all contribute to the conductance. The current flows down the backbone of the cluster, indeed 'backbone' is defined to mean the connected cluster of bonds that do carry current. The form of the backbone and its relation to the whole percolating cluster is another substantial problem in percolation theory (Shlifer *et al* 1979). Figure 9.3 shows some results of numerical computation (Kirkpatrick 1973, Nakamura 1984).

The indices t and s, like those in percolation, are expected to be universal for a given dimensionality. At the time of writing their values are still under discussion but $s = t \sim 1.3$ in 2-D and $s \sim 0.7$, $t \sim 2.0$ in 3-D (Sahimi 1984, Roux 1985). Whether and how t and s are related to the percolation indices β, ν are interesting questions apparently unsolved.

It is not intended to pursue these network problems here although they are simple models, in relative terms at least, for real matter. Inhomogeneous media are really more complicated. For example, even if they are macroscopically homogeneous, random mixtures of conducting and insulating solids, whether in films or bulk, will probably have distributions of grain size, grain shape and grain conductivity and these distributions will in general not be well known. The problem of calculating their overall conductivity theoretically, while clearly related to percolation theory, involves added uncertainty. There may also be contact resistances between touching pairs of metal grains or dielectric breakdown across insulating bridges between separated pairs. According to the mode of preparation the disposition of grains may be correlated.

Nevertheless, a large amount of empirical information on composite metal–insulator systems is available and careful studies of the structure by transmission electron microscopy (TEM) and of the conductivity for various compositions reveal behaviours clearly connected with percolation. For example, the 0.15 volume fraction threshold for continuum percolation in 3-D referred to in the previous section has been confirmed for Al–Ge, Ag–KCl and other mixtures. Measured values of t are in the range predicted for resistor networks and therefore support the view that such indices have universal validity, i.e., are independent of the details of structure or interactions. From thin-film TEM photographs it is even possible to analyse cluster sizes and the geometry of percolating clusters and backbones with results to which percolation theory is clearly relevant. All these properties of composite metal–insulator systems are discussed extensively by Deutscher and colleagues (Deutscher *et al* 1983, Deutscher 1981).

The problem of mixed media had a technological significance long before percolation theory and composite conductivity was discussed by James Clerk Maxwell. An approximate treatment based on classical electromagnetic theory is available and the simplest case is that of a random mixture of regions which have either $\sigma = \sigma_1$ or $\sigma = \sigma_2$. Following Landauer (1952) we take any particular region P, with σ_1, shaded in figure 9.4, and idealise it to a sphere of radius large enough for macroscopic electrical theory to apply; thus P is not an atom but it could be a metallic grain. The rest of the medium is deemed to be uniform with the conductivity σ_m of the mixed medium as a whole. This assumption excludes any correlation of properties in the immediate neighbourhood of P. These assumptions are strongly reminiscent of the single-site CPA and EMA (§§7.2, 7.3) and in adopting them we are setting up effective-medium theory (EMT).

The electrostatics of a sphere embedded in a uniform medium is well known and can be treated either as a current flow problem using σ and i or as a polarisation problem using ε and D since the field lines of i and D are identical and both follow from Laplace's equation. Suppose the uniform

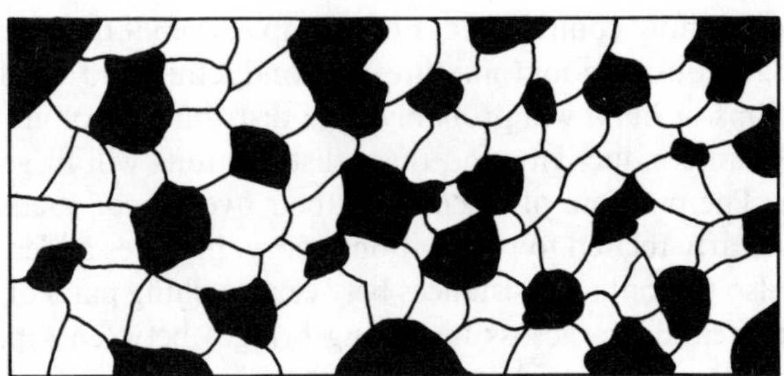

Figure 9.4 A medium with two kinds of conducting matter in randomly distributed regions.

applied field that would exist in the effective medium of $\sigma = \sigma_m$ is E_m when an external potential difference is applied. Then the dipole moment of the embedded sphere is proportional to $E_m(\sigma_1 - \sigma_m)/(\sigma_1 + 2\sigma_m)$ by electrostatics. Now every region is treated in the same way as if they were all acting independently so that if C, $(1 - C)$ are the volume fractions of material with σ_1, σ_2 then the total dipole moment per unit volume is

$$P \propto CE_m(\sigma_1 - \sigma_m)/(\sigma_1 + 2\sigma_m) + (1 - C)E_m(\sigma_2 - \sigma_m)/(\sigma_2 + 2\sigma_m). \quad (9.9)$$

If a sphere of the inhomogeneous medium is now imagined to be embedded in a volume of a uniform effective medium of conductivity σ_m subject to the uniform field E_m, the embedded sample must be electrostatically indistinguishable since its $\sigma = \sigma_m$; so $P = 0$. In other words σ_m is the value that makes $P = 0$, namely,

$$\sigma_m = \tfrac{1}{4}[\![(3C - 2)\sigma_2 + (3C - 1)\sigma_1 + \{[(3C - 2)\sigma_2 + (3C - 1)\sigma_1]^2 + 8\sigma_1\sigma_2\}^{1/2}]\!]. \quad (9.10)$$

For the metal–insulator case put $\sigma_2 = 0$. Then

$$\sigma_m = (\tfrac{3}{2}C - \tfrac{1}{2})\sigma_1 \qquad C > \tfrac{1}{3} \quad (9.11)$$

so that σ_m is linear in C with a percolation threshold at $\frac{1}{3}$.

The assumptions of EMT are such that it will fail if the size of the inhomogeneities is so small and σ_1 so different from σ_2 that the flow lines of i are deflected by one grain on to another, i.e. the grains do not contribute independently to the whole. The EMT has a corresponding resistor network version (see figure 9.3, and Kirkpatrick (1973) and Nakamura (1984)) and works well except near the threshold.

The EMT is sufficiently useful and tractable to have been generalised to other transport properties. For the Hall coefficient R_H, and the Hall mobility μ_H, Cohen and Jortner (1974a,b) give

$$\mu_{H,m}/\mu_{H,1} = [1 - b(1 - xy)]^{-1} \quad (9.12a)$$

$$R_{H,m}/R_{H,1} = [1 - b(1 - xy)]f^{-2} \quad (9.12b)$$

$$f \equiv a + (a^2 + \tfrac{1}{2}x)^{1/2} \quad (9.12c)$$

$$a \equiv \tfrac{1}{2}[(\tfrac{3}{2}C - \tfrac{1}{2})(1 - x) + \tfrac{1}{2}x] \quad (9.12d)$$

$$b \equiv \frac{(2f + 1)^2(1 - C)}{(2f + 1)^2(1 - C) + (2f + x)^2 C} \quad (9.12e)$$

$$x \equiv \sigma_2/\sigma_1 \qquad y \equiv \mu_{H,2}/\mu_{H,1}. \quad (9.12f)$$

Unlike σ_m and $R_{H,m}$, the thermoelectric power will not depend on the geometric form of the percolation path if $\sigma_2 = 0$; if there is a flow path at all, $\alpha_m = \alpha_1$. The EMT formulae have proved useful in applications to metal–insulator transition theory in cases where local conductivity is made inhomogeneous by density or concentration fluctuations (see §11.12).

A microscopic process for which a percolating resistor network could serve as a model is conduction by electron hopping. If electrons can occupy alternative localised states a field will bias the transition probability and cause a net flow of electrons by successive hopping transitions. If the sites of localisation are disordered a percolation path is needed for DC conductivity (Ambegaokar *et al* 1971). The subject will be taken up again after discussing localisation.

9.5 Localisation—an introduction

The wavefunction of an electron in a hydrogen-like state centred on a donor impurity in an otherwise perfect crystal falls off exponentially with some decay or localisation length, α^{-1}. At absolute zero the electron does not move away even under a small electric field. It is in a state readily conceded to be localised and clearly different from one-electron Bloch states which have a periodic amplitude throughout the crystal. Changes in the boundary conditions at far away faces of the crystal will have negligible effects on the localised state and the number of lattice sites at which its mean-square amplitude exceeds an arbitrary finite size will be a small fraction of the total in a crystal of linear size L sufficiently greater than α^{-1}. Since the impurity is an interruption of structural order one could say the electron was localised by disorder.

Localisation by more generalised forms of disorder as in amorphous matter or a random variation of well depths on a lattice is less easily conceivable. It is not intuitively obvious whether or when it would occur but it is clearly an interesting matter to investigate because the behaviours of localised and non-localised electrons differ. What is this difference or, in other words, what are the characteristics of a localised state?

A localised state does not diffuse away from its site even in infinite time. Stated more precisely, if at $t = 0$ the particle has an exponential envelope to its wavefunction with a maximum amplitude at $\boldsymbol{R}_0$, then at absolute zero there is a finite probability of finding the particle at $\boldsymbol{R}_0$ even as $t \to \infty$ in an infinite system. The reader may be aware that the classic paper in the field, that of Anderson (1959), was entitled 'Absence of Diffusion in Certain Random Lattices', and localisation by disorder is often called Anderson localisation.

Diffusion is connected with conductivity (see equation (7.38)). If the localised state has energy E its contribution to $\sigma(E)$ in equation (7.43c) is zero for $T = 0$ and zero frequency. The exponential fall-off of ψ and the vanishing diffusion and conductivity are alternative definitions of localised states. The latter two are more general, for ψ might fall off as $r \to \infty$ other than exponentially, e.g., according to a power law.

In searching for localised states either theoretically or by numerical work, α, $D(E)$ and $\sigma(E)$ could be examined but other criteria also serve as hallmarks of localisation. One is the participation ratio or its inverse. The participation ratio was suggested by Bell and Dean as a measure of the fraction of atoms out of a system of N atoms that participated in the vibrations of a normal mode of frequency ω; if all participated equally the ratio would be unity. Applied to the TBA for electrons this general idea suggests the quantity

$$P_n \equiv \sum_i |a_i^{(n)}|^4 \left(\sum_i |a_i^{(n)}|^2\right)^{-2}$$

for a particular eigenstate $|n\rangle = \Sigma_i a_i^{(n)} |\alpha_i\rangle$. An alternative definition for a wavefunction $\psi_n(\boldsymbol{r})$ is:

$$P_n \equiv \int |\psi_n|^4 \, \mathrm{d}\boldsymbol{r} \left(\int |\psi_n|^2 \, \mathrm{d}\boldsymbol{r}\right)^{-2}.$$

P_n or similar quantities, or their reciprocals, can be used as measures of localisation because they vary significantly between the extremes of extended and localised states. For example in the TBA, $P_n = N^{-1}$ and unity respectively for Bloch states and states localised at a single site; it is called the inverse participation ratio.

Insensitivity to changes in boundary conditions is another characteristic of localised states which are remote from the boundaries. In numerical work it is possible to compute the change in energy of a state due to a switch from, say, periodic to antiperiodic boundary conditions. The change is relatively much less in localised states and can be used to distinguish them from extended states (Edwards and Thouless 1972). In the literature the reader will find other definitions and criteria of localisation.

An extended state may be defined as one that is not localised; potentially it is a current-carrying state. It is often taken for granted that the same system could not have a localised state with the same energy eigenvalue as an extended state but the point is a subtle one and there appear to be counter-examples (Shapir *et al* 1982). Localised and extended states with different energies may well coexist however.

That extended states would not maintain their character indefinitely as scattering increases is strongly hinted at by the Joffe–Regel criterion which dates from 1960. This says that mean-free paths, Λ, for electron propagation with wavenumber k cease to make sense if $k\Lambda < 1$. This idea is connected with those of §7.1 where scattering by disorder was shown to be incompatible with the use of k as a precise quantum number and with a sharp Fermi surface. If this is pressed to the Joffe–Regel limit a new regime might be expected to set in but it is not easy to calculate Λ as the limit is approached and the criterion is not a precise indicator of localisation.

In a region of disordered potential a classical electron of energy E might be hemmed in by insurmountable potential barriers. For some $E_c > E$ it could percolate (§9.3). This is indeed a case of electron localisation but quantum mechanical electrons can tunnel and for both tunnelling electrons and those with $E > E_c$ localisation can still occur when the matter is considered quantum mechanically. This localisation is wavemechanical in essence. It is not simply a phenomenon of electrons and any waves in a sufficiently disordered medium might be localised; indeed it has been suggested that building suitable irregularities into architectural structures might localise acoustic disturbances and limit the propagation of noise (Hodges and Woodhouse 1986). By the same token lattice vibrations may be localised but this is a more classical phenomenon. Indeed it has been argued that if p_c is the classical site percolation threshold for a random distribution of heavier atoms in a lattice they share with lighter ones, and if p is the concentration of heavier atoms, then there will be a frequency separating localised and extended modes (i.e. a 'mobility edge'—see below) if $p > p_c$ and $1 - p < p_c$ (Carnisius and van Hemmen 1985).

This section has been concerned with how localised states might be conceived and recognised. It remains to discuss under what conditions they occur. The rather remarkable conclusion of two decades of study is that in the presence of disorder all electron eigenstates in 1-D and 2-D are localised and that in 3-D localisation is possible under certain conditions. A numerical demonstration of the localised state in 2-D is shown in figure 9.5.

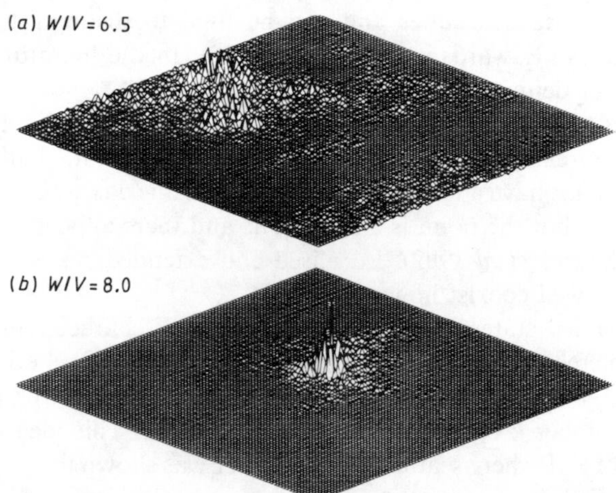

Figure 9.5 Localised states in 2-D resulting from numerical computations by Yoshino and Okazaki (1977). The computations are based on equation (9.13) for a square lattice with periodic boundary conditions and the energies are near the band centre.

9.6 Some simple, if not compelling, ideas

Much discussion of localisation is based on a tight-binding model with a single band which in a crystalline array has a bandwidth B equal to $2zV$ where z is the coordination number and V is the transfer integral. V is taken to be constant for nearest-neighbour pairs and zero otherwise. The integrals V distribute the probability density through the crystal making the states into extended Bloch states. The band is centred on a single-site energy, ε_0. Suppose now that random increments are given to the potential wells so that the energy of an electron on site j becomes $\varepsilon_j \neq \varepsilon_0$ and ε_j is uniformly distributed over $\varepsilon_0 \pm \frac{1}{2}W$. If the difference $\Delta\varepsilon = \varepsilon_j - \varepsilon_i$ between two neighbouring wells is less than γB, where γ is of order unity, we assume that the transfer integral continues its function of sharing the electron between the sites. If $\Delta\varepsilon > \gamma B$ we shall provisionally regard the sharing as ruled out. A set of nearest neighbours all connected by $\Delta\varepsilon < \gamma B$ is a cluster in the sense of bond percolation (§9.2). If the cluster is infinite there is an extended electron state; if not, the states are localised on finite clusters. Which happens depends on the size of W because the larger it is the smaller the probability of $\Delta\varepsilon < \gamma B$. A criterion for the presence of an infinite cluster would have to connect the fraction of bonds having $\Delta\varepsilon < \gamma B$ with the percolation threshold. This line of argument, which is a rough paraphrase from an interesting article by Efros (1978), can be refined somewhat but because of the unacceptable crudity of the distinction between blocked and unblocked bonds it serves only to raise three possibilities:

(i) that the existence of extended states may require the disorder not to be too great (i.e. W small enough);

(ii) that there may be some numerical relation between W and B at which an extended to localised transition occurs;

(iii) that the percolation concept may be relevant to the discussion.

In fact, all these suggestions turn out to be highly pertinent.

Another provocative and interesting argument was given by Allen (1980). The nearest-neighbour transfer integral is interpreted to represent hopping. As a rough measure of hopping rate we take $B/2\hbar$ from the uncertainty relation and bandwidth and then imagine a particular electron to perform a random walk by hopping from site to site. In time t it performs $n(t) = Bt/2\hbar = zVt/\hbar$ hops. Since it may recross its path it visits only $N(t)$ distinct sites and $N(t) \leqslant n(t)$. The average relation between N and n depends on the dimensionality, d, so let $\overline{N(t)} = f_d(zVt/\hbar)$. Now suppose the electron starts with energy ε. The crucial assumption is now made that the walk is forbidden (i.e. the electron is confined or localised) if the uncertainty in the electron energy, of amount $\hbar/t$, is not big enough to bracket the difference, $\Delta\varepsilon$, between ε and the nearest site energy ε_i available. The criterion separating allowed and disallowed walks is therefore $\hbar = t\Delta\varepsilon$. For $\Delta\varepsilon$ we write the rough value $W/N(t)$ since the $N(t)$ sites visited have energies spread uniformly over W; so the criterion is $f_d(zVt/\hbar) = tW/\hbar$. Random-walk theory supplies f_d; $f_1(n)$, $f_2(n)$ and $f_3(n)$ are, respectively, $(8n/\pi)^{1/2}$, $\pi n/\ln n$ and $\sim 0.659n$. It therefore follows that in 3-D there is a definite criterion, $W/Vz \sim$ unity, separating localised and unlocalised states or, in other words, localisation sets in only if W is big enough. In 1-D and 2-D, however, a limit to the duration of a walk is always found however small W is and this is clearly connected with the greater likelihood in lower dimensions that the path will self-intersect. In random-walk theory it can be demonstrated that for $d \leqslant 2$, walks always return to the neighbourhood of their origin.

There are more comparable arguments in the literature and the reader will enjoy the one by Hodges (1981). Nevertheless these arguments are only suggestions and the acceptance of localisation criteria depends on numerical calculations, on experiment and on deeper theory, either general in the sense that it covers all dimensions in a relatively model-free way or specially concerned with well defined models of definite dimensionality.

9.7 Localisation in 1-D

In 1961 a very clear indication that the eigenstates in a disordered chain might well be localised was given by Mott and Twose (1961) during a discussion of impurity conduction in semiconductors. After further influential papers in the early sixties by Borland, by Makinson and Roberts and others this conclusion has been accepted. It is a comment on the difficulty of the problem even in 1-D that two decades later papers still emerge to correct, modify, or make mathematically more rigorous the original ideas (see, e.g. Goda 1982, Kunz and Souillard 1980, Delyon *et al* 1985).

To illustrate some of the points we consider a chain of equal δ-function potential wells. A regularly spaced chain is a special case of the Kronig–Penney model but we are interested in the spatially disordered case (DKP model). The following are typical questions to be considered. How can $g(E)$ be found? Are the states localised? Do the energy gaps of the KP model persist in the DKP? To specify the disorder let the distance between a well and the next well on its left be $a(1 + \varepsilon\gamma)$ where $\varepsilon < 1$ and γ is a random variable. Various distributions of γ could be chosen, e.g., $(1 + \varepsilon\gamma)$ could be given a cut-off parabolic distribution with mean value unity and standard deviation σ; σ then measures the amount of disorder.

The density of states in the DKP has been studied numerically and theoretically. A valuable theorem for this purpose is that the integrated density of states up to E equals the number of zeros in the real wavefunction. Node counting by computer along very long chains gave some of the early results. One such is that energy gaps present in the KP spectrum persist in the DKP spectrum for a certain amount of short- or long-range disorder but close up as the disorder increases; the highest energy gaps close up sooner (Makinson and Roberts 1960). Borland was later able to prove this result rigorously provided that the distribution of γ allowed only a finite range of separation between neighbouring wells, the range depending on the well depth (Borland 1961a,b).

It is well known that the wavefunctions on either side of the energy gap in the KP model have their peaks spaced regularly with respect to the δ-wells. The periodicity of ψ is 'locked' (in Makinson's terminology) to the well spacing. If there were a range of energy E_1 to E_2 such that locking persisted throughout, then the number of nodes of ψ would be constant. But this number is proportional to the integrated density of states; therefore E_1 to E_2 would be an energy gap. The question: 'Is there an energy gap in the DKP model?' can therefore be rephrased: 'Does ψ lock on to a disordered well spacing for a range of energy?' or yet again: 'Do the wells always coincide with the same, or nearly the same, phase of ψ?'.

Figure 9.6 shows the effects of a δ-well at x_n. ψ, which would have been the sine wave DBC, has its slope changed at x_n and continues as if it were the sine wave D′B′C′. The distance CC′ measures the amount by which the phase of ψ has been altered by the well; this effect is called 'pull'. The phase change between two successive wells, a distance u apart, would be $2\pi u/\lambda$ if there were no pull, but the actual phase change will also contain the 'pull' contribution due to the first well. If these two contributions cancel, and if the lattice is regular, locking will persist throughout. This is possible for a range of energy (the Kronig–Penney gap) within which, as the energy is varied, the pull and the $2\pi u/\lambda$ contributions alter by equal and opposite amounts. This is elucidated in detail by Borland (1961a,b) and Roberts and Makinson (1962) who go on to prove that, even if the chain is disordered, locking can still occur over a range of E if the wells are deep enough and if the quantity $(1+\varepsilon\gamma)$ is confined within maximum and minimum limits.

The rise of B′ above B is another effect: the amplitude has increased. If the wave is nearly locked successive wells will have similar effects and the amplitude will be strongly amplified. What the calculations show is that for an arbitrary energy solutions of Schrödinger's equation which satisfy boundary conditions at the left- and right-hand ends of a long chain have envelopes which increase exponentially with distance from the end and in general do not match in the interior. The eigenenergies are those for which matching does occur giving a localised peak in the eigenfunction at, say, x_m falling off exponentially to the left and right (Borland 1961a,b). This conforms to the original idea of Mott and Twose. If the disorder itself were localised at one place, this will be x_m; but in a uniformly disordered chain x_m may be at any point. The localisation length α^{-1} depends on E, the amount of disorder and on the well depth.

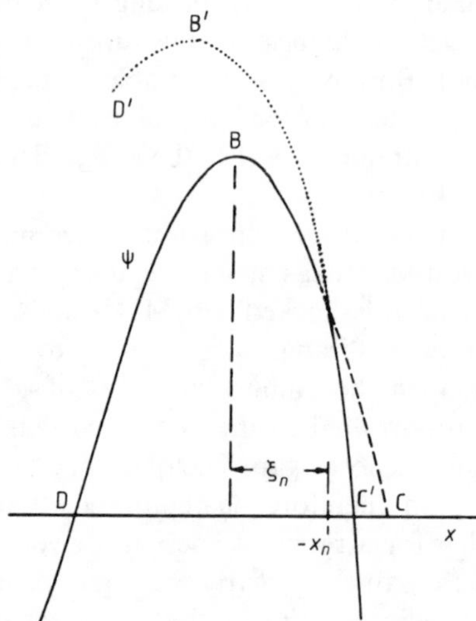

Figure 9.6 Schematic diagram to show the effect of a δ-well in the DKP model (from Cusack 1963).

Early opinions that the effect of an arbitrary atomic potential could be represented by a δ-well flanked by two intervals of zero potential, and that consequently the results for the DKP model would carry over into atomic chains, seem now not to be held (Erdos and Herndon 1982). Nevertheless the general conclusion about localisation does not depend on the shape of the wells.

The existence of localised eigenstates does not imply that electrons cannot be transmitted down a finite disordered chain. If the chain is regarded as a 1-D object in which electron waves of specified energy impinge from the left it is possible to consider reflection and transmission coefficients r and t, and to connect these with the resistance by the formula $R = (2\pi\hbar/e^2)(r/t)$. If the states are localised we should expect R to diverge as the length of the chain goes to infinity. There are arguments of some considerable subtlety to this effect (Anderson *et al* 1980) and this point, together with all other aspects of electrons in 1-D disordered systems, were discussed in 1982 by Erdos and Herndon (1982) (see also §9.9 below).

9.8 The Anderson model

The 3-D tight-binding model first considered by Anderson in 1958 was that referred to at the beginning of §9.6. Its Hamiltonian is

$$H = \sum_{i}^{N} \varepsilon_i |i\rangle\langle i| + V \sum_{i}^{N} \sum_{j \neq i}^{z} |i\rangle\langle j| \tag{9.13}$$

where $\{\varepsilon_i\}$ are the random energies distributed uniformly over $\pm\frac{1}{2}W$, V is the transfer integral which is constant between site i and its z nearest neighbours j and zero otherwise; the $|i\rangle$ are basis states of an atomic nature localised on sites $\{i\}$.

The argument leading from equation (9.13) to a criterion for localisation in 3-D is both long and subtle. It involves identifying a mathematical property of the Green function which signifies localisation and then discovering under what circumstances it has this property. Since G is expressed as an infinite series, and a random variable is involved, detailed discussion of the statistical properties of the terms is required as well as arguments (which involve percolation concepts) about the convergence. The conclusion was the important one that all states in the band are localised if W/zV is greater than about 11. The interest and the difficulty of this stimulated a number of expositions, critiques and developments (Ziman 1969, Thouless 1970, Licciardello and Economou 1975).

Meanwhile the uncertain status of some of the assumptions in the theory prompted attempts to solve equation (9.13) numerically and to answer the localisation question by calculating one or other of the characteristics of localisation mentioned in §9.5. Authorities on numerical procedures have not disguised the difficulties of the computing approach. One is clearly that the number of sites in the sample is limited by computer capacity and may not be big enough to distinguish extended states from localised ones with large α^{-1}; dubious extrapolations to infinite sample size may be unavoidable. Further, the properties calculated depend on the probability distribution of the disorder and are statistical quantities which may show large fluctuations. A fairly recent calculation proceeded through a numerical integration of the time-dependent Schrödinger equation to a calculation of σ from the Kubo–Greenwood formula (Kramer *et al* 1981). σ was seen to fall with increasing disorder suggesting a zero value at $W/zV \sim 2.5$ for a simple cubic lattice of about 3000 sites—notably lower than given by the original theory. Numerical work by others with comparable, though different, methods gave a similar value for the diamond lattice. To obtain more detailed information about how σ approaches zero, Kramer *et al* estimated that samples of at least 10^8 sites would be needed.

Both theoretical and numerical arguments have given a localisation criterion for 2-D; for example, the computation by Kramer *et al* gave $W/zV \sim 1.5$ for the square lattice but it appears that this does not separate localised from extended states but weakly from strongly localised ones. After outlining scaling arguments in §9.9 numerical work will be referred to again.

This section has referred to the model in equation (9.13). It would be possible to complicate an already very difficult matter by adding randomness into V (usually called 'off-diagonal disorder') or into the site positions themselves; or other distributions of ε_i could be invented. Such matters have been investigated to some extent but it has often simply been assumed that any kind of homogeneous disorder will have localising effects qualitatively similar to those of the basic Anderson model.

9.9 Scaling arguments

During the mid seventies a number of papers notably by Thouless contributed to an important new way of considering localisation. The arguments are conducted without overt initial reference to Hamiltonians describing particular models and without the computational solutions of Schrödinger's equation. Ideas more characteristic of phase transition theory are involved and were embodied in a significant and much quoted contribution by Abrahams *et al* (Abrahams *et al* 1979, Anderson *et al* 1979).

First let us rehearse an argument by Thouless (1977). Consider a block of metal of volume L^3; it is at zero temperature but has finite conductivity because of some unspecified disorder. Somewhere near the centre is an electron wavepacket of energy E with a spread small compared with L. While it is about $L/2$ distant from a surface, the boundary conditions hardly affect it but by the time, Δt, it has diffused to the surface it will have become subject to any change in boundary conditions and its energy may be altered thereby, say by ΔE. A rough use of the uncertainty relation gives $\Delta E \sim \hbar(\Delta t)^{-1}$. ΔE could be called the sensitivity of the energy to the boundary conditions. With a diffusion coefficient D, $\Delta t \sim L^2/4D$. Now a little manipulation of equation (7.38) shows that $\sigma = 2e^2 D W_L^{-1} L^{-3}$ where W_L is the spacing between the levels near E_F. It follows that

$$\frac{\Delta E}{W_L} \sim \left(\frac{2\hbar}{e^2}\right)\frac{1}{R} \tag{9.14}$$

where R is the resistance across L. The upshot is that the ratio of level spacing to the sensitivity to boundary conditions is proportional to the resistance of the system and that in this context there is a quantum unit of resistance $\hbar/e^2$—about 4000 Ω. Later arguments by Anderson and Lee suggested that $\Delta E/W$ should appear squared but this does not affect the following argument.

Now imagine two blocks which are similar but not identical; there may be a different configuration of disorder, and the energy levels, though similarly spaced on the average, will not coincide. Now suppose they are joined face to face. The quantity $\Delta E/W_L$ is now given a major significance as a measure of the coupling strength between the states in one block and those in the other. If R is very large, $\Delta E \ll W_L$ and energy matching between the states in the two blocks will be poor; conversely, for small R, $\Delta E \gg W_L$ and coupling is no problem. In the latter case, new energy levels will form for the combined system which can be treated just like the original block except that now W_L is one half and Δt four times as great. So from equation (9.14),

R has doubled or, in other words, the resistances in series have added in the normal way. In the opposite case, coupling is weak and the states in one block penetrate only exponentially into the next being localised with $\alpha^{-1} \sim L$. Since the condition for this is $R \gg \hbar/e^2$ it appears that adding wires of $R > 4000\ \Omega$ in series increases the resistance exponentially by progressively reducing the amplitude of any localised state at either end. This appears to show that long thin wires, like true one-dimensional systems (§9.7), have localised states if disorder is present.

This rather extraordinary conclusion is hard to accept at first sight. But it is essentially a zero-temperature argument. Suppose a wavepacket is made up of superposed localised states which have α^{-1} much greater than the size of the packet. Except at $T \simeq 0$, if the packet diffuses it will be scattered by phonon interaction into another packet long before it has diffused a distance α^{-1}. Phonon-controlled mobility is therefore essentially unaffected by the localisation. As phonon scattering decreases at low temperature its mean-free path will become comparable with α^{-1} and localisation begins to matter. Thus the predicted effect of localisation on R has a chance of showing up only if wires of $R \gg 4000\ \Omega$ are examined at $T \lesssim 1$ K (Thouless 1977). Before considering experiments we turn to the hypothesis of Abrahams *et al.*

Since the quantity in equation (9.14) contains the reciprocal of the resistance of a block of side L it is a non-dimensional conductance to be denoted by $g(L)$. We shall now suppose that L is the linear dimension of a regular figure of size L^d. Dimensionalities $d > 3$ have been considered in the literature but we shall stay with $d = 1, 2, 3$ while, for convenience, always referring to 'cubes'. Suppose b^d cubes are joined side by side. We now have one on a larger scale with side bL and the hypothesis is that $g(bL)$ depends only on b and $g(L)$. Thus

$$g(bL) = f(b, g(L)) \tag{9.15a}$$

or, as a differential equation,

$$\frac{\mathrm{d}\ln g}{\mathrm{d}\ln L} = \beta_d(g) \tag{9.15b}$$

where $g = g(L)$ and β_d is some scaling function to be determined. While it could hardly be said that this hypothesis is 'intuitively obvious' it is certainly hard to see what alternative there is to equation (9.15) if the argument of Thouless is accepted.

Abrahams *et al* established β by considering large g first. When g is large, ordinary conduction theory holds, i.e. $R \propto L^{2-d}$ or $g \propto L^{d-2}$. So lim β as $g \to \infty = 2 - d$. For very small g, exponential localisation is to be expected according to the arguments of Thouless so $g = g_0(d) \exp(-\alpha L)$ where the pre-exponential could depend on d. In that case, lim β as $g \to 0$ can be written as $\ln(g/g_0(d))$ which means that $\beta_d \to -\infty$ as $g \to 0$. Abrahams *et al* were able to show that as g rises from zero the slope of β versus $\ln g$ increases at first. Using this and plausible physical assumptions that β_d is a continuous monotonic function, curves of the shape shown in figure 9.7 result. Although the whole function β_d has not been derived some conclusions may be drawn, namely, that in 1-D and 2-D, β_d is always negative which means that as $L \to \infty$, $g \to 0$. This is also true in 3-D if g is less than a certain critical value g_c for which $\beta = 0$. For $g > g_c$ in 3-D, $\beta > 0$ which means g increases with L once the conductance is high enough. The interpretation is that the eigenstates in 1-D and 2-D are always localised and that in 3-D they are localised if the disorder is great enough. The behaviour in 2-D was particularly unexpected since numerical work had suggested a localisation criterion at a particular degree of disorder. This now appears to mark a changeover from exponential or severe localisation at low g to a much milder or logarithmic localisation where the $d = 2$ curve approaches the ln g axis and d ln g/d ln L, though negative, is very small. The changeover occurs smoothly in the region $R \sim 10^4\ \Omega$ per square and the lower resistance regime is said to exhibit weak localisation. In such ways 2-D is particularly interesting. As Pendry (1984) puts it '. . . the 2-D case escapes the conducting regime only by a hair's breadth'. Extra terms in the Hamiltonian such as an external magnetic field or the spin–orbit interaction may radically change the behaviour. At the time of writing there seems more to be learned about this.

Scaling arguments have had their effect on numerical computations. By combining a scaling hypothesis very much in the spirit of equation (9.15) with computations involving $\sim 10^9$ atoms, MacKinnon and Kramer (1983) were able greatly to extend the accuracy of numerical calculations of the

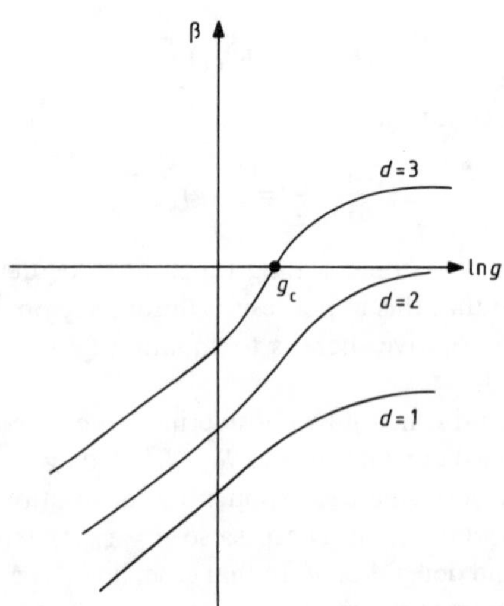

Figure 9.7 Scaling functions showing localisation in 1-, 2- and 3-D.

localisation length α^{-1} in the Anderson model and to make statements about localisation in 1-D, 2-D and 3-D. This not only confirmed the conclusions in the preceding paragraphs but offered evidence about critical behaviour, namely, the approach $\alpha^{-1} \to \infty$ as localised states go over to extended states in 3-D. As with the percolation equations like (9.8), α^{-1} might behave like $(w - w_c)^{-\nu}$ when $w \equiv W/zV$ approaches its critical value from above. This appears to be so with $\nu = 1.5 \pm 0.05$ but, like a number of matters in localisation theory, the value of ν has been subject to much discussion and disagreement (see, e.g., Sarkar and Domany 1981, Bulka *et al* 1985).

Because localisation theory is both challenging and interesting various theoretical approaches to it have been, and continue to be, tried. For example the mode-coupling theory of memory functions briefly referred to in connection with current fluctuations in classical liquids (§8.4) and the glass transition (§10.4) can be developed for conduction and the Anderson transition in fluids of fermions in disordered potentials. This and some other studies, usually too complex for this introduction, are given in Gotze (1979), Vollhardt and Wolfle (1980), Pendry (1984), Morgan and Hickey (1985), Belitz *et al* (1981) and Fröhlich and Spencer (1984).

9.10 Mobility edges and gaps

Looked at from the point of view of multiple scattering, electrons of energy E are sufficiently scattered by the disorder under conditions of Anderson localisation that the scattered wavelets interfere destructively at distances large compared with α^{-1} and collectively reduce the amplitude $\psi_E(r)$ beneath the exponential envelope. When W/zV falls just below its critical value in 3-D the localisation does not switch off suddenly throughout the band; states in the centre of the band become extended first. In general for an intermediate degree of disorder less than the critical value there will be two energies which separate localised from extended states as shown in figure 9.8(*a*). This was first emphasised by Mott and the energies E_A, E_B came to be known as mobility edges after an important application to amorphous semiconductors by Cohen *et al* (1969). In a long and influential series of papers and books Mott and collaborators applied the mobility edge idea to the analysis of a great many observations on a variety of materials (Mott and Davies 1979).

For a small degree of disorder only the states in the band tails are localised. The tails themselves stretch into the energy gap only because ranges of additional states become available through the disordered potential distribution. At either side of an energy gap in a disordered insulator or semiconductor tails of localised states penetrate into the gap (figure 9.8(*b*)); they may or may not overlap. If they do, $g(E) > 0$ throughout, though it may be very

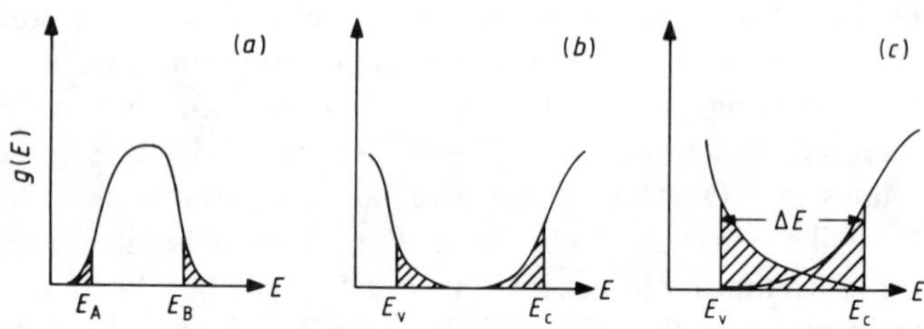

Figure 9.8 Mobility edges.

small. Because of localisation in the tails there will be a range $\Delta E = (E_c - E_v)$ where the mobility is zero and this is called a mobility gap (figure 9.8(*c*)). Where there is a deep dip in $g(E)$, but not a zero, Mott refers to a pseudogap and all the states in a pseudogap may or may not be localised depending on the nature and degree of disorder.

Mott (1974) also emphasised the possibility of metal-to-non-metal (MNM) transitions—called Anderson transitions—if by some operations such as doping or applying pressure the Fermi level E_F can be made to cross a mobility edge. In figure 9.8(*a*), for example, if $E_F > E_A$ the material is metallic and conducts at $T = 0$; if $E_F < E_A$ it will not conduct unless, at $T > 0$, electrons can be thermally excited from one localised state to another or to an extended state. If $g(E_F)$ is high enough for the electron gas to be degenerate and E_F lies in a mobility gap the material is called a Fermi glass.

As $E \to E_c$ from below or $E \to E_v$ from above, the localisation length $\alpha^{-1} \to \infty$. $\alpha^{-1}(E)$ resembles $\alpha^{-1}(Wz/V)$ in the previous section in that $\alpha^{-1} \sim (E_c - E)^{-\nu'}$. Like ν, ν' has a much discussed value which may be the same as ν (Sarkar and Domany 1981, Bulka *et al* 1985).

An interesting question is: at $T = 0$, if W/zV approaches its critical value from below, or E_F approaches E_c, how does σ vary? There was an argument due to Mott which said that σ decreased to σ_{min} and then went discontinuously to zero as localisation set in (Mott and Davies 1979). A little manipulation of equation (7.44) using the quantum mechanical relation

$$|\langle \varphi_i | v | \varphi_{i'} \rangle|^2 = a^2 V^2 / \hbar^2$$

where V is the transfer integral between nearest neighbours gives

$$\sigma = \frac{2\pi}{3}\left(\frac{e^2}{\hbar a}\right) a^6 z V^2 (g(E_F))^2. \tag{9.16}$$

With large disorder, N levels are spread over an energy width $\sim W$ so $\Omega g(E_F) \simeq N/W$ or $g(E_F) \simeq (Wa^3)^{-1}$, so that

$$\sigma \sim \frac{2\pi}{3z}\left(\frac{e^2}{\hbar a}\right)\left(\frac{zV}{W}\right)^2. \tag{9.17}$$

As zV/W falls to its critical value, σ also falls to a value $\sigma_{\min}$ at which point the states localise and σ goes discontinuously to zero. The value of $\sigma_{\min}$ depends on z, a and the critical value of zV/W and Mott's estimate for $a = 3$ Å was $\sim 200\ \Omega^{-1}\ \text{cm}^{-1}$ (Mott and Davies 1979). Insofar as states in 1-D and 2-D are localised, the discontinuity is a concept for 3-D only and for some years experimental and theoretical evidence appeared for and against it. There is enough contrary evidence to show it cannot have the universal application it first looked like having. Indeed it now seems accepted that a continuous fall of σ to zero should be expected. It has been found in a number of experiments (see below) and also in theory and is an implication of figure 9.7 (see the end of §7.7 and Economou *et al* 1985, Gotze 1979, Vollhardt and Wölfle 1980, Pendry 1984, Morgan and Hickey 1985, Belitz *et al* 1981, Frohlich and Spencer 1984, Lee and Ramakrishnan 1985, Mott and Kaveh 1985).

9.11 Experimental tests—some preliminaries

In 2-D it looks as if a straightforward test of localisation would be to see if the resistance per square increases with sample size near $T = 0$. Apart from the practical difficulties it is necessary to complicate the physics further before definitive experiments become possible. It has been remarked in §9.9, and also in §7.7 in connection with $2k_F$ scattering, that inelastic collisions alter the electron energy and vitiate the phase relations that secure localisation. The time, τ_i, between inelastic collisions, and the distance an electron can move in τ_i, provide upper limits above which localisation could not apply. Since $\tau_i(T) \propto T^{-p}$, the effect of localisation reveals itself as a characteristic T-dependence of σ for fixed sample size; in effect $L_i(T)$ is controlling the length scale for the localisation theory. $L_i(T)$ is the diffusion length between inelastic collisions which is approximately $(D\tau_i)^{1/2} \simeq (l_i l_e)^{1/2}$, where l_e, l_i are mean-free paths for elastic and inelastic scattering and $v_F\tau_i = l_i$. In Abrahams *et al* (1979) and Anderson *et al* (1979) the corresponding length and temperature formulae in the weak localisation regime in 2-D were shown to be

$$\sigma(L) = \sigma_B(L_0) - \frac{Ae^2}{\pi^2\hbar} \ln (L/L_0) \qquad (9.18a)$$

$$\sigma(T) = \sigma_B(T_0) + \frac{Ap}{2}\frac{e^2}{\pi^2\hbar} \ln (T/T_0) \qquad (9.18b)$$

where A is of order unity and the suffix B refers to an arbitrary state of the sample in which σ_B is given by the normal Boltzmann treatment of conductivity by extended states. Equation (9.18) implies that as T rises towards T_0 inelastic scattering effectively reduces the length scale and makes the $\ln L$ term smaller; so σ rises logarithmically to a value unaffected by localisation.

What determines p is also an issue: either electron–phonon or electron–electron events may dominate with different values of p. In 2-D, p is thought to be two for electron–electron collisions and three for electron–phonon collisions.

This might set the scene for definitive experiments except that a ln T term also emerges from a completely different cause, namely electron–electron interactions which have been ignored in all discussion of localisation theory so far (Altshuler *et al* 1980, Kaveh and Mott 1981). The electron–electron interactions, without localisation theory, give

$$\sigma(T) = \sigma(T_0) + (2 - \tfrac{3}{2}\tilde{F}) \frac{e^2}{4\pi^2 \hbar} \ln T/T_0 \tag{9.19}$$

instead of equation (9.18*b*), where $\tilde{F}$ is a complicated electron density function varying between zero and about one for long and short screening lengths. Thus a means of distinguishing the two types of logarithmic variation was needed and was found in magnetoresistance.

An applied magnetic field is itself not an instant key to the situation. If it is perpendicular to the 2-D sample the electrons are induced to move in cyclotron orbits. If the smallest orbit has linear dimension $l_{\text{cyc}} \ll L_{\text{i}}$ then it, and not the latter, determines the length scale. Being independent of T, l_{cyc} removes the ln T dependence due to localisation; in fact H affects the phase of the electron waves and destroys localisation. Negative magnetoresistance should therefore be observable. Further increases in H bring the electron–electron interactions into play and a more general formula can be written rather inexplicitly:

$$\begin{aligned}\sigma(T, H) = \sigma(T_0, 0) &+ \Delta_{\text{loc}}\sigma(T) + \Delta_{\text{loc}}\sigma(H) \\ &+ \Delta_{\text{ee}}\sigma(T) + \Delta_{\text{ee}}\sigma(H)\end{aligned} \tag{9.20}$$

where the 'loc' terms are from localisation theory and the 'ee' terms are from many-electron theory. The $\Delta\sigma$ are somewhat complicated functions of the quantities already introduced, namely, A, p, l_{i}, l_{e}, l_{cyc}, H, T and F plus a relaxation time for spin–orbit interaction if that is important (see, e.g., Uren *et al* 1980, 1981, Bishop *et al* 1982, Lee and Ramakrishnan 1985, Mott and Kaveh 1985 for these expressions).

These various effects were gradually elucidated through the interplay of theoretical and experimental contributions since 1979 and are reviewed in Lee and Ramakrishnan (1985) and Mott and Kaveh (1985).

9.12 Experiments—a few examples

In 2-D, experimenters have made use of thin films and inversion layers in the range $\sim 0.05\ \text{K} < T < \sim 10\ \text{K}$. Films might be about 100 Å thick and evaporated onto cooled glass substrates. Whether low- or high-Z metals are used controls the importance of SO effects. Inversion layers are a facility provided by transistor technology. In a field effect transistor shown schematically in figure 9.9 a positive voltage on the gate bends the energy bands down in energy near the Si surface until the conduction band edge is just below E_F while majority carriers are repelled away from the surface. Minority carriers (electrons in this case) can enter the conduction band in a narrow layer—the inversion layer—close to the Si surface where $E_{cond} < E_F$; their number density depends linearly on the gate voltage. Their component of motion perpendicular to the surface is quantised and for each allowed mode the parallel components have a large density of levels making motion parallel to the surface essentially free. Since the oxide layer and the interface contain charged defects the potential in the inversion layer is disorderd. For more details about these devices and their use see Adkins (1978) or Pepper (1985). Here we simply assume that effectively 2-D electron gases can be created and controlled by this technique and that they will have localised states if scaling theory is correct.

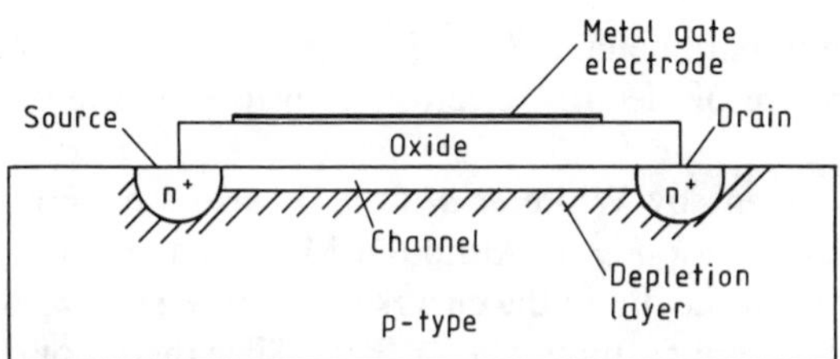

Figure 9.9. Schematic diagram of a field effect transistor used for inversion layer studies.

Some of the first inversion layer experiments by Uren *et al* (1980, 1981) and Bishop *et al* (1982) gave evidence for the smooth change in behaviour at about 10^4 Ω per square. For higher resistances the temperature dependence was exponential and for lower resistances it was logarithmic, consistent with equation (9.18). With small magnetic fields negative magnetoresistance occurred and the logarithmic variation with T was suppressed—which are both consistent with weak localisation theory. More extensive investigations brought out the coexistence of localisation and many-electron effects by results exemplified in figure 9.10. Here the low fields reduce the resistance because of l_{cyc}— the $\Delta_{loc}\sigma(H)$ term in equation (9.20). Higher fields bring an increase due to the $\Delta_{ee}\sigma(H)$ term which appears not to depend on whether H is parallel to perpendicular to the layer. This suggests a spin rather than a cyclotron orbit effect and it is the factor $(2-\frac{3}{2}\tilde{F})$ which is modified by spin polarisation. Fitting the results to equation (9.20) was possible leading to the

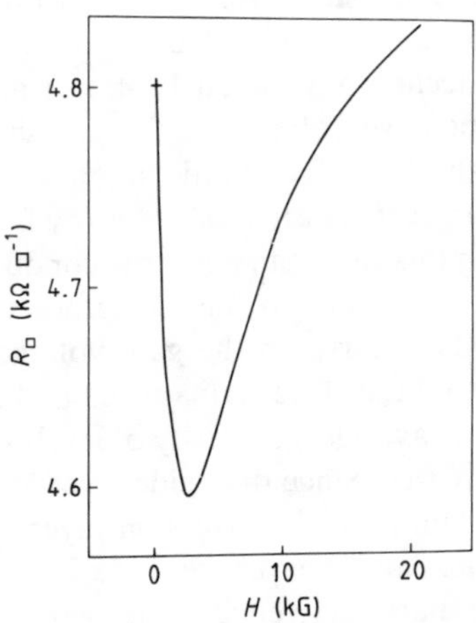

Figure 9.10 Magnetoresistance effect in an inversion layer with H perpendicular to the 2-D electron gas, 1.2×10^{12} electrons cm^{-2}. (From Bishop *et al* 1982.)

evaluation of the parameter $\tau_i(T) \propto T^{-p}$; $p \sim 1$ near 1 K, strongly suggesting the dominance of electron–electron collisions as the origin of inelastic scattering.

A film experiment by White *et al* (1984) (also see Bergman 1983) used 100 Å Mg films with $k_F l_e > 1$. Although Mg has a relatively low Z the so effect was visible in the dip in the curves of figure 9.11 at zero field. This was masked at higher temperatures when $\tau_i < \tau_{SO}$. The theory of $\sigma(H, T)$ includes τ_{SO} which could be separated out from τ_i by fitting these curves to the 'loc' terms of equation (9.20), the electron–electron terms being negligible at the low fields. It was concluded that $p \sim 1$ (electron–electron collisions) for $T < 4$ K above which electron–phonon collisions began to take over and increase the value of p.

From these few examples the reader may conclude that in 2-D there is strong evidence for the existence of both localisation and many-electron effects and for the ability of current techniques (early 1980s) to separate them. It is probably a valid generalisation that wherever localisation is suspected many-electron effects had better be looked for also. Anderson transitions could be expected in 2-D if T was high enough to eliminate localisation, say 0.5 K (Mott *et al* 1975).

By advanced design of inversion layer devices it is possible to make two dimensions small, with their components of motion quantised, instead of one. In this way Dean and Pepper (1982) have given evidence for a changeover from 2-D to 1-D localisation phenomena as the temperature falls below 1 K and L_i exceeds two sample dimensions. It has also proved possible to make very thin wires by depositing metals in steps formed by ion bombard-

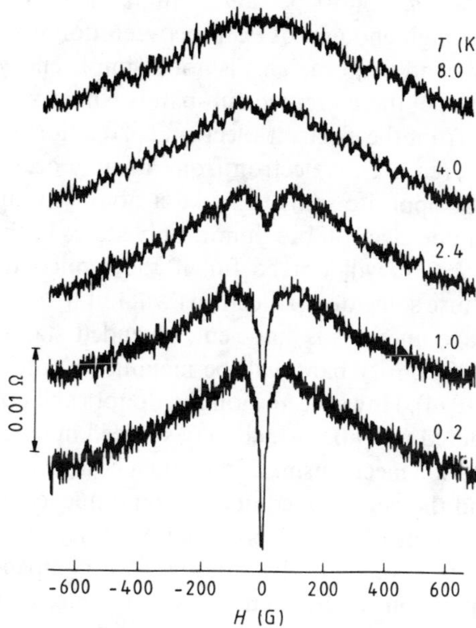

Figure 9.11 Magnetoresistance in 2-D film of Mg at various temperatures (from White *et al* 1984).

ment of insulating substrates. Platinum of cross sectional area of about $1.7 \times 10^{-12}\,\text{cm}^2$ was used by Masden and Giordano in the range $\sim 1\,\text{K} < T < \sim 10\,\text{K}$ where it was found that the resistance, $R_0 + \Delta R(T)$, rose as T fell. If the impurity resistivity in the absence of localisation effects is ρ_e (corresponding to R_0), the 1-D equivalents of equations (9.18) and (9.19) predict $\Delta R/R_0 \propto \rho_e^{1/2}/A$ (Masden and Giordano 1985).

Although these and similar experiments certainly revealed localisation and electron–electron effects, quantitative agreement with the theory was not found and at the time of writing they are not fully understood (Lee and Ramakrishnan 1985, Mott and Kaveh 1985).

There are many disordered 3-D systems in which MNM transitions occur; they include liquid and solid alloys as composition is varied, liquid semiconductors, expanded liquid metals, amorphous semiconductors and impurity bands and some of these are described in Chapter 11. In a number of such cases $\sigma(T)$ rising with T is suggestive of excitations to extended states from localised states or of hopping between the latter. It is not always obvious, indeed it is sometimes very difficult, to demonstrate that the transition is a case of Anderson localisation though that may be a plausible hypothesis. Detailed discussion of many examples is given in the book by Mott and Davies (1979) who took the view that some of the strongest evidence for the Anderson transition comes from doped semiconductors.

If Ge or Si contain n_d donors per unit volume the properties are sensitive to n_d. Then if n_d is high enough, overlap between donor wavefunctions will cause an impurity band to form. This is not a simple entity (Fritsche 1978). In the Hubbard model there are two sub-bands whose separation in energy derives ultimately from the electron–electron correlation energy which is the energy required to remove an electron from a singly occupied site and pair it with another (of opposite spin) to make a doubly occupied site. At low temperature, with one electron per donor, only states in the lower sub-band will be occupied but it will not be full if the semiconductor is partially compensated because some donor electrons will fall to acceptor levels. When the overlap of wavefunctions is sufficient, extended states occur at E_F and conduction in the impurity band will be metallic in nature. This is a Mott transition (Mott 1974). However, the ionised donors and acceptors are bound to create a disordered potential which may succeed in localising some of the states by the Anderson mechanism. E_F may then be above or below a mobility edge. Thus, n_d and the degree of compensation influence the impurity bandwidth and the positions of E_c and E_F. The possibility opens up therefore of creating an Anderson transition by varying n_d or compensation or both. If $E_F < E_c$, the conduction might occur by various mechanisms: at very low temperatures by hopping between localised states near E_F; at higher temperatures by excitation to extended states at E_c; and at still higher temperatures by excitation to extended states in the conduction band. The hopping option contains both 'nearest neighbour' and 'variable-range' versions (see next section).

Many studies of doped compensated semiconductors down to milliKelvin temperatures leave little doubt of the crossover from semiconducting to metallic behaviour as n_d increases. In the work of Allen and Adkins with compensated Ge having $n_d \sim 10^{17}$ cm^{-3} and temperatures down to about 0.15 K, the variable-range hopping and the two excitations to extended states could all be recognised making the evidence for Anderson localisation strong (Allen and Adkins 1972). More recently uncompensated phosphorus-doped Si was examined to even lower temperatures with the results shown in figure 9.12 (Rosenbaum *et al* 1980). Such results look very decisive but raise, as well as settle, both theoretical and experimental questions. While there is no doubt that a MNM transition occurs, it is not certain from the graph that the transition is continuous as well as sharp because the apparent continuity may be rounding due to sample inhomogeneity. Such experiments have been pursued by varying the compensation, temperature, magnetic field and also uniaxial stress which can cause a sample of fixed n_d and T to undergo a MNM transition. Evidence is strong for both localisation and electron–electron effects and for a continuous descent of $\sigma(T=0)$ to zero as n_d or stress is varied, but exact agreement with the results of scaling theory has not been found (Rosenbaum *et al* 1983). A continuous variation of $\sigma(T=0)$ is also observed in Au_xGe_{1-x} as x is varied. A review of this work is given in Lee and Ramakrishnan (1985) and Mott and Kaveh (1985).

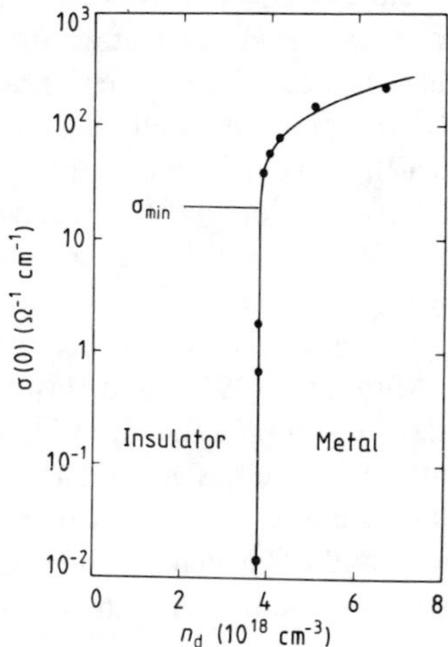

Figure 9.12 σ extrapolated to zero temperature versus n_d for phosphorus-doped Si. (From Rosenbaum *et al* 1980.)

By 1985 it was clear that localisation by disorder in 1-D, 2-D and 3-D was complicated by electron interaction effects but that it was well on the way to being understood. This question recurs in Chapter 12 in connection with metallic glasses.

9.13 Hopping conductivity

Whatever the cause of localisation an electron in a localised state cannot contribute to DC conductivity at zero temperature. At $T > 0$ it may be thermally excited to an extended state—a process familiar in doped semiconductors which need not concern us. Another possibility is a transition to an empty localised state. Such transitions can occur reversibly but an applied field will bias the motion and create a current down the field. This is called hopping conduction and was first appreciated by Conwell and by Mott in 1956. If it can be recognised by some characteristic T-dependence it serves to reveal the presence of localised states.

Hopping transitions are most easily envisaged between two neighbouring sites. ('Site' is used here to mean the place where an electron can be localised; it may or may not be a site in a crystal lattice.) In §10.7, in another connection, a two-level system is discussed briefly. Hopping is a property of this. A hop is a change in position and energy; more precisely electron hopping is the phenomenon in which a change of potential caused by the thermal displacement of atoms in the vicinity is the perturbation which stimulates a transition between two states. The electron and thermal motions jointly conserve energy, e.g., a phonon is emitted or absorbed. The states differ by the amount electron charge is located near one site rather than the other, and to cause this asymmetry the potential wells near the two centres must be different (see figure 10.12). If sufficient is known or assumed about the wavefunctions and the perturbation the transition rate can be calculated from perturbation theory. We assume this has been done and denote by R_{ij} the transition rate from site i to site j.

The calculation of R_{ij} and its use to set up a steady-state equation were put forward by Miller and Abrahams (1961) and this approach has frequently been followed as the basis for hopping theory (Pollak 1978). At the site i we need the following quantities as well as R_{ij} and R_{ji}: $\varepsilon_i + U_i$ and ε_i which are the energies of electrons localised at site i with and without an external electric field switched on; f_i and $f_i^{(0)}$ which are the occupation probabilities of site i under the same two circumstances. f is the Fermi function. The steady-state equation is then:

$$\frac{\mathrm{d}f_i}{\mathrm{d}t} = \sum_i [f_j(1-f_i)R_{ji} - f_i(1-f_j)R_{ij}]. \tag{9.21}$$

This is a Boltzmann equation and products like $f_j(1-f_i)$ are valid if the probabilities of occupation of i and j are independent. Detailed balancing will be assumed, namely,

$$R_{ij}\exp[-\beta(\varepsilon_i + U_i)] = R_{ji}\exp[-\beta(\varepsilon_j + U_j)]. \tag{9.22}$$

Following the exposition by Butcher (1980) we define φ_i by

$$f_i = f_i^{(0)} - \frac{\mathrm{d}f_i^{(0)}}{\mathrm{d}\varepsilon_i}\varphi_i \tag{9.23}$$

and linearise the rate equation (9.21) by substituting for R and f from equations (9.22) and (9.23) and retaining first-order terms in U and φ. The algebra leads to

$$C_i\frac{\mathrm{d}}{\mathrm{d}t}(V_i + U_i/e) = \sum_i g_{ij}(V_j - V_i) \tag{9.24a}$$

where:

$$C_i = -e^2\frac{\mathrm{d}f_i}{\mathrm{d}\varepsilon_i} \tag{9.24b}$$

$$g_{ij} = e^2\beta f_i^{(0)}(1-f_j^{(0)})R_{ij}^{(0)} \tag{9.24c}$$

$$-eV_i = U_i + \varphi_i. \tag{9.24d}$$

There is an elegant interpretation of these equations. From inspection it appears that φ_i and $-eV_i$ are, respectively, the chemical and electrochemical potentials at site i (the μ and ζ of §6.11). Thus the current flowing from i to

j is proportional to $V_{ij} \equiv V_i - V_j$ where the V are voltages. In terms of the rate of charging of a capacitance, C_i, equation (9.24a) is expressing the rate of accumulation of charge at i caused by currents resulting from all voltages V_{ij} being applied to the corresponding conductances $g_{ij} = g_{ji}$.

The formula for the conductivity may be obtained by calculating the power when a sinusoidal voltage is applied, equating this to the macroscopic expression, and then letting $\omega \to 0$, thus:

$$\tfrac{1}{2}\sum_{i,j}\tfrac{1}{2}|V_{ij}|^2 g_{ij} = \tfrac{1}{2}\Omega E_0^2 \sigma$$

where a factor $\frac{1}{2}$ on the left avoids double counting. This still lacks the vital step of configuration averaging for an infinite system and this requires further physical assumptions. A reasonable assumption is that g_{ij} depends only on ε_i, ε_j and $r \equiv |\boldsymbol{r}_i - \boldsymbol{r}_j|$ and must therefore be averaged over these variables. Another assumption is that the site positions are uncorrelated so the probability of finding a site j distant r to $r + \mathrm{d}r$ from site i at the origin is $4\pi n_s r^2\,\mathrm{d}r$ from equation (2.12), where n_s is the number density of sites. Writing $g(\varepsilon)$ for the density of states per unit volume, the combined probability that ε_i, ε_j lie in the usual infinitesimal ranges is $g(\varepsilon_i)g(\varepsilon_j)\,\mathrm{d}\varepsilon_i\,\mathrm{d}\varepsilon_j/n_s^2$. Thus if we integrate over $\mathrm{d}\varepsilon_i$, $\mathrm{d}\varepsilon_j$ and $\mathrm{d}r$ to include all sites j for a fixed i and then multiply by $n_s\Omega$ for the number of i sites we find

$$\sigma = \frac{1}{2E_0^2}\int g(\varepsilon_i)\,\mathrm{d}\varepsilon_i \int g(\varepsilon_j)\,\mathrm{d}\varepsilon_j \int 4\pi r^2 g_{ij}(\varepsilon_i, \varepsilon_j, r)\langle V_{ij}^2\rangle\,\mathrm{d}r \qquad (9.25)$$

in which $\langle V_{ij}^2\rangle$ is the configuration average of V_{ij} over any variables other than the ε and r. This applies in the limit $\Omega \to \infty$. Further progress requires more discussion of g_{ij}.

Calculating g_{ij} for realistic models, e.g. between donor sites in crystalline Ge, is not simple (Miller and Abrahams 1961). However g_{ij} contains $R_{ij}^{(0)}$ and we may suppose the transition rate will have an exponential dependence of the form $\exp(-2\alpha r)$ where α^{-1} is the localisation length as before. This is the form of the transfer integral between states centred r apart and it could be computed exactly for specified wavefunctions. Suppose therefore that

$$g(r) = g_0 \exp(-2\alpha r) \qquad (9.26)$$

where α is assumed constant. It has to be considered whether there will be a further factor with an explicit dependence on ε_i and ε_j. This turns out to be important at very low temperatures, but temporarily we defer this by assuming T to be high enough that any activation energy term can be lumped in with a constant g_0. Since r is a random variable $g(r)$ will be also, with a spread of values over many orders of magnitude. The problem is now reduced to finding the conductance of a vast network of resistors joined at the sites; current across the system will presumably take 'the line of least resistance', whatever that is.

It is a plausible guess that a percolation problem is involved but it will not be the relatively simple bond percolation on a lattice as in §9.4. The sites have been assumed uncorrelated so there is no well defined coordination number or nearest-neighbour distance. Resistors forming bonds between pairs of sites have widely different values depending on r. One pertinent type of percolation problem can be envisaged as follows. Suppose the sites are given for a large finite network and that the resistors are ranked in order of size ready to be joined between sites. Now imagine they are successively connected into the network starting with the lowest resistances, i.e., shortest r. As many resistors go in, the network becomes more and more multiply connected until on inserting resistors of a critical conductance g_c (or length r_c) a percolating cluster will have been reached. r_c is thus the shortest length such that site-to-site distances $\lesssim r_c$ form a percolating cluster. There will now be a certain average number, N_c, of bonds per site. N_c has been computed by the Monte Carlo method for random site distributions and is 2.7 in 3-D and 4.5 in 2-D (Pike and Seager 1974). A little reflection shows that

$$\tfrac{4}{3}\pi r_c^3 n_s \simeq N_c. \tag{9.27}$$

At the percolation threshold let the conductance of the system be G_c. The percolation hypothesis for DC hopping conduction is that the conductance of the entire network is well approximated by G_c and therefore depends on N_c. An argument to justify this was given by Ambegaokar *et al* (1971) along the following lines. In the whole network there will be clusters in which all sites are joined to others in the cluster by $g_{ij} \gg G_c$. Some of the clusters will be *joined to each other* by links having $g_{ij} \sim G_c$ collectively making a spanning cluster across the system; this is the situation at the threshold when it is arrived at by the process in the preceding paragraph. All resistors with $g_{ij} \gg G_c$ could now be shorted out, i.e. $g_{ij} \to \infty$, with little effect on the overall conductance because those in isolated clusters will not contribute to the current-carrying backbone anyway and those within the backbone are connected to it by $g_{ij} \sim G_c$ through which the current will have to flow. The links so far not mentioned are the ones with $g_{ij} \ll G_c$. These will also not affect the overall G much; indeed if they were broken, i.e. $g_{ij} \to 0$, their infinite resistance would still be bypassed by the spanning cluster with $G \sim G_c$. This is an order of magnitude argument to show that the important quantity defining the DC hopping conductance of a network is the conductance at the percolation threshold as defined above.

It still remains to estimate G_c and the corresponding σ. Reverting to equation (9.25) and the treatment by Butcher we insert equation (9.26) to obtain

$$\sigma = \frac{2\pi n_s^2}{E_0^2} \int_0^\infty r^2 P(r)\, dr$$

where

$$P(r) \equiv g(r)W^2(r) = I^2(r)/g(r).$$

Here $W(r)$ is $\langle V_{ij}^2 \rangle^{1/2}$ and $P(r)$ is the average power dissipated in $g(r)$. Therefore $I(r) \equiv W(r)g(r)$ is the RMS current. If E_0 is in the x-direction, $V_{ij}^2 = E_0^2(x_i - x_j)^2$ and $W^2 = \frac{1}{3}E_0^2 r^2$. If r is large, here interpreted as $r > r_c$, $P(r) = \frac{1}{3}g(r)E^2r^2$ and its r-variation is mainly controlled by the exponential factor $\exp(-2\alpha r)$ in $g(r)$. If r is small, i.e. $r < r_c$, the resistance is small and its current does not depend on r but is controlled by the conductances feeding it. This current is written $I^2(r) = \frac{1}{3}g^2(r_c)E^2r_c^2$ making $P(r)$ dependent on the exponential $\exp(2\alpha r)$ in $g^{-1}(r)$. With these approximations, $P(r)$ may be substituted into equation (9.25) which can be integrated from $0 \to r_c$ and $r_c \to \infty$ leading to

$$\sigma = 2\pi n_s^2(r_c^4/3\alpha)g_0 \exp(-2\alpha r_c) \tag{9.28a}$$

where r_c is given by equation (9.27). This equation is most accurate for $r_c \gg \alpha^{-1}$ and is best for a low density of sites. For the opposite limit

$$\sigma = 2\pi n_s^2 \left(\frac{1}{4\alpha^5}\right) g_0. \tag{9.28b}$$

Equation (9.28) has successfully survived detailed checking against computed results for all values of αr_c by Butcher and McInnes but does not represent a complete solution of the problem because $g(r)$ has no dependence on temperature.

If $|\varepsilon_i - \varepsilon_j|$, ε_i and ε_j are all several $k_B T$ from the Fermi level, the transition rate and the occupation probabilities will all acquire factors of the form $\exp(-\beta\varepsilon)$ which will introduce a temperature dependence into g_{ij}. It is quite possible to treat this case with percolation arguments on the preceding lines, indeed Ambegaokar *et al* introduced their percolation approach in order to do so. The references given cover this case and also the calculation of g_0 which must contain a parameter, representing the strength of the interaction between the electrons and the phonons, which causes the transition and another factor deriving from the integration of equation (9.25) over the density of states. To avoid the complexities of a complete treatment some simpler arguments which preceded the percolation approach will be taken from Mott and Davies (1979).

There is more than one regime for temperature-dependent hopping. The original concept of Miller and Abrahams was of transitions chiefly impeded by the distance r through the factor $\exp(-2\alpha r)$. Hopping is to sites with the smallest r and is therefore sometimes called nearest-neighbour hopping. For a hop $i \to j$ with $\varepsilon_j > \varepsilon_i$ a factor $\exp[-2\alpha r - \beta(\varepsilon_j - \varepsilon_i)]$ would be expected in the transition rate and when this is averaged over the sample the conductivity acquires an activation energy factor $\exp(-\beta\varepsilon)$. ε was originally calculated by Miller and Abrahams as a function of doping and compensation and was of the order of 10^{-3} eV in n-type Si.

The assumption here is that T is high enough to make hopping through $(\varepsilon_j - \varepsilon_i)$ to a near neighbour the most likely event. If T is lower this may not be so; the increased probability of a jump to a site j with energy nearer to ε_i may more than compensate for the decrease in probability due to j's greater spatial distance. This idea was first proposed by Mott in 1968 and the process is known as 'variable-range hopping'. Mott's argument is as follows (see Mott and Davies 1979). Suppose the electron hops a distance r from site i; then sites in a volume $4\pi r^3/3$ are available for it. It is most likely to jump to the site nearest in energy so we may put $(\varepsilon_j - \varepsilon_i)$ equal to the level spacing $(4\pi r^3 g(\varepsilon_F)/3)^{-1}$. The mean jump distance in a sphere of radius r is $3r/4$ assuming all points are available, so the transition rate should be proportional to

$$\exp[-\tfrac{3}{2}\alpha r - 3\beta/4\pi r^3 g(\varepsilon_F)].$$

On maximising this with respect to r, assuming the pre-exponential factor does not depend on r, it readily appears that

$$\text{transition rate} \propto \exp(-B/T^{1/4}) \tag{9.29a}$$

where

$$B \equiv 2\left(\frac{3}{2\pi}\right)^{1/4}\left(\frac{\alpha^3}{k_B g(E_F)}\right)^{1/4}. \tag{9.29b}$$

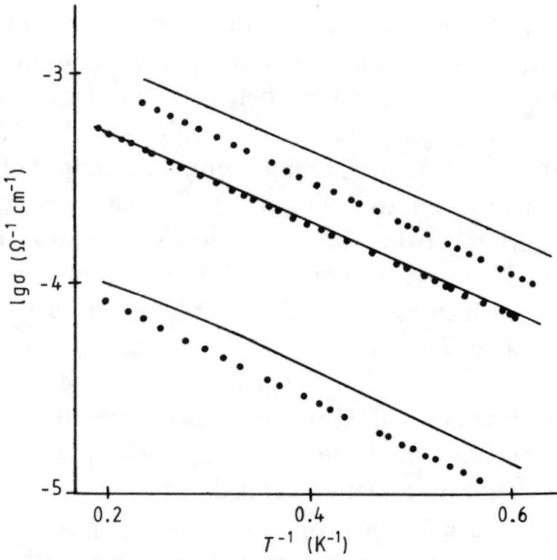

Figure 9.13 Comparison of observed DC hopping conductivity (points) in doped GaAs with a theory by Summerfeld and Butcher (1982) (lines). (From Summerfield and Butcher 1983.)

The $T^{1/4}$-law is characteristic of variable-range hopping and it is now known to follow from percolation arguments also (Ambegaokar *et al* 1971). It becomes $\exp(-C/T^{1/3})$ in 2-D.

There are detailed studies of other aspects of hopping, notably AC hopping and the theory of the pre-exponential factor in conductivity expressions like $\sigma = \sigma_0 \exp(-\beta\varepsilon)$ or $\sigma = \sigma_0 \exp(-B/T^{1/4})$. An attempt at a unified theory of AC and DC hopping and a comparison with conceptually different approaches to the problem are given by Summerfield and Butcher (1982, 1983). Hopping was originally identified between localised impurity states in doped crystalline semiconductors. 'Nearest neighbour' and $T^{1/4}$ hopping have been observed in a number of such systems and amorphous semiconductors (Mott and Davies 1979). A comparison of recent hopping theory with observation in n-type Si and GaAs and amorphous Ge was given by Summerfield and Butcher (1982). Figure 9.13 shows an example. The DC conductivity is rather well explained by the theory but AC conductivity less so leaving some theoretical questions unresolved. Figure 9.14 illustrates variable-range hopping in a 2-D inversion layer (Mott *et al* 1975).

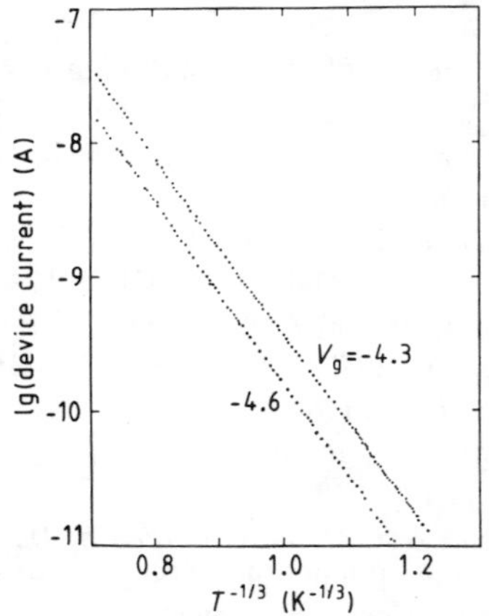

Figure 9.14 Variable-range hopping in a 2-D inversion layer at two different values of gate voltage (from Mott *et al* 1975).

References

Abrahams E, Anderson P W, Licciardello D C and Ramakrishnan T V 1979 *Phys. Rev. Lett.* **42** 673

Adkins C J 1978 *J. Phys. C: Solid State Phys.* **11** 851

Allen F R and Adkins C J 1972 *Phil. Mag.* **26** 1027

Allen P B 1980 *J. Phys. C: Solid State Phys.* **13** L667

Altshuler B L, Aronov A G and Lee P A 1980 *Phys. Rev. Lett.* **44** 1288
Ambegaokar V, Halperin B I and Langer J S 1971 *Phys. Rev.* **4** 2612
Anderson P W 1959 *Phys. Rev.* **109** 1492
Anderson P W, Abrahams E and Ramakrishnan T V 1979 *Phys. Rev. Lett.* **43** 719
Anderson P W, Thouless D J, Abrahams E and Fisher D S 1980 *Phys. Rev.* B **22** 3519
Balian R, Maynard R and Toulouse G (ed) 1979 *Ill-Condensed Matter* (Amsterdam: North-Holland)
Belitz D, Gold A and Gotze W 1981 *Z. Phys.* B **44** 273
Bergman G 1983 *Phys. Rev.* B **28** 515
Bishop D J, Dynes R C and Tsui D C 1982 *Phys. Rev.* B **26** 773
Borland R E 1961a *Proc. Phys. Soc.* **77** 705
—— 1961b *Proc. Phys. Soc.* **78** 926
Bulka B R, Kramer B and MacKinnon A 1985 *Z. Phys.* B **60** 13
Butcher P N 1980 *Phil. Mag.* B **42** 79
Carnisius J and van Hemmen J L 1985 *J. Phys. C: Solid State Phys.* **18** 4873
Castellani C, Di Castro C and Peliti L (ed) 1981 *Disordered Systems and Localisation* (Berlin: Springer) p 783
Cohen M H, Fritsche H and Ovshinsky S R 1969 *Phys. Rev. Lett.* **22** 1065
Cohen M H and Jortner J 1974a *Phys. Rev.* **10** 978
—— 1974b *J. Physique Coll.* **35** C4 345
Cusack N E 1963 *Rep. Prog. Phys.* **26** 361
Dean C C and Pepper M 1982 *J. Phys. C: Solid State Phys.* **15** L1287
Delyon F, Lévy Y and Souillard B 1985 *J. Stat. Phys.* **41** 375
Deutscher G 1981 *Disordered Systems and Localisation* ed C Castellani, C Di Castro and L Peliti (Berlin: Springer)
Deutscher G, Kapitulnik A and Rappaport M 1983 *Percolation Structures and Processes* ed G Deutscher, R Zallen and J Adler (Bristol: Adam Hilger)
Economou E N, Soukoulis C M and Zdetsis A D 1985 *Phys. Rev.* B **31** 6483
Edwards J T and Thouless D J 1972 *J. Phys. C: Solid State Phys.* **5** 807
Efros L A 1978 *Sov. Phys.–Usp.* **21** 746
Erdos P and Herndon R C 1982 *Adv. Phys.* **31** 65
Essam J W 1980 *Rep. Prog. Phys.* **43** 833
Fritsche H 1978 *The Metal Non-Metal Transition in Disordered Systems* ed L F Friedman and D P Tunstall (Edinburgh: SUSSP) p 193
Frohlich J and Spencer T 1984 *Phys. Rep.* **103** 9
Gaunt D S and Sykes M F 1983 *J. Phys. A: Math. Gen.* **16** 783
Goda M 1982 *Prog. Theor. Phys. Suppl.* **72** 232
Götze W 1979 *J. Phys. C: Solid State Phys.* **12** 1271
Hammersley J M 1983 *Percolation Structures and Processes* ed G Deutscher, R Zallen and J Adler (Bristol: Adam Hilger) p 47
Hodges C H 1981 *J. Phys. C: Solid State Phys.* **14** L247
Hodges C H and Woodhouse J 1986 *Rep. Prog. Phys.* **49** 107
Jug G 1986 *J. Phys. A: Math. Phys.* **19** 1459
Kaveh M and Mott N F 1981 *J. Phys. C: Solid State Phys.* **14** L183
Kirkpatrick S 1973 *Rev. Mod. Phys.* **45** 574
Kramer B, MacKinnon A and Weaire D 1981 *Phys. Rev.* B **23** 6357
Kunz H and Souillard B 1980 *Commun. Math. Phys.* **78** 201
Landauer R 1952 *J. Appl. Phys.* **23** 779

Last B J and Thouless D J 1971 *Phys. Rev. Lett.* **27** 1719
Lee P A and Ramakrishnan T V 1985 *Rev. Mod. Phys.* **57** 287
Licciardello D C and Economou E N 1975 *Phys. Rev.* B **11** 3697
MacKinnon A and Kramer B 1983 *Z. Phys.* B **53** 1
Makinson R E B and Roberts A P 1960 *Aust. J. Phys.* **13** 437
Masden J T and Giordano N 1985 *Phys. Rev.* B **31** 6395
Miller A E and Abrahams E 1961 *Phys. Rev.* **120** 745
Morgan G J and Hickey B J 1985 *J. Phys. F: Met. Phys.* **15** 2473
Mott N F 1974 *Metal–Insulator Transitions* (London: Taylor and Francis)
Mott N F and Davies E A 1979 *Electronic Processes in Non-Crystalline Materials* (Oxford: Clarendon)
Mott N F and Kaveh M 1985 *Adv. Phys.* **34** 335
Mott N F, Pepper M, Pollitt S, Wallis R H and Adkins C J 1975 *Proc. R. Soc.* A **345** 169
Mott N F and Twose W D 1961 *Adv. Phys.* **10** 107
Nakamura M 1984 *Phys. Rev.* B **29** 3691
Pendry J B 1984 *J. Phys. C: Solid State Phys.* **17** 5317
Pepper M 1985 *Contemp. Phys.* **26** 257
Pike G E and Seager C H 1974 *Phys. Rev.* B **10** 1421
Pollak M 1978 *The Metal Non-Metal Transition in Disordered Systems* ed L F Friedman and D P Tunstall (Edinburgh: SUSSP) p 95
Roberts A P and Makinson R E B 1962 *Proc. Phys. Soc.* **79** 509
Rosenbaum T F, Andres K, Thomas G A and Bhatt R N 1980 *Phys. Rev. Lett.* **45** 1723
Rosenbaum T F, Milligan R F, Paalanen M A, Thomas G A, Bhatt R N and Lin W 1983 *Phys. Rev.* B **27** 7509
Roux S 1985 *C. R. Acad. Sci., Paris* B **301** 369
Sahimi M 1984 *J. Phys. A: Math. Phys.* **17** L601
Sarker S and Domany E 1981 *Phys. Rev.* B **23** 6018
Shapir Y, Aharony A and Harris A B 1982 *Phys. Rev. Lett.* **49** 486
Shlifer G, Klein W, Reynolds P J and Stanley H E 1979 *J. Phys. A: Math. Phys.* **12** L169
Stauffer D 1979 *Phys. Rep.* **54** 1
Straley J P 1983 *Percolation Structures and Processes* ed G Deutscher, R Zallen and J Adler (Bristol: Adam Hilger) p 353
Summerfield S and Butcher P N 1982 *J. Phys. C: Solid State Phys.* **15** 7003
—— 1983 *J. Phys. C: Solid State Phys.* **16** 295
Thouless D J 1970 *J. Phys. C: Solid State Phys.* **3** 1559
—— 1977 *Phys. Rev. Lett.* **39** 1167
Uren M J, Davies R A and Pepper M 1980 *J. Phys. C: Solid State Phys.* **13** L985
—— 1981 *J. Phys. C: Solid State Phys.* **14** L531
Vollhardt D and Wölfle P 1980 *Phys. Rev.* B **22** 4666
White A E, Dynes R C and Carno J P 1984 *Phys. Rev.* B **29** 3694
Yoshino and Okazaki 1977 *J. Phys. Soc. Japan* **43** 415
Zallen R 1983 *Percolation Structures and Processes* ed G Deutscher, R Zallen and J Adler (Bristol: Adam Hilger)
Zallen R and Scher H 1971 *Phys. Rev.* B **4** 4471
Ziman J M 1968 *J. Phys. C: Solid State Phys.* **1** 1532
—— 1969 *J. Phys. C: Solid State Phys.* **2** 1230

10

INSULATING GLASSES

There has been some discussion of what is acceptable as the definition of a glass (§1.9). Recognising that there are different usages for the word we shall here take a glass to be the non-crystalline solid form of a system capable of undergoing a glass transition. This shifts the need for definition to 'glass transition' and this chapter will attempt to deal with that. Although they have properties in common, metallic, insulating and semiconducting glasses are in different chapters for convenience. Computer glasses (see §4.6) are not excluded from discussion.

As is well known glasses are not systems which are in thermodynamic equilibrium. Nevertheless conventional measurements can be made of thermodynamic properties such as the heat capacity. To put this in a reasonable context it is helpful to take a molecular and structural view and recognise that there are various internal processes that can occur with different relaxation times. Two pertinent processes are structural, or configurational, reorganisation and the vibration of molecules about equilibrium positions. A process will contribute to the result of a measurement unless its relaxation time is much greater than the duration of the measurement. Thus we should expect measured volumes or heat capacities to be affected by both configurational and vibrational degrees of freedom if both motions take place during the measurement. In a glass, although configurational rearrangement into lower-energy crystalline states is conceivable, it takes place so extremely slowly that such motion does not affect the measurements. Vibrations are still active, however, and in thermal equilibrium with the heat bath containing the sample. Orientational and magnetic alignments may be active but configurational changes are inhibited by lack of enough thermal energy to overcome the potential barriers.

These ideas occur in various contexts. For example, one statement of the third law of thermodynamics is: 'the contribution to the entropy of a system by each aspect which is in internal thermodynamic equilibrium tends to zero as the temperature tends to zero' (Adkins 1983). 'Aspect' means some internal process which is sufficiently decoupled from the rest that it makes an independent contribution to the properties of the system. In a glass this is again making a distinction between active and inactive processes; vibrations are an equilibrium aspect and configuration changes are not.

Supercooled liquids are cases of metastable thermodynamic equilibrium. Some authorities say the same of glasses but it seems more reasonable to call a glass unstable though it alters its state exceedingly slowly. During the glass transition active processes become inactive as their relaxation times increase by large factors, making glass seemingly stable.

The technical processes of glass manufacture are designed to bring about this condition. The practical utility of the products needs no emphasis and the strength, dielectric behaviour, chemical resistance and many other important properties are explained in Rawson (1980). While it is arguable that anything may be made glassy, given appropriate treatment, the majority of familar non-metallic glasses are based on oxides such as SiO_2, B_2O_3 and P_2O_5 which form glasses in their own right but often have their properties modified in important ways by Na_2O, Li_2O, PbO, CaO and others. For reasons of space and mutual relevance the examples below are normally taken from oxide glasses—which is not to imply that a great deal of interesting physics cannot be found in chalcogenide glasses based on S, Se or Te (see Chapter 11), organic glasses like glycerol, or in ionic and aqueous glasses.

10.1 Empirical description of the glass transition

Let us suppose one of the common glasses is cooled at constant pressure from its liquid state at a rate fast enough to avoid crystallisation—about $10\ \mathrm{K\ s^{-1}}$ would suffice. During cooling its volume (V) and its change of enthalpy (ΔH) are measured as functions of T from which α_p and C_p can later be calculated. Figure 10.1 shows $V(T)$ schematically and, in particular, the significant change of slope that occurs after supercooling has proceeded

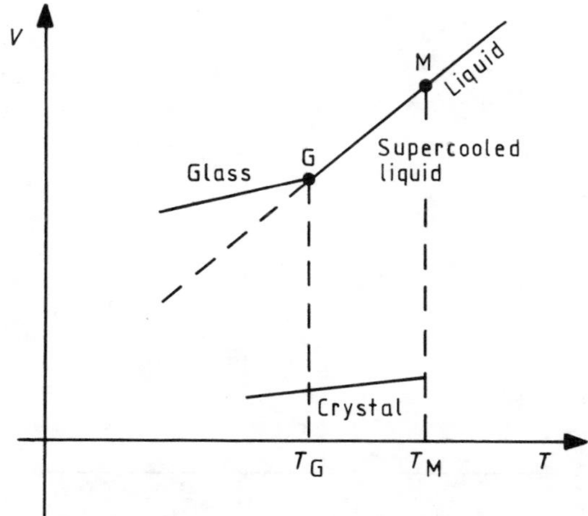

Figure 10.1 Schematic drawing of the glass transition.

to the temperature T_G. It will be important to see what happens to the shear viscosity η between G and the crystal melting point M but before this some more properties of $V(T)$ must be mentioned. T_G is not a unique temperature; slower cooling lowers T_G and differences of the order of 10 % of the absolute temperature are quite possible. Furthermore if T is held constant below T_G, $V(T)$ falls with time and approaches at an ever decreasing rate the value on the dotted extrapolation of MG. This change will be imperceptibly slow if T is far below T_G. The curve $V(T)$ may be reversible if the heating and cooling rates are equal. Reheating too slowly will make the return curve bend down and merge into the broken line below T_G. Reheating too fast will carry the return curve past G so that it has to bend upwards to regain MG from below.

The change in slope at G separates the supercooled liquid, existing from M to G, from the glass which exists for $T < T_G$. T_G is therefore called the glass transition temperature. Because there is no discontinuity of slope and because the location of the bend depends on the rates of cooling or heating it begs fewer questions to refer to a *transition range* of temperature of which T_G is the central or representative point. T_G is a few hundred °C for common oxide glasses. Outside the transition range, the supercooled liquid and the glass are in metastable and unstable states respectively. What is happening inside the range is discussed below.

Figure 10.2 shows α_p and C_p schematically. The rises through the glass transition are sharp but it would be exaggerating to call them discontinuities. Clearly, however, passage through the transition strongly affects the mechanisms controlling expansion and specific heat.

From M to G the shear viscosity rises by many orders of magnitude. Special measuring techniques are needed (Rawson 1980). For $T_G < T < T_M$ many viscosities follow the Vogel–Fulcher equation over wide ranges of

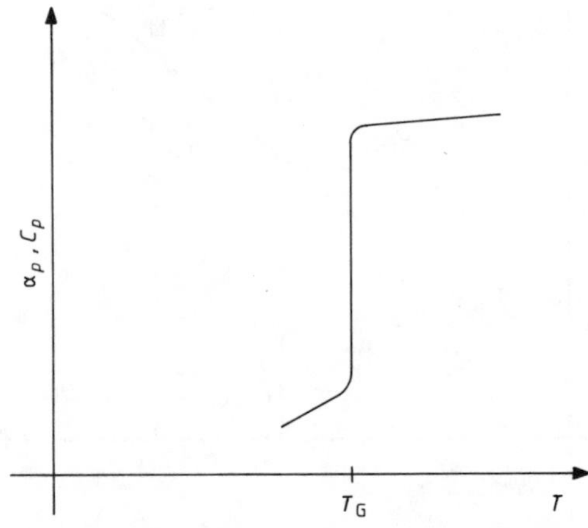

Figure 10.2 C_p and α_p at the glass transition.

temperature. This equation is

$$\eta(T) = \eta_0 \exp\left(\frac{A}{k_B(T - T_0)}\right) \tag{10.1}$$

where η_0 and A are constants and T_0 is positive but less than T_G. The Vogel–Fulcher equation reduces to Arrhenius's form when $T_0 = 0$ and this fits some materials better, especially as T approaches T_G. At T_G, η is typically 10^{12} Pa s (10^{13} P) making viscous flow negligible under everyday stresses—which is to say that glasses are solids. It remains very difficult to interpret the apparent singularity in η as $T \to T_0$.

At this point we recall the concepts of viscoelasticity and Maxwell relaxation time in §8.4 and identify the glass transition with the changeover from viscous behaviour in the supercooled melt to elastic behaviour in the glass. $\tau_m \equiv \eta/G$ and with $G \sim 10^{10}$ Pa and $\eta \sim 10^{12}$ Pa s, $\tau_m(T_G) \sim 100$ s. Putting $\omega\tau_m \sim 1$ for the viscoelastic transition, $\omega \sim 0.01$ is the lowest frequency at which an elastic response might be expected to a periodic shear stress at T_G. τ_m is an average; in a real glass there must be a number of mechanisms or local configurations with a spread of relaxation times.

Glass transitions at constant temperature can be induced by pressure changes.

10.2 Some thermodynamic and statistical considerations

In a supercooled liquid slightly below T_M structural rearrangements will occur which are not substantially different from those a little above T_M. They can in principle be investigated by the methods of Chapter 8. At $T < T_M$ freezing may occur if nucleation of a crystal occurs spontaneously or is provided. The nucleation probability depends on such factors as the enthalpy and entropy of fusion and the interfacial tension of crystal and melt. Nucleation is an important phenomenon in its own right but will not be pursued here (see, e.g., Toschev (1973)). Let us simply assume that heat is extracted rapidly enough to pre-empt nucleation by reducing the probability of creating and maintaining suitable crystalline arrangements. This probability is progressively lowered by cooling until in the transition range the system becomes unable in the available times to explore those regions of configuration space which represent structural change. This cannot be an absolute statement; presumably among the very large variety of local arrangements and the vast number of conceivable rearrangements there will always be changes requiring so little energy that their probability is not completely negligible. However, for $T < T_G$ extracting heat is chiefly a matter of reducing the vibrational amplitude, not of altering the structure. In this sense, glasses resemble solids more than liquids. Another way of expressing the same idea is that for $T < T_G$ the relaxation time for structural, as opposed to vibrational, modes of motion is very large on the timescale of measurements. This is a general view inferred from the phenonema shown in figure 10.2 together with the dramatic decrease in fluidity and corresponding changes in other parameters such as dielectric relaxation times.

The question of whether thermodynamics and statistical mechanics are applicable to glasses has often been raised; some recent contributions have been made by Jäckle (1981, 1984). If the state of the system can be specified by thermodynamic variables such as p, T there is no special problem in applying thermodynamic arguments. This clearly applies to the stable equilibrium states of the crystal below, and the liquid above, T_M. It also applies to the states of the supercooled liquid above the transition range and to the glass below T_G, where the observational timescale is very short compared with the relaxation time of structural change but very long compared with that for vibrational motion. It does not apply in the glass transition region. Consequently the thermodynamic equation $\Delta S = \int_{\Delta T} C_p \, dT/T$ can only be used if ΔT does not overlap any part of the transition range even if experimental values of C_p are calculated from observations in that range. In particular, if $S_C(T_2)$ is known to be the absolute value of the entropy in the crystal with $T_2 < T_G$, a unique absolute value, $S_G(T_2)$, for the glass cannot be obtained from

$$S_C(T_2) + \int_{T_2}^{T_M} (C_p)_C \frac{dT}{T} + \Delta S_M - \int_{T_2}^{T_M} (C_p)_G \frac{dT}{T} = S_G(T_2)$$

where ΔS_M is the entropy of melting and $(C_p)_G$ is for the supercooled liquid or the glass. Simon pointed out this difficulty in the 1930s.

The inappropriateness of thermodynamics over the transition range makes it unlikely that the glass transition could conform to the equations given by Ehrenfest for a second-order phase transition, though this has been suggested because such a transition does involve jumps in C_p and α_p like those in figure 10.2. Ehrenfest's equations are (Adkins 1983):

$$\frac{dp}{dT} = \frac{1}{VT} \frac{\Delta C_p}{\Delta \alpha_p} = \frac{\Delta \alpha_p}{\Delta \kappa_T} \tag{10.2a}$$

or

$$\frac{\Delta \kappa_T \, \Delta C_p}{TV(\Delta \alpha_p)^2} \equiv R = 1. \tag{10.2b}$$

R is called the Prigogine–Defay ratio and dp/dT refers to the slope of the phase boundary.

In practice $\sim 2 < R < \sim 5$ for glass transitions (Jäckle 1986, Wong and Angell 1976) and it appears invalid to treat the glass transition as a second-order phase change. To provide more flexibility in applying thermodynamics to glasses it was proposed to specify the state of a glass by an additional variable besides p and T: thus $V = V(p, T, z)$ where z is an unspecified parameter which presumably relates to some structural or other internal aspect of the system (Davies and Jones 1953). It turns out that equation (10.2*b*) formally follows from this hypothesis; therefore, if parameters like z are to be invoked, a single parameter is insufficient to secure agreement with experiment.

It is not impossible to apply formulae from statistical mechanics to glasses but there is not a unique way to do so (Jäckle 1981, 1984). In discussing this the word 'structure' is used to mean the spatial array of equilibrium positions of vibrating molecules. A particular glass sample has a fixed structure which is the one it froze into as it left the transition range. To make this process more definite, if more hypothetical, suppose the sample is held at a temperature T_F in the transition range long enough for it to reach the broken line in figure 10.1. It is then a supercooled liquid in a metastable condition capable of exploring, albeit slowly, an ensemble of different structures and having entropy $S_L(T_F)$. Suppose it is now quenched instantaneously to $T < T_F$, thus retaining the structure it happened to have when it was quenched. It keeps this structure indefinitely long on the timescale of observations and it is this structure which specifies the forces that control the vibrations. An ensemble can be imagined of which the members all have this fixed structure but various vibrational coordinates. The ensemble could be used for statistical mechanical calculations. *Inter alia*, the entropy would vanish at $T = 0$ just as for a crystal which also has a fixed structure. $S_G(T_F)$ could be calculated in principle and would be less than $S_L(T_F)$ by a finite amount, ΔS_F, because the liquid at T_F, unlike the glass, can explore a variety of structures, i.e., it has a much bigger phase space. T_F is sometimes called a fictive temperature; like the instantaneous quenching and the 'entropy of softening', ΔS_F, T_F is hypothetical because glass forming or softening actually proceeds continuously over the transition range.

In another way of approaching the matter the ensemble used for calculating the glass properties would include not only the vibration possibilities but also the set of structures that the sample might have frozen into. This set is the one available to the liquid at T_F because any one of them might have been the one which was quenched instantaneously. The probability of each structure would have to be known for the calculation. The entropy does not then have a discontinuity at T_F, nor does it vanish at absolute zero. At absolute zero there is still an ensemble of configurations even if the vibrational entropy has vanished. In fact the entropy expected to be connected with the range of possible disordered structures of the glass enters *either* as the 'entropy of softening' *or* as the residual entropy at zero temperature, depending on the way it is looked at. It still remains an intractable problem to calculate the entropy from first principles though there have been interesting attempts to do it (Rivier and Duffy 1982).

An experiment in 1985 demonstrated very clearly the removal of effective modes of motion as $T \to T_G$ from above. Birge and Nagel (1985) injected sinusoidal heat flow of frequency ν into glycerol, having devised a method of detecting its periodic temperature variation as a linear response and thus measuring $C_p(\nu)$. As figure 10.3 shows, there is a high-ν limit to C_p at which fast modes are being excited, and a low-ν limit where all modes respond; $C_p(0)$ is the ordinary specific heat. The curves show the removal of effective high-frequency modes as T falls. T_G for glycerol is about 180 K.

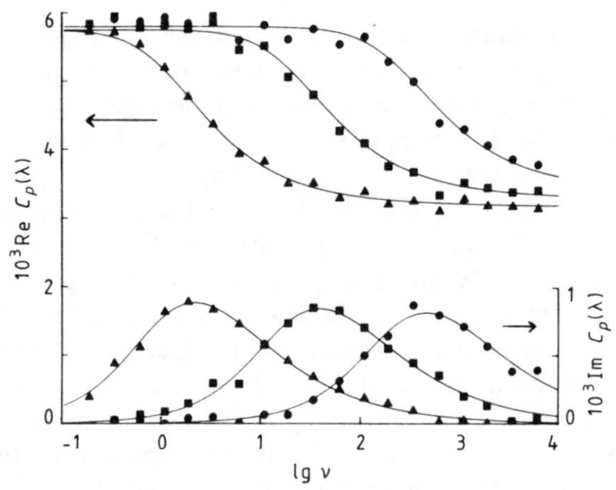

Figure 10.3 Real and imaginary parts of $C_p(\lambda)$ in $J^2\ cm^{-4}\ s^{-1}\ K^{-2}$ in glycerol, λ, the thermal conductivity, varies very slowly. ▲, −203.9 K, ■, −211.4 K, ●, −219.0 K. (From Birge and Nagel 1985.)

In this context we may recall the long-wavelength limit of the structure factors (see §§3.2, 3.12). For a monomolecular system in thermodynamic equilibrium $S(0) = n_0 k_B T \kappa_T$ which is a measure of the number density fluctuations. What is this for a glass? Presumably the number density fluctuations are frozen in so the limit $Q = 0$ measured in the glass will be that of the supercooled liquid at T_F rather than $n_0 k_B T \kappa_T$ of the glass. Similar arguments should apply to $S_{ij}(0)$, $S_{cc}(0)$ etc though at the time of writing this appears not to have been tested experimentally.

Owing to the extremely slow exploration of its configuration phase space by a glass, the time average of a property observed over a feasible timescale is not the same as the ensemble average. The system is said to be non-ergodic and the discussion of its statistical mechanics raises deep problems only touched upon above. They can be pursued in Palmer (1982).

10.3 Can all materials be vitrified?

With the advent of splat cooling, melt spinning, evaporation onto cold substrates and laser glazing, practical methods for very-high-speed cooling became available, say 10^7 K s^{-1} with splat cooling and about 10^{13} K s^{-1} with laser glazing (Breinan *et al* 1976). Consequently, many previously unrealised vitreous systems have been made, notably metallic glasses. It still remains difficult or impossible to vitrify pure metals or noble gases but many authorities would hold that, in principle, any fluid could be turned into a glass if the cooling were rapid enough. It is not possible to identify by its composition or bonding any class of materials incapable of vitrification.

This view is strengthened by the preparation of computer glasses (§4.6). But it is not obvious that systems reached by virtually instantaneous quenching in a simulation can be said to have undergone a glass transition. The necessary and sufficient conditions for this have been discussed by Angell *et al* (1981) and by Jäckle (1986).

If a real quench were halted at some temperature T, the system ought to be able to last long enough without crystallisation to equilibrate into either a metastable supercooled liquid or a glass, depending on $(T - T_G)$. To count as a glass transition a simulation should have this property. Of course the glass may not remain amorphous indefinitely unless kept at a very low temperature but this is true in the laboratory also. Furthermore at all temperatures the vibrational modes should remain in temperature equilibrium with the heat bath which is controlling the fall in temperature. Above T_G the configurational modes should do this also. The simulation should be consistent with this so that the concept of temperature is preserved through the transition. It appears that the computer-formed Ar glasses referred to in §4.6 and §8.5 do meet these conditions with cooling rate of about 10^{12} K s^{-1} and it seems very valuable to study computer glasses as valid hypothetical representatives of the class if not real ones (Jäckle 1986).

10.4 A glass phase transition? Models

All the preceding sections have expressed and supported the view that the glass transition has to do with the relaxation time of configuration changes becoming first comparable with and then, at lower temperatures, much longer than measurement times.

As a crystal melts the liquid receives ΔS_M of entropy. If the liquid is now supercooled to T near T_G the entropy decreases by

$$\Delta S_L = \int_T^{T_M} C_p \, dT/T.$$

Observed values of ΔS_M and ΔS_C are such that values of T can be found at which ΔS_L is almost as big as ΔS_M. This is sometimes quite striking as in lithium acetate (Wong and Angell 1976). It suggests that, if by careful and slow cooling to maintain metastable equilibrium T could be made somewhat lower still, the supercooled liquid would give up all its entropy of melting and more, and then have less entropy than the crystal at some finite temperature. Since this seems unacceptable it has been suggested that the supercooled liquid phase must end in a phase transition to another state either crystalline or glassy. This argument was put first by Kauzmann (1948) whose own view was that as the effort was made to reduce T very slowly the probability of crystalline nucleation would always overtake that of other structural rearrangements and crystallisation would pre-empt the hypothetical phase transition. The prevailing view is that the glass transition is a relaxation phenomenon, not a thermodynamic phase transition. Whether the relaxation pre-empts or in some way overlies what would otherwise have been a phase transition seems difficult to decide with certainty.

In order to clarify the glass transition theoretically a number of models have been proposed which were reviewed by Jäckle (1986). New models continue to emerge. All that is attempted here is to outline briefly one such model, the free-volume theory of Grest and Cohen. This is interesting for the variety of ideas it brings to bear on the matter and because it provides for both phase change and relaxation. The arguments are both detailed and extended and justice will only be done to the model by reference to the original papers (Grest and Cohen 1981, Cohen and Grest 1983).

The idea of a free-volume theory is that each of the N molecules has a space of its own which can be identified with its Voronoi polyhedron (§2.2). Because of the disorder the cell volumes have a distribution $P(v)$. There is supposed to be a critical cell volume v_c such that a cell with $v = v_c + v_f$ is said to contribute v_f of free volume which is capable of redistribution including amalgamation with other v_f into voids big enough to allow molecular movement. This aspect of the theory leads to expressions for D and η which was its original intention as a theory of dense liquids.

To deal with the statistical thermodynamics of glass transitions it is supposed that each cell makes a local contribution of free energy, $f(v)$, dependent only on the cell volume. This does not mean that $\int P(v)f(v)\,dv$ is the total free energy; on the contrary, a term $Nk_BT \int P(v) \ln P(v)\,dv$ is necessary for the configuration entropy of the cells (compare the $c \ln c$ term for mixing entropy in a binary system). There is also a term $-TS_c$ where S_c is the communal entropy deriving from the possibility that a molecule in a liquid has to move throughout the volume. Thus, so far

$$F = N \int P(v)(f(v) + k_B T \ln P(v)) \, dv - TS_c.$$

It is not obvious what the function $f(v)$ is but it will be a measure of the energy involved in putting a molecule into its volume and, as a function of v, it will be similar in shape to the pair potential $\varphi(r)$ as a function of r, i.e., large and positive for small v, small and negative for large v with a minimum v_{min}. Grest and Cohen give arguments for modelling $f(v)$ between v_{min} and v_c by a quadratic form joining continuously onto a linear form for $v > v_c$ (see figure 10.4). This is important to the theory because v_c is a critical cell volume enabling a distinction to be made between cells with $v < v_c$, called solid-like (s-) and those liquid-like (l-) cells with $v > v_c$.

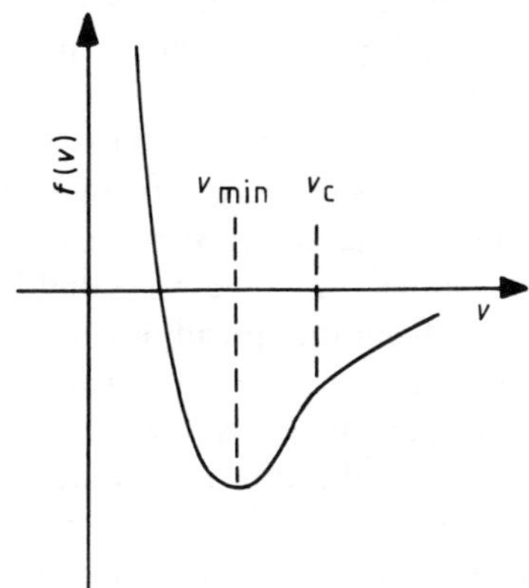

Figure 10.4 (After Grest and Cohen 1981.)

Therefore

$$\bar{v}_f = \int_{v_c}^{\infty} (v - v_c) P(v) \, dv \left(\int_{v_c}^{\infty} P(v) \, dv \right)^{-1}$$

and the fraction of l-cells is

$$p = \int_{v_c}^{\infty} P(v) \, dv.$$

An l-cell that has several others as neighbours can redistribute free volume among them without disturbing any neighbouring s-cells. Since $f(v)$ is linear for the l-cells, and the s-cells are unchanged, there is no change in the local amount of free energy. Isolated l-cells could not interchange free volume without involving a free-energy change for the s-cells. This indicates the importance for free-volume redistribution of clusters of l-cells. Clusters of several touching l-cells become more probable as p increases and when p reaches a critical value p_c there will be an infinite cluster reaching throughout the system. The reader will recognise this as a percolation problem (see Chapter 9). It is a feature of the free-volume model that a glass transition is associated with the critical percolation fraction p_c: $p < p_c$ corresponds to a solid glass, $p > p_c$ to the supercooled liquid. In order to consolidate this idea it is necessary to derive an expression for S_c contributed by all the liquid volume, namely, that in the finite clusters and the infinite one if present, and from this to derive $P(v)$ and p. It is also necessary to add another term to F because of the surface entropy of clusters. The details of these arguments are given in Grest and Cohen (1981) and Cohen and Grest (1983). The interesting outcome is that the glass transition turns out to be a first-order phase transition and since the theory involves a complete—though approximately modelled—statistical mechanics, all the thermodynamic quantities such as C_p and α_p can be derived. The free-volume theory also succeeds in fitting the viscosity and diffusion data.

However, the notion that the glass transition is actually first order does not agree with observations. Therefore the free-volume theory accommodates the relaxation nature of the observed process by treating the adjustment of p to changes in temperature as a relaxation phenomenon using a distribution of τ's derivable from the spread of cluster sizes given by percolation theory. Since p is the fraction of l-cells this amounts to saying that the structural rearrangements which alter this fraction do so at a rate comparable with measurement times in the transition range. We have therefore a model in which relaxation overlies a phase transition and of which the results show encouraging agreement with observations (Grest and Cohen 1981, Cohen and Grest 1983).

The mode-coupling theory referred to in §8.4 has been applied to temperatures near the glass transition and many features of the transition follow from it. It is a somewhat complex theory and can be studied in Bengtzelius *et al* (1984) and Bengtzelius and Sjögren (1984). For more discussions of glass transition models and comments on the free-volume theory see Anderson (1979), Kirkpatrick (1985) and Shukla (1983).

10.5 Constraints and stability

While it may be that any system can be made amorphous if the conditions are right it is certainly the case that some substances form covalently bonded networks more readily than others. There have been many attempts to explain this and a recent method of considerable interest was introduced by Phillips (1979). The basic idea is to relate the number of mechanical degrees of freedom of an atom, denoted by n_d, to the number of constraints imposed on it by the forces from other atoms. To simplify the enumeration of the constraints it is assumed that very short-range bonding interactions are involved; long-range Coulomb forces would greatly complicate the issue. Let $\langle r \rangle$ stand for the average number of bonds per atom in a glass. Then

$$\langle r \rangle \equiv \sum_r rN_r \left(\sum_r N_r\right)^{-1} = N^{-1} \sum_r rN_r$$

where N is the total number of atoms and N_r is the number of atoms with r bonds. In a binary system of A and B atoms

$$\langle r \rangle = r_A x_A + r_B(1 - x_A)$$

where x_A is the concentration of A atoms.

Suppose now that the number of mechanical constraints per atom is n_c which will be some function of $\langle r \rangle$. Phillips proposed that glass formation was favoured by the condition

$$n_c = n_d$$

which expresses the idea that the structure forms in such a way that the number of constraints just uses up, or cancels out, the degrees of freedom. This suggests a criterion for the stability or rigidity of the structure. Systems with $n_c > n_d$ or $n_c < n_d$ are called over- and under-constrained respectively.

Before relating this hypothesis to observations or calculations n_c must be estimated. The underlying idea for this is that small displacements of atoms are controlled by forces expressed by the potential

$$V = \frac{1}{2} \sum_{i,j} \alpha_{ij} (\Delta r_{ij})^2 + \frac{1}{2} \sum_{i,j,k} \beta_{ijk} (\Delta \theta_{ijk})^2$$

in which Δr_{ij} is the change in length of the i–j bond and the corresponding forces are called α-forces. $\Delta \theta_{ijk}$ is the change in angle between the bonds i–j and j–k and is controlled by β-forces. Each bond is a constraint so there will be $\frac{1}{2}r$ such constraints per atom for atoms with r bonds; these are due to the α-forces. For an atom with $r = 2$, the β-force leads to one extra constraint on the bond angle. For an $r = 3$ atom, the additional bond brings in two more angular constraints because it implies the specification of two more angles, namely, those with the two existing bonds. Likewise, a fourth bond adds two more constraints because angles with two of the previous bonds will be determined. Thus the formula $(2r - 3)$ covers the number of β-force constraints for $r = 2, 3, 4$, namely, n_c (β-forces) $= 1, 3, 5$ respectively.

A useful quantity in these arguments is the fraction, f, of modes which involve displacements costing no energy, called zero-frequency modes. To estimate f for a network glass of N atoms we need:

Number of degrees of freedom $= 3N$

Number of α-force constraints $= \sum_r n_r r/2$

Number of β-force constraints $= \sum_r n_r(2r - 3)$.

The number of zero-frequency modes is $3N$ minus the number of constraints, so $f = 2 - \frac{5}{6}\langle r \rangle$. Cancelling out the degrees of freedom by the constraints means putting $f = 0$ or $\langle r \rangle = 2.4$. This result has been confirmed by numerical simulations on a diamond lattice from which bonds were removed to give a distribution of atoms with $r = 2, 3, 4$ (Thorpe 1984, 1985). The proposition, originally due to Phillips, could therefore be stated without undue overprecision as follows: covalent networks are most likely to form if there are two to three bonds per atom. There are certainly many real glasses with $\langle r \rangle$ in this range. The glasses in this chapter and the chalcogenide glasses of §11.5 are examples, whereas the amorphous tetrahedrally bonded semiconductors in §11.4 are overconstrained.

The underlying idea that the condition for mechanical stability of a network is significant for the formation of a glass can be connected in an interesting way with percolation theory. Stability of hinged connected frameworks is also a macroscopic mechanical problem in its own right and exercised Maxwell in the 1860s as Thorpe (1984, 1985) has pointed out. If b bars are connected at j hinged joints in 2-D, Maxwell showed that the condition $b = 2j - 3$ separates rigid or overconstrained frameworks from non-rigid, underconstrained or floppy ones. Starting with the smallest rigid framework—a triangle with $b = 3 = j$—it is not difficult to arrive at Maxwell's conclusion by induction but the result, though accurate in practice, is not rigorous because rigidity depends not only on the number of bars but also on which joints they connect; b may equal $2j - 3$ in a structure which is overconstrained in some parts and floppy in others. What is needed in the argument is the number of constraints and this is approximately, but not exactly, equal to the number of bars. In 3-D, the corresponding condition is $b = 3j - 6$ where the 6 accounts for the three translational and three rotational degrees of freedom of the structure as a whole.

Site and bond percolation were discussed in §9.2. By imagining two triangles hinged at one common vertex in 2-D it becomes clear that connectedness of sites does not guarantee rigidity of structure; a further bond between two other vertices would be required. By the same token, a percolating cluster in site or bond percolation will be connected but not in general rigid. Thorpe (1984, 1985), who developed the idea of rigidity percolation, gives the following example. A 2-D triangular net of atoms is connected by central pair forces between nearest neighbours. Suppose bonds are removed randomly. Connectivity vanishes at the percolation threshold $p_c = 0.35$ but the elastic constants c_{11} and c_{44} vanish earlier at a threshold $p^* \sim 0.67$. c_{11} and c_{44} decrease linearly with p towards p^* from their values at $p = 1$; this has been shown by numerical simulation studies. There is also an effective-medium theory of rigidity percolation (compare §9.4) which gives the same linear result. For $p < p^*$ rigid and floppy regions or clusters coexist, and the glass transition can be conceived as a rigidity percolation threshold interesting both for the similarities and the contrasts with the free-volume-percolation theory in §10.4.

Rigidity percolation will no doubt undergo further developments. It is in some ways more complex than ordinary site or bond percolation. The latter depends on connectedness at one place: either a site or bond is or is not occupied. Rigidity demands connectedness at one or more distant places as well—a requirement which can be called non-locality. This can be seen from the example given above of two triangles hinged at one vertex.

If we now imagine a binary glass A_xB_{1-x} where A has $r = 4$ and B has $r = 2$, then almost pure B will have $\langle r \rangle < 2.4$ and consist of non-rigid chains of two coordinated atoms, e.g., Se. In practice crosslinks between chains caused by A atoms or van der Waals' weak forces will stop the structure from being intrinsically fluid but its elastic moduli will be low. High concentrations of A (say A = Ge) will make $\langle r \rangle > 2.4$ and raise the moduli. At the time of writing experiments demonstrating this behaviour of moduli as a function of x are being performed and more evolution of the rigidity percolation concept of glass formation can be expected (Halfpap and Lindsay 1986). The domains drawn in figure 10.9 may relate to this concept.

10.6 The structure of oxide glasses

A priori a glass may be a mass of small crystallites; they would have to be very small otherwise their presence would be revealed by diffraction studies. On the other hand, a glass might be a continuous random network. Considering how glasses are formed a compromise view might commend itself, namely that in some regions the local order resembles rather closely that of one or other of the crystalline polymorphs of the material but that in between the structure is more random. It has proved difficult to settle this matter by a definitive diffraction experiment and in any case there are many different glasses and various ways of making the same glass so the class may defy generalisation (Gaskell *et al* 1982).

Structures have been described and studied by the methods of Chapters 2 and 3. One purpose of this section is to indicate that although the CRN concept of Zachariasen (§2.6) is of major significance there is much more to be said and discovered about the atomic arrangement in covalently bonded glasses. A second purpose is to illustrate the value of techniques other than diffraction for elucidating SRO, notably NMR and EXAFS.

In order to interpret experimental results models are very helpful. Ball and spoke models of vitreous SiO_2 by Evans and King and by Bell and Dean have been referred to (§§2.6, 8.5) and there are others in the literature. Modelling continues, some of it starting with the CRN concept and building the system subject to specific restraints on bond angles, torsional angles or, to a lesser extent, bond lengths (Tarlros *et al* 1984). In simple cases the building block is a well defined structural unit such as the SiO_4 tetrahedron in SiO_2 or the boroxol ring—a three-membered ring of BO_3 triangles—in

B_2O_3. The disorder derives from the way they are joined together and the choice of this determines the short- and medium-range order.

In the conceptual construction of a crystal, a basis is located at each Bravais lattice point. A corresponding idea for an amorphous solid is to place a structural block at each unit of a topological network (Wright *et al* 1980). A topological unit is an object with a certain connectivity, i.e., a certain number of points at which it is connected to neighbouring units. A unit with connectivity three is clearly representable in 2-D by a triangle of any shape. In 3-D, four-connected units could be tetrahedra joined at their vertices.

The network itself is most easily illustrated in 2-D using triangles (see figure 10.5). Since, by hypothesis, the only feature that matters to the network is connectedness, not any particular shape or size of triangle, the network can be defined by giving the serial numbers of every unit's neighbours, e.g., unit 4 has neighbours 5, 3, 8; 5 has 4, 9, 6 etc. As with the analogous problem of specifying the coordinates of atoms in a disordered array this can only be done for small finite models; for real macroscopic samples fairly crude statistical measures must suffice. For atoms one such is the pair distribution function; for topological networks ring statistics play a comparable role. In neither case does the statistic completely describe the structure.

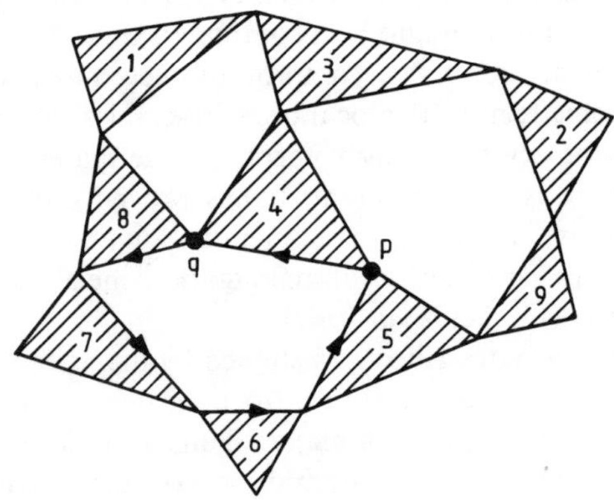

Figure 10.5

A glass may be conceived by postulating a network of topological units and decorating each one with a structural block. Figure 10.6 is from Wright *et al* (1980) and shows the same topological network decorated in two different ways. In figure 10.6(*a*) each topological unit (triangle) has a single atom and the resulting hypothetical element has a CRN of atoms each joined to three others. In figure 10.6(*b*), the structural block is a triangle of B-type atoms with an A-type atom in the middle; each AB_3 joins three others making a material of formula A_2B_3. More elaborate structures can be envisaged. Apart from moving into 3-D, it is possible to construct topological networks with more than one kind of topological unit, or to keep one kind of unit but have two or more types of block for the decoration which might therefore be performed in an ordered or disordered way; or these complications could be combined. SiO_2 can be envisaged as SiO_4 tetrahedra decorating tetrahedral topological units—similar to those of figure 10.6(*b*) but in 3-D.

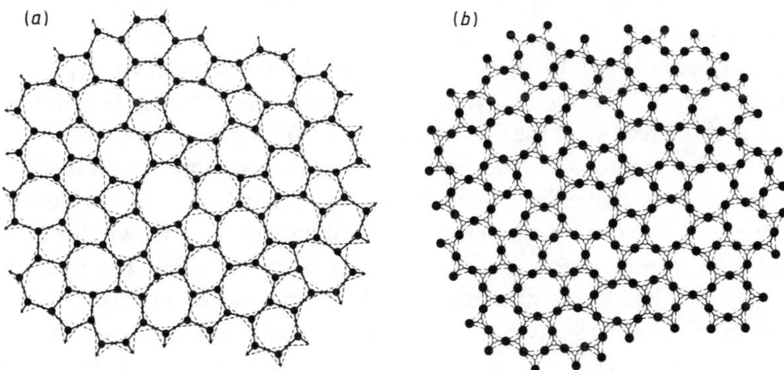

Figure 10.6 Two decorations of the same topological network (from Wright *et al* 1980).

If a model is required to be plausible for some theoretical purpose, such as computing a vibrational spectrum, it is not sufficient to postulate a topological network and decorate it. It may also be necessary to relax the model (§2.5). A suitable potential will be required and the energy minimised by altering the bond angles and lengths with constant topological connections. If the relaxation leads to a density and a pair distribution consistent with experimental observations the model can plausibly be used for further computations.

Intermediate- or medium-range order was referred to in §1.7 as a difficult subject of increasing interest. In CRN it is an expression of the topology of the network. Ring statistics may be used to quantify it and they were first investigated in SiO_2 following a suggestion by Bernal (King 1967). Let *m* be defined as follows: if p, q are two connection points of the same topological unit (e.g. two vertices of a triangle in figure 10.5), *m* is the smallest number of topological units which jointly make a closed ring containing p and q. For example, for p, q in figure 10.5, *m* is 5, involving units 4, 8, 7, 6 and 5. *M*(*m*) is the number of all the *m*-membered shortest rings obtained by taking every pair of vertices p, q on each topological unit in turn divided by the total number of topological units. *M*(*m*) is thus the number of *m*-membered shortest rings per topological unit. The distribution of 3-, 4-, 5-, 6-, . . . membered rings is then a property of the model; the proportion of even and odd membered rings may have significance for the character of the allowed chemical bonds.

It has proved difficult to measure ring statistics in real matter. Indeed it is a considerable problem to elucidate the actual distribution of bond or torsional angles or the nature of structural blocks even in comparatively simple materials like SiO_2 and B_2O_3. There is also the question of the nature and frequency of defects in whatever is the typical structure of the material, though we shall not pursue that question here (see §§11.4, 11.5).

SiO_2 has often been studied by diffraction (Wright and Leadbetter 1976). One well known x-ray study was made by Mozzi and Warren (1969) (also see Tanaka *et al* 1985). Figure 10.7(*a*) is from this work and its interpretation is straightforward for the first two peaks. The first is at the Si–O distance of 1.62 Å and the second at the O–O distance of 2.65 Å. These lengths are in the right ratio for an SiO_4 tetrahedron. The Si–Si distance depends on the Si–O–Si bond angle, β, and is in fact $2(1.62 \sin \frac{1}{2}\beta)$ Å, assuming a negligible variation in Si–O. The third peak marks the Si–Si distance and its width suggests a spread in β. Corresponding remarks can be made about the next three peaks which relate to the distances from Si atom to second O atom, O atom to second O atom, Si atom to second Si atom, only successively more parameters, such as the torson angle α in figure 2.8, need distributing in order to provide for the variation in these distances. Mozzi and Warren used an optimisation procedure which made assumptions about the spread of β and the other variables, and adjusted the assumed distributions until the computed pair distribution agreed best with the measured one. Figure 10.7(*b*) shows their conclusion about the Si–Si distance and bond angle variations which are, of course, expressions of the disorder in the CRN. The mean value of β was about 144°.

This samples the kind of inference that can be made by fitting x-ray data. The difference curve in figure 10.7(*a*) illustrates the difficulty of deriving totally satisfactory structure information from a single diffraction experiment. It is not surprising therefore that other experiments have been tried.

The EXAFS technique was introduced in §3.13 and was applied to silicate glasses by Greaves *et al* (1981). This method reveals the environment of one species at a time and the conclusion about the Si environment, apart from confirming the O tetrahedron, was that $\beta = 160 \pm 20°$.

The NMR literature shows that the various anisotropic forces, such as the nuclear magnetic dipole interaction, that broaden the NMR resonance line in crystals are removed in liquids because the relative motion of atoms averages them out. By the same token, if a solid sample is spun rapidly about an axis making an angle γ with the static applied magnetic field, the rotation reduces the broadening by a factor $\frac{1}{2}(3\cos^2\gamma - 1)$ which vanishes for $\gamma = \cos^{-1} 3^{-1/2}$. Spinning at this angle results in much sharper spectra

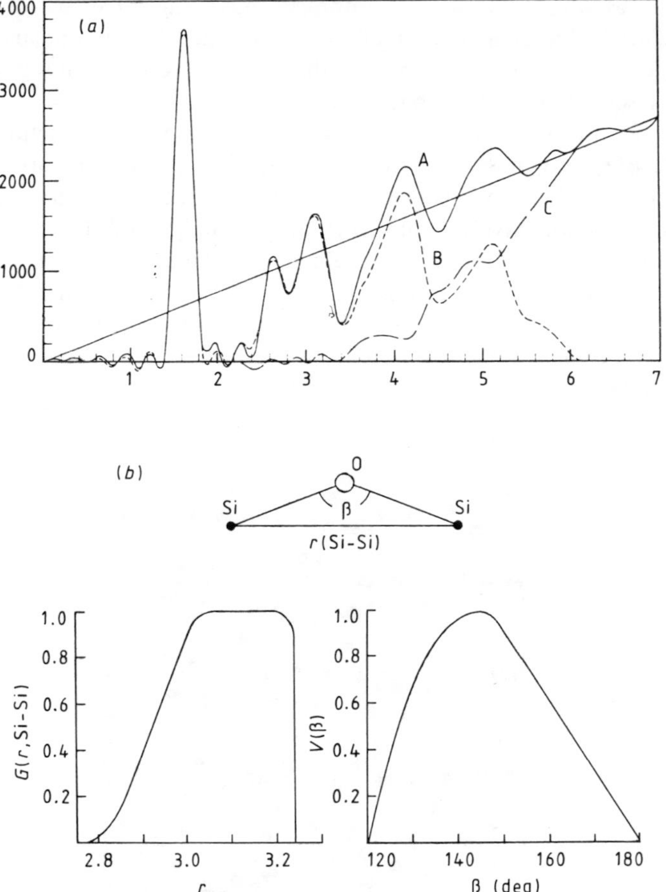

Figure 10.7 (*a*) Pair distribution function of SiO_2. A, x-ray results; B, computed function from first six interparticle distances; C, difference between A and B. (See Mozzi and Warren 1969.) (*b*) Some inferences from figure 10.7(*a*). Distributions of the distance and the angle defined in the figure.

from solids and since its invention by Andrew and more widespread exploitation in the eighties has become the established technique of 'magic-angle spinning' (MAS). The sharpness of MAS spectra enabled it to be shown that the Si–O–Si angle in seven crystalline forms of SiO_2 (quartz, cristobalite, etc) correlated with the small shift in NMR frequency due to the different structural environments. The MAS–NMR lineshape in vitreous SiO_2 is therefore a product of the built-in distribution of Si environments and Dupree and Pettifer (1984) were able to interpret their observed lineshape in terms of the spread of bond angles in the glass. Their inferred distribution was broad enough—about $150 \pm 20°$—to embrace those suggested by diffraction, EXAFS, model building and MD, all of which disagreed with one another. At the time of writing it cannot be said that the details of the SiO_2 glass structure are settled beyond reasonable doubt.

Another much studied glass is B_2O_3. Candidates for the structural blocks include BO_3 triangles linked by their vertices and a ring of three of these called a boroxol group. Postulated or modelled structures containing one, or both, of these will have a variety of B–B, O–O and B–O distances which will determine the diffraction patterns. Reviewing the evidence from x-ray and neutron diffraction in 1982, Johnson *et al* (1982) concluded that it was most consistent with a model in which about 0.75 of the B atoms were members of boroxol rings and the rest were in BO_3 triangles leading to a structure of which figure 10.8 is a 2-D representation.

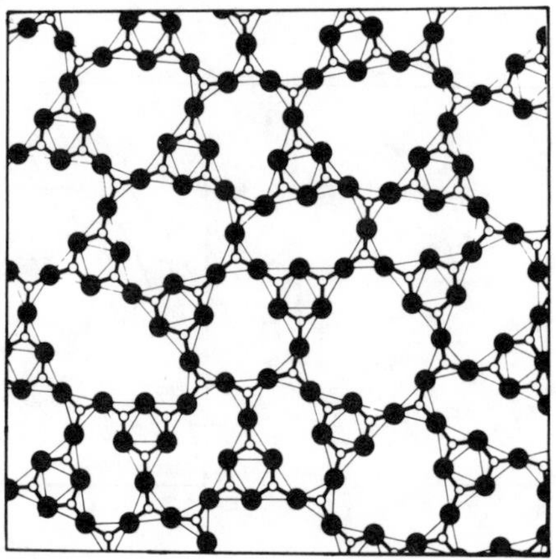

Figure 10.8 2-D representation of B_2O_3 structure: open circles, boron (from Johnson *et al* 1982).

Such a conclusion is not based solely on diffraction. For example the Raman spectrum contains a line considered from other studies of boron-containing compounds to characterise the boroxol group. The NMR experiments also contribute valuable evidence because ^{10}B, ^{11}B and ^{17}O nuclei have quadruple moments which make their NMR spectra notably sensitive to the electric field gradient from the neighbours (see also §8.6). The gradient depends on the electric charge distribution and is therefore an indicator of local geometry and bonding. Suppose there is reason to postulate that B exists with possible environments E_1, E_2, . . . , having spectra S_1, S_2, ... Then it is possible to compute the compound spectrum obtained by superposing S_1, S_2, . . . , with weightings proportional to the fractions of B nuclei in

$E_1, E_2, \ldots$. Computations can be made to agree with the observed spectra with remarkable accuracy and details and examples are given by Bray *et al* (1982). In B_2O_3 the evidence strongly supports the presence of boroxol rings and of O atoms in bridging sites between rings. But it is interesting that MD simulations of B_2O_3 do not necessarily lead to boroxol structures, though this may simply be a feature of the potential employed (Soules 1982).

When Na_2O, Li_2O or similar compounds are added to SiO_2 or B_2O_3 important property changes occur. For example, Na_2O in SiO_2 reduces the viscosity and greatly affects the softening temperature and expansion coefficient (Rawson 1980). Alkali oxides are examples of 'network modifiers' and alter the structure. Some of the excess O atoms attach themselves to one Si atom instead of two thus leading to the categorisation of 'non-bridging' (nb) and 'bridging' (b) oxygens—the latter being those that link two SiO_2 tetrahedra at a common vertex. The Na^+ cations are thought to be attracted to nb oxygens; there is diffraction and MD evidence for this and extensive discussion in the glass literature. The modifiers create new environments for Si and B, and this is exactly what NMR can detect. In fact, Bray *et al* (1982) show from the effect of the quadrupole interaction on the NMR spectrum of ^{10}B how the various concentrations of modifier both affect the proportion of B atoms in boroxol rings and also introduce other forms of threefold and fourfold O coordination round B, together with different proportions of b and nb oxygens.

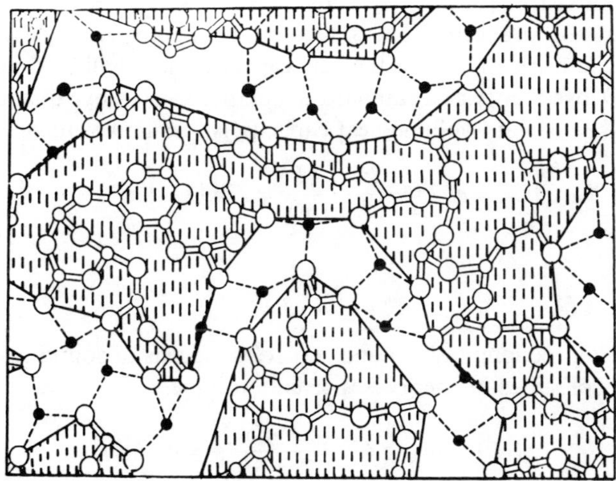

Figure 10.9 Schematic 2-D diagram of the structural function of a modifer (black dots) in an oxide glass suggested by EXAFS studies (from Greaves 1985). Full lines, covalent bonds; broken lines, ionic bonds. The glass has formula $M_2O_3(G_2O_3)_2$ where M = modifier (e.g. Na) and G = glass former (e.g. B). Note the b and nb oxygen atoms (large circles).

The MAS-NMR and EXAFS experiments also show interesting effects of Na_2O in SiO_2. The MAS spectra of ^{29}Si are sufficiently sharp to distinguish between Si with four b oxygens, those with three b and one nb and those with two b and two nb. It appears that b oxygens are gradually replaced as Na_2O is introduced; at 33.3 mol. %, all Si atoms have three b and one nb and at 50 mol. % they have two b and two nb. Similar observations on glassy and crystalline forms of the same composition showed no evidence for regions of crystallinity in the glass (Dupree *et al* 1984). Greaves *et al* interpret their EXAFS spectra from sodium disilicate glass ($Na_2Si_2O_5$) as evidence for two interpenetrating subnetworks, a covalent one of SiO_2 and an ionic one of Na_2O, linked together by what in another context are called non-bridging oxygens (Greaves 1985, see figure 10.9). At the time of writing thorough integration of views on soda–silica glasses appears incomplete.

The reader may well conclude that considerable progress has been made with the glass structure problem in the 1980s but that much remains to be discovered especially for the many glasses less thoroughly investigated than SiO_2 and B_2O_3.

10.7 Low-temperature properties and two-level systems

In §10.2 it was remarked that even at $T < T_G$ an absolute inhibition of configurational change seemed very unlikely. Changes involving energy differences of $k_B T$ at about 1 K, i.e. about 10^{-4} eV, appear to be involved in some remarkable properties possessed at low temperatures by glasses but not by crystals. The first of these is the heat capacity and it will be convenient to recall the Debye theory as a standard of comparison. Debye quantities have suffix D. The constant-volume heat capacity is, per unit volume,

$$C_D = c_D T^3 = \frac{2\pi^2}{5} \frac{k_B^4 T^3}{\hbar^3 v_D^3} = 234 \left(\frac{T}{\theta_D}\right)^3 nk_B \qquad (10.3)$$

where

$$1/v_D^n = \tfrac{1}{3}(1/v_l^n + 2/v_t^n)$$

v_l and v_t being the longitudinal and transverse speeds of sound respectively. The density of states in energy is, per unit volume,

$$g_D(E) = \hbar^{-1} g(\omega) = \frac{3}{2\pi^2} \frac{(\hbar\omega)^2}{\hbar^3 v_D^3} = b_D(\hbar\omega)^2. \qquad (10.4)$$

The Debye theory applies rather well to pure crystalline solids at $T < 10^{-2}\,\theta_D$. Above this, departures from the T^3-law are explicable by the differences of $g_D(E)$ from the real phonon spectrum. $10^{-2}\,\theta_D$ is typically about 1 K and the dominant phonon wavelengths might be of the order of 100 lattice spacings.

It emerged clearly during the 1970s that glasses do not behave like this. Figure 10.10 taken from Stephens (1976) shows the behaviour of vitreous silica. Even among insulating glasses SiO_2 is sometimes unusual but this non-Debye behaviour is shared by many other glasses including Se, B_2O_3, borosilicates, chalcogenides, polymers and metals. The observations lead to the generalisation (Pohl in Phillips (1981))

$$C = c_1 T + c_3 T^3 \qquad 0.1 < T(\mathrm{K}) < 1. \tag{10.5}$$

Confining our attention to insulating glasses the noteworthy features are the term $c_1 T$, often called the linear specific heat anomaly, the fact that $c_3 > c_D$ and the remarkable universality of the behaviour. The anomaly is not small; for $T < 0.1$ K, C can exceed C_D by a factor of 100 or more. A simple equation cannot embrace all the relevant facts, e.g., the first term sometimes reads T^n where $n \sim 1.45$ in B_2O_3. C is also influenced by impurities, but without the implication that the anomaly is entirely due to them. The inference seems to be that there are some excitations peculiar to glasses that lead to an approximately linear specific heat at low temperatures. This contradicts the rather natural assumption that the Debye continuum model with long-wavelength phonons ought to apply to glasses if it does to crystals.

This is not an isolated phenomenon. It soom became clear that properties involving phonon propagation also showed marked peculiarities and we refer to the thermal conductivity. Figure 10.11 is from Zeller and Pohl (1971). These and other results show that glasses of different compositions

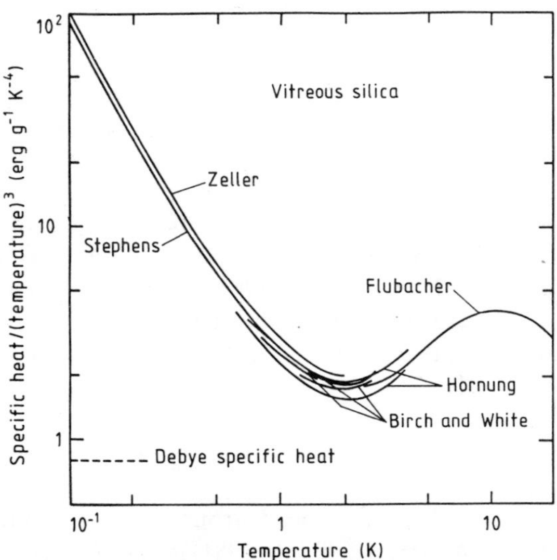

Figure 10.10 Departures from Debye theory detected by several workers in the specific heat of vitreous silica (from Stephens 1976).

have remarkably similar thermal conductivities and that, below a plateau of roughly constant value around 10 K, λ falls with temperature to vary as $\beta T^{1.9 \pm 0.1}$ for $T < 1$ K with $\beta \sim 10^{-3}$ to 10^{-4} W cm^{-1} K^{-1}. All these properties are in marked contrast with those of crystals in which λ is larger by several orders of magnitude and varies significantly from one material to another. It has a maximum, not a plateau, below which it decreases approximately as T^3 like the specific heat.

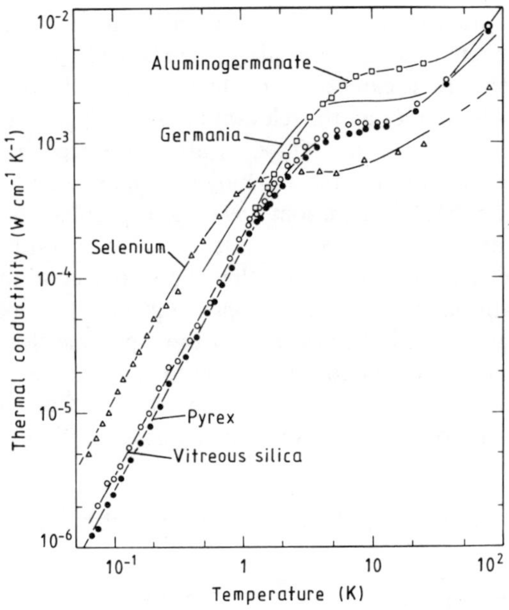

Figure 10.11 Thermal conductivities of several glasses (from Zeller and Pohl 1971).

It is possible to estimate a phonon mean-free path, l, on the assumption that the thermal conduction in glass is by phonons. In the Debye theory the number of quanta in dω is $\omega^2\,d\omega[\exp(\hbar\omega/k_B T) - 1]^{-1}$ which has a maximum at $\hbar\omega_{dom} \simeq 1.6 k_B T$. Simplifying the picture to one of heat transport by phonons of dominant frequency, ω_{dom}, the formula can be taken from the kinetic theory, namely,

$$\lambda(T) = \tfrac{1}{3} C_D v l = \frac{4.08 \times 10^{10} T^3 l}{v_D^2} \tag{10.6}$$

which can be used if l does not depend strongly on ω. An interesting experiment referred to by Anderson (Phillips 1981) *defined* l by a dispersion of tiny holes in the glass sample. Knowing l, T and also v from acoustic measurements, $\lambda(T)$ was calculated from equation (10.6) and found to agree rather well with the measured $\lambda(T)$ for $T < 0.2$ K. The suggestion is therefore that thermal conduction is by phonons propagating with the velocity of sound and that any special excitations required for explaining the linear specific heat anomaly are non-propagating. At higher temperatures, around the plateau, the observed $\lambda(T)$ is incompatible with so simple a picture and points to the further assumption that l is strongly dependent on ω, falling from $\sim 10^{-3}$ to $\sim 10^{-7}$ cm between 10^{10} and 10^3 Hz (Anderson in Phillips (1981)).

Phonon propagation can also be investigated by experiments on ultrasonic attenuation and dispersion. Sufficiently high frequencies can be used, say $\omega \simeq 10^{10}$, to overlap with those of the thermal phonons invoked in discussing thermal conduction. Again there is a pattern of behaviour common to many different glasses (see e.g., Hunklinger and Schickfus in Phillips (1981)). α, the absorption coefficient, or reciprocal of the mean-free path, is much greater than in crystals and in general the same for longitudinal and transverse waves. It has a broad maximum at $T_1 \sim 100$ to 200 K and another maximum, considerably smaller, at $T_2 \sim$ a few kelvins and both vary with ω. The peak at T_2 falls off as T^3 on the low-temperature side but at $T < 1$ K what happens depends somewhat unexpectedly on the intensity of the acoustic signal. Low intensities (say 10^{-1} to 10^{-2} μW cm^{-2}) show α passing through a minimum ($\sim$0.6 K in SiO_2) but high intensities (say 10^2 to 10^3 μW cm^{-2}) cause α to decrease with T towards zero as if the absorption mechanism was overwhelmed or 'saturated' and ceased to function.

This by no means exhausts the rather remarkable phenomena in glasses at low temperature and the reader will find an extensive account of other relevant experiments in Phillips (1981) and Hunklinger (1982). Dielectric relaxation phenomena should be mentioned briefly because, although their behaviour is not wholly identical with the acoustic absorption, it does parallel it to a marked degree. If $\ln \omega$ is plotted against $1/T_1$ the same straight line results for both acoustic and dielectric absorption, which strongly suggests that they are both the result of the same thermally activated relaxation process with

$$\tau(T) = \tau_0 \exp(\Delta E / k_B T).$$

The Debye equations for relaxation phenomena give $\alpha \propto \omega^2 \tau / (1 + \omega^2 \tau^2)$ and $\alpha(T)$ should follow by inserting $\tau(T)$. This deos not fit the observed $\alpha(T)$ unless it is assumed that whatever the relaxation mechanism is it occurs with a spread of ΔE and therefore a spread of $\tau(T)$.

A considerable number of hypotheses were put forward to explain one or all of these phenomena. An important one, proposed independently by Phillips and by Anderson, Halperin and Varma in 1972, proposes the existence in glasses of systems with double potential wells allowing the possibility of quantum mechanical tunnelling states. It is not obvious what these systems are. Perhaps particular atoms in favoured places can shift between two positions; perhaps a whole SiO_2 tetrahedron has two orientations. Possibly larger groups of atoms are involved. The free-volume theory (§10.4) associates tunnelling centres with small concentrations of free volume, (Grest and Cohen 1981, Cohen and Grest 1983). Whatever the structure, it is required that two closely spaced energy levels per system should exist and that there should be a wide spectrum of separations, ΔE, between the levels. Because of the common features in the behaviour of different glasses the two-level system (TLS) should be an intrinsic feature of disordered glassy structures and not the property of some special chemical circumstance. For a model of the TLS the potential diagram in figure 10.12 is used but the coordinate is not necessarily a linear distance, it could be a rotation or a generalised coordinate to represent some complicated configurational adjustment influenced by two potential minima.

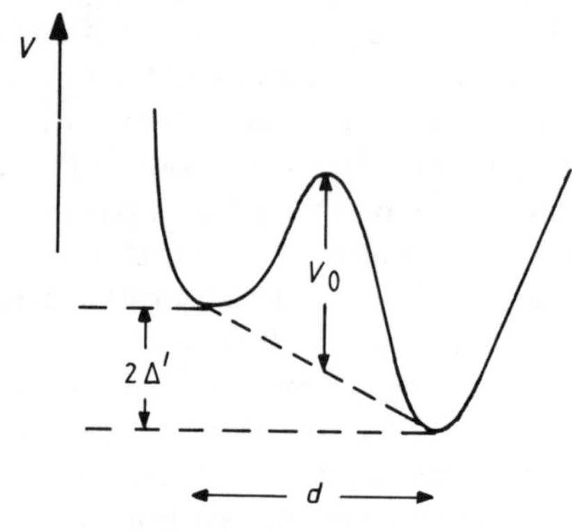

Figure 10.12

The Hamiltonian is

$$H = \frac{-\hbar^2}{2m} \nabla^2 + V = H_1 + V_2 = H_2 + V_1$$

where H_1, V_1, H_2 and V_2 are for the wells 1 and 2 separately and m is the effective mass of the particle, i.e., of whatever moves. The matrix elements of H require a choice of basis states. For finding the ground-state energy an evident assumption for the basis states would be $|1\rangle$ and $|2\rangle$ representing, respectively, the ground states of the particle in V_1 or V_2 separately. H_{11}, for example, would be $\langle 1|H_1 + V_2|1\rangle$ or $E_1 + \langle 1|V - V_1|1\rangle$ where E_1 is the ground-state energy in well 1. Actually the second term is negligible compared with E_1 because V is close to V_1 in the first well. Proceeding in this way the basis states and matrix elements can be represented as

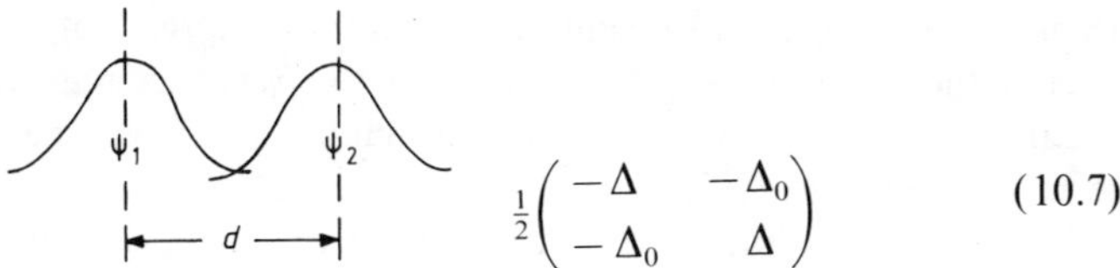

$$\frac{1}{2}\begin{pmatrix} -\Delta & -\Delta_0 \\ -\Delta_0 & \Delta \end{pmatrix} \tag{10.7}$$

where $\Delta_0 \equiv -2\langle 1|H|2\rangle$ and $\Delta = -E_1$. By choosing the energy zero to be $\frac{1}{2}(E_1 + E_2)$ we make $\Delta = +E_2$. Δ is the Δ' in figure 10.12 if the two wells have identical shapes. Δ_0 is often written

$$\Delta_0 \simeq -\hbar\omega_0 \exp\left(\frac{-d}{\hbar}(2mV_0)^{1/2}\right) \tag{10.8}$$

where d and V_0 are as in figure 10.12 and $\hbar\omega_0$ is of the order E_1 or E_2. This can be proved by taking particular forms of potential well, or guessed intuitively from the idea that a particle in one basis state could tunnel into the other potential well with an exponentially lowered amplitude depending on d and V_0 according to the usual tunnelling formula.

Another option for basis states is linear combinations of $|1\rangle$ and $|2\rangle$ which make eigenfunctions for the Hamiltonian after it has been diagonalised. The two diagonal matrix elements are then the two energies. Schematically

$$\begin{aligned} \psi_1 &= |1\rangle \cos \tfrac{1}{2}\delta + |2\rangle \sin \tfrac{1}{2}\delta \\ E_1 &= \tfrac{1}{2}(\Delta^2 + \Delta_0^2)^{1/2} \\ \psi_2 &= -|1\rangle \sin \tfrac{1}{2}\delta + |2\rangle \cos \tfrac{1}{2}\delta \\ E_2 &= -\tfrac{1}{2}(\Delta^2 + \Delta_0^2)^{1/2} \end{aligned} \tag{10.9}$$

where $\tan \delta = \Delta_0/\Delta$.

The energy splitting between these two states is $\Delta E = (\Delta^2 + \Delta_0^2)^{1/2}$ and for thermal properties at very low temperatures only TLS with very small ΔE need be considered.

A simple application of statistical mechanics to an ensemble of systems with two levels ΔE apart shows that

$$C_v \text{ per system} = 4k_B\left(\frac{\Delta E}{k_B T}\right)^2 \operatorname{sech}^2\left(\frac{\Delta E}{2k_B T}\right) \tag{10.10}$$

which gives a peak in C_v called a Schottky anomaly. In a glass this would have to be integrated over all the ΔE weighted with a factor $p(\Delta E)\,\mathrm{d}(\Delta E)$ which is the probability of a tunnelling TLS having an energy splitting ΔE to $\Delta E + \mathrm{d}(\Delta E)$. Moreover in a glass a wide distribution of local configurations permitting a large and essentially continuous spread of ΔE would be expected and a hypothesis of constant $p(E) = p_0$ is very plausible. In that case the integral of C_v over ΔE is proportional to $p_0 T$ in explanation of the linear term in equation (10.5).

In order to exploit the TLS for phonon propagation phenomena a hypothesis is needed for the TLS phonon interaction. The perturbation which causes transitions between ψ_1 and ψ_2 is the disturbance of local configurations by the passage of long-wavelength phonons. This is an elastic deformation which alters the potential $(V_1 + V_2)$ and adds the terms $(\partial H/\partial e_i)\delta e_i$ to the Hamiltonian where e_i is a component of strain. TLS's irradiated by phonons can react in various ways. A TLS can offer resonant absorption to phonons with $h\nu = \Delta E$. This leads to a mean-free path $l \propto T^2$ which goes a long way towards accounting for the low-temperature behaviour of thermal conductivity. Integration with a weighting of $p(\Delta E)\,\mathrm{d}(\Delta E)$ would be required again of course. Resonance also contributes towards the absorption coefficient of phonons of comparable wavelength and the low intensity in acoustic experiments near the maximum at T_2 (see above). The upper state will have a finite decay time and if the intensity rises to the point where the upper and lower levels are equally populated this absorption mechanism ceases to be effective and shows the saturation referred to at very low temperatures.

Acoustic absorption is more complicated than this however. Near the absorption maximum at T_1 the absorption probably requires a thermally activated, as opposed to a tunnelling, excitation between states in double-well potentials and this is a relaxation phenomenon. Relaxation processes in glasses and experiments designed to explore them have been intensively discussed (see, e.g., Golding and Graebner in Phillips (1981)).

As a final example we return to the specific heat of SiO_2 because it illustrates not only the low-temperature properties but also the general nature of glass inferred in §10.2 and elsewhere, namely that the value of an observed property depends on the processes that have time to occur during the observation period. Loponen *et al* (1980) devised a method for measuring C_p at $T \lesssim 1$ K in SiO_2 samples into which heat pulses of duration 10^{-6} s could be sent. C_p was referred from the observed rise in temperature. The rapid initial rise implied a short-time ($\lesssim 1$ ms) heat capacity of the form $c_3 T^3$. This was attributed to the excitation of phonons. As time went on ($\gtrsim 10$ ms) the phonons appeared to begin interacting with the tunnelling states of the TLS which increased the heat capacity by an amount proportional to T. The total heat capacity therefore followed equation (10.5) but it appears that the two terms have separate origins in processes with different relaxation times.

This section has served to introduce a few of the properties of glasses for which a model using the TLS has proved illuminating. It would not be correct to imply however that all the observations have been reconciled by one simple model still less that the microscopic origin of the TLS is clear

(Guttman and Rahman 1986). The references give more detail about all these points, and they include the corresponding phenomena in metallic glasses which will be taken up again in Chapter 12.

References

Adkins C J 1983 *Equilibrium Thermodynamics* (Cambridge: Cambridge University Press)

Anderson P W 1979 *Ill-Condensed Matter* ed R Balian, R Maynard and G Toulouse (Amsterdam: North-Holland)

Angell C A, Clarke J H R and Woodcock L V 1981 *Adv. Chem. Phys.* **48** 397

Bengtzelius U, Götze W and Sjölander A 1984 *J. Phys. C: Solid State Phys.* **17** 5915

Bengtzelius U and Sjögren L 1986 *J. Chem. Phys.* **84** 1744

Birge N O and Nagel S R 1985 *Phys. Rev. Lett.* **54** 2675

Bray P J, Geissberger A E, Bucholtz F and Harris L A 1982 *J. Non-Cryst. Solids* **52** 45

Breinan E M, Kear B H and Banas C M 1976 *Phys. Today* **29** 45

Cohen M H and Grest G S 1983 *Proc. 5th Int. Conf. on Liquid and Amorphous Metals* vol 5, ed C N J Wagner and W L Johnson (Amsterdam: North-Holland) (*J. Non-Cryst. Solids* **61/62** 749)

Davies R O and Jones G O 1953 *Adv. Phys.* **2** 370

Dupree R, Holland D, McMillan P W and Pettifer R F 1984 *J. Non-Cryst. Solids* **68** 39

Dupree R and Pettifer R F 1984 *Nature* **308** 523

Gaskell P H, Parker J M and Davies E A (ed) 1982 *The Structure of Non-Crystalline Materials* (London: Taylor and Francis)

Greaves G N 1985 *J. Non-Cryst. Solids* **71** 203

Greaves G N, Fontaine A, Lagarde P, Raoux D and Gurman S J 1981 *Nature* **293** 611

Grest G S and Cohen M H 1981 *Adv. Chem. Phys.* **48** 455

Guttman L and Rahman S M 1986 *Phys. Rev.* B **33** 1506

Halfpap B L and Lindsay S M 1986 *Phys. Rev. Lett.* **57** 847

Hunklinger S 1982 *J. Physique Coll.* **43** C9 461

Jäckle J 1981 *Phil. Mag.* **1344** 533

—— 1984 *Physica* **127B**

—— 1986 *Rep. Prog. Phys.* **49** 172

Johnson P A V, Wright A C and Sinclair R N 1982 *J. Non-Cryst. Solids* **50** 281

Kauzmann W 1948 *Chem. Rev.* **43** 219

King S V 1967 *Nature* **213** 1112

Kirkpatrick 1985 *Phys. Rev.* A **31** 939

Loponen M T, Dynes R C, Narayanamurti V and Garno J P 1980 *Phys. Rev. Lett.* **45** 457

Mozzi R L and Warren B E 1969 *J. Appl. Crystallogr.* **2** 164

Palmer R G 1982 *Adv. Phys.* **31** 669

Phillips J C 1979 *J. Non-Cryst. Solids* **34** 153

Phillips W A (ed) 1981 *Amorphous Solids, Low Temperature Properties* (Berlin: Springer)

Rawson H 1980 *Properties and Applications of Glass* (Amsterdam: Elsevier)

Rivier N and Duffy D M 1982 *J. Phys. C: Solid State Phys.* **15** 2867

Shukla P 1983 *Z. Phys.* B **52** 179

Soules T E 1982 *J. Non-Cryst. Solids* **49** 40

Stephens R B 1976 *Phys. Rev.* B **13** 852

Tanaka Y, Ohtomo N and Katayama 1985 *J. Phys. Soc. Japan* **54** 967

Tarlros A, Kleinin M A and Lucovsky G 1984 *J. Non-Cryst. Solids* **64** 215

Thorpe M F 1984 *J. Non-Cryst. Solids* **57** 355

—— 1985 *J. Non-Cryst. Solids* **76** 109

Toschev T S 1973 in *Crystal Growth: An Introduction* ed P Hartmann (Amsterdam: North-Holland)

Wong J and Angell C A 1976 *Glass Structure by Spectroscopy* (New York: Dekker)

Wright A C 1982 *J. Non-Cryst. Solids* **49** 63

Wright A C, Connell G A N and Allen J W 1980 *J. Non-Cryst. Solids* **42** 69

Wright A C and Leadbetter A J 1976 *Phys. Chem. Glasses* **17** 122

Zeller R C and Pohl R O 1971 *Phys. Rev.* B **4** 2029

11

AMORPHOUS AND LIQUID SEMICONDUCTORS

The reference to liquid semiconductors in the chapter heading stretches the definition to some extent. It is not intended to deal with the liquid state of Ge or Si which is metallic, nor very much with molten semiconducting glasses because, although their properties are by no means identical with those of the glassy form, similar concepts can be used in their study (Mott and Davies 1979, Cutler 1977). The liquids in this chapter are certain special alloys and expanded liquid metals which, under appropriate conditions of temperature, pressure or composition, undergo, or almost undergo, metal–non-metal transitions and enter semiconducting phases. These interesting systems bring out new problems not introduced elsewhere in the book.

By the mid-1980s a considerable number of commercial products exploiting solid non-crystalline semiconductors were on the market and their development has stimulated a great deal of fundamental research and will no doubt continue to do so. This chapter will be concerned with this physics and not with technical devices based on it. For applications of amorphous semiconductors the reader should consult the Device Physics Sections of 1985 *J. Non-Cryst. Solids* **77/78** 1363ff.

Amorphous semiconductors are not a homogeneous group. They include, *inter alia*, chalcogenide glasses made from melts and tetrahedrally bonded elements produced in films by evaporation and other methods. For some purposes in general discussion, metallic and insulating glasses might enter alongside semiconductors and the treatment of these three groups in separate chapters is only an expedient for introducing a vast and growing amount of interconnected information. With semiconductors there is a focus of interest on electronic properties especially insofar as technological applications have been found for them.

There has been, and continues to be, such a prolific output of research into these systems that here, perhaps more than elsewhere, the problems of selection are particularly difficult. The reader will appreciate therefore that the following paragraphs merely sample the activity in a major scientific field. However it need not be a random sample because the work of the 1970s and 1980s, while leaving many unsolved problems, has also achieved some clarifications. To limit the scope while introducing a variety of properties, silicon and chalcogenide glasses will be used as examples for the most part.

11.1 Amorphous semiconductors—an introduction

Much of the emphasis in these pages has been on new, or different, properties consequent upon disorder. It is also worth emphasising in the context of amorphous semiconductors how similar they are in some ways to the corresponding crystals. As with SiO_2 (§10.6), the SRO is dictated by chemical bonding and closely resembles that in the crystalline forms, tetrahedral coordination being the rule in both a-Si and c-Si for example. It is a consequence—not originally obvious but one that emerged—that the electronic structure is similar in its gross features also. One piece of evidence for this is from UPS (see §7.6). The spectra in figure 11.1 show that the effect of disorder is to obliterate sharp features, not to transform the situation completely.

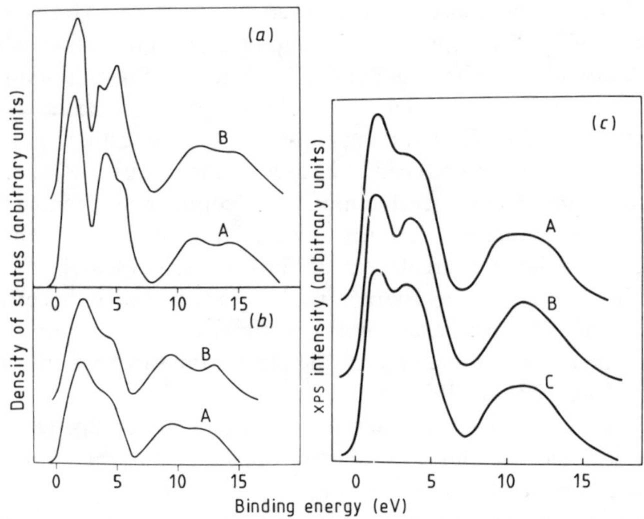

Figure 11.1 Photoemission spectra to show overall similarity between electron spectra in crystalline and amorphous solids: (*a*), selenium; (*b*), arsenic (A, amorphous; B, crystalline); (*c*), As_2Te_3 (A, amorphous (25 °C); B, crystalline (200 °C); C, powdered crystal).

Much of the work concerns amorphous Si, amorphous Ge, their amorphous alloys and also III–V compounds like amorphous GaAs. For brevity, and unless the statements are very specific, they could collectively be called a-IV semiconductors. Some of the following sections exemplify the investigations that have led to a certain conception of a-IV systems. But it will be convenient to sketch a picture of them first and fill in some evidence later. A semiconducting crystal has a regular structure and well defined band edges and energy gap. If the structure were distorted into an ideal CRN the band edges would tail into the energy gap, with at least some of their states localised. As in a crystal, E_F would be somewhere in the gap. By accident or design, however, a real structure will not be an ideal CRN; there will be inhomogeneities, defects and impurities which will introduce gap states.

The properties of a semiconductor depend on the circumstances of its preparation. This is obviously important for practical reasons but it raises matters of principle as well. A liquid in thermodynamic equilibrium will be specified by its composition, density and temperature. In principle its structure, given by $g(r)$, $g^{(3)}(r_1, r_2, r_3)$ etc, can be calculated by statistical mechanics as can its electronic properties. With an amorphous semiconductor this is not in general true. Even for a given temperature and density its structure may depend on how it was made and the structural input into a theory of, say, optical absorption, is almost certain to come from an experiment on a particular sample or from a model, not from statistical mechanics. This is a characteristic feature of the physics of amorphous solids.

The density, $g(E)$, of gap states is an important function, though difficult to discover, and the position and behaviour of the Fermi level and the electronic properties will be strongly influenced by it. E_F is found by putting the integral of $f(E, T)g(E)$ over all energies equal to n, the total number of electrons. If $g(E_F)$ is small, or zero as in intrinsic crystalline semiconductors, E_F can be significantly altered by small amounts of donor or acceptor impurities which change n. This is not the case if $g(E_F)$ is large as in metals. In intermediate cases $g(E_F)$ may not be small enough to allow E_F to be varied by doping and in a-IV materials the size of $g(E)$ in the gap is a significant controller of this. That $g(E_F)$ in the gap can be compatible with effective doping in a-Si was an important discovery which was made by Spear and Le Comber (1975). It does not follow however that the doping process in a-Si and c-Si is the same (see §11.4 below).

Whether doping is indeed possible turned out to depend on the composition of the sample and the method of preparation. Whatever the method, the product will not be an unflawed CRN. Perfection does not exist in SiO_2 where the pair of bonds to bridging oxygen atoms lends some flexibility to the structure. In a-IV even this element of flexibility is lacking and many departures from perfection will be present. Of these we shall refer to at least three: defects in pure a-IV; incorporated hydrogen; and dopants from groups III and V.

It will be seen later that doping is difficult or impossible in chalcogenide glasses and an interesting difference in the nature of defects in these and a-IV materials is relevant to this.

11.2 Preparation

There are already numerous ways of making a-IV materials and the product is normally a thin layer up to a few tens of micrometres thick.

The glow discharge (GD) method will be outlined first because it reached the stage in the early 1980s of production on an industrial scale, apart from which its output is hydrogenated a-Si—a material of great intrinsic interest. The essence of the method is to create a plasma in silane gas (SiH_4), disilane (Si_2H_6) or trisilane (Si_3H_8) by using radiofrequency power to maintain a glow discharge. The plasma is a complicated system including neutral and charged species such as SiH, SiH_2, SiH_2^+, . . . , and many others, and the deposition process remains incompletely understood. In equipment, of which figure 11.2 shows a laboratory example, a-Si (which contains hydrogen) is deposited on a substrate which has a deposition temperature $T_D \sim 300$ °C. By opening valves to admit phosphine (PH_3) or diborane (B_2H_6), doping with P or B can be achieved. These simple statements obscure many variables and much empirical knowledge. The properties of the deposit can be influenced by T_D, by deposition rate, by equipment geometry, gas purity, RF input power, gas pressure and flow rate, and other parameters. The article by Spear and Le Comber in Joannopoulos and Lucovsky (1984) gives details and further reading. The method is sometimes relabelled PECVD—meaning plasma-enhanced chemical vapour deposition. UV and IR lasers have also been used since about 1979 for laser-assisted CVD.

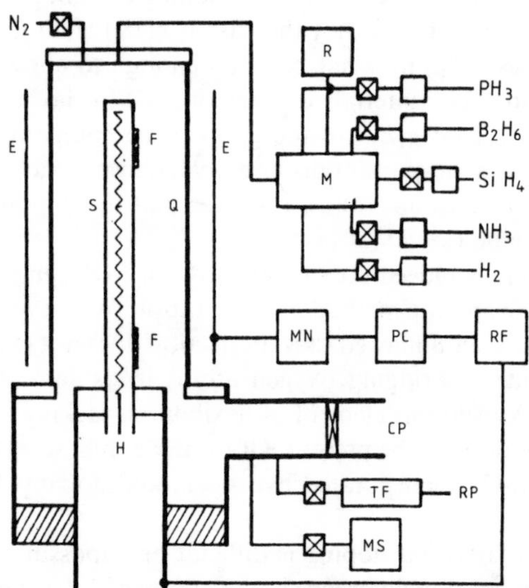

Figure 11.2 Schematic drawing of glow discharge equipment used by Spear and colleagues. E, RF electrodes; S, specimen holder; H, heater; F, rotatable flaps; RF, PC, MN, RF source, power controller, matching network; CP, RP, cryo- and rotary pumps; MS, mass spectrometer; TF, furnace; M, mixing chamber; R, premixing reservoir. (From Joannopoulos and Lucovsky 1984.)

The RF power is not necessary to decompose silane; a sufficiently high temperature will do. If the temperature is less than 600 °C the deposited Si will be amorphous containing from about 0.1 to about 10 % of hydrogen for temperatures ranging from about 600 to about 100 °C. The process is called chemical vapour deposition (CVD) (Joannopoulos and Lucovsky 1984). It is possible to increase the hydrogen content by treating the amorphous film with a hydrogen plasma or to dope by adding phosphine or diborane to the initial gas.

Sputtering is also a possibility. In both DC and RF sputtering, ions from the sputtering gas, typically Ar, eject atoms from a target and these proceed to a substrate on which the deposit accumulates. As with GD and CVD, conditions in the plasma and at the solid–gas interfaces are complicated. The Si and Ge deposited on the substrate will acquire some impurities from the plasma and the material has different properties from GD samples. However, it was shown that hydrogen added to the sputtering Ar resulted in a hydrogenated deposit with properties close to the GD material. An article by Thompson discusses sputtering (see Joannopoulos and Lucovsky 1984).

Ion implantation and thermal diffusion are two more doping techniques, and fluorine is an alternative additive to hydrogen. If it were desired to produce a-Si or a-Ge devoid of dopants, hydrogen and fluorine, clean evaporation in UHV onto a cold substrate leads to a nominally pure sample. A method not open for exploitation is cooling from the melt through a glass transition; this appears not to be possible. The various preparation methods are exhaustively discussed in Pankove (1984) and Phillips (1986).

Semiconducting chalcogenide glasses are mixtures or compounds containing one or more of S, Se and Te. Some of the simpler glasses are binary compounds of these elements with As, Ge, Si, Tl and a number of other elements. Much studied examples are As_2S_3, As_2Se_3, Ge_xSe_{1-x} . . . . Commonly these are made by sealing pure powdered constituents under vacuum, or an inert gas, in a silica tube, and rotating it for some hours at, say, 900 °C. The glass is then formed by cooling or quenching through a glass transition (§10.1). For optical use special purifications to remove oxides may be required. The glass transition temperature may depend on the thermal history of the liquid phase because that can affect the configurations—chains, rings, etc—in which chalcogenide atoms variously arrange themselves.

11.3 Composition and structure

What are the products of these preparative techniques like? To begin to answer this we consider the grosser features and the structure, before going to the electronic and optical consequences.

Pure evaporated a-Si or a-Ge are hightly strained materials in which some strain relief is achieved by forming small low-strain islands and a network of voids on a scale of ~ 10 to $\sim 10^3$ Å. This is evident from electron microscopy and the material may be up to 10 % less dense than the crystalline form.

Much a-Si is hydrogenated during preparation and the product, symbolised a-Si:H, may contain up to as much as 50 at. % of H depending on the mode of production. a-Si:H made for its useful electronic properties often has from about 5 to 12 at. % hydrogen. The amount can be estimated by heating the material in a closed cell and measuring the pressure of the hydrogen that is driven out. IR spectroscopy, proton NMR and infrared absorption are alternative H detectors which can be made quantitative with care and calibration.

Samples are not necessarily homogeneous. Examining them for inhomogeneities is a good field for transmission and scanning electron microscopy. The considerable literature shows that by varying the deposition conditions of GD or sputtered a-Si:H, samples can be made either with, or apparently without, structural inhomogeneities on the scale of ~ 10 to $\sim 10^2$ Å. Nodules, granules or columnar features in inhomogeneous films can be detected. There is evidence that the material between columnar features is richer in hydrogen than the columns themselves. There are also detectable relations between the structures and the infrared spectra of Si–H bonds, the Si–O bonds that develop on oxidation, and the hydrogen evolution processes that appear to make use of a network of internally connected channels. Proton NMR is able to detect and investigate the location of H atoms. Such matters, and the general question of inhomogeneity in a-Si, can be pursued in Joannopoulos and Lucovsky (1984), Pankove (1984) and Phillips (1986).

Naturally the structure of a-IV solids on the atomic scale has been explored by the methods of Chapter 3. Neutron diffraction studies of pure evaporated a-Ge led to the results shown in figure 2.5(b) which indicate better agreement with relaxed CRN models than with quasicrystalline ones. A recent computer modelling programme for CNR's achieves even better agreement (Wooten *et al* 1985). The distribution of H atoms in a-Si:H is not entirely clear. An EXAFS experiment in 1984 was interpreted as showing that, in a sample with 4.9 at. % H, H bonded to Si on the surface of voids of which the estimated size was 30 Å (Bouldin *et al* 1984). H atoms are particularly difficult to locate by x-rays because of the small scattering, and by neutrons because of the large inelastic scattering correction.

The distinction, already referred to several times, between pure and hydrogenated a-IV materials is very significant. $g(E)$ in the gap is about 100 times greater in the pure materials which, unlike the hydrogenated materials, cannot be doped. The temperature dependence of the conductivity is different, so are the electron spin resonance (ESR) and the optical absorption. Hydrogenated material is more photoconductive and more luminescent. It is the hydrogenated a-Si that is suitable for technical devices. These differences are explicable in general terms by reference to defects and the relation of H atoms to them. Many observations of electronic properties have helped to elucidate the defect concept, and vice versa.

As for chalcogenides, even the structure of pure a-Se has not been easy to determine. The trigonal crystalline form of Se is made of two-coordinated atoms in parallel helical chains held together by van der Waals' forces. After detailed x-ray diffraction, review of the literature and computer modelling, Corb *et al* (1982) concluded that a-Se is less dense than the crystalline form and consists of disordered chains coiled round each other, probably with some closed rings of atoms as well. However, within about 4 Å the order is similar in a-Se and c-Se.

As has been emphasised before many models are consistent with the same pair distribution. This is true of a-Se and another example is provided by the diffraction study of $As_{100}S_{160}$ glass by Apling *et al* (1977). Assuming that local valence requirements are satisfied with a As–S bond of 2.25 Å and a S–As–S angle of 98°, at least two models with a similar $S(Q)$ can be envisaged. One is a random packing of roughly spherical As_4S_6 molecules; the other is a quasicrystalline model with covalently bonded layers. Either case can be built up with AsS_3 pyramids and the structure in $S(Q)$ for $Q \gtrsim 3$ Å^{-1} is dominated by this fact. The detailed discussion also makes clear that bulk and thin-film glass differ somewhat in structure. As a final example we may take the concentration of neutron diffraction, with and without isotopic substitution, EXAFS, Raman and infrared spectroscopy on to the P_xSe_{1-x} glasses (Price *et al* 1984). Diffraction gives the peak positions (r_n) in the radial distribution function, the average number of nearest neighbours (~ 2) and the effective bond angle ($2\sin^{-1}r_2/2r_1 \simeq 105°$, see figure 10.7(*b*)). EXAFS using the Se absorption edge gives the P and Se distances from Se atoms and also the ratio of P to Se neighbours possessed by a Se atom: this rises from 0 to about 2.4 as the P concentration rises to 0.5. In interpreting these data it is assumed that all Se atoms and some P atoms obey the $8 - N$ rule (see below) with two and three bonds respectively. But a fraction, y, of P atoms are supposed to bond to four Se atoms. There are also P–P bonds in numbers increasing with the concentration of P, and the fraction of them per P atom is z. The values of y and z are known for certain crystal configurations and the glass data can be used to infer that y in the glass is about 0.25 and that z rises from 0 to 0.86 as the P concentration rises to 0.5. The general picture is of units P_4Se_n with $n = 3, 4, 5$ such as are found in some crystal structures, embedded in Se-rich matrices. The presence of particular postulated bonds in certain concentration ranges, e.g. –Se–Se– for P concentrations <0.4, is confirmed by Raman and infrared spectroscopy which have peaks which are characteristic of specific bonds. The reader will find many more discussions of chalcogenide glass structures in the literature and the measurement of structure as the temperature approaches the glass transition is of considerable interest (Busse and Nagel 1981, Busse 1981). Nevertheless, as with all glasses, it is difficult to arrive at structures which are both unique and convincing.

11.4 Defects and dangling bonds in a-Si

It was remarked in §1.6 that crystal defects cannot obviously be invoked for amorphous solids. But the concept of CRN contains perfection of bonding or of coordination and a departure from perfection can reasonably be called a defect in this context. It would be remarkable if defects were not built in when layers are deposited or glasses undergo the glass transition. Their number might be further increased by particle bombardment or reduced by annealing. In a-Si, hydrogen or dopant atoms or other chemical impurities will constitute different kinds of defect site and, if hydrogen is driven out by heating, defect centres will be left behind.

Absences of atoms from a-Si or a-Ge, the equivalent of vacancies or polyvacancies in crystals, may well occur and atoms nearby may have three or two bound neighbours instead of four. To take a particularly important case, suppose a neutral three-coordinated Si atom has an unpaired electron, called a dangling bond. D^0 symbolises this defect and E_{D^0} stands for the electron energy, which will be in the energy gap. Many authorities regard the dangling bond as the most important defect in a-IV material. It has a net spin and is a paramagnetic centre; in pure a-Si there may be 10^{18} to 10^{20} cm^{-3} of them. If another electron pairs up with it, the correlation energy, U, from the mutual electric repulsion will increase E_{D^0} to an energy $E_{D^-} = E_{D^0} + U$ and the centre, now called D^-, is charged but spinless. E_{D^-} also lies in the energy gap. Because of the disorder, D^0 and D^- have broadened, not sharp, levels (see figure 11.4(*b*)). Empty dangling bonds, positively charged and spinless, could also exist. A hydrogen atom in a-Si:H may attach itself to a dangling bond; this creates another spinless defect and when this occurs during deposition the sample is without the corresponding D^0 states and has a much reduced overall strain. The dangling bond is said to be passivated by the hydrogen. Hydrogenated material accordingly has a much lower $g(E)$ in the gap and the dangling bond density may be only 10^{16} cm^{-3} or less. The electron states bonding the H have energies below the gap. Si atoms at the surface of a small void do not necessarily retain all their unpaired electrons as dangling bonds. Two otherwise dangling bonds may join to form a weak spinless bond—another category of defect. Weak bonds might also break up and acquire H atoms, and for this and other reasons a-Si:H contains in general a far larger percentage of H atoms than pure a-Si has dangling bonds. Si atoms with one, two or three attached hydrogens can be recognised by their infrared spectra and $(SiH_2)_n$ material is thought to be common when columnar inhomogeneities are present.

Substitutional doping in crystalline Si creates donor and acceptor levels and is familiar. The doping in amorphous systems is not necessarily analogous. To appreciate the problem let us recall the $8-N$ rule which refers to the ability of the elements to form covalent bonds with their eight s- and p-valence states or hybrid combinations thereof. The summarising diagram, figure 11.3, is helpful and three examples from it will suffice. In monovalent elements, one occupied orbital permits one bond leading to a diatomic molecule. With the number of valence electrons $N \leqslant 4$, N singly occupied states are possible and the number of potential bonds is N. Ge and Si are well known cases having $N=4$ with the bonds formed by sp^3 hybrid states spatially oriented 109.5° apart. If $N=5$, $8-N$ p orbitals are singly occupied and able to form $8-N$ bonds. The doubly occupied s orbital and any unoccupied orbitals cannot do so. In general, the number of possible bonds is $8-N$ for $N>4$ and N for $N \leqslant 4$.

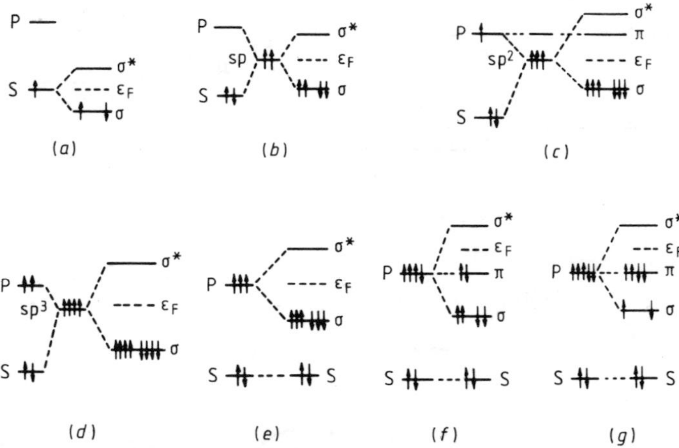

Figure 11.3 Schematic diagram showing, from left to right in each column of the periodic table, levels in the free atom, hybridisation possibilities and bonds in condensed matter. σ, σ^*, bonding and antibonding states; π, lone pairs; ε_F, Fermi level. (From Robertson 1983.)

In an ideal covalent CRN all the atoms have their valence requirements satisfied in accordance with the $8-N$ rule. It is a common finding that the $8-N$ valence rule does hold, and a hypothesis that it is a general feature of amorphous semiconductors and non-metallic glasses. A P atom in a-Si would therefore be expected to be electrically inactive with $8-N=3$ bonds, whereas the periodicity in c-Si requires the dopant to have four bonds and a donor level readily emptied by thermal excitation. Nevertheless doping in a-Si does occur and an interesting conception of it was proposed by Street (1982). This was that during the incorporation of, say, n_P P atoms, n_D of them enter the structure as P^+ ions obeying the $8-N$ rule with $N=4$, accompanied by the generation of n_D dangling bonds on Si atoms which then become charged D^- centres by acquiring the electrons detached from the P^+. There will be $n_P - n_D$ P atoms left to obey the $8-N$ rule with $N=3$ without any accompanying dangling bonds. The postulated equilibrium between the neutral and ionised P centres could be written according to the law of mass action as

$$P_3^0 \leftrightharpoons P_4^+ + D^- \tag{11.1a}$$

$$(n_P - n_D)/n_D^2 = \text{constant.} \tag{11.1b}$$

It is supposed that this equilibrium sets in to keep E_F below the band of donor levels and above the D^- states so the former is empty and the latter

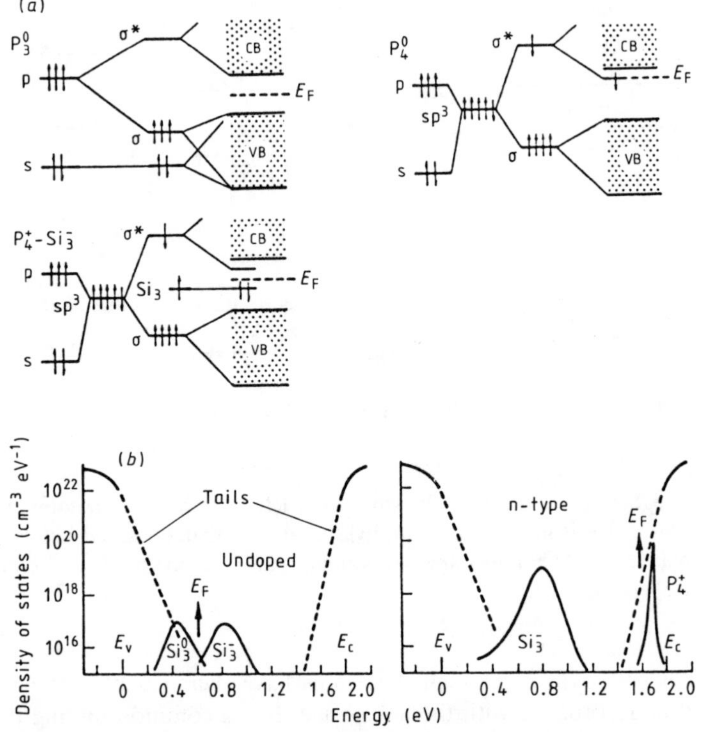

Figure 11.4 (*a*) As figure 11.3 for the special case of phosphorus in silicon. P_3^0, neutral atom obeys $8-N$ rule; P_4^0, four-bonded state found in Si crystal doping; P_4^+–Si, ionised P obeys $8-N$ rule in collaboration with dangling bond—see text. (From Robertson 1984.) (*b*) Schematic density of gap states in a-Si showing doping with P. Broken lines are band tails, peaks are dangling-bond states. (From Robertson in Device Physics Section of 1985 *J. Non.-Cryst. Solids* **77/78**.)

occupied. There is experimental support for this: for instance, equation (11.1*b*) predicts that the doping efficiency, defined as n_D/n_P, will be proportional to $n_P^{-1/2}$ for $n_D \ll n_P$ and this is observed. Furthermore, ESR experiments do not detect any occupied donor levels. Corresponding statements can be made about group III doping. Other suggestions have been made about dopant incorporation, and it probably remains true to say that exactly how dopant atoms enter is not entirely clear but it is certainly not the familiar process imposed by the fourfold coordination in the diamond structure. Figure 11.4 illustrates the preceding ideas (Street 1985, Robertson 1983, 1984).

11.5 Defects in chalcogenide glasses

In chalcogenide glasses, as in a-IV materials, there are defects, but their nature and the properties they give rise to are different. There is evidence from electronic and optical phenomena that chalcogenide glasses have a significant density of gap states with E_F pinned near the centre of the energy gap (Mott and Davies 1979). Doping is ineffective. Singly occupied states near E_F would be expected but, in contrast to a-IV behaviour, no ESR or paramagnetism confirms their presence. It appears that if dangling bonds occur then they must be doubly occupied and the implication is that the correlation energy required to overcome the mutual repulsion is more than compensated by some other adjustment such as energetically favourable displacements of atoms. This was emphasised by P W Anderson in 1975 and expressed in an effective U which was overall *attractive* between electrons of opposite spin at a particular bonding location. An important extension of this idea by Street and Mott revealed that many properties of As_2Se_3 could be rationalised if structural distortions were postulated which made it energetically favourable for a pair of D^0 sites to rearrange themselves into $D^+ + D^-$ both of which are spinless gap states (Street and Mott 1975). We shall now outline yet another development of these suggestions made shortly after by Kastner *et al* (1976).

Following Kastner *et al* we first symbolise the possible bonding arrangements a chalcogenide might have. This relates to figure 11.3 and it is important that, in bonding, two of the four p electrons normally enter a lone-pair non-bonding orbital leaving two others to form a pair of bonds according to the $8 - N$ rule. The first row shown in table 11.1 is a symbol of which the superscript is the charge and the subscript the number of bonds. The second row symbolises the number of p electrons and their spin directions in a hypothetical isolated atom or ion and in bonding combinations in condensed matter. The third row is a crude energy estimate taking the lone-pair energy as zero, giving the bonding orbital (σ) an energy $-E_b$ and the antibonding

Table 11.1

C_2^0	C_3^0	C_3^+	C_3^-	C_1^0	C_1^-
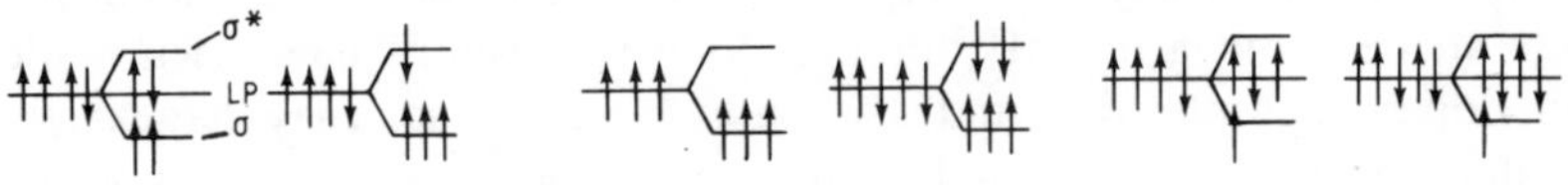					
$-2E_b$	$-2E_b+\Delta$	$-3E_b$	$-E_b+2\Delta$ $+U_{\sigma^*}$	$-E_b$	$-E_b+U_{LP}$

orbital (σ^*) an energy $+E_b+\Delta$ where $\Delta>0$ as generally accepted in theoretical chemistry.

To take some examples: C_2^0 is the normal situation where Se, say, is joined to two other atoms; it has the lowest energy of the neutral configurations. The lowest-energy neutral *defect* is the three-coordinated atom C_3^0 where the lone-pair electrons have moved to σ and σ^* orbitals. If an electron is removed, C_3^+ becomes possible; if one is added, C_3^- can occur but, with two electrons in a σ^* state, an amount, U_{σ^*}, of correlation energy has to be supplied. C_1^0 is a dangling bond. C_1^- implies a correlation energy, U_{LP}, for the lone pairs which is not necessarily equal to U_{σ^*}.

It is an essential feature of the argument that even the lowest-energy uncharged defect, C_3^0, will spontaneously change by the exothermic reaction

$$2C_3^0 \rightarrow C_3^+ + C_1^- \tag{11.2}$$

which represents a bonding rearrangement into an associated, electrostatically attracting, pair of defects C_3^+ and C_1^- called a valence alternation pair (VAP).

The formation of the various defect species, requiring different amounts of formation energy, will presumably occur during the glass transition but the higher-energy ones, including C_3^0, will in general be less preferred. The glass structure for $T<T_G$ will retain the thermodynamic equilibrium numbers of defects appropriate to $T=T_G$. It is thought that the VAP is the predominant one. The right-hand side of reaction (11.2) has $(U_{LP}-2\Delta)$ more energy than the left-hand side according to table 11.1; but this should be reduced to $(U_{LP}-2\Delta-W)$ because an amount, W, of energy will be regained in the relaxations that occur after the charge transfer. The quantity $(U_{LP}-2\Delta-W)$ is expected to be negative: in other words, VAP should be energetically preferable to C_3^0 states. It is thought that the energy of formation of a VAP defect in Se is about 0.8 eV which gives a concentration in thermal equilibrium of about 10^{-6} at $T_G=310$ K (Kastner *et al* 1976). The relatively low energy of formation is consistent with the fact that an overcoordinated (C_3^+) atom and an undercoordinated one (C_1^-) conserve the total number of bonds when formed simultaneously.

It remains to consider the energy levels and processes conceivable with VAP and figure 11.5 illustrates this. Before explaining the figure it should be remarked that optical transitions occur on a timescale much faster than atomic movements and therefore involve amounts of energy uncompensated by the relaxation processes already referred to. These occur when an atom loses or gains an electron and energy W is released as the environment adjusts itself to the new charge distribution by emitting phonons.

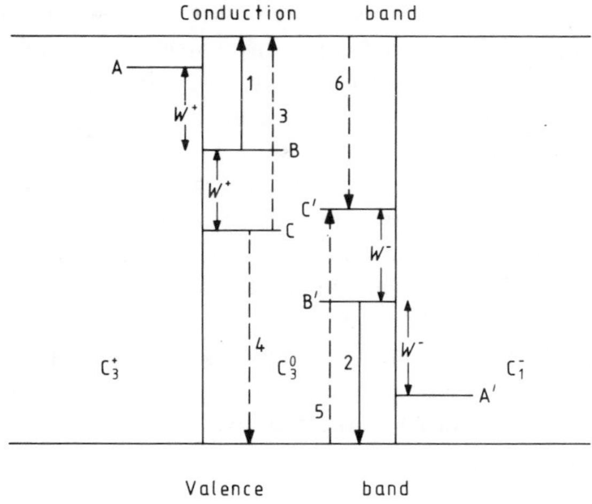

Figure 11.5 Schematic diagram to show states and transitions associated with a VAP—see text.

If a C_3^+ centre acquired an electron and nothing else happened, the electron would enter an antibonding level. It would not enter the conduction band which is formed out of the antibonding states because of the attraction of the positive C_3^+ defect. Its energy level, labelled A, lies just below the conduction band, forming a donor level. Level A is thus open to receive optically excited electrons, e.g., from the valence band. Analogously, the level A′ would result if C_1^- lost an electron without relaxation, e.g., by optical excitation to the conduction band. If we now consider the relaxation after a C_3^+ defect has captured an electron (wherever from) then $C_3^+ + e \rightarrow C_3^0$ and some relaxation energy is released. The level A is now not relevant for picturing the energy required to excite the electron thermally to the conduction band: this energy is given by the arrow denoted by 1 upwards from level B. Level B is W^+ below A and W^+ derives from the relaxation. The analogous C_1^- defect behaviour is the capture of a hole followed by adjustments creating the level B′ a distance W^- above A′. The arrow denoted by 2 represents thermal excitation of the hole into the valence band. Unlike W^+, W^- embraces bond rearrangement and not simply relaxation.

How could the optical transitions of C_3^0 be represented? Excitation of an electron from C_3^0 to the conduction band without benefit of relaxation requires W^+ more energy than shown by arrow 1 and can be represented by dashed arrow 3. The electron could also be removed by descent to the valence band but in the absence of relaxation the photon energy available in such a radiative recombination with a hole is only that represented by dashed arrow 4. Analogously, if an electron *enters* C_3^0 by optical excitation from the valence band or radiative transition from the conduction band, the processes are those shown by dashed arrows 5 and 6. Without configurational changes a correlation energy $U_{\sigma*}$ is needed for this which is why level C′ lies above level C in figure 11.5. The C_3^- centre so formed is of higher energy than C_1^- so bonding changes will subsequently occur to create C_1^- again.

These arguments show that the C_3^0 state can act as either a donor or an acceptor. They also illustrate the value of chemical bonding ideas in the discussion of these defects. Since subtractions and additions of electrons can convert C_3^+ and C_1^- into one another, their relative numbers can alter if electrons and holes become available from other sources such as other kinds of donors and acceptors. The model represented by figure 11.5 receives some credence because it helps one to understand the activation energies and spectra observable in several different kinds of experiment as will shortly be seen. The prevailing disorder in the glass will prevent the levels in the diagram from being sharp.

11.6 ESR and photoluminescence

The descriptive statements so far made about amorphous semiconductors are inferences from a very large number of experiments. It will not be possible to refer here to every one of the rather remarkable range of spectroscopies that have contributed. So a limited selection will be made, beginning in this section with electron spin resonance (ESR) and photoluminescence (PL).

An ESR resonance line signals the energy difference between spin-up and spin-down electrons in an applied field H_0; the energy difference is $g\mu_B H_0$. Measuring the resonant frequency determines g which is 2.0023 for free electrons, but otherwise departs from this value by small but measurable amounts depending on the orbital motion of the electron. The intensity of the signal depends on the number of unpaired spins. The spin density, n_s, in a sample can be measured by comparing the intensity with that from a sample of known spin population. Omitting technical details we assume that ESR can *detect* unpaired electrons, *resolve* between spins in different environments and *measure* their numbers.

ESR was detected in a-Si by Brodsky and Title in 1969. The value $g = 2.0055$ has come to be associated with neutral dangling bonds throughout many subsequent experiments through comparing it with the value of g for other situations, in which dangling bonds are definitely responsible for the signal, such as c-Si–a-SiO_2 interfaces and samples in which the Si–H bonds have been broken by driving H out with heat. The g-value and ESR lineshape vary little with the mode of sample preparation, but n_s shows the wide variations referred to in §11.4. The inference is that deposition conditions influence the population, but not the nature, of the resonating particle which is therefore a highly localised paramagnetic centre. g is a tensor quantity with different components for H_0 in different directions relative to the orbit; 2.0055 is an average applicable to disordered a-Si (Beigelsen 1980).

Photoluminescence in solids can be described in general terms as follows. A photon of energy $\hbar\omega$ is absorbed by exciting an electron transition. The electron will have an interaction with the ions present, e.g., it will have an electron–phonon interaction with a lattice, and the probability of excitation from the ground state will vary with ω, the shape of the absorption line depending on the nature of the ground state. On a simple model of excitation to a discrete level the spectrum is Gaussian as shown on the left-hand side of figure 11.6. Excitation to a continuum of levels is shown schematically by the absorption curve (broken curve).

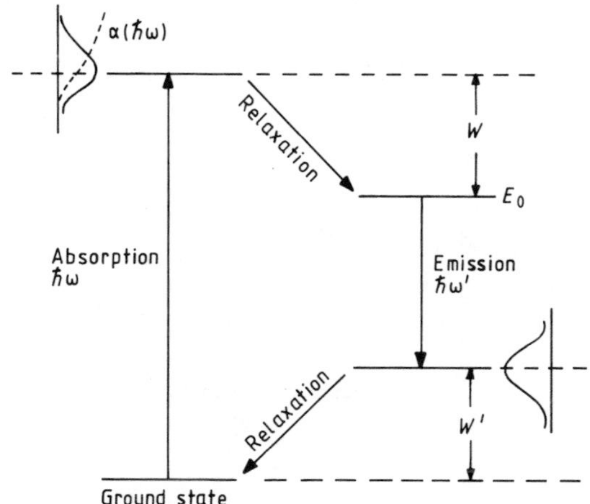

Figure 11.6 Schematic diagram to show transitions in luminescence.

The electronic transition takes much less time than is needed for atomic rearrangements. The latter now occur, reducing the energy of the excited centre to E_0 by removing an amount W by means of one or several phonons. The relaxed centre now emits a photon, $\hbar\omega'$, leaving the centre at energy W' above the original ground state which is regained by further phonon emission. On a simple model the emission line is also Gaussian as shown on the right-hand side of figure 11.6. The energy difference $\hbar(\omega - \omega')$ between the absorbed and emitted lines is the Stokes shift. It is not impossible that the state of energy E_0 decays directly to the ground state without phonon emission but this so-called zero-phonon line may be very weak or undetectable (Street 1976). There is a distribution of delay times between absorption and emission which can be measured by transient techniques (Street 1976, 1981).

Photoluminesence is widespread in amorphous semiconductors and has been detected in a-Si:H with a small Stokes's shift of about 0.4 eV. Sample luminescence spectra are shown in figure 11.7 which also shows that, like ESR, PL has an intensity strongly dependent on deposition conditions. In some contexts this is very troublesome, but a virtue can be made of it by systematically altering the preparation parameters and measuring both PL and ESR for the same samples. In one experiment on GD material, the SiH_4 concentration, the RF power, the substrate temperature and the gas pressure were all varied (Street *et al* 1978). As figure 11.8 shows, it turned out that there was a general negative correlation of dangling-bond density with PL intensity irrespective of the conditions of preparation. In other words, dangling bonds encourage competing non-radiative processes of de-excitation and quench luminescent photon emission especially for $n_s > 10^{17}$ cm^{-3}. The

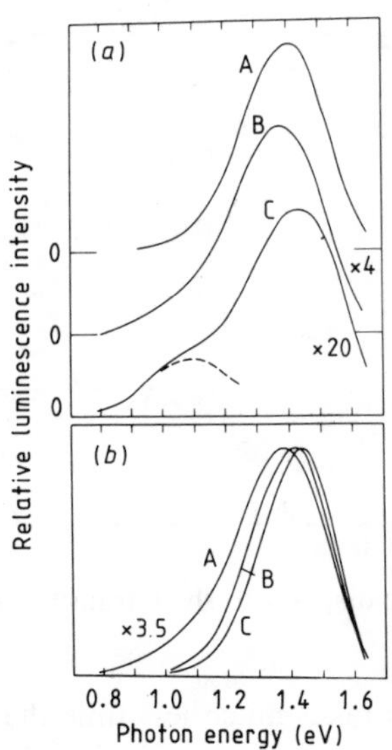

Figure 11.7 Luminescence in a-Si. (*a*) Main peak with 30 W discharge power and various percentages by volume of silane in the silane–argon gas: A, 100 %; B, 10 %; C, 1 %. (*b*) Power variation at constant concentration (A, 10 %, 30 W; B, 10 %, 5 W; C, 10 %, 0.5 W). (From Street *et al* 1978.)

interpretation is that the initial excitation creates an electron–hole pair in localised states in the band tails. Luminescent emission subsequently occurs by recombination unless n_s is so great that electron tunnelling to pair up with a neutral dangling bond becomes more probable. As the temperature is increased, PL decreases and this is explained by the thermal separation of the electron–hole pairs and their subsequent capture at dangling bonds. This general picture is compatible with further experiments on luminescent decay times (Street *et al* 1978), and with the absence of PL in unhydrogenated a-Si which has a high density of dangling bonds.

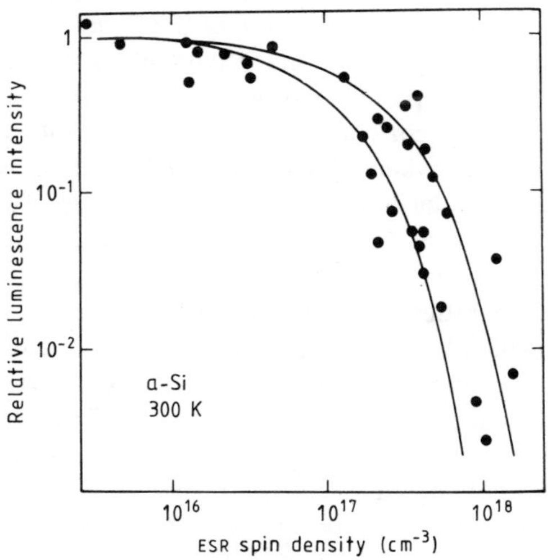

Figure 11.8 Correlation of luminescence and n_s. The lines are from a model discussed in Street *et al* (1978). (From Street *et al* 1978.)

Further extensive observations of ESR and PL have been made in doped and undoped a-Si:H, with and without compensation (Austin *et al* 1979, Street *et al* 1981, Dersch *et al* 1981). One of the main observations is that both p- and n-type doping reduce luminescent intensity and render it undetectable at $\gtrsim 1$ mol. % of dopant gas in silane. This is caused by numerous additional dangling bonds and it appears that the incorporation of dopant atoms entails the creation of associated defects as already suggested in §11.4. Compensation decreases n_s and restores the PL intensity and ESR while introducing new states near the valence band edge. The reader will find many more associated effects in the literature and should note that the PL peak at about 1.4 eV discussed above is the main, but not the only, one. Another peak at about 0.9 eV is visible under certain circumstances and is due to the de-excitation of electrons bound in defects (Bhat *et al* 1983).

As for ESR, it can be detected not only in equilibrium but also transiently in paramagnetic centres created by photon irradiation—a property called light-induced ESR (LESR). LESR indicates that light can generate neutral dangling bonds and also centres with two other g-values. These have been attributed to unpaired electrons in localised tail states of the conduction and valence bands—the very states responsible for the main PL transition. This identification follows the observation that the LESR and the decay time of the PL have the same temperature variation (Street and Beigelsen 1980). Apparently light excites carriers out of charged dangling bonds into tail states revealing many neutral dangling bonds by their ESR.

An extensive study of the three g-values by Dersch *et al* (1981) gave the results shown in figure 11.9. The position of the Fermi level can be estimated from the activation energy of the conductivity and is shifted by doping. This significantly alters the unpaired spin density as figure 11.9 shows. We may now refer back to the correlation energy U in §11.4. The peak of the $g = 2.0055$ curve occurs when all dangling bonds contain one electron; E_F is then midway between the peaks D^0 and D^- in figure 11.4(*b*). Raising or lowering E_F decreases n_s by increasing double occupation or emptying dangling bonds respectively. The width of the curve at half maximum measures the energy separation $D^- - D^0$ giving $U \sim 0.1$ eV. Comparable results in a-Ge give $U \sim 0.1$ eV. Doping does not simply move E_F in an otherwise static system; it introduces more dangling bonds as indicated earlier but this effect is small over the U-range of about 0.4 eV. From this experiment it appears that D^0 and D^- lie about 1.1 and 0.7 eV, respectively, below the conduction band mobility edge in a-Si. In spite of this, and the evidence in support given below, the magnitude and even the sign of U are uncertain and the theory uncompleted (Bar-Yam and Joannopoulos 1985).

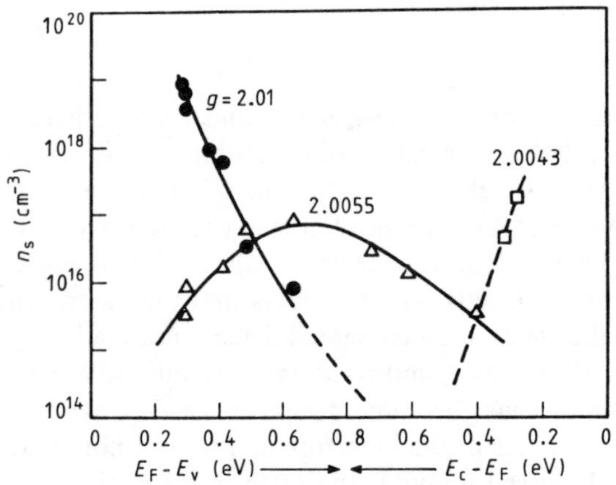

Figure 11.9 Spin densities versus E_F for variously doped samples of a-Si. (From Dersch *et al* 1981.)

In chalcogenides PL, but not ESR, is a well displayed phenomenon and figure 11.10 illustrates this. The luminescence spectrum has a considerable Stokes shift and neither its shape nor its temperature dependence varies with the excitation frequency or intensity; the PL reflects the nature of the emission process. The integrated intensity, normalised to that of the incident beam, *does* depend on the incident frequency and the excitation curve displays this. Its position indicates that $\hbar\omega$ and the band gap are about equal while the absorption coefficient (also shown in figure 11.10) depends on the shape of the band tail which in a glass will contain localised states. The excitation peak falls off on the left because of the fall in $\alpha(\hbar\omega)$ but its fall on the high-ω side appears to be due to at least two circumstances. One is that a rise in $\hbar\omega$ increases the probability that the excited electron–hole pair separate and get trapped at non-radiative recombination centres. Another is that the probability of PL emission is decreased unless the electron–hole pair is created near a radiative recombination centre, and it is supposed that absorption near the centre increases more slowly with $\hbar\omega$ than absorption in general, thus lowering the normalised luminescent intensity.

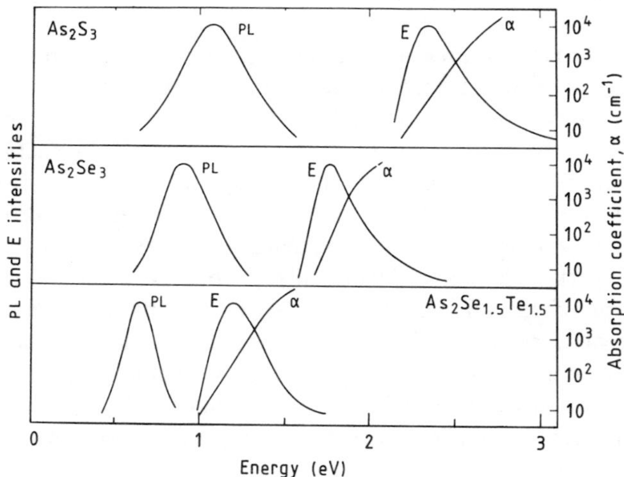

Figure 11.10 Excitation (E) and luminescence (PL) spectra and absorption edges in three chalcogenide glasses. (From Street 1976.)

The radiative recombination centres are generally believed to be the characteristic chalcogenide defects discussed in §11.5. The interpretation of PL according to the model goes as follows—see figure 11.5. An electron is excited to the conduction band from the C_1^- state (level A′) or possibly from the valence band in which case, if the electron–hole pair is near the defect, the hole can be captured non-radiatively leaving the same overall result.

Reorganisation then occurs to C_3^0. The electron from the conduction band now descends to level C′ emitting the photon $\hbar\omega'$ (arrow 6) and C_1^- reforms by further reorganisation. The relative positions of the levels A′ and C′, basically due to relaxation and bonding changes, accounts for $\hbar\omega'$ being about $\frac{1}{2}E_G$ and thus for the large Stokes shift ($\simeq$0.5 to $\simeq$1.5 eV). Corresponding phenomena with C_3^+ may occur. Increasing the temperature would help the electron to escape from the defect before recombination and this would decrease the PL intensity and also leave an uncharged centre which (unlike C_3^+ and C_1^-) has an unpaired spin which should show LESR. Both these effects are observed.

This section illustrates the notable insights to be gained from PL, ESR and LESR. There are other associated phenomena such as photoconductivity, with and without simultaneous ESR, which add more detail to the picture and confirm, *inter alia*, the importance of dangling-bond defects in α-IV solids (Dersch *et al* 1983). These methods do not however directly measure $g(E)$ in the gap.

11.7 Density of states in the gap

A sign that $g(E)$ in the gap is a solved problem would be the presentation of a $g(E)$ curve for a well defined sample accurately based on several different experiments convincingly analysed. At the time of writing this is not possible and attempts to measure $g(E)$ in the mobility gap continue to emerge usually accompanied by descriptions of the lack of agreed conclusions from other methods. Since the matter is important for understanding the properties it remains of great interest. When measurements disagree at least four causes may be at work: $g(E)$ may be truly different because the samples are not identically prepared; different experiments may use different assumptions in the course of extracting $g(E)$ from the observations; more detail may be inferred than is really warranted by the data, so minor features (e.g. small peaks) may be artefacts; inhomogeneities—quite probably present—may influence different techniques in different ways. There is some evidence in the literature for all four circumstances.

The major difficulty lies in the train of arguments and assumptions connecting the observed quantities and $g(E)$ for all the methods are very indirect. A few examples will now be given in outline only. The reader will find other experiments and much detail in specialist works.

The pioneering study was carried out by Spear and colleagues in the early 1970s with the field effect (FE) method (Spear and Le Comber 1983). A device like that shown in figure 11.11 was used to measure I_{SD} for a fixed voltage V_D as a function of the gate voltage V_G, and the problem is to see how this quantity leads to $g(E)$. Suppose V_G is applied with such a sign that the energy levels in the sample are bent downwards in energy by an amount

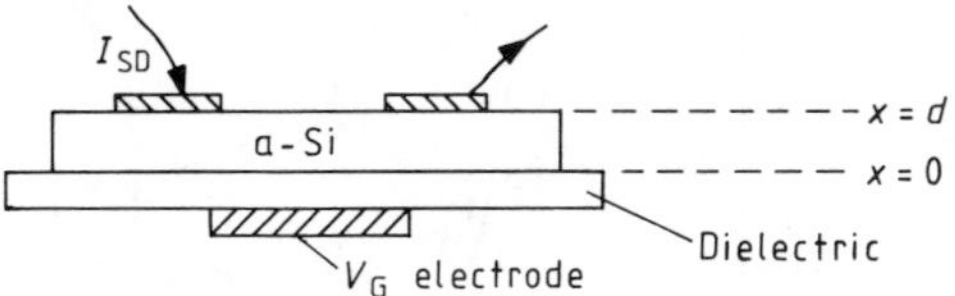

Figure 11.11 Schematic diagram of sample geometry used by Madan *et al* (1976) for FE measurements of $g(E)$.

$eV(x)$. Since E_F is constant, increasing V_G sweeps the energy E in $g(E)$ downwards past E_F with the effect that empty gap states fill up as $E \leqslant E_F$. The space charge induced in the semiconductor by V_G is $n(x)$ and must be related to $V(x)$ by Poisson's equation. $V(x = d) = 0$ is one boundary condition and we may put $-\partial V/\partial x$ at $x = 0$ equal to the known field strength. It is now assumed that the space charge is wholly that of electrons which entered previously empty localised gap states as the levels were pulled down past E_F. Thus for any V at position x:

$$n(V) = \int g(E)(f(E - eV) - f(E))\, \mathrm{d}E \qquad (11.3)$$

where f is the Fermi function. $n(V)$ would vanish for unbent bands which we provisionally assume to occur when $V_G = 0 = V(x)$. $V(x)$ pulls the mobility edge down and the number of free carriers is thereby enhanced by a Boltzmann factor. If $I_{SD}(0)$ is the current for the flat-band condition ($V(x) = 0$),

$$I_{SD}(V_G) = I_{SD}(0)d^{-1} \int_0^d \exp(eV(x)/k_B T)\, \mathrm{d}x.$$

This relates I_{SD} to an integral of $V(x)$ which is itself related to $g(E)$ through the space charge and Poisson's equation. In fact $n(V)$ can be computed from the observed $I_{SD}(V_G)$ and the equation (11.3) is deconvoluted to obtain $g(E)$. The evolution of data processing methods and of corrections, e.g., for the influence of surface states on the flat-band condition is explained in Spear and Le Comber (1983). Figure 11.12 shows some results. These were the first to indicate the order of magnitude of $g(E)$ and to reveal structure.

Spear and Le Comber emphasised from the beginning that the experiment does not by itself guarantee that $g(E)$ so obtained is a bulk, and not a surface layer, property. Another interesting point of physics is that the use of $f(E)$ in equation (11.3) evades any consideration of the correlation energy controlling the double occupation of gap states. Insofar as the D^- dangling-bond state affects the FE observations, the analysis might be expected to involve the correlation energy, U (Grünewald *et al* 1981, Schweitzer *et al* 1981). For such reasons the shape and origin of structures in $g(E)$ attracted many attempts to observe bulk properties less susceptible to interface effects. Some of these make use of capacitance measurements on p–n junctions and

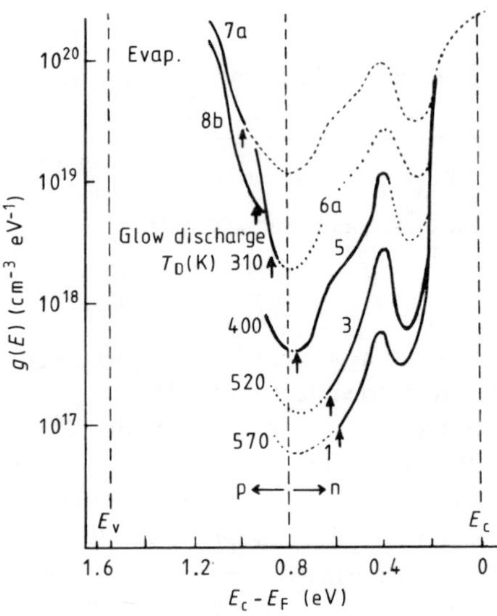

Figure 11.12 Some early results by the FE method giving $g(E)$ for various samples prepared under different circumstances. (From Madan *et al* 1986.)

we now refer to isothermal capacitance transient spectroscopy (ICTS) (Okushi 1985).

$g(E)$ is studied in the n region of a p^+–n junction in which it is supposed that gap states capture electrons from, and emit them to, the extended states. Capture and emission rates will be denoted by $c(E)$ and $e(E)$ with the suffixes e and h for electrons and holes; they are connected in equilibrium by

$$e_e(E) = c_e(E)N_c \exp[-(E_c - E)/k_B T] = \nu(E) \exp[-(E_c - E)/k_B T]$$

where N_c is the effective density of extended states in the conduction band. $\nu(E)$ is an attempt-to-escape frequency. From the device literature we take the capacitance under reverse bias, V_R, of a junction of area A and built-in voltage V_D, in a material of permittivity ε, namely,

$$C^2(\infty) = \frac{e\varepsilon A^2 N(\infty)}{2(V_D + V_R)} = BN(\infty)$$

where e is the electronic charge and $N(\infty)$ is the density of ionised gap states in the depletion region. If a burst of additional ionisation is injected by a voltage pulse which terminates at $t = 0$, N and C subsequently vary with time returning at $t = \infty$ to $N(\infty)$ and $C(\infty)$. The density of states enters through the equation representing the decay, namely,

$$\Delta N(t) \equiv N(t) - N(\infty) = \int_{E_v}^{E_c} (F_0(E) - F_\infty(E))g(E) \exp[-(e_e + e_h)t] \, dE$$

where F_0, F_∞ are the occupation probabilities of a level at E for $t=0$ and $t=\infty$. Letting n and p stand for the electron and hole densities at $t=0$, a little reflection shows that

$$F_0(E)=\frac{nc_e(E)}{nc_e(E)+pc_h(E)}$$

while in equilibrium when electron and hole emissions from a level occur in equal numbers

$$F_\infty(E)=\frac{e_h(E)}{e_e(E)+e_h(E)}.$$

Taking the special case that the pulse injects majority carriers, $n \gg p$ so $F_0(E)=1$ and assuming that for electron traps $e_e \gg e_h$, the ICTS signal, $S(t)$, is

$$S(t)\equiv tB\frac{\mathrm{d}\Delta N(t)}{\mathrm{d}t}=-B\int_{E_V}^{E_c} g(E)te_e(E)\exp(-e_e(E)t)\,\mathrm{d}E.$$

This quantity is measurable.

Now for a given t, $te_e\exp(-e_et)$ peaks at $e_et=1$ which defines an energy $E_p(t)$ since e_e is a function of E. Now $te_e\exp(-e_et)$ can be approximated by the δ-function $k_BT\delta(E-E_p(t))$ so

$$g(E_p(t))=-(Bk_BT)^{-1}S(t). \tag{11.4a}$$

To find what the energy $E_p(t)$ is, we solve the expression for e_e giving

$$E_c-E_p(t)=k_BT\ln[\nu(E_p(t))t]. \tag{11.4b}$$

Thus each instant of the signal gives the density of states at an energy calculable from equation (11.4*b*) provided the function $\nu(E)$ is known. An auxiliary experiment determines $\nu(E)$ by varying the initial voltage pulse width; it is linear in E and of order $10^{-9}\ \mathrm{s}^{-1}$. A density of states result is in figure 11.13(*a*) and the broad peak is centred at 0.52 ± 0.02 eV below E_c.

This experiment cannot be said to have settled the matter. The interested reader will want to pursue the allied technique of deep-level transient spectroscopy (DLTS) which is much used for studying trapping levels in crystalline semiconductors and which exploits the temperature dependence of the quantity $C(t_1, T)-C(t_2, T)$ where C is the capacitance transient referred to in the ICTS method. In an intensive study of a-Si:H by DLTS and related methods (Lang *et al* 1982), supplemented by ESR in a depletion layer of variable width (Cohen *et al* 1982), $g(E)$ of the form shown in figure 11.13(*b*) was found. The hump at about 0.85 eV below E_c was identified as the D^- state of the dangling bond and lies about 0.3 eV away from the presumably equivalent hump in the ICTS result possibly because of differences in the estimates of $\nu(E)$.

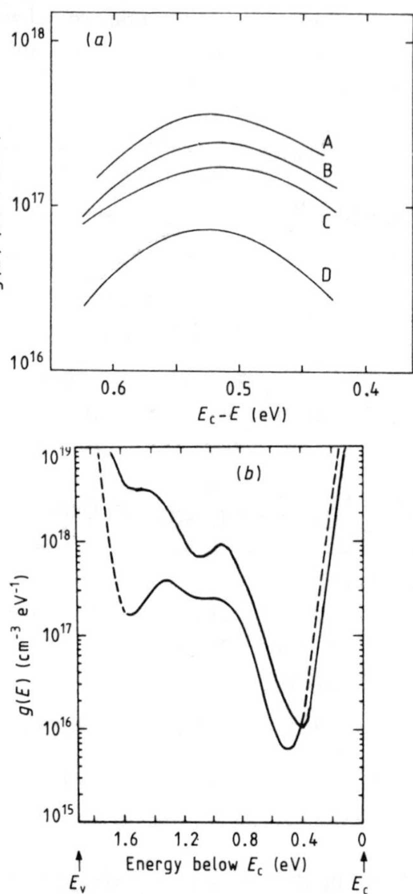

Figure 11.13 (*a*) $g(E)$ in the gap of a-Si according to an ICTS experiment—see text. PH_3/SiH_4: A, 10^{-3}; B, 3×10^{-4}; C, 10^{-4}; D, 10^{-5}. (From Okushi 1985.) (*b*) $g(E)$ from a DLTS experiment—see text. (From Lang *et al* 1982.)

Another example of these studies is the attempt to pin down the position of the D^- hump using measurements of electron and hole lifetimes (Spear *et al* 1984). These are relevant on the hypothesis that the lifetime is largely determined by capture at dangling-bond sites. The sample is lightly doped a-Si:H sandwiched between heavily doped n and p layers and electrons and holes are generated in it by a light pulse. There exists a delayed-field time-of-flight method for measuring τ_e and τ_h (see §11.9) and we shall assume that done. By measuring the conductivity of the unilluminated sample $(E_c - E_F)$ can be found from the activation energy while $(E_c - E_v)$ is assumed to be 1.8 eV from other work. Repeating this for different samples leads to the results shown in figure 11.14(*a*). Clearly $\tau_h \gg \tau_e$ but the near mirror symmetry of the curves is striking and suggests a single cause. An explanation is required for the rise and fall of τ with doping. On the hypothesis that dangling bonds are involved, it is relevant to know the probabilities, P^+, P^0, P^-, that a dangling bond is in the D^+, D^0, D^- states. To calculate these U must be known. Using $U = 0.35$ eV, the P were calculated for various positions of E_F relative to E_{D^0} subject to the obvious condition $P^+ + P^0 + P^- = 1$ and the results are shown in figure 11.14(*b*).

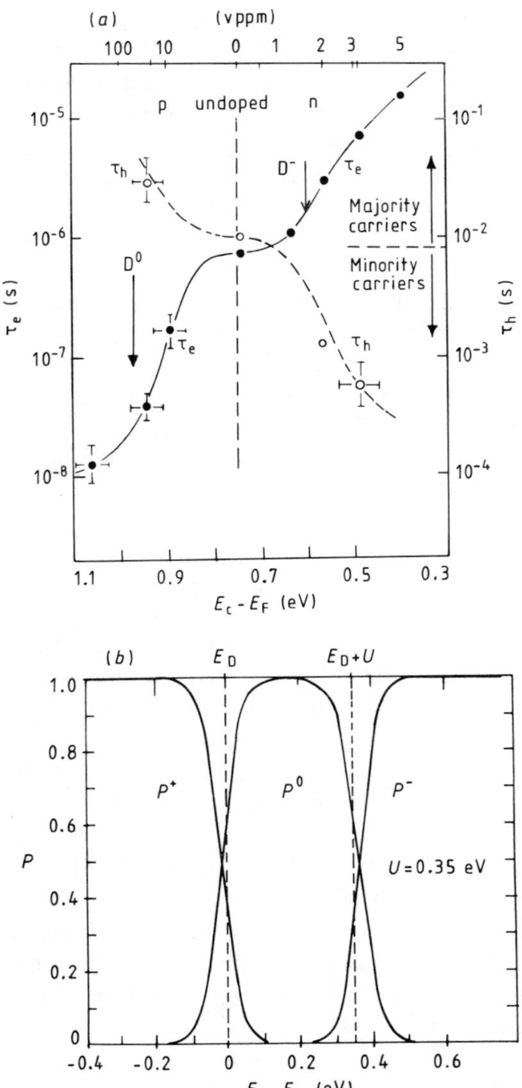

Figure 11.14 (*a*) Electron and hole lifetimes in doped a-Si:H, obtained by the delayed-field method. (See also § 11.9.) (*b*) Calculated probabilities of the dangling-bond states. (From Spear *et al* 1984.)

Where P^- is rising and P^0 is decreasing, τ_e will be expected to rise because electrons are much less likely to be captured at bonds already doubly occupied. When doping moves E_F towards the valence band converting D^0 to D^+ states, the attraction of the positively charged states will decrease τ_e. With converse arguments for holes, figure 11.14(*b*) contains an explanation of figure 11.14(*a*). Now the undoped sample has $E_c - E_F = 0.75$ eV and its τ are centrally situated between the rises and falls. This must correspond to the position between the fall of P^+ and the rise of P^-, namely, $E_F - E_{D^0} = 0.2$ eV. So $E_c - E_{D^0} = 0.95$ eV and D^- lies $(0.95 - U) = 0.6$ eV below E_c. It is the D^0 singly occupied states that control τ_e in undoped samples. This forms a rather consistent picture and the D^- position is not far from that of the ICTS method.

It might be thought that an obvious way to explore $g(E)$ is by optical absorption spectroscopy using photon energies lower than the gap width. There are at least two difficulties with this. One is experimental and the technique of photothermal deflection spectroscopy (PDS) seems well adapted to solve the problem. In this the sample is irradiated periodically with a laser beam of fixed $\hbar\omega$ and the periodic temperature fluctuations created by the absorbed energy cause similar oscillations in the refractive index of the gas adjacent to the sample. A probe beam skimming the sample surface undergoes rhythmic deflections because of the refractive index gradient and these are detectable. $\alpha(\hbar\omega)$, the absorption coefficient, can be inferred (Jackson *et al* 1981). If the initial state is a gap state within $g(E)$ and the final state is in the extended-state distribution, $g_c(E)$, then

$$\alpha(\hbar\omega) \propto \int g(E) g_c(E + \hbar\omega)\, \mathrm{d}E$$

assuming that the transition probability is independent of E. Deconvoluting this integral to get a unique $g(E)$ has proved difficult (Payson and Guha 1985) but PDS experiments are consistent with D^0 and D^- states 1.25 and 0.9 eV below E_c (Jackson 1982).

Numerous other techniques, including xerography (Imagawa *et al* 1984) and space-charge-limited currents (SCLC) (Mackenzie *et al* 1982), have been applied to the problem but enough examples have probably been given to show that the gap state problem, though not quantitatively solved beyond doubt, has been considerably clarified by the mid-1980s through the recognition that the dangling-bond states play the major role. Nevertheless it cannot be said that $g(E)$ in the gap is definitively known nor that the energies of the three dangling-bond states relative to E_c are settled (Fritzsche 1985, Le Comber and Spear 1986).

11.8 Bands, tails and optical gaps

The preceding sections have been much concerned with defects and gap states; the existence of the underlying gap has been taken for granted. The calculation of the energy spectrum in an amorphous semiconductor raises some of the same problems that occurred in Chapter 7 with metals. Band structure calculations for crystalline semiconductors are extremely well developed but in amorphous semiconductors the spectrum will be complicated by the topological disorder of the ideal CRN, by various defects and, in a-Si:H, by hydrogenation.

In the first place let defects and hydrogen be ignored so that we may consider the band structure of ideal four-coordinated a-Si. From the point of view of the TBA it is necessary to involve the overlap of the four sp^3 hybrid orbitals per atom. Even ignoring other than nearest neighbours there are six interactions of various strengths symbolised schematically in 2-D in figure 11.15. It looks plausible, and is known from the theory of the crystals, that $|V_2| > |V_1|$ and both are substantially stronger than the others. It was a notable advance when Weaire and Thorpe (1971) showed that valuable conclusions could be deduced from an even further simplified model in which V_3 to V_6 are made zero and V_1 and V_2 assumed to be the same for all atoms in the network.

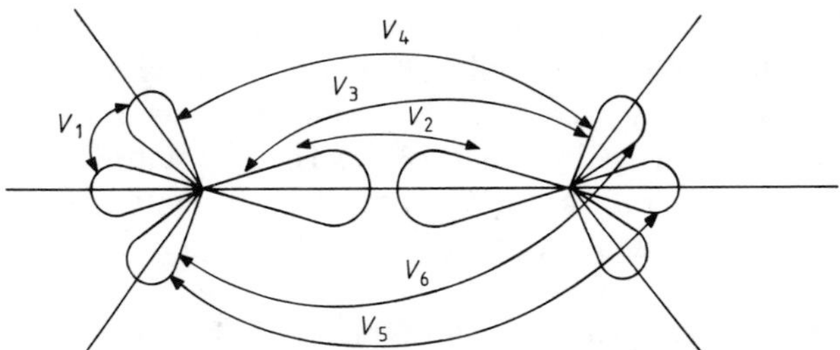

Figure 11.15 Schematic diagram showing the six nearest-neighbour interactions between various pairs of bonds.

Let $|i\mu\rangle$ stand for an sp^3 orbital from the atom at site i which contributes to the bond μ and let the set of orbitals $\{|i\mu\rangle\}$ be orthonormal. What is being assumed is the tetrahedral bonding and the CRN connectedness. Distortions of bond lengths and angles—which involve V_3 to V_6 and imply randomness in V_1 and V_2—are not taken into account but, although the diamond structure is included as a special case, crystallinity is not a requirement. The Hamiltonian representing the Weaire–Thorpe model is then

$$H = \sum_i \sum_{\mu \neq \mu'} |i\mu\rangle V_1 \langle i\mu'| + \sum_i \sum_\mu |i\mu\rangle V_2 \langle i_\mu \mu|$$

where i_μ denotes a nearest neighbour of i connected by the bond μ, and the first and second terms represent intra-atomic and interatomic interactions respectively. An important deduction made rigorously from this was that the band structure depends in detail on V_1/V_2 but that in general there is an energy gap between the valence band derived from bonding states and the conduction band derived from antibonding states (Weaire and Thorpe 1971). This demonstration did much to remove some suspicion that the topological disorder of the CRN must *of itself* destroy the energy gap and to encourage the now accepted view that band structure is more a question of SRO than LRO.

It is possible that the band edges shown to exist by Weaire and Thorpe could be modified and given tails and localised states by further disorder introduced by fluctuations in V_1 and V_2. To investigate this the parameters in the TBA could be determined by fitting the band structure calculated for c-Si to a well established one derived with pseudopotential theory. Then the variation in parameters with bond length and angle would also have to be derived from crystal calculations; and finally some estimate of typical bond length and angle fluctuations, Δr and $\Delta\theta$, could be taken from ball and spoke models or diffraction data (§2.6). From this the effect of Δr (say 1 %) and $\Delta\theta$ (say 10° in 109°) on the band edges could be found. This does not end the matter because fluctuations in the dihedral angle (φ in figure 11.16) also occur and to estimate their effect V_4 to V_6, which are constraints on φ, have to be introduced. The programme of considering fluctuations in V_1 to V_6 was carried through by Yonezawa and Cohen (1981) and by Singh (1981) with the result that Δr and $\Delta\theta$ were relatively unimportant while $\Delta\varphi$ could smear out the valence band edge by about 0.3 eV. The ring statistics (§10.6) could also be shown to affect the outcome especially at the bottom of the conduction band where odd-numbered rings randomly distributed imply band tailing and localisation.

A handmade or computed structural model (§2.6, also see Wooten *et al* 1985) and a chosen atomic potential can provide the system potential, $\Sigma_i V(\boldsymbol{r} - \boldsymbol{R}_i)$, for a numerical calculation by the recursion method or similar

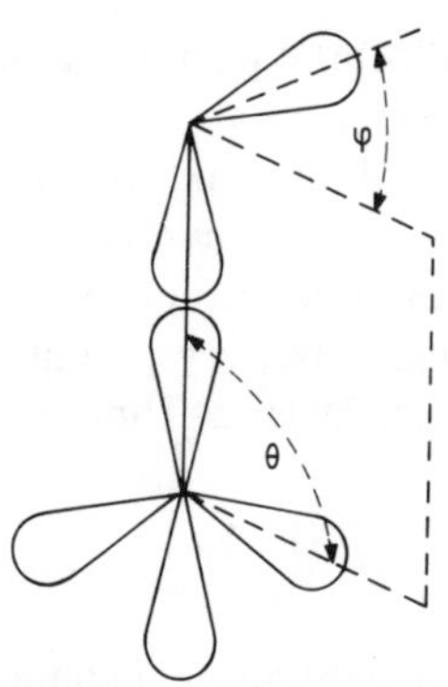

Figure 11.16 Definition of bond (θ) and dihedral (φ) angles.

techniques (Hickey *et al* 1985). Alternatively, stress can be laid on the relatively large departures from local tetrahedral coordination shown by the simple tetragonal form of c-Ge with twelve atoms per unit cell (called 'ST 12'). For an ST 12 crystal a pseudopotential method will give $g(E)$. Figure 11.17 shows the x-ray photoemission spectrum of crystalline and amorphous Si and Ge compared with an ST 12 crystal calculation of $g(E)$, and a recursion calculation of ST 12 and two amorphous models (Joannopoulos and Cohen 1983, Kelly and Bullett 1976). The latter are with (PB) and without (CT) odd-membered rings. The ST crystal also has odd rings. The inferences are that the main peak in the crystal spectrum survives in the amorphous state but other features are smeared out while preserving the overall form; that the centre of gravity of the main peak is moved towards E_F by disorder; that the ring statistics play an important role in determining $g(E)$ and the presence of odd rings seems essential for reproducing the observed spectrum, even a crystal, provided it is complex enough locally (ST 12), being able to approximate the general shape of the amorphous spectrum. Later work questions this, emphasising the dihedral angle disorder over the odd rings (Hayes *et al* 1985).

Since real a-Si contains defects, states occur in the energy gap and, if the material is substantially hydrogenated, the width of the gap and the form of the valence band are unlikely to remain unaffected. One method of tackling this theoretically has been tried by Zdetsis *et al* (1985). Since it is very difficult to include everything at once, the hydrogenation investigation omitted topological disorder by modelling the system as a lattice on which Si atoms were randomly placed with a probability $(1-c)$. A proportion, c, of vacancies therefore existed at which up to four H atoms could reside in positions between the vacant site and the adjacent atoms. The calculation method was the TBA version of the CPA (§7.2) which showed that, with four H atoms per vacancy passivating all the dangling bonds, the energy gap was about 1.4 eV which is 0.4 eV wider than in the unhydrogenated case. The valence band density of states receives the electrons in the Si–H bonds whereas, with less complete passivation, dangling-bond states encroached upon and eventually obliterated the gap. In practice dangling bonds near vacancies might well reconstitute into weak bonds (§11.4) and this case was investigated by a one H per vacancy model with the other three dangling bonds being effectively reconstituted with Si atoms. This also entailed a recession of the valence band edge leading to a wider gap and to a qualitatively similar band structure suggesting that the general features of the gap, and of $g(E)$ in the bands, are not dependent on the exact location of the H atoms. The widening of the gap by the recession of $\lesssim 1$ eV of the valence band edge and the appearance of additional Si–H bond peaks in the valence band is clearly revealed by the difference between the UPS spectra of hydrogenated and pure a-Si (von Roedern *et al* 1970).

There has been less study of the conduction band but a time-reversed

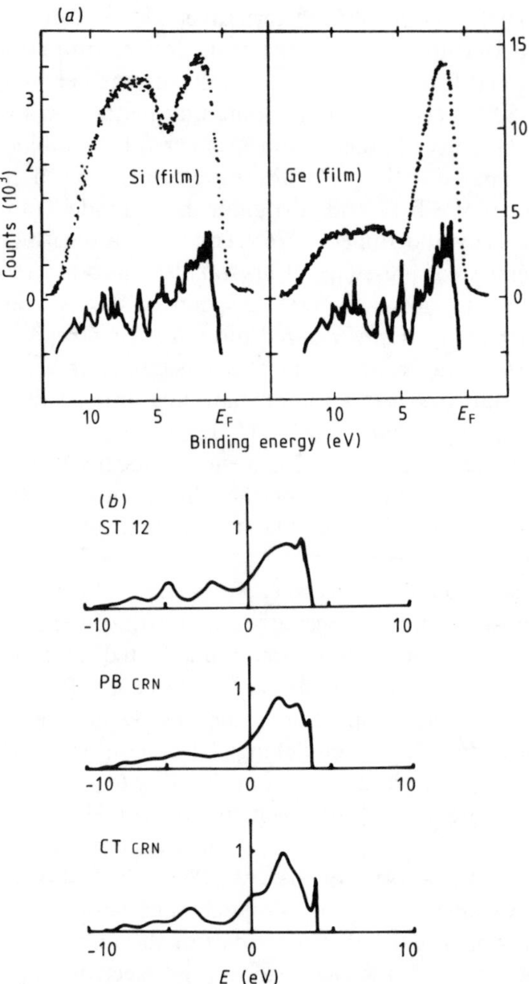

Figure 11.17 (*a*) $g(E)$ in a-Si and a-Ge from x-ray photoemission (top curves) and from pseudopotential calculations on ST 12. (From Joannopoulos and Cohen 1983.) (*b*) $g(E)$ in a-Si from recursion calculations on ST 12 and two continuous random networks. (From Kelly and Bullett 1976.)

version of XPS called bremsstrahlung isochromat spectroscopy (BIS) is available for the purpose. In this, monoenergetic electrons enter excited states in the sample and generate x-rays by descending to empty levels in the conduction band. The number of photons of a fixed frequency is counted as the incident electron energy is varied. This number is porportional to $g(E)$ at the receiving end of the downward electron transition. With BIS the conduction band of a-Si:H has been shown to have a pronounced peak about

3 eV wide just above the band edge; it decreases on driving out H and is attributed to antibonding states of Si–H_2 and SiH_3 sites (Jackson *et al* 1985). The density of states distributions in a large number of different amorphous semiconductors have been reviewed extensively by Robertson (1983).

Given any sort of disorder it is very plausible that band edges should lose their sharpness and tail into the band gap. Theories of the shape of band tails and of localisation in them have nevertheless proved difficult to establish. A number of experiments, including the study of trapping of carriers and the absorption of light (see below), suggest that the tails of bands fall off exponentially as

$$g(E) \propto \exp(-E/E_0) \tag{11.5}$$

where E_0 is the exponential tail width. General arguments from localisation theory indicate that the tail will contain localised states. It has not proved easy to demonstrate exponential tailing theoretically but arguments by Soukoulis *et al* (1984) support the following picture. The conduction band will be discussed but, *mutatis mutandis*, the arguments apply to the valence band. Descending in energy through the extended states we reach localisation at the mobility edge, E_c. For $E_1^{(c)} < E < E_c$, a range to be called the 'near tail', the states have a localisation length α^{-1} which diverges at $E = E_c$ (§9.5 *et seq*). For lower energies, $E < E_1^{(c)}$, there remain only a very small number of states which are those trapped in fairly large-scale potential wells, say a few atomic spacings or more, which are there by chance spatial fluctuations of the disordered structure. To cause localisation, the wells must be deep enough to confine electrons despite the greater kinetic energy implied by confinement. Soukoulis *et al* calculated the probability of such confinement, taking into account the shorter-range disorder that still exists inside the wells. It is this consideration that leads to an exponential tail of width $E_0^{(c)}$ for $E < E_1^{(c)}$; thus $E_1^{(c)}$ means the energy where localistion by large-scale fluctuations ceases to be effective. Such arguments refer to systems and disorder in general, not to particular kinds of them; nor are they restricted to 3-D. The difference $(E_c - E_1^{(c)})$ turns out to be of the order of $W/10$ and $E_0^{(c)} \sim W/100$ where W^2 is the variance of the disordered potential (Soukoulis *et al* 1984). Such concepts have relevance to any electronic properties, such as optical absorption, that involve tail states.

The optical properties of amorphous semiconductors have been widely studied and there are two plausible reasons for expecting them to differ in frequency dependence from those in crystals. Firstly the $\boldsymbol{k}$-conservation rules arising from the Bloch nature of the electron states should no longer apply because $\boldsymbol{k}$ is no longer a relevant quantum number; and secondly the disorder will have changed $g(E)$. Generally speaking, sharp peaks in crystal spectra are smeared out and edges caused to tail or blur. Optical properties in general are widely covered in Mott and Davies (1979). Here we take the particular case of the interband absorption edges because it is relevant to the band gap.

It is not perfectly straightforward to detect an absorption edge at frequency ω and establish an energy gap by $E_G = \hbar\omega$. There are at least three ranges over which the absorption coefficient, $\alpha(\hbar\omega)$, rises from low values as ω increases. A slow rise at first, sometimes seen in chalcogenide glasses, is attributed to excitations from the valence band to defect states. α then rises exponentially with ω over a few orders of magnitude and its dependence can be written

$$\alpha(\hbar\omega) = \alpha_0 \exp[(\hbar\omega - E_1)/E_0(T, X)] \tag{11.6a}$$

where $E_0(T, X)$ is a parameter to be discussed and α_0 and E_1 are constants. This exponential dependence is called an Urbach edge and expresses the main increase in α as $\hbar\omega$ approaches the band gap. Thirdly, the rise becomes less steep and follows

$$\hbar\omega\alpha(\hbar\omega, T) = B(\hbar\omega - E_G(T))^n \tag{11.6b}$$

where $n = 2$ frequently, though not invariably. The second and third of these ranges are shown for some chalcogenide glasses in figure 11.18.

A quantity referred to as the optical gap, E_G, has been extracted from equation (11.6b) for a number of chalcogenide glasses by extrapolating the straight lines obtained by plotting $(\hbar\omega\alpha)^{1/2}$ against $\hbar\omega$ with $n = 2$. For example, at room temperature E_G(eV) = 1.76 for As_2Se_3, 2.32 for As_2S_3, 0.7 for GeTe with the quantity B typically 10 μm^{-1} eV^{-1} (Mott and Davies 1979).

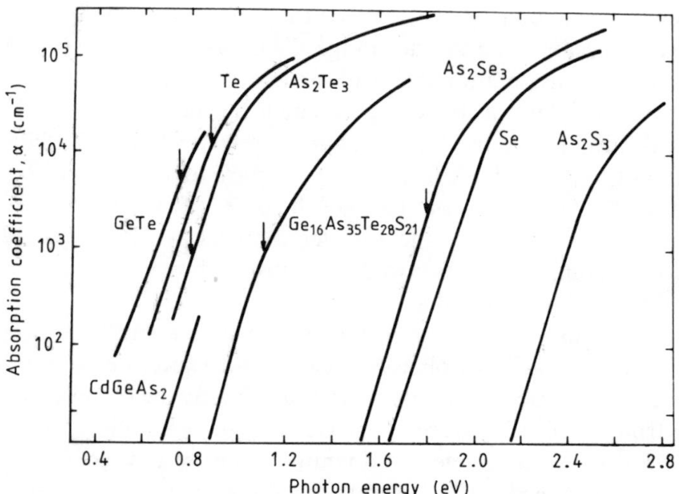

Figure 11.18 Optical absorption in some chalcogenide glasses. The arrows mark twice the conductivity activation energy. (From Mott and Davies 1979.)

In chalcogenides, the quantity E_0 is not very temperature dependent until near the glass transition.

From equations (6.61) and (7.43a), with E_n in the valence band and E_m in the conduction band, we infer

$$\hbar\omega\alpha(\hbar\omega) \propto \int g(E)g(E+\hbar\omega)\,\mathrm{d}E \tag{11.7}$$

provided that the matrix element is constant over the range of ω involved in the discussion. Equation (11.6b) with $n = 2$, often associated with the name Tauc, follows from this if both bands have the form $g(E) \propto |\text{const} - E|^{1/2}$ though it can be derived from other assumptions too (Mott and Davies 1979).

The Urbach edge occurs in many crystals and a favoured one among several explanatory hypotheses is that it is due to thermal disorder. In amorphous semiconductors it might therefore be that thermal and structural disorder jointly cause the exponential shape of α (Dunstan 1982). This approach was used to analyse their experiments by Cody *et al* (1981) and made the subject of theoretical interpretation by Cohen *et al* (1984c). Figure 11.19 displays the Urbach region of a-Si:H in which the full symbols show how

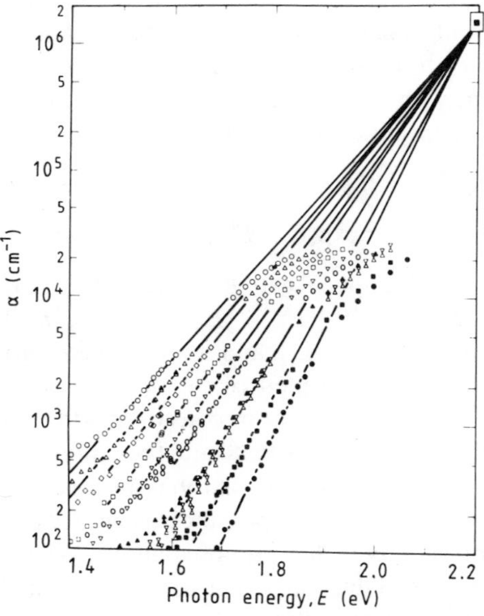

Figure 11.19 Urbach edges in a-Si:H for various amounts of disorder. T_H (°C) (⧗, 293 K; ◯, 500; ▽, 525, □, 550; ◇, 575; △, 600; ○, 625) is the temperature at which a spell of heating increased structural disorder by hydrogen evolution before the optical observation at 293 K. The full symbols denote measurement temperatures (K): ●, 12.7; ■, 151; ▲, 293. (From Cody *et al* 1981.)

increasing thermal disorder lowers the slope while the open symbols, all for 293 K, show how increasing structural disorder does the same. These results can be expressed by equation (11.6*a*) where $E_0(T, X)$ is the width of the Urbach tail, typically several tens of meV and α_0 and E_1 are fitting parameters equal to $1.5 \times 10^6\ \mathrm{cm}^{-1}$ and 2.2 eV respectively. For higher ω, equation (11.6*b*) with $n = 2$ yielded a temperature dependent E_G between 1 and 2 eV and $B = 48\ \mu\mathrm{m}^{-1}\ \mathrm{eV}^{-1}$. The interesting conclusion was that, whether temperature or structural order was taken as the parameter, E_G was always the same linear function of E_0, namely,

$$E_G = E_G^0 - (\mathrm{const}) \times E_0 \tag{11.8}$$

with $E_G^0 = 2.1$ eV.

Cody *et al* explained this on the basis that thermal and structural disorder simply aggregated their effects each contributing proportionally to a quantity $\langle \Delta R^2 \rangle$ where ΔR is the displacement, thermal or structural, of atoms from their ideal crystalline sites. X is then a number representing a normalised value of $\langle \Delta R_{STR}^2 \rangle$.

It still remains to say what excitations lead to equation (11.6). At the time of writing this seems not to have been entirely clarified but in Cohen *et al* (1984c, 1985) reasons are given for taking the matrix element to be constant and therefore for relying on equation (11.7). Excitation can be envisaged from extended or 'near-tail' states in the valence band to those in the conduction band. Provided $g(E) \propto |\mathrm{const} - E|^{1/2}$, equation (11.6*b*) should follow with $n = 2$, in agreement with experiment. Throughout the conduction band tail for example, and requiring continuity at $E_1^{(c)}$, we suppose

$$g(E) = A_c(E - E_1^{(c)} + \tfrac{1}{2}E_0^{(c)})^{1/2} \qquad E > E_1^{(c)} \tag{11.9a}$$

$$= A_c(\tfrac{1}{2}E_0^{(c)})^{1/2} \exp(E - E_1^{(c)})/E_0^{(c)} \qquad E < E_1^{(c)} \tag{11.9b}$$

where A_c is a constant. The optical gap, E_G, obtained by extrapolating the first of these to $g(E) = 0$, would be $[(E_1^{(c)} - \tfrac{1}{2}E_0^{(c)}) - E_1^{(v)} + \tfrac{1}{2}E_0^{(v)})]$. This is not the same as the mobility gap, $(E_c - E_v)$, and does not necessarily have the same temperature dependence. Attempted correlations of optical and transport properties should therefore take that into account. On the other hand, exciatations between the near tail of one band and the exponential tail of the other lead to the Urbach edge when equation (11.9) is inserted into equation (11.7).

The Urbach edge observations support the existence of exponential tails but do not prove their origin to be that proposed above. Whether they are really due to large-scale potential fluctuations or to steeper, more localised, wells is not clear (Cohen *et al* 1984c, 1985).

11.9 Some electron transport properties

Conductivity is of the essence of semiconductors and selected data from the vast accumulation will be presented in this section. Explaining it has not been altogether simple and as late as 1984 a leading authority found it 'somewhat depressing to note that perhaps the least understood property is the DC transport' (Shapiro and Adler 1984, 1985). We begin by collecting a few definitions and formulae.

An electron in an extended state above the mobility edge, $E > E_c$, is supposed to respond linearly to an applied electric field, $\boldsymbol{\varepsilon}$, by reaching a steady velocity $\boldsymbol{v}$. The microscopic mobility tensor $\boldsymbol{\mu}(E)$ is defined by

$$\boldsymbol{v}(E) = \boldsymbol{\mu}(E)\boldsymbol{\varepsilon}. \qquad (11.10)$$

For simplicity we go over to scalar quantities for the x-axis of isotropic matter. The electron conductivity can then be written

$$\sigma = -|e| \int \mathrm{d}E\, g(E) f(E) \mu(E) \qquad (11.11)$$

where $f(E)$ is the occupation probability and $v_x(E) = -\mu(E)\varepsilon_x$. Recalling the Kubo–Greenwood formula, equation (7.43c), we may integrate it by parts thus:

$$\sigma = -\int_0^\infty \mathrm{d}E\, \sigma(E) \frac{\mathrm{d}f}{\mathrm{d}E} \qquad (7.43c), \quad (11.12a)$$

$$= \int_0^\infty \mathrm{d}E \frac{\mathrm{d}\sigma(E)}{\mathrm{d}E} f(E) \qquad (11.12b)$$

because $f(\infty) = 0$ and the energy zero can be chosen at such a low energy that $\sigma(0) = 0$. $f(E, T)$ is known to be the Fermi function, but in an interesting discussion Cohen *et al* emphasised that setting up equations (11.11) and (11.12) from first principles does not rely on this and $f(E)$ can be arbitrary— in which case these equations imply equality of integrands or

$$|e| g(E) \mu(E) = -\frac{\mathrm{d}\sigma(E)}{\mathrm{d}E}. \qquad (11.13)$$

For a full band $f(E) = 1$ so that

$$\sigma = 0 = \int \mathrm{d}E\, g(E) \mu(E) \qquad \text{(full band)}. \qquad (11.14)$$

Equation (11.13) relates the microscopic mobility of electrons to the slope of $\sigma(E)$ which is positive for E above E_c. The slope is negative just below a valence band mobility edge making $\mu(E)$ positive and fulfilling the expectation that extended states near the top of the valence band imply hole-like carriers (Cohen *et al* 1984a,b). Let $\mu_c(E) \equiv -\mu(E)$ in the conduction band and $\mu_v(E) = \mu(E)$ in the valence band; near the mobility edges both μ_c and μ_v are positive. The probability of a hole is $f_v(E) \equiv 1 - f(E)$. Equations (11.11) and (11.14) then lead to

$$\sigma = \sum_{B=c,v} |e| \int \mathrm{d}E\, g_B(E) f_B(E) \mu_B(E). \qquad (11.15)$$

Assuming that ideas familiar in crystal physics apply to amorphous materials, equation (11.15) would have followed intuitively.

In the case that $E_c - E_F \gg k_B T$ and that $\sigma(E)$ close to the mobility edge may be written σ_0 and taken outside the integral, equation (11.12*a*) gives

$$\sigma_c = \sigma_0 \exp[-(E_c - E_F)/k_B T]. \tag{11.16}$$

For σ_0, σ_{min} of §9.10 could be inserted or, as a result of more detailed studies, $\sigma_0 = \sigma_{min}$ times other factors depending on the exact behaviour of the extended states as they go over into localised states at E_c (Cohen *et al* 1984a,b). The average or conductivity mobility, $\bar{\mu}_c$, is defined by $\sigma_c = |e|\bar{\mu}_c n$ where

$$n = \int dE\, g_c(E) f(E).$$

Corresponding formulae would apply if σ were dominated by holes and $E_F - E_v \gg k_B T$.

These formulae are for extended-state conduction and are derived from the Kubo–Greenwood formula which omits both inelastic scattering by phonons and electron–electron effects. The formulae do not take account of hopping conductivity of the varieties referred to in §9.13 which might occur between localised gap or tail states in suitable ranges of temperature; e.g. at very low temperatures, variable-range hopping near E_F is a possibility. In general an observed conductivity will be the sum of those in all the active mechanisms at the prevailing temperature.

A general relation for the thermoelectric power, α, can be found from the electrical and heat current densities. Let j_x be a flux or current density with superscript C, U, F, S, H, N to denote electric charge, energy, free energy, entropy, heat, particle number. $j_x^{(H)}$ is usually called Q_x (see, e.g. equation (6.47)). Recalling that $dQ = T\,dS$, Q_x may be defined as $Tj_x^{(S)}$. From $F = U - TS$ in thermodynamics, $Q_x = j_x^{(U)} - j_x^{(F)}$. Furthermore, for $j_x^{(F)}$ we may write $E_F j_x^{(N)}$ because E_F is the chemical potential of electrons or their free energy per particle, so

$$j_x^{(C)} = \int j_x^{(C)}(E)\, dE = \varepsilon\sigma = -\varepsilon \int dE\, \sigma(E) \frac{df}{dE}$$

$$Q_x = \int Q_x(E)\, dE = -\frac{\varepsilon}{|e|} \int dE\, \sigma(E) \frac{df}{dE}(E - E_F).$$

But from equation (7.36), $Q_x = T\alpha j_x^{(C)}$ from which

$$\sigma\alpha = \frac{k_B}{|e|} \int dE\, \sigma(E) \frac{df}{dE} \frac{E - E_F}{k_B T}. \tag{11.17}$$

This can be derived rigorously from Kubo's equations and applied to amorphous and crystalline systems. Since $\sigma(E)$ is positive and $\mathrm{d}f/\mathrm{d}E$ negative in equilibrium, α is positive (hole behaviour) for $E < E_F$ and negative (electron behaviour) in the opposite case. With the same assumptions that led to equation (11.16) the expression for α follows as

$$\alpha_c = -\frac{k_B}{|e|}\left(\frac{E_c - E_F}{k_B T} + A\right) \qquad A = 1. \tag{11.18}$$

There is a corresponding expression for holes; and for holes and electrons together

$$\alpha = \frac{\sigma_e \alpha_e + \sigma_h \alpha_h}{\sigma_e + \sigma_h} \tag{11.19}$$

In the event that the activation energies have a temperature dependence adequately represented by the linear relation $E_c - E_F = E(0) - \gamma T$, then

$$\sigma_c = \sigma_0 \exp(\gamma/k_B) \exp(-E(0)/k_B T) \tag{11.20}$$

$$\alpha_c = -\frac{k_B}{|e|}\left(\frac{E(0)}{k_B T} + A - \frac{\gamma}{k_B}\right). \tag{11.21}$$

Measurement of γ from the latter equation looks possible but depends on the correctness of the other constant A. This may not be exactly unity but will depend on how $\sigma(E)$ approaches E_c (Cohen *et al* 1984a,b). By combining equations (11.20) and (11.21) to eliminate the temperature dependent $(E_c - E_F)$ we find

$$Q \equiv \ln \sigma_c - \frac{|e|\alpha_c}{k_B} = \ln \sigma_0 + A$$

where σ is usually expressed in $\Omega^{-1}\ \mathrm{cm}^{-1}$.

In a number of chalcogenide glasses, experiments show that

$$\sigma = C \exp(-E_\sigma/k_B T) \tag{11.22}$$

with an activation energy, E_σ, about half the optical gap (see figure 11.18). C lies in the range 10^3 to $10^4\ \Omega^{-1}\ \mathrm{cm}^{-1}$. Plots of $\ln \sigma$ against T^{-1} over long ranges often show slight curvature. α is commonly positive which implies that the holes dominate the conduction. Equation (11.21) shows that an activation energy, E_α, can be found by plotting α against T^{-1}, and E_α might be expected to equal E_σ though in fact it is often about 0.1 eV less. The α-plots are also curved. Figure 11.20 shows some observations.

The curvatures may indicate that two or more mechanisms with different activation energies are contributing. An obvious possibility is that extended-state conduction by holes is supplemented by hopping along the localised states in the valence band tail with a hopping conductivity

$$\sigma_{hop} = \sigma_1 \exp(-E_1/k_B T). \tag{11.23}$$

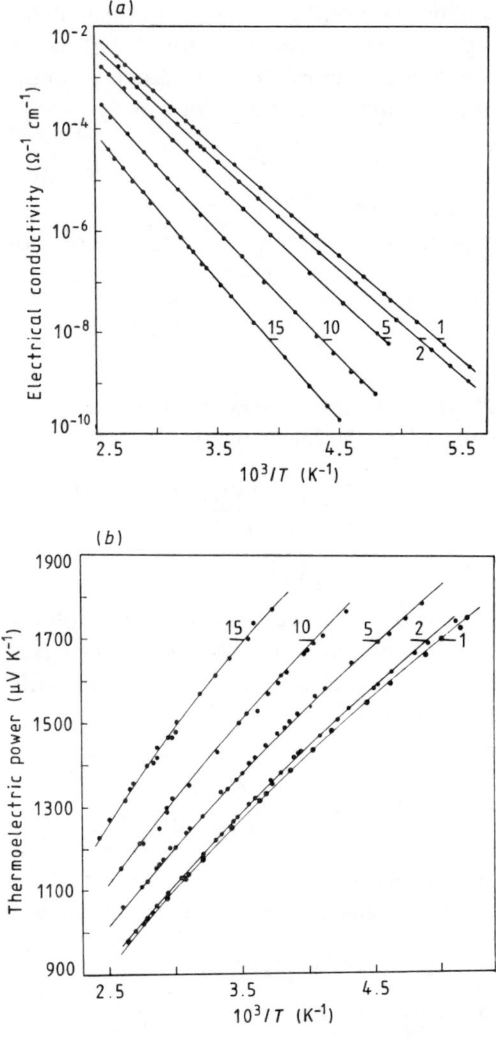

Figure 11.20 (*a*) Conductivity and (*b*) thermoelectric power of $AsTe_{1.5}Si_x$ chalcogenide glasses. The numbers are atomic percentages of Si. (From Nagels *et al* 1974.)

In this case $E_1 = E_F - E_L + W$ where E_L is an energy about which hopping is centred and W is the activation energy for a hop. σ_1 is difficult to calculate. In §9.13 it was pointed out that σ_1 will depend on $g(E)$, the electron–phonon interaction and the spatial distribution of localised states. For near and exponential tails anything like those of equation (11.9), σ_{hop} will depend on E_1, E_c, E_0 and possibly other quantities (Cohen *et al* 1984a,b). Many authorities have argued that $\sigma_1 \ll \sigma_0$ (Mott and Davies 1979). In Nagels *et al* (1974) extensive observations of σ and α in $AsTe_{1.5}Si_x$ were fitted very satisfactorily on the assumption that extended and localised states were providing two conduction channels in parallel. Equations of type (11.20) and (11.23) were summed for σ and one of the form (11.19) used for the combined α. A typical result was: for extended states, $E(0) \simeq 0.5$ eV with a pre-exponential factor $3800\ \Omega^{-1}\ \mathrm{cm}^{-1}$; for hopping, $E_F - E_L + W \simeq 0.418$ eV for $T = 0$ with a pre-exponential factor $92\ \Omega^{-1}\ \mathrm{cm}^{-1}$.

A similar principle of multiple conduction channels accounts satisfactorily for the quite complicated behaviour of evaporated a-Si and a-Ge which show both positive and negative α depending on preparation, annealing and temperature. These materials have many gap states and except at high temperatures conduct approximately according to the $T^{-1/4}$ law (§9.13). This channel is at $E \sim E_F$ and has a small $\alpha \sim 60\ \mu\mathrm{V\ K}^{-1}$. At $T \gtrsim 300$ K contributions from hopping by carriers excited to both band edges arise and much larger σ and α result. Very good fitting of the data requires electron and hole activation energies for each sample and plausible assumptions about the quantities A and $(E_\sigma - E_\alpha)$ (Beyer and Stuke 1974).

Hydrogenated a-Si is very different and has some highly interesting properties. In the first place it responds to doping with very large changes in conductivity as shown in figure 11.21(a) from the work of Spear and Le Comber (see Joannopoulos and Lucovsky 1984, Jones *et al* 1977). This gives technical control over σ for practical applications. Doping with P and B gives n- and p-type as revealed by the sign of α. In the 1970s many measurements of σ and α in a-Si:H were made for various dopings (Jones *et al* 1977, Bayer and Overhof 1979). The accumulated data show some remarkable regularities which the formulae used so far do not readily account for. One

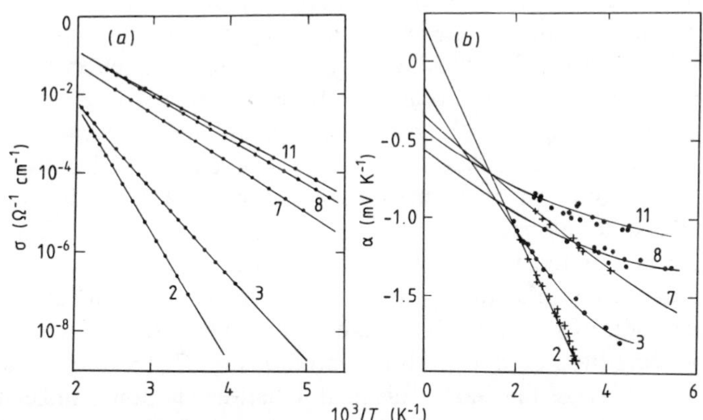

Figure 11.21 (a) Conductivity and (b) thermoelectric power of phosphorus-doped a-Si:H. The numbers 2 to 11 refer to samples with successively higher doping from light to heavy. The very large effect of doping on σ is evident. (From Jones *et al* 1977.)

is the Meyer–Neldel rule which is a correlation between E_σ and C (equation 11.22) for materials in which E_σ can be varied by doping, exposure to light or other means, namely,

$$\ln C = A + BE_\sigma. \tag{11.24}$$

Another is the behaviour of Q at $T \gtrsim 250$ K for numerous materials, n- and p-type, including a-Si and chalcogenides. This behaviour is

$$Q \simeq L - M/k_B T \tag{11.25}$$

where $L \sim 10$ universally, while M is sample dependent. This makes it look as if the large variations of C have to do with the temperature dependence of the activation energies (as suggested by equation (11.20)) since these have been eliminated by definition from Q, of which the behaviour is much more regular.

To take an illustrative experiment, some results of Jones *et al* on n-type a-Si:H are shown in figure 11.21. It appears that with light doping, except at the lowest temperatures, one activation energy for excitation to $E > E_c$ is adequate to explain both α and σ using equations (11.16) and (11.18). With stronger doping, the curvature of the α-plots requires a two-channel hypothesis and the fitting gave all the parameters in equations (11.20) to (11.23) and the curves in the diagrams. Increased doping reduced $E(0)$ from about 0.6 to about 0.23 eV while $(E_c - E_L)$ fell from 0.22 to about 0.13 eV. W was about 0.1 eV. The picture is one of conduction in extended states above, and in localised states below, E_c with the position of the hopping channel being influenced by the doping level. Values of the quantity $(A - \gamma/k_B)$ also emerge from the fitting and are about -0.5 eV. Without reliable knowledge of A it is difficult to evaluate and explain γ, but it may be presumed to depend partly on any intrinsic dependence of energy levels and gap widths on temperature and partly on the changes in E_F as varying T repopulates the available levels. Some authors have laid particular emphasis on the latter effect and indeed Overhof (1984) gives arguments and computer modelling to show that Fermi level movement, together with an assumed spatial modulation of the mobility edge by long-range random fluctuations of potential due to charged centres, can account for both equations (11.24) and (11.25) with conduction in extended states alone.

Nevertheless at the time of writing models of transport in amorphous semiconductors continue to appear, some of them emphasising an exponential distribution of tail states (equation (11.9)) rather than extended states and depending little or not at all on variations of E_F (Shapiro and Adler 1984, 1985). It is possible that the distinction between hopping and extended-state conduction is too simple and that a treatment which, unlike the Kubo–Greenwood equation, includes electron–phonon interaction from the start is required to deal with conduction near a mobility edge (Mueller and Thomas 1983, Cohen *et al* 1984b).

The general difficulty of understanding transport is not eased by observing the Hall effect. The Hall coefficient, R_H, is a valuable tool in crystal physics indicating, by its sign and magnitude, the sign and number density of carriers in single-carrier transport. In chalcogenide glasses R_H is usually negative even when carriers are predominantly positive according to the sign of α, while samples of a-Si:H judged to be p- and n-type by their thermopower have negative and positive R_H respectively.

The observations of σ and α quoted above were accompanied by measurements of R_H and therefore of Hall mobility $\mu_H \equiv |R_H \sigma|$ which is shown in figure 11.22. In both cases it was possible to model the curves with the same two-channel hypothesis of conduction in extended and hopping states that was used for σ and α but with additional specific assumptions about μ_H, namely,

$$\mu_H = \frac{\sigma_c \mu_{H,c} + \sigma_L \mu_{H,L}}{\sigma_c + \sigma_L} \qquad (11.26)$$

where L indicates the hopping contribution and $\sigma_L \mu_{H,L}$ was taken to be very small. In both cases $\mu_{H,c}$ turned out to be independent of temperature and of order 0.1 $cm^2\,V^{-1}\,s^{-1}$ (Le Comber *et al* 1977).

The sign of R_H presents a difficult problem. For diffusive motion in extended states near a mobility edge, say E_c, the RPM given in §7.7 was used by Friedman for R_H in semiconductors as well as metals. Instead of equation (7.45) the result is

$$\mu_H \cong 2\pi \left(\frac{ea^2}{\hbar}\right) Ba^3 \left(\frac{\eta \bar{z}}{z^2}\right) g(E_c) \qquad (11.27)$$

where B = bandwidth, a = nearest-neighbour distance, z = coordination number, $\bar{z}$ = the average number of triangular paths starting at one site and passing through two nearest neighbours, η = a geometrical factor $\sim \frac{1}{3}$. With $z = 6 = \bar{z}$, Friedman's own estimate was $\mu_H \sim 0.1\ cm^2\,V^{-1}\,s^{-1}$. This is the same as the experimental result but depends on a crude estimate of $g(E_c)$ like that used in connection with equation (9.17). As for the sign, it depends on the path provided for an electron by a sequence of transfer integrals. To get a Hall effect at all there must be alternative paths for which the relative probability is influenced by the magnetic field. So at least two sites must be considered as possible routes for an electron starting at a third. For three-membered ring paths, Friedman and others showed that both electrons and holes have $R_H < 0$, which provides one reason why an n-type Hall behaviour is compatible with a p-type thermopower. The sign problem was considered in more detail by Emin (1977), and the crucial criterion was the sign of the product of transfer integrals round a closed path. The sign depends on (*a*) whether the path is an odd- or even-membered ring and (*b*) the nature, s- or p-like, bonding or antibonding, of the orbitals in the transfer integrals. Thus sign anomalies could in principle be explained given sufficient microscopic information.

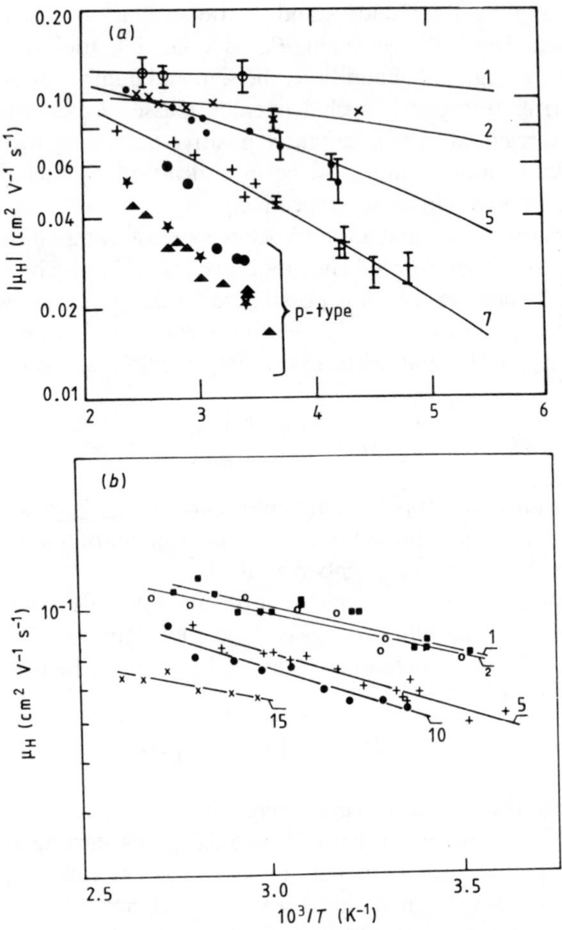

Figure 11.22 Hall mobilities in (*a*) a-Si:H with n and p dopings (from Le Comber *et al* 1977, compare figure 11.22) and (*b*) chalcogenide glasses (from Nagels *et al* 1974).

Many studies of transport during transient phenomena have been made since drift mobility experiments were developed by Spear in 1957 and thereafter. To conclude this section on transport some of the important quantities will be defined and characteristic results quoted. Apart from the chance that transient experiments offer of studying trapping and the density of states, the measurement of mobility and lifetime of excess carriers is essential for characterising materials for technical use. Electrons and conduction bands will be referred to but corresponding arguments, though with different numerical outcomes, would apply to holes. The assumption is always that the effects measure bulk properties though contact-controlled results have sometimes been suspected.

The quantity μ_c is the mobility of electrons in extended states in equilibrium. If a non-equilibrium excess of carriers is generated near an electrode, probably nowadays with a laser light pulse, and if the electrons are then drawn across to another electrode in a measurable transit time, t_t, by a known uniform field, an inference can be made of their drift mobility, μ_d; this is defined as their speed per unit field. $\mu_d \neq \mu_c$ because some of the electrons get trapped in empty tail states and re-emitted, possibly many times, and are thereby delayed. If τ_f, τ_t are the average times spent free and trapped then

$$\mu_d = \frac{\tau_f}{\tau_f + \tau_t}\,\mu_c.$$

Some electrons may descend into deeper traps from which re-emission during the experiment is improbable. Their contribution to the transient current is lost and their lifetime, τ_e, may be defined as the decay time of the number of excess carriers. This view implies a model of transient transport as being due to current in extended states impeded by trapping which may be temporary and which may entail some hopping. We refer to it as the multiple-trapping model.

There has been, and will probably continue to be, a considerable variety of opinion about what mechanisms control drift and on how anything can be inferred from the observations. Many of these ideas were reviewed by Marshall (1983, 1985), and the experiments by Spear (1983). It is difficult to interpret experiments without a clear hypothesis as to the mechanisms and, in what follows, examples of this will be given in relation to a-Si:H and As_2Se_3 without the implication that alternative treatments are out of the question.

A sample in a recent a-Si:H experiment might be undoped or lightly doped, several microns thick, and sandwiched between thin p^+ and n^+ heavily doped regions backed by electrodes. One electrode will transmit a tuned laser light flash to irradiate the p^+ layer at time $t = 0$. A collecting field pulse has to be applied in the appropriate sense and with sufficient duration and it draws electrons to the other electrode enabling the current pulse to be displayed. The schematic diagram 11.23(*a*) shows a simple pulse shape from which t_t is identifiable for obtaining μ_d, e.g., $\mu_d \sim 0.8$ $cm^2\ V^{-1}\ s^{-1}$ for electrons and $\mu_d \sim 10^{-3}$ for holes. For technical reasons, it is often desirable to delay or interrupt the collecting field by a quiescent period Δ (figure 11.23(*b*)) which enables $Q(\Delta)$, the charge still collectable after Δ, to be measured (Spear *et al* 1984). It is discovered that $Q(\Delta) = Q\exp(-\Delta/\tau_e)$ from which τ_e is derived, having typical values $\tau_e = 0.5\ \mu s$ at 300 K or 2.5 μs at 230 K, independent of sample thickness. There is confirmatory evidence that the decay of Q is due to deep trapping, not to surface recombination or other cause. Measurements like this were

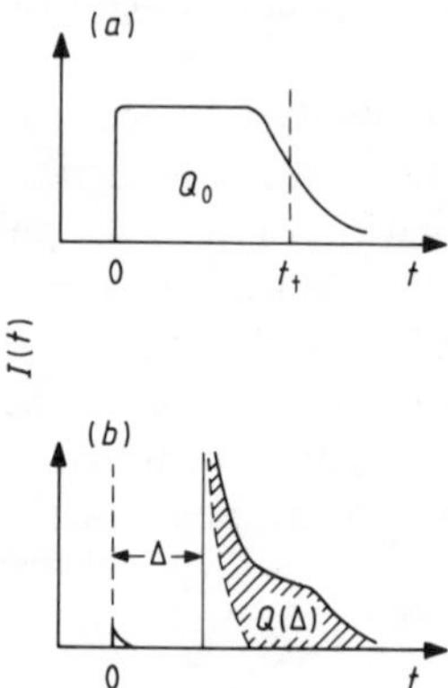

Figure 11.23 Schematic drawing of pulse shapes in drift experiments (after Spear *et al* 1984). Current versus time: (*a*), with sweeping field on (t_t is the mean transit time); (*b*), with field delayed by Δ (*Q* is the charge collected). The shaded area is the current integrated with the exclusion of the switching transient at $t = \Delta$.

quoted in §11.9 which illustrated one use for the τ_e-measurements. τ_h can be obtained similarly (see figure 11.14).

The simplest application of the multiple-trapping model would be to assume a layer of traps at a definite energy below E_c but not so deep that thermal equilibrium cannot be set up—or at least used as a good approximation—for times less than t_t (Spear 1983). A more elaborate hypothesis is necessary for pulse shapes less readily interpretable than figure 11.23(*a*). In fact many suggestions have been made to explain why some $I(t)$ curves have a variety of slope changes making it difficult to identify a definite t_t. This phenomenon is called dispersive transport and presumably implies that the mean μ_d varies with time. An attractive hypothesis is that there is actually a range of localised states acting as traps at different depths and an analysis of pulse shapes on this basis was performed by Orenstein and Kastner (1981) for As_2Se_3 and by Tiedje *et al* (1981) for a-Si:H.

To show how this hypothesis leads to $g(E)$ in the tail we follow the argument of Orenstein and Kastner. Here the phenomenon in question is the transient photoconductivity (TPC) induced by a laser flash in a small plate of the polished glass. Two electrodes in the top surface monitor the photocurrent $I(t)$ for times up to 10^{-2} s after the flash. $I(t)$ was found to decrease as t^{-1+T/T_0} where T_0 is a constant and T is the temperature. Recombination of electrons and holes is negligible for the duration of the experiment, so the decrease of I can be hypothetically attributed to trapping in a tail of localised states. If this is so, then at $t = 0$, excess electrons start to be captured into various states $E < E_c$. (We continue the description with electrons, though chalcogenide glasses are commonly p-type—see above.) However, thermal re-emission from the shallower traps is probable, followed sooner or later by trapping into deeper levels from which escape before the end of the observation is less likely. At any time, t, the excess electrons have energies in a distribution peaked at, say, $E_p(t)$. For $E < E_p$, the peak falls away because of the rapid fall of $g(E)$. In fact, $E_p(t)$ may be defined as the energy such that the lifetime, $\tau(E_p(t))$, in a trap at this energy, is t; electrons of higher energy are likely to have been released and collected, those of lower energy are trapped. As time goes on, $E_p(t)$ gradually falls further into the tail of states. Eventually all electrons are in states deep enough to prevent further emission before the end of the observation and $I(t)$ has become imperceptible.

$I(t)$ is carried by the fraction of the excess electrons still high enough in the wing of the distribution centred on $E_p(t)$ so assuming they are close to thermal equilibrium,

$$I(t) \propto \mu_c \frac{g(E_c)}{g(E_p(t))} \exp[-(E_c - E_p(t))/k_B T].$$

If $\nu(E)$ is an attempt-to-escape frequency from a trap at E, then

$$\tau^{-1}(E) = \nu(E) \exp[-(E_c - E)/k_B T].$$

Inserting this into the previous equation

$$I(t) \propto \frac{\mu_c}{\nu_0 t} \frac{g(E_c)}{g(E_c - k_B T \ln \nu_0 t)}$$

where $\nu(E)$ has been replaced by a representative fequency $\nu_0 \sim 10^{12}$ to 10^{13} Hz.

This equation suggests that the shape of $I(t)$ allows an inference of $g(E)$. In fact an exponential tail

$$g(E) = g_0 \exp[-(E_c - E)/k_B T_0]$$

leads to $I(t) \propto t^{-1+T/T_0}$ in agreement with experiment. The quantity $k_B T_0$ is thus the width of the exponential tail and corresponds to E_0 in equation (11.9*b*). This elegant experiment and analysis lent credence to the exponential tail and to the multiple-trapping model and it was found that $k_B T_0 = E_0 \simeq 50$ meV in As_2Se_3. $k_B T_0$ in a-Si:H was 27 and 43 meV for the conduction and valence bands respectively.

This general picture is enhanced by the observation during TPC of another phenomenon, photoinduced absorption (TPA). The time evolution of the optical absorption showed that it was caused by the excitation out of traps of the same carriers originally freed by the laser flash. As time went on absorption required higher energy as the carriers descended to deeper traps.

Little more will be said here about transient experiments except to observe that the multiple-trapping model just used, though not necessarily the exponential tail itself, has been subject to objection both in principle and from experiment. The objection in principle concerns the use of thermal equilibrium relations between E_c and E_p (Marshall 1983, 1985). The experimental objection is that at $T < 200$ K for holes in both a-Si:H and As_2Se_3 the relation $I(t) \propto T^{-1+T/T_0}$ fails. Another version of the multiple-trapping model appears to accommodate this but only by postulating hopping transport among the tail states instead of trap-controlled drift in extended states at E_c (Monroe 1985). This bears witness to the substantial obstacles encountered in fully understanding amorphous semiconductors.

11.10 A few more properties of amorphous semiconductors

The size of the band gap, the nature of the defects, the localised tail states and hopping, the possibility of doping and the structural variations compatible with the overall disorder, all provide amorphous semiconductors with an immense richness of properties. Almost any external influence—photons, heat, particle bombardment, stress, magnetic and electric fields and even time—does something interesting, and the different effects frequently influence each other as when photoconductivity is not divorced from optical absorption or luminescence. For this reason there are many interesting properties which cannot be dealt with in a chapter like this for reasons of time and space. In this section a few more will be drawn to the reader's attention without discussion; more detail and further references can be found in general treatises or collections like Mott and Davies (1979), Joannopoulos and Lucovsky (1984), Pankove (1984), Phillips (1986), Spear and Le Comber (1983), Yonezawa and Cohen (1981) and the Device Physics Sections of 1985 *J. Non-Cryst. Phys.* **77/78** 1363ff.

Since a-Si:H can be doped, p–n junctions can be made with all that this implies for photovoltaic effects, electroluminescence, transistor action and memory switching between 'on' and 'off' states of diodes. Apart from being research tools (see §§11.7, 11.9), p–n junctions are the obvious source of many device possibilities (1985 *J. Non-Cryst. Phys.* **77/78** 1363ff). a-Si:H itself can be influenced by light in many ways. An effect named after its discoverers, Staebler and Wronski, and abbreviated the SWE, is the production of metastable defects in undoped or lightly doped material by irradiation with photons. The defects are thought to be dangling bonds though perhaps not exclusively so and, interestingly, they are produced in numbers which increase more slowly than the product of intensity and exposure time. The epithet metastable means that the defects can be removed by annealing at, say, 200 °C, the corresponding activation energy being of the order 1 eV. It is thought that after the initial photoexcitation, electron–hole recombination occurs in a way which involves the rupture of weak Si–Si bonds (§11.4) and the stabilisation of the resulting local structure, perhaps by the relocation of nearby H atoms. The SWE is important for the behaviour of photovoltaic cells and at the time of writing is a challenging problem.

Light also produces many interesting phenomena in chalcogenide glasses. One is photodarkening (PD). First discovered in As_2S_3, and found in many chalcogenide glasses, PD is a movement of the optical absorption edge to lower frequency on irradiation with photons of band gap energy. Band gap light on these glasses also induces both irreversible and reversible photo-structural (PS) changes which have been detected by x-ray and neutron diffraction and by EXAFS and Raman spectroscopy; the volume also alters and changes in SRO are implicated. Yet another phenomenon is photodoping or photodissolution in which films of silver or silver-containing materials deposited on chalcogenide glass disappear or dissolve when they are exposed to light. Technical processes including imaging and photoresist processing have been based on this. All these light-induced responses require considerably more research before they are fully understood.

One of the stimuli for research in chalcogenide glasses in the seventies was their ability to exist in high- and low-current-carrying states between which they could be switched reversibly with great rapidity, $\sim 10^{-10}$ s. In the high-resistance low-current phase the field might rise to $\sim 10^5$ V cm^{-1} before the rapid transition to a high-current lower-voltage phase is made. There is a large scientific and technical literature on this phenomenon and quite complex chalcogenides such as the Te–As–Si–Ge system have been studied.

Before leaving solid amorphous systems for liquids it should perhaps be re-emphasised that the preceding sections sample an exceedingly active field in which agreement has not been reached by competent authorities on a number of issues.

11.11 Compound-forming liquid alloys

Chapter 6 showed that simple metals and alloys can be described very reasonably by the NFE model using pseudopotentials and a hard-sphere reference system. It was remarked in §6.9 that there are systems in which bonding or negative-ion formation is suggested by the properties and in which the electron–ion interaction looks too strong for the NFE theory to deal with. A fundamental difficulty in constructing theories of these phenomena is that, by definition, the materials are not metals nor ionic materials or molecular fluids in the full senses of those expressions. They are intermediate, or are in the process of undergoing metal–insulator transitions as the concentration is varied. *A priori* it is not obvious whether an advance on the NFE theory will serve or whether it should be abandoned in favour of the TBA. Both approaches are in use and in the following introduction the first will be emphasised in relation to Li–Pb and the second with regard to Cs–Au—which is not to suggest that either approach is out of the question with any of these compound-forming alloys. One of the problems it to understand when, and when not, to employ the languages of covalent, metallic or ionic bonding.

To take Li–Pb first: it was discovered by Nguyen and Enderby (1977) that liquid alloys of the Li–M type behaved as shown in figure 11.24. Accompanying the high peak of resistivity are other phenomena which must be related to it, e.g. a kink or even reversal of sign in the thermopower and, in Li_4Pb, a deep negative minimum in $\partial\rho/\partial T$.

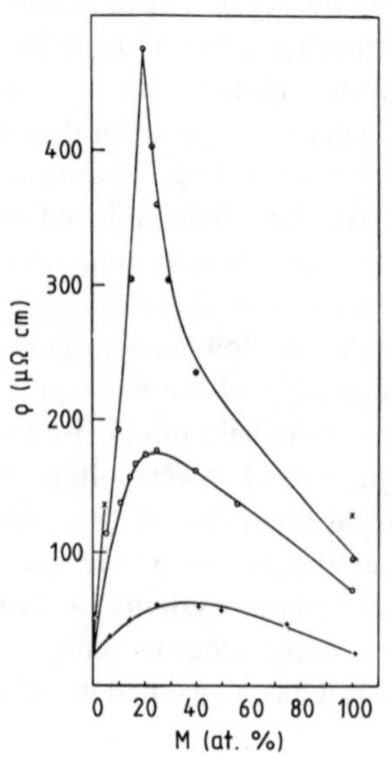

Figure 11.24 Resistivity versus composition. ×, Li–Bi; ○, Li–Pb; □, Li–Tl; +, Li–Mg. (From Nguyen and Enderby 1977.)

It is not only electron transport which shows special features at a particular concentration. There are accompanying anomalies or extrema in thermodynamic properties such as ΔV or $S_{cc}(0)$ which depart from their ideal form of solution (§6.9), and in the Knight shift, the structure and magnetic behaviour. There are a number of systems which are similar to, though not exactly the same as, Li–M. They include alloys of other alkali metals with other members of Group IV (e.g. Na–Sn, Cs–Pb), Mg–Bi, Tl–Te. The significant features are that there is a special composition (Li_4Pb, Mg_3Bi_2, Tl_2Te etc), or perhaps two compositions, at which the anomalies appear most intense and that these compositions have stoichiometric proportions which suggest some chemical relation between the components. There are usually crystalline compounds of the same composition. In fact these liquids are often called compound-forming alloys, though that begs the question. There are so many cases known that complete coverage here is impossible; a review and references are given by van der Lugt and Geertsma (1984). We take as our example Li–Pb.

It was suggested early on that what happens is perhaps the formation of complexes or associations of chemically determined composition which are most in evidence when the overall concentration conforms to the stoichiometric proportions, e.g. Li_4Pb. These complexes are not to be thought of as stable molecules; they are generally supposed to have some finite lifetime. They represent chemical SRO where the electron states are locally modified with the result that some electrons may be withdrawn completely from the conduction process thus increasing the resistivity—an effect intensified by electron scattering off the complexes. Heating breaks up the complexes and a large negative value of $\partial\rho/\partial T$ results. Clear expositions of that approach can be found in Ratti and Bhatia (1975) and Hoshino and Young (1980) where, after a certain amount of parameter fitting, the resistivity of Li–Pb in figure 11.24 can be reproduced.

In any such treatment the concentrations of Li, Pb and Li_4Pb complexes have to be known. In an influential paper, Bhatia and Hargrove (1974) showed how a chemical thermodynamic model will provide for this. All the thermodynamic properties follow from the assumptions of the model which is a generalisation of that represented by equation (6.44). An essential concept is the complex of atoms, $A_\mu B_\nu$, assumed to form with lifetime short compared with that of a stable molecule and to be in equilibrium with its decomposition products A and B. μ, ν are small integers and $A_\mu B_\nu$ influences by its presence the $g_{ij}(r)$ of the system and therefore also the thermodynamic properties which can be calculated from $g_{ij}(r)$ in principle (§5.2). It is supposed that the complexes have a Gibbs free energy of formation $g(T)$ per mole. It is not certain that any such compounds will resemble either a crystal structure because there is no requirement for LRO or a gas molecule because the effective interatomic forces will be modified by the condensed environment. Generalising equation (6.44) we have for the total free energy of a binary mixture of N moles in volume V

$$G = \sum_{i=1}^{3} n_i G_i^{(0)} + RT \sum_{i}^{3} n_i \ln \varphi_i + \tfrac{1}{2} \sum_{i \neq j}\sum n_i \varphi_j \chi_{ij} \tag{11.28a}$$

$$\sum_{i=1}^{3} n_i G_i^{(0)} = NcG_1^{(0)} + N(1-c)\,G_2^{(0)} - n_3 g \tag{11.28b}$$

where the terms have the following meanings: $i = 1, 2, 3$ denote atoms A and B and complex $A_\mu B_\nu$. As in §6.9, χ_{ij}, φ_i are, respectively, an interaction energy and the quantity $n_i v_i/V$ where n_i is the number of moles and v_i the partial molar volume of species i. c is the fraction of A atoms in the mixture as a whole. $G^{(0)}$ is the free energy per mole of a pure constituent. If $g = 0$, $i = 3$ becomes irrelevant and the equation reduces to (6.44). The condition for chemical equilibrium is

$$\left(\frac{\partial G}{\partial n_3}\right)_{T,p,N,c} = 0. \tag{11.29}$$

Although no detailed demonstration is given here, equations (11.28) and (11.29) lead to formulae for all the thermodynamic mixing properties ΔH, ΔS, $S_{cc}(0)$ etc in terms of χ_{ij}, v_i, g, μ and ν. The deductions are fairly complicated and can be found in Bhatia and Hargrove (1974) and Gray *et al* (1980). Fitting such a model to observed values of all the thermodynamic properties requires plausible suppositions as to the values of μ and ν, e.g., the smallest group of correct stoichiometry could be chosen. After injecting any further assumptions or information from elsewhere, the fitting gives χ_{ij} and g from which the $n_i(c)$ can be calculated. $n_i(c)$ can then be used in electron transport calculations if required. As shown in the references quoted, such fits can be quite successful but they are subject to the comments that there are rather many parameters to fit, that they do not establish the reality of the complexes and that they are phenomenological only.

Any SRO ought to be demonstrable by structure or motion studies of the kind discussed in Chapters 3 and 8. Li_4Pb can be made into a zero alloy—see §3.11—using ^7Li. In an original application of this to inelastic scattering, Solwisch *et al* (1983) measured $S_{cc}(Q, \omega)$ and inferred the diffusion coefficients of Li and Pb and the lifetime of concentration fluctuations. $S_{cc}(Q)$ has its main peak at about 1.8 Å^{-1} at which value the lifetime of the concentration fluctuation has a pronounced maximum of about 1.5 ps. If the peak in $S_{cc}(Q)$ is held to be evidence of complexes then they would have the enhanced lifetime. But other studies lead to the inference that there is strong and relatively long-ranged SRO in Li_4Pb, though decreasing as T rises, and this would be compatible with a somewhat salt-like structure with A atoms preferring B neighbours and vice versa (Ruppersberg and Reiter 1982). The enhanced lifetime would then characterise this. It cannot be said that $S_{cc}(Q)$ and $S_{cc}(Q, \omega)$ studies confirm the hypothesis of long-lived localised complexes but by combining the thermodynamic theory of complexes with the observation of R_Q (see §8.6) Elwenspoek *et al* (1983) inferred a lifetime of about 10^{-11} s for complexes in InSb.

We now draw equations and concepts from previous chapters and apply them to the Li_4Pb problem in an attempt to make an advance on the hard-sphere pseudopotential approach of Chapter 6. The aim is to derive the structure and thermodynamic properties. The calculation is typical of approaches being made in the 1980s to problems of chemical SRO and is due to Copestake *et al* (1983), Hafner *et al* (1984) and Pasturel *et al* (1985).

The pairwise term of equation (5.2) can easily be generalised to a binary system making the structure-dependent energy per particle:

$$U_{\text{str}} = 2\pi n_0 \sum_{i,j=\text{A,B}} c_i c_j \int_0^\infty g_{ij}(r)\varphi_{ij}(r) r^2 \, \mathrm{d}r. \tag{11.30}$$

To obtain the partial structure factors the MSA of §5.7 is invoked and $\varphi_{ij}(r)$ is chosen to be that of hard spheres with screened Coulomb tails, thus

$$\varphi_{ij} = \infty, \qquad r < \sigma \tag{11.31a}$$

$$= \frac{Q_i Q_j e^2}{r} \exp(-\kappa r) \qquad r > \sigma. \tag{11.31b}$$

Really the bare ions and their electron clouds have a complicated interaction, probably especially so at the stoichiometric concentration. The model can be looked at as intermediate between the neutral pseudoatoms of §6.4 and a molten salt in which one or more electrons are wholly transferred, leaving positive and negative ions with a Coulomb interaction. Here we have what could be called pseudoions and all their interactions are modelled by the screened potentials. Q_1, Q_2 and κ are parameters but if the Q are looked on as effective ionic charges then it becomes plausible to impose charge neutrality by requiring

$$c_A Q_A + c_B Q_B = 0 \tag{11.32}$$

and this is what is done.

The MSA equations for a binary system are then

$$c_{ij}(r) = -\beta\varphi_{ij}(r) \qquad r > \sigma \tag{11.33a}$$

$$g_{ij}(r) = 0 \qquad r < \sigma. \tag{11.33b}$$

In addition there is the Ornstein–Zernike equation (5.23) which generalises to

$$g_{ij} - 1 = h_{ij}(r) = c_{ij}(r) + \sum_k n_k \int c_{ik}(\mathbf{r}')h_{kj}(\mathbf{r} - \mathbf{r}')\, d\mathbf{r}'. \tag{11.34}$$

It is useful now to form linear combinations of the φ_{ij}:

$$\varphi_{cc} \equiv c_A c_B(\varphi_{AA} + \varphi_{BB} - 2\varphi_{AB}) \tag{11.35a}$$

$$\varphi_{NN} \equiv c_A^2\varphi_{AA} + c_B^2\varphi_{BB} + 2c_A c_B\varphi_{AB} \tag{11.35b}$$

$$\varphi_{Nc} \equiv 2c_A c_B[c_A\varphi_{AA} - c_B\varphi_{BB} + (c_A - c_B)\varphi_{AB}]. \tag{11.35c}$$

If $\varphi_{AA} + \varphi_{BB} > 2\varphi_{AB}$ around the nearest-neighbour distance, unlike neighbours will be preferred (ordering); conversely, like neighbours will be favoured (clustering). φ_{cc} is therefore called an ordering potential and implies ordering when positive. φ_{NN} is a weighted average potential. The suffixes are those of the Bhatia–Thornton structure factors of §3.12. As a consequence of equation (11.32), φ_{NN} vanishes outside the core and $\varphi_{cc} \propto (Q_1 - Q_2)^2$.

The $g_{ij}(r)$ and $S_{ij}(Q)$ are mutual Fourier transforms as in equation (3.15). Similarly it is possible to relate S_{NN}, S_{Nc} and S_{cc} to pair distributions as follows (with the arguments omitted):

$$
\begin{aligned}
g_{\mathrm{NN}} &\equiv 1 + (8\pi^3 n_0)^{-1} \int (S_{\mathrm{NN}} - 1) \frac{\sin Qr}{Qr} 4\pi Q^2 \,\mathrm{d}r \\
&= c_{\mathrm{A}}^2 g_{\mathrm{AA}} + c_{\mathrm{B}}^2 g_{\mathrm{BB}} + 2c_{\mathrm{A}} c_{\mathrm{B}} g_{\mathrm{AB}} \\
g_{\mathrm{Nc}} &\equiv (8\pi^3 n_0 c_{\mathrm{A}} c_{\mathrm{B}})^{-1} \int S_{\mathrm{Nc}} \frac{\sin Qr}{Qr} 4\pi Q^2 \,\mathrm{d}Q \\
&= c_{\mathrm{A}}(g_{\mathrm{AA}} - g_{\mathrm{BB}}) - c_{\mathrm{B}}(g_{\mathrm{BB}} - g_{\mathrm{AB}}) \\
g_{\mathrm{cc}} &\equiv (8\pi^3 n_0)^{-1} \int \left(\frac{S_{\mathrm{cc}}}{c_{\mathrm{A}} c_{\mathrm{B}}} - 1 \right) \frac{\sin Qr}{Qr} 4\pi Q^2 \,\mathrm{d}Q \\
&= c_{\mathrm{A}} c_{\mathrm{B}} (g_{\mathrm{AA}} + g_{\mathrm{BB}} - 2g_{\mathrm{AB}}).
\end{aligned}
\tag{11.36}
$$

In each case the second line follows from equations (3.30) and (3.15). After this, a little more algebra and the use of equation (3.31) transform the expression for the pairwise energy into

$$U_{\mathrm{str}} = 2\pi n_0 \int_0^\infty (g_{\mathrm{NN}}\varphi_{\mathrm{NN}} + g_{\mathrm{Nc}}\varphi_{\mathrm{Nc}} + g_{\mathrm{cc}}\varphi_{\mathrm{cc}}) r^2 \,\mathrm{d}r. \tag{11.37}$$

These transformations lead to a useful expression for the ordering energy

$$\Delta H_{\mathrm{ord}} \equiv 2\pi n_0 \int_0^\infty g_{\mathrm{cc}}\varphi_{\mathrm{cc}} r^2 \,\mathrm{d}r$$

which is one contribution to the heat of mixing and which differs from zero only if the ordering potential is not zero. The other contribution, ΔH_{eg}, comes from the structure-independent electron gas energy which is a function of electron density and therefore of both concentration and any change of volume, ΔV, on mixing. ΔH_{eg} can be calculated in terms of r_s (see §6.6) so the measured ΔV, which is negative and considerable, with a minimum near Li_4Pb, is needed for this.

The remaining problem is to calculate the partial pair distributions. It was mentioned in §5.7 that exact solutions are sometimes possible in the MSA. The present is a case in point though we refer the reader to the references for the proof. What is needed is a solution of the three integral equations (11.34), given also equations (11.32) and (11.33). If c_{ij} and h_{ij} are transformed into c_{NN} etc by linear equations analogous to (11.36), it turns out that they can be obtained analytically and in particular that $h_{\mathrm{Nc}} = c_{\mathrm{Nc}} = \varphi_{\mathrm{Nc}} = 0$. ΔH_{ord} can therefore be evaluated, but only in terms of the parameters Q_1, Q_2, κ and σ in the potentials. The core diameter can be obtained from the volume and packing fraction η. The latter is interpolated between its values for the pure components which are in turn found from equation (4.15) and the measured entropies of Li and Pb (see, e.g., §6.8). κ is made arbitrarily but plausibly of a size to screen off φ_{cc} effectively outside the nearest-neighbour distance. Finally the Q are settled by fitting $(\Delta H_{ord} + \Delta H_{eg})$ to the observed value of ΔH.

The calculation thus requires two empirical functions, $\Delta V(c)$ and $\Delta H(c)$. From the now fully determined theory, ΔS and the structure factors can be deduced. Some results are shown in figure 11.25 and show that in spite of its manifest simplifications—such as having the same value of σ for Li and Pb—the theory is qualitatively satisfactory.

The reader can find other discussions of Li–Pb alloys in the literature: it is not yet a solved problem (Copestake *et al* 1983, Hafner *et al* 1984, Pasturel

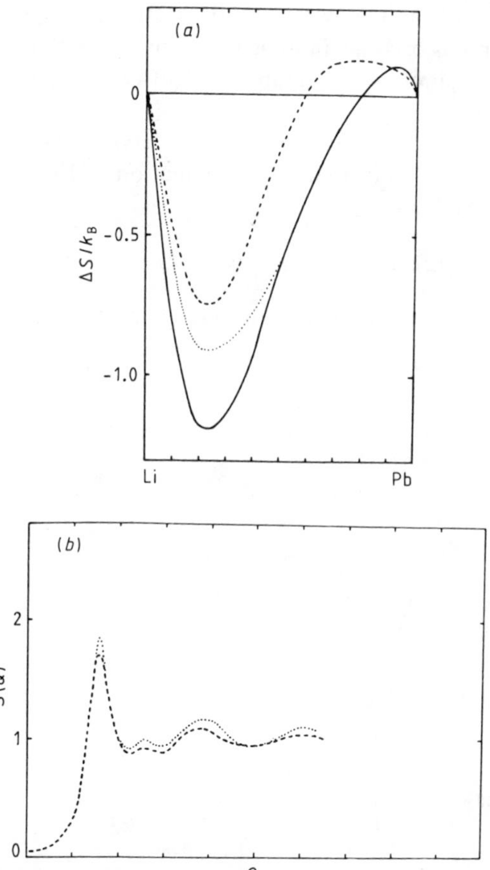

Figure 11.25 (*a*) Entropy of mixing of Li–Pb. Broken curve, observed at 932 K; full curve, theory using observed ΔH; dotted curve, theory assuming $Q_1 - Q_2$ is constant. (*b*) Total neutron structure factor $S(Q)$ for $Li_{0.62}Pb_{0.38}$. Dotted curve, observed at 775 K; broken curve, theory for 1000 K. (From Hafner *et al* 1984.)

et al 1985). The parameters Q_1 and Q_2 are not of course equal to 1 and 4, respectively, for Li and Pb; Copestake *et al* in their discussion suggest 0.5 and -2.0 at Li_4Pb which implies a transfer of electronic charge from Li to Pb and leads to mutual attraction. At the time of writing charge transfer is not a very fully clarified idea and the assumption in equation (11.32) is not wholly compelling. No doubt more understanding of the actual distribution of electronic charge and the more complete theories will evolve together. They should facilitate an acceptable explanation of figure 11.24. Computer simulations will also help (Jacucci *et al* 1985).

Cs_xAu_{1-x} is perhaps even more striking in its behaviour as shown by figure 11.26(*a*). The precipitous dip in conductivity which is repeated in thermopower, together with observations that CsAu is diamagnetic, has a very small NMR shift and shows electrolysis, strongly suggest that CsAu at this composition has become an ionic insulator. Au with other alkali metals is not necessarily similar—see figure 11.26(*b*) (Schmutzler *et al* 1976, Nicoloso *et al* 1978).

A model with the TBA is obviously worth trying and was developed in detail by Franz *et al* (1980, 1982). From equation (7.18) or by analogy with equation (7.26) we have

$$G_{ii}(z) \equiv \langle i|(z-H)^{-1}|i\rangle = (z-\varepsilon_i-\Delta_i)^{-1}$$

where z is the complex energy, ε_i is the energy level of an electron in orbital

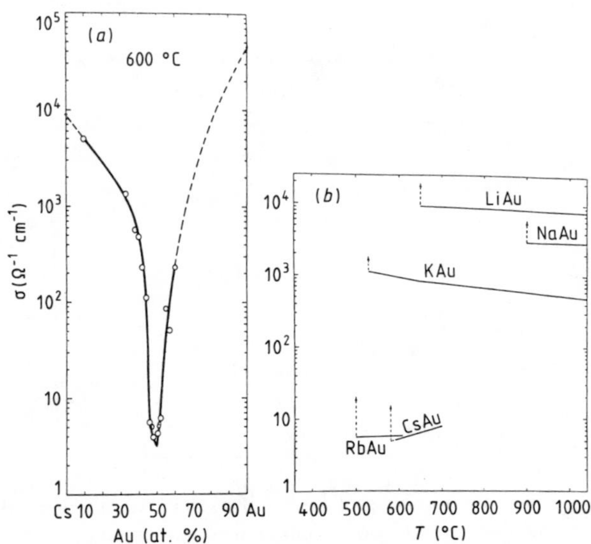

Figure 11.26 (*a*) Conductivity of liquid Cs–Au (from Schmutzler *et al* 1976). (*b*) $\sigma(T)$ for liquid Au–alkali-metal alloys (from Nicoloso *et al* 1978).

$|i\rangle$ on the ith site, and Δ_i is a self-energy. This is formally exact but Δ_i has to be expressed as an expansion or continued fraction which cannot be evaluated without fairly drastic assumptions. The simplification used, that of the Cayley tree, is explained in the references cited and amounts to treating the system as one in which every atom is attainable from every other atom by nearest-neighbour steps forming paths that do not self-intersect. One consequence is that the bandwidth is significantly smaller than in a TBA for a complete lattice. From the approximate G_{ii}, $g(E)$ is found by equation (7.24*b*).

For a binary system of A, B atoms, G_{ii} has to be averaged using

$$G_{ii} = cG_{ii}^{(A)} + (1-c)G_{ii}^{(B)}.$$

A system like CsAu is interesting because of its ionicity and the charge transfer this implies and, as it is a liquid, there will be some degree of disorder; these factors must be allowed for. The disorder is specified by a SRO parameter η (see §2.7) chosen to minimise the free energy for a given concentration; this requires a calculation of $U - TS$ and therefore an assumed expression for S. The effect of charge transfer is to change ε_i from its isolated-atom value ε_i^0 to $\varepsilon_i^0 + \Delta\varepsilon_i(c, \eta)$. As a first approximation ε_i^0 is used to calculate the local density of states at A or B sites and the charges can be found by integrating this over energy. This is used as a first step to recalculate ε_i by iterating for self-consistency taking the Coulomb interaction of the displaced charge into account.

This elegant but complex scheme led to several conclusions including the location of maximum SRO at $c = 0.5$, irrespective of temperature, accompanied by a deep minimum in $g(E_F)$. At concentrations away from 0.5, $g(E_F)$ was much higher. Using the RPM (§7.7) the conductivity was calculated and fell catastrophically at 50 at. %. This is all consistent with the onset of ionic bonding at the composition CsAu as revealed by experiments. It is no accident that these extrema coincide. The maximum SRO corresponds to salt-like behaviour (§3.11) and the self-avoidance on the part of the like ions reduces overlap of their orbitals and narrows the energy bands inducing a gap or pseudogap.

The success of this was repeated at least qualitatively for the other alkali–Au systems and later extended to cover compound-forming alloys like Li_xPb_{1-x} or Cs_xSb_{1-x} in which other than s states had to be assumed, e.g., s and p states in Pb. Since Li–Pb remains a metal even at Li_4Pb the application of the TBA is to some extent suspect. Nevertheless these methods were able to find a SRO parameter which implied high local order, high resistivity and a sharp density of states minimum at Li_4Pb. Presumably at this stoichiometric composition the charge transfer, though significant, is not complete and the mixture just fails to become ionic.

The details of this theory will not be taken further but it should be added that in spite of the notable success of both the pseudopotential and the TB approaches the subject is still open at the time of writing. It is possible for example to argue that, to a first approximation, the alkali cations simply keep the others apart and it is the interactions of anions (Au, Sb etc) that have the greater significance. In some equiatomic liquid mixtures such as NaSn and KPb the anions may form tetrahedral clusters as they are known to do in the crystalline form. In Au-containing systems the d states may have a role. These matters can be pursued in Holzhey *et al* (1982) and Geertsma and Dijkstra (1985).

11.12 Expanded liquid metals

It is a highly interesting fact that liquid semiconductors can be created by inducing pure metals to undergo a metal-to-non-metal transition (MNMT). Although the experiments chiefly concern Cs and Hg the number of observations is considerable and by the mid-1980s has provoked interesting theoretical ideas but not yet a generally accepted explanation. Figure 11.27(*a*) shows why semiconductors must exist. The path pq near the triple point implies a change from liquid metal to insulating vapour. p to q across the saturated vapour curve traverses a first-order phase transition; p to q by the long route round the critical point does not, although somewhere along the line a MNM transition must occur. At X the liquid is less dense than at the triple point and conceivably the line XY is the locus of points where the density is low enough for metallic binding to cease and semiconduction to set in. A priori the transition may be sharp or gradual but it may not occur there at all. An alternative possibility is the line LM in the vapour phase if the density is high enough for metallic properties to appear above LM. A third possibility is that change occurs where the path pq crosses the line CC′ which is the prolongation of the curve SVP connecting points where the density equals the critical density (i.e. the critical isochore).

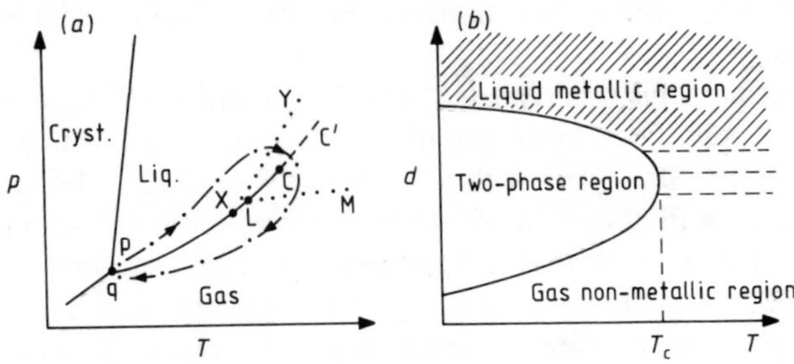

Figure 11.27 Schematic phase diagram: (*a*), *p–T* plane showing triple and critical points and possible transition loci; (*b*), *d–T* plane showing isochores (broken lines) and hypothetical division into metallic and non-metallic regions. T_c is the critical temperature.

There is an obvious way to explore the matter experimentally. All we need do is to attach electrodes to a liquid metal sample and measure its conductivity while simultaneously measuring its density as a function and T and p. Lines of constant conductivity could be drawn on figure 11.27(b) and the regions of high (metallic) and low (semiconducting) conductivity identified. This can now be done: the liquids are expanded by heating and prevented from boiling by pressure and the conductivity falls dramatically. But about two decades of experimental development were needed to overcome the technical difficulties caused by the very high critical temperatures and pressures of metals, their chemical reactivity, and the divergence of the compressibility and specific heat which occur at the critical point and make thermal stability in the neigbourhood very problematical (see table 11.2).

Table 11.2

	p_c (MPa)	T_c (K)	d_c (gm cm^{-3})	d_{TP} (gm cm^{-3})
Rb	12.5	2017	0.292	1.48
Cs	9.3	1924	0.379	1.85
Hg	167	1760	6	13.6

A limited selection of important facts about expanded Hg and Cs is as follows. Theory and experiment up to about 1982 with all the important references can be found in *Prog. Theor. Phys. Suppl.* 1982 **72** 1 and later references are given below. The equation of state and location of the critical point of Cs is shown in figure 11.28 (Jüngst *et al* 1985, Hensel *et al* 1986). There is comparable data for Hg (Schönherr *et al* 1979). Figure 11.29 shows how three electronic properties vary with density in Hg. The rise of R_H, the fall in σ and especially the vanishing of the Knight shift (Warren and Hensel 1982) which is proportional to $g(E_F)$, all indicate that at $d \simeq 9$ gm cm^{-3} liquid Hg ceases to be a metal. Since $d_c \simeq 5.75$ gm cm^{-3}, the transition region—not a sharp line—follows XY in figure 11.27(a) not CC′ or LM.

If it is true that Hg is a semiconductor for densities less than about 9 gm cm^{-3} some kind of activated conduction would be anticipated. In fact experiments show that in the range $6 \lesssim d \lesssim 8$ gm cm^{-3}

$$\sigma = \sigma_0 \exp(-\Delta E/k_B T) \tag{11.38a}$$

$$\alpha = -\frac{k_B}{|e|}\left(\frac{\Delta E}{k_B T} + \text{const}\right). \tag{11.38b}$$

This gives good ground for believing that low-density liquid Hg is a liquid semiconductor, but the quantities σ_0 and ΔE are strong functions of pressure

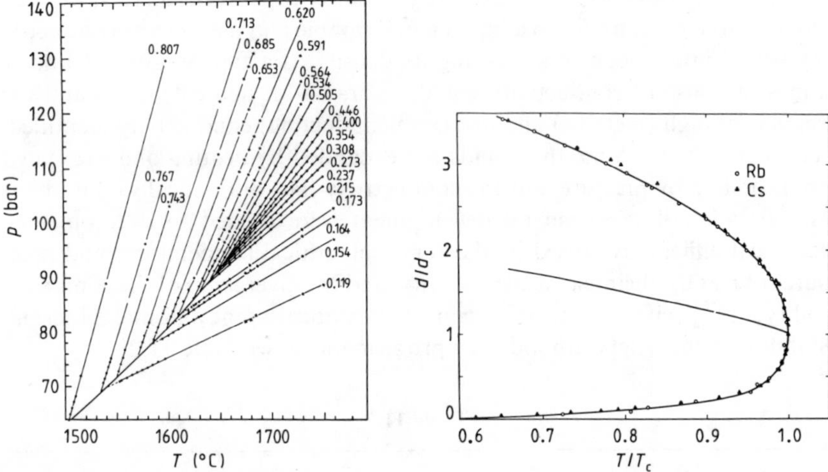

Figure 11.28 Equation of state data in expanded Cs and Rb. Note that in the d–T plane the 'rectilinear diameter', $\frac{1}{2}(d_{\text{liq}} + d_{\text{gas}})$, is not straight near the critical point. The Cs isochore labels are in gm cm^{-3}. ●, from isochoric measurements; ×, from isothermal measurements. (From Hensel *et al* 1986, Jüngst *et al* 1985.)

and temperature and the situation is more complicated than in a semiconductor of fixed density. Semiconducting behaviour occurs in Cs also but sets in at $d \sim d_c$. This already indicates a difference in the behaviours of Cs and Hg and by adducing more experimental facts it is possible to speculate about this difference with only a short preliminary excursion into MNMT concepts.

When a divalent metal is sufficiently expanded we expect its overlapping energy bands to draw apart. In a crystal a real energy gap appears and

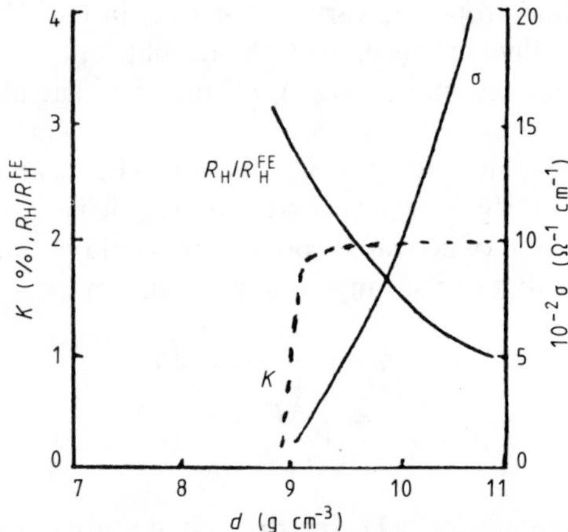

Figure 11.29 Hall coefficient (R_H), conductivity (σ) and Knight shift (K) in expanded liquid Hg.

widens as the density falls. In a liquid the bands could be calculated by one of the Green function methods in Chapter 7 and the disorder replaces the sharp band edges by tails. The tails overlap in the region of E_F and it is probable that the $g(E)$ is never precisely zero—a circumstance called by Mott a pseudogap (see figure 9.8). The pseudogap will be more or less pronounced depending on density and the states in it may or may not be localised by the disorder. If the pseudogap is what is giving rise to ΔE in equation (11.38)—which seems entirely plausible—and if $g(E_F)$ is substantial, the states will presumably be localised so there is a mobility gap (§9.10). ΔE can then be identified with $(E_c - E_F)$ because α is strongly negative signifying electrons as carriers. But it is possible that $g(E)$ is negligible in the gap and in that case the optical properties ought to be compatible with the opening up of an energy gap as $d \to 9$ g cm^{-3}. Figure 11.30 shows the real part, $\sigma(\omega)$, of the optical conductivity measured by reflectivity in the visible at various densities (Ikezi *et al* 1978). Dense liquid Hg has a behaviour close to that of equation (6.63) showing a Drude-like $\sigma(\omega)$ deriving from the free electrons. This gradually changes with density until by 9 gm cm^{-3} the curve is characteristic of materials with energy gaps. A demonstration of the compatibility of these results with a real gap in $g(E)$ was given by Krohn and Thompson (1980). The optical evidence and the low value of the Knight shift together support the view that $g(E)$ in the gap is actually very low for $d \lesssim$ 9 g cm^{-3}. At even lower densities, down to about 1.0 gm cm^{-3} in the vapour, optical absorption edges are observed with forms compatible with equation (11.6). A theory of these edges has been provided by Bhatt and Rice and others, though not without further hypotheses about the existence

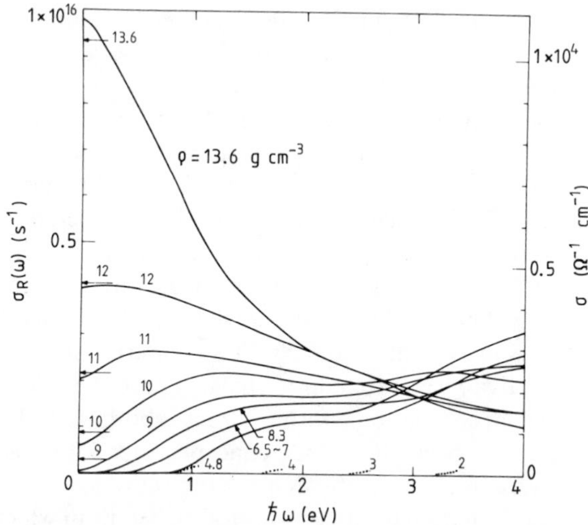

Figure 11.30 Real part of $\sigma(\omega)$, the optical conductivity, in liquid Hg at various densities. (From Ikezi *et al* 1978.)

in the vapour of atom clusters with a distribution of sizes (Bhatt and Rice 1979, Uchtmann *et al* 1981).

Attempts to calculate $g(E)$ for expanded Hg have been made and certainly show a gap opening up at low density between states derived from 6s and 6p electrons but with no very close correlation with observations.

The preceding paragraph has the tacit assumption that in whatever way $g(E_F)$ falls with density the system can be treated as homogeneous. But when $g(E_F)$ and σ are sensitive to density—as they are in expanded liquid metals—density fluctuations may become significant. To put the matter in an extreme form: if the system near the MNMT consists of a random array of small volumes, some metallic and some semiconducting as determined by their local density, the observed overall σ will depend on the percolation of current through the more highly conducting regions. A very interesting theory on this basis was devised in 1974 by Cohen and Jortner for the MNMT in liquid Hg and the EMT equations of §9.4 found a valuable application there. The development of metallic properties as liquid Te is heated can be discussed in similar terms (Cohen and Jortner 1976). Concentration fluctuations in binary systems may have a comparable role in metal–insulator transitions in liquid semiconductors and theories of this kind have been put forward for Ga_2Te_3, the Se–Te system, In_2Te_3 and other systems, e.g. by Tsuchiya and Seymour (1985).

The difficult experiments in expanded Cs show that for densities down to about twice the critical density (d_c), the conductivity and Hall effect can be described by the NFE theory of metals but near d_c semiconducting behaviour sets in. The problem is to understand by what mechanism the MNMT occurs and Cs is probably more complicated than Hg because at least three different concepts seem relevant. One is localisation by disorder which can in principle be contained in a one-electron theory (Chapter 9), though not with the Anderson model itself because the relevant disorder is certainly the topological kind with the off-diagonal randomness referred to in §9.8. Then there is the significant fact that alkali metals have half full bands and a continuously expanded crystal would be expected to become a semiconductor through the operation of electron correlations as invoked in the theory of the Mott transition (Mott 1974). A role for conduction through percolation channels which are blocked at the MNMT is also conceivable and the theory ought to take into account the existence at high temperature and low density of neutral or ionised dimers or other clusters. It is very difficult to design a theory which takes in everything and clarifies the part, if any, played by each mechanism. Conceivably all mechanisms have interacting roles. The localisation and the electron correlation approaches have both led to highly interesting theories, each ignoring the other concept, and both succeeding to a considerable degree in deriving an MNMT at the right density and predicting a notable peak in the paramagnetic susceptibility per atom which was discovered experimentally by Freyland. A valuable description of the properties was given by Freyland (1981) and theories of both kinds can be found in papers by Rose (1981), Franz (1984) and Warren and Matthiess (1984).

To give the flavour of some of the arguments we recall figure 3.6 and the proposition that as liquid Ar approaches its critical density, it is the coordination number (n_1) that decreases not the nearest-neighbour distance that increases. Neutron diffraction from expanded Rb suggests the same for liquid metals. The preceding section indicated that n_1 and the transfer integral t_{ij} were important parameters in the TBA applied to liquid Cs–Au when it was discussed in the Green-function–Cayley-tree approximation. Suppose therefore that low-density Cs is regarded as a Cs–X alloy in which X is simply absent—then only one (constant) transfer integral and the variable n_1 are involved. The density takes the place of the concentration through an effective concentration $x_e \equiv \langle n_1 \rangle / n_1(\max)$, $n_1(\max)$ being the highest value of n_1, ~ 9 for liquid Cs. If the Green function is calculated by the Cayley-tree method, with n_1 lying randomly between 0 and $n_1(\max)$, then ensemble averages of the spectral density and $g(E)$ become available and so does σ through an equation of the form (7.43a). This was pursued in detail by Franz and among the numerous results were the conductivities shown in figure 11.31. The decrease of σ is caused by the opening up of a pseudogap, accompanied by localisation of states, at E_F; this occurs at about d_c. There is qualitative agreement with experiment but the extra conductivity observed at the lowest densities is serious and may be due to free electrons from ionised Cs_2 molecules.

The reader is referred to the theoretical papers for further details such as the density variation of $g(E_F)$ per atom which may account for the peak in the paramagnetism. However, the latter is not the only spin effect to have been observed. The Knight shift and susceptibility combined give the quantity ζ which is the ratio of the charge density of electrons of Fermi energy to the charge density in the free atom, both densities taken at the nucleus. ζ might be expected to tend to unity in the low-density limit but in fact it reaches a plateau of about 0.5 and then falls for $d \lesssim 1.3$ g cm^{-3}. It seems hard to explain this by a one-electron model and some authorities hold that susceptibility, Knight shift and allied quantities favour a strong role for electron correlations in the MNMT of alkali metal liquids.

The intrinsic difficulty of both experiments and theories in the expanded metal field has meant that progress in understanding has not been rapid, but it has not been without surprises. As a final example we return to Hg. In the dense vapour (below the saturated vapour curve in figure 11.27(a)) Hefner and Hensel (1982) discovered a dielectric anomaly at a density of about 3 g cm^{-3}. It consisted of a sharp peak in the real part of the dielectric constant as the pressure was increased over a small range near 1400 bar at constant temperature. It appears that before the vapour can be compressed into a liquid near the critical point it first undergoes some other transition sometimes called a plasma transition. Additional optical absorption at

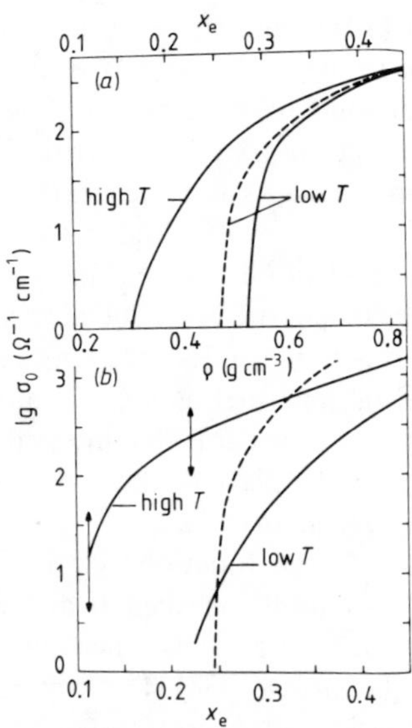

Figure 11.31 Observed and calculated conductivity in expanded liquid Cs. (*a*), calculated; (*b*), observed. High T refers to liquid Cs. Low T refers to different but related experiments in solid Cs–Xe mixtures. $x_e \equiv$ mean actual number of nearest neighbours divided by maximum number. (From Franz *et al* 1980.)

infrared frequencies sets in at the transition. An explanation suggested by the discoverers was that the transition was from a slightly ionised homogeneous vapour to an inhomogeneous phase containing dense charged droplets. At the time of writing neither the theoretical nor the experimental situation is fully resolved and there exists an interesting discussion of the two transitions in Hg (one in the liquid and one in the vapour) on the basis of a new hypothesis: that fluid Hg at high temperatures and pressures is an excitonic insulator (Turkevich and Cohen 1984).

This section has referred to the use of high temperatures and pressures to turn liquid metals into semiconductors. Although the matter will not be discussed here, the comparable phenomenon of converting liquid semiconductors into metals has also been studied. Much information about structural, thermodynamic, electronic and magnetic changes with temperature and pressure in molten Se, Se–Te, transition-metal mixtures, etc can be found in Endo (1983), Takeda *et al* (1985), and the Liquids Section of *J. Non-Cryst. Solids* 1985 **77/78** 1343ff.

References

Apling A J, Leadbetter A J and Wright A C 1977 *J. Non-Cryst. Solids* **23** 369

Austin I, Nashashibi T S, Searle T M, Le Comber P G and Spear W E 1979 *J. Non-Cryst. Solids* **32** 373

Bar-Yam Y and Joannopoulos J D 1985 *J. Non-Cryst. Solids* **77/78** 99
Beigelsen D K 1980 *Sol. Cells* **2** 421
Beyer W and Overhof H 1979 *Solid State Commun.* **31** 1
Beyer W and Stuke J 1974 in *Amorphous and Liquid Semiconductors* ed J Stuke and W Brenig (London: Taylor and Francis)
Bhat P K, Rhodes A J, Searle T M and Austin I G 1983 *Phil. Mag.* B **47** L99
Bhatia A B and Hargrove W H 1974 *Phys. Rev.* B **10** 3186
Bhatt R N and Rice T M 1979 *Phys. Rev.* B **20** 466
Bishop 1974 *Phys. Rev.* B **12** 1567
Bouldin E, Stern E A, van Roedern B and Azoulay J 1984 *J. Non-Cryst. Solids* **66** 105
Busse L E 1981 *Phys. Rev.* B **29** 3639
Busse L E and Nagel S R 1981 *Phys. Rev. Lett.* **47** 1848
Cody G D, Tiedje T, Abeles B, Brooks B and Goldstein Y 1981 *Phys. Rev. Lett.* **47** 1480
Cohen J D, Harbison J R and Wecht K W 1982 *Phys. Rev. Lett.* **48** 109
Cohen M H, Economou E N and Soukoulis C M 1984a *Phys. Rev.* B **30** 4493
—— 1984b *J. Non-Cryst. Solids* **66** 285
Cohen M H and Jortner J 1976 *Phys. Rev.* B **13** 5255
Cohen M H, Soukoulis C M and Economou E N 1984c *AIP Conf. Proc.* **120** 371
—— 1985 *J. Non-Cryst. Solids* **77** 171
Copestake A P, Evans R, Ruppersberg H and Schirmacher W 1983 *J. Phys. F: Met. Phys.* **13** 1993
Corb B W, Wei W D and Averbach B L 1982 *J. Non-Cryst. Solids* **53** 29
Cutler M 1977 *Liquid Semiconductors* (London: Academic)
Dersch H, Schweitzer L and Stuke J 1983 *Phys. Rev.* B **28** 4678
Dersch H, Stuke J and Beichler J 1981 *Phys. Status Solidi* B **105** 265
Dunstan D J 1982 *J. Phys. C: Solid State Phys.* **15** L419
Elwenspoek M, Maxim P, Brinkman R, Weihreiter E and Quitman D 1983 *Phys. Lett.* **96A** 435
Emin D 1977 *Phil. Mag.* **35** 1189
Endo H 1983 *J. Non-Cryst. Solids* **59/60** 1047
Franz J R 1984 *Phys. Rev.* B **29** 1565
Franz J R, Brouers F and Holzhey C 1980 *J. Phys. F: Met. Phys.* **10** 235
—— 1982 *J. Phys. F: Met. Phys.* **12** 2611
Freyland W 1981 *Comment. Solid State Phys.* **10** 1
Fritzsche H 1985 *J. Non-Cryst. Solids* **77** 273
Geertsma W and Dijkstra J 1985 *J. Phys. C: Solid State Phys.* **18** 5987
Gray P, Cusack N E, Tamaki S and Tsuchiya Y 1980 *Phys. Chem. Liq.* **9** 307
Grünewald M, Weber K, Fuhs W and Thomas P 1981 *J. Physique Coll.* **42** C4 523
Hafner J, Pasturel A and Hicter P 1984 *J. Phys. F: Met. Phys.* **14** 1137
Hayes T M, Allen J W, Beeby J L and Oh S-J 1985 *Solid State Commun.* **56** 953
Hefner W and Hensel F 1982 *Phys. Rev. Lett.* **48** 1026
Hensel F, Jüngst S, Knuth B, Uchtmann H and Yao M 1986 *Physica* **139/140** B 90
Hickey B J, Morgan G J, Weaire D L and Wooten F 1985 *J. Non-Cryst. Solids* **77** 95
Holzhey C, Brouers F, Franz J R and Schirmacher W 1982 *J. Phys. F: Met. Phys.* **12** 2601
Hoshino K and Young W H 1980 *J. Phys. F: Met. Phys.* **10** 1365
Ikezi H, Schwarzenegger K, Simons A L, Passner A L and McCall S L 1978 *Phys. Rev.* B **18** 2494

Ishida Y, Asano A and Yonezawa F 1980 *J. Physique Coll.* **41** C8 81
Imagawa O, Akiyama T and Shimakawa K 1984 *Appl. Phys. Lett.* **45** 438
Jackson W B 1982 *Solid State Commun.* **44** 477
Jackson W B, Amer N M, Boccara A C and Fournier D 1981 *Appl. Opt.* **20** 1333
Jackson W B, Tsai C C and Kelso S M 1985 *J. Non-Cryst. Solids* **77/78** 281
Jacucci G, Ronchetti M and Schirmacher W 1985 *Proc. 3rd Int. Conf. on the Structure of Non-Crystalline Materials* ed C Janot (Paris: Les Editions Physique)
Joannopoulos J D and Cohen M L 1983 *Proc. R. Soc.* B **7** 2644
Joannopoulos J D and Lucovsky G (ed) 1984 *Hydrogenated Amorphous Silicon* vol 1 (Berlin: Springer)
Jones D I, Spear W E and Le Comber P 1977 *Phil. Mag.* **36** 541
Jüngst S, Knuth B and Hensel F 1985 *Phys. Rev. Lett.* **55** 2160
Kastner M, Adler D and Fritzsche H 1976 *Phys. Rev. Lett.* **37** 1504
Kelly M J and Bullett D W 1976 *J. Non-Cryst. Solids* **21** 155
Krohn C E and Thompson J C 1980 *Phys. Rev.* B **21** 2619
Lang D K, Cohen J D and Harbison J R 1982 *Phys. Rev.* B **25** 5284
Le Comber P G, Jones D I and Spear W E 1977 *Phil. Mag.* **35** 1173
Le Comber P G and Spear W E 1986 *Phil. Mag.* B **53** L1
Linke K, Morán-Lopez J L and Benneman K 1983 *Phys. Rev.* B **27** 7348
van der Lugt W and Geertsma W 1984 *J. Non. Cryst. Solids* **61** 187
Mackenzie K D, Le Comber P G and Spear W E 1982 *Phil. Mag.* B **46** 377
Madan A, Le Comber P G and Spear W E 1976 *J. Non-Cryst. Solids* **20** 250
Marshall J M 1983 *Rep. Prog. Phys.* **46** 1235
—— 1985 *J. Non-Cryst. Solids* **77** 425
Matthiess L F and Warren W W 1977 *Phys. Rev.* B **16** 624
Monroe D 1985 *Phys. Rev. Lett.* **54** 146
Mott N F 1974 *Metal–Insulator Transitions* (London: Taylor and Francis)
Mott N F and Davies E A 1979 *Electronic Processes in Non-Crystalline Materials* (Oxford: Clarendon)
Mueller M and Thomas P 1983 *Phys. Rev. Lett.* **51** 702
Nagels P, Callaerts R and Denayer M 1974 in *Amorphous and Liquid Semiconductors* ed J Stuke and W Brenig (London: Taylor and Francis)
Nguyen V T and Enderby J E 1977 *Phil. Mag.* **35** 1013
Nicoloso N, Schmutzler R W and Hensel F 1978 *Ber. Bunsenges. Phys. Chem.* **82** 621
Okushi H 1985 *Phil. Mag.* B **52** 33
Orenstein J and Kastner M 1981 *Phys. Rev. Lett.* **46** 1421
Overhof H 1984 *J. Non-Cryst. Solids* **66** 261
Pankove J I (ed) 1984 *Semiconductors and Semimetals* vol 21 (London: Academic)
Pasturel A, Hafner J and Hicker P 1985 *Phys. Rev.* B **32** 5009
Payson J S and Guha S 1985 *Phys. Rev.* B **32** 1326
Phillips J C 1986 *J. Appl. Phys.* **59** 383
Price D L, Misawa M, Susman S, Morrison T I, Shenoy G K and Grimsditch M 1984 *J. Non-Cryst. Solids* **66** 443, 1986 *Ibid.* **68** 80
Ratti V K and Bhatia A B 1975 *J. Phys. F: Met. Phys.* **5** 893
Robertson J 1983 *Adv. Phys.* **32** 361
—— 1984 *J. Phys. C: Solid State Phys.* **17** L349
van Roedern B, Ley L, Cardona M and Smith F W 1970 *Phil. Mag.* B **40** 433
Rose J H 1981 *Phys. Rev.* B **23** 552
Ruppersberg H and Reiter H 1982 *J. Phys. F: Met. Phys.* **12** 1311

Schmutzler R W, Hoshino H, Fischer R and Hensel F 1976 *Ber. Bunsenges. Phys. Chem.* **80** 707

Schönherr G, Schmutzler R W and Hensel F 1979 *Phil. Mag.* B **40** 411

Schweitzer L, Grünewald M and Dersch H 1981 *Solid State Commun.* **39** 355

Shapiro F R and Adler D 1984 *J. Non-Cryst. Solids* **66** 303

—— 1985 *J. Non-Cryst. Solids* **74** 189

Singh J 1981 *Phys. Rev.* B **23** 4156

Solwisch M, Quitman D, Ruppersberg H and Suck J B 1983 *Phys. Rev.* B **28** 5583

Soukoulis C M, Cohen M H and Economou E N 1984 *Phys. Rev. Lett.* **53** 616

Spear W E 1983 *J. Non-Cryst. Solids* **59/60** 1

Spear W E and Le Comber P G 1975 *Solid State Commun.* **17** 1193

—— 1983 in *The Physics of Amorphous Silicon and Its Applications* ed J Joannopoulos and G Lucovsky (Berlin: Springer) (and references therein)

Spear W E, Steemers H L, Le Comber P G and Gibson R A 1984 *Phil. Mag.* B **50** L33

Spear W E, Steemers H L and Mannsperger H 1984 *Phil. Mag.* B **48** L49

Street R A 1976 *Adv. Phys.* **25** 397

—— 1981 *Adv. Phys.* **30** 593

—— 1982 *Phys. Rev. Lett.* **49** 1187

—— 1985 *J. Non-Cryst. Solids* **77/78** 1

Street R A and Beigelsen D K 1980 *J. Non-Cryst. Solids* **35/36** 651

Street R A, Biegelsen D K and Knights J C 1981 *Phys. Rev.* B **24** 969

Street R A, Knights J C and Biegelsen D K 1978 *Phys. Rev.* B **18** 1880

Street R A and Mott N F 1975 *Phys. Rev. Lett.* **35** 1293

Takeda S, Okazaki H and Tamaki S 1985 *J. Phys. Soc. Japan* **54** 1891

Tiedje T, Cebulka J M, Morel D L and Abeles B 1981 *Phys. Rev. Lett.* **46** 1425

Tsuchiya Y and Seymour E F W 1985 *J. Phys. C: Solid State Phys.* **18** 4771

Turkevich L A and Cohen M H 1984 *J. Non-Cryst. Solids* **61/62** 13

Uchtmann H, Popielawski J and Hensel F 1981 *Ber. Bunsenges. Phys. Chem.* **85** 555

Warren W W and Hensel F 1982 *Phys. Rev.* B **26** 5980

Warren W W and Matthiess L F 1984 *Phys. Rev.* B **30** 3103

Weaire D and Thorpe M F 1971 *Phys. Rev.* **4** 2508, 3518

Wooten F, Winer K and Weaire D 1985 *Phys. Rev. Lett.* **54** 1392

Yonezawa F and Cohen M H 1981 in *Fundamental Physics of Amorphous Semiconductors* ed F Yonezawa (Berlin: Springer)

Zdetsis A D, Economou E B, Papaconstantopoulos D A and Flytzanis N 1985 *Phys. Rev.* B **31** 2410

12

THE STRUCTURE AND ELECTRONIC PROPERTIES OF METALLIC GLASSES

Metal physics acquired a new and rapidly growing branch when it became easy in the 1970s to fabricate metallic glasses and when technical applications were found for them. Since alloy glasses can be made with all kinds of metals, amorphous systems with many kinds of properties can be studied, for example, a wide variety of magnetic behaviour is obtainable.

So many metallic glasses have been studied that various attempts to classify them have been made. No particular classification scheme will be emphasised in the sections below, but for some purposes a systematisation can be valuable. In a major review, a division by magnetic properties was used by Mizutani (1983), whose five classes were: ferromagnetic, weakly ferromagnetic, spin glass and Kondo-effect types, paramagnetic, weakly paramagnetic and diamagnetic. The magnetic and all the other properties must depend on the nature of the constituents, and grouping by chemical composition is also very common. Let us designate classes of pure metals by letters: S, simple metal (in the sense of Chapter 6); T, transition metal; M, metalloid†; R, rare earth; N, noble metal. Then some binary glass groupings, with examples, would be: S–S (Ca–Mg, Zn–Al); T–M (Ni–B, Fe–P); N–T (Cu–Ti, Au–Co); T–T (Fe–W, Pd–Zr); R–T (Ni–Dy, Fe–Hf). Different values of x in A_xB_{1-x} will lead to different properties, indeed ferromagnetism or even metallic conductivity may be present for some concentrations and not others. Consequently, property and composition classifications cut across each other and at certain concentrations a T–M glass might have properties characteristic of the S–S group. Many glasses, including some of practical utility, will be ternary, or more complicated still.

As with amorphous semiconductors, the realisation of technical applications has stimulated research into glassy alloys. Mechanical and anticorrosion properties are valuable but it would probably be true to say that up to the mid-eighties most commercial applications have exploited the magnetism in such devices or processes such as in recording, recorder heads, transformer cores, permanent magnets, shielding, transducers and sensors. This chapter will not deal with the technical applications; many pertinent papers can be found in Steeb and Warlimont (1985).

†In chemistry a metalloid is an element with properties intermediate between those of metals and non-metals. The class is not perfectly well defined but usually includes B, P, Ge, Si, As, Sb, Te.

12.1 Formation of metallic glasses

It is not easy to make amorphous metallic elements. By receiving evaporated Bi or Ga atoms on a cooled substrate, non-crystalline solid films were made by Bückel in the fifties, and Wright and colleagues did this with Co in 1972. Liquid quenching of glassy Ni was reported by Davies *et al* in 1973. But it is much easier to make liquid alloys into glasses, and a major step was taken with the discovery of splat cooling by Duwez and collaborators in 1959. This technique uses the impact of liquid droplets on a cooled highly conducting target to produce rapid quenching, say 10^6 K s^{-1}, resulting in a small foil of metallic glass. It soon became clear that many alloys could be vitrified in this way and, as interest grew in their properties, other quenching techniques evolved.

Several families of methods exist for making non-crystalline solid alloys. One is deposition 'atom by atom' from gaseous or liquid phases as in sputtering, evaporation or electrodeposition. Another is ion implantation or particle bombardment. Mechanical cold working of foils or powders can also be effective and a fourth approach is to cause amorphisation reactions to occur in the solid state; it even appears possible to convert $Cr_{1-x}Ti_x$ from amorphous to crystalline and back by thermal cycling (Blatter and von Allman 1985). However, the most widely exploited method is quenching from the melt and this family includes splat cooling, melt spinning and laser glazing (Beck and Güntherodt 1983, Luborsky 1983).

Laboratory equipment suitable for melt-spinning research samples is shown in figure 12.1. An air or helium stream impinging on the buckets drives the roller to a rim surface speed of, say, 30 m s^{-1} normally at room temperature. Then the He injection pressure operates to squirt the alloy—now made molten by the induction winding—against the copper rim of the roller where a ribbon of amorphous alloy forms (Pavuna 1981). In other versions the molten metal is aimed at the narrow gap between two cold oppositely rotating rollers or at a chilled metal belt at the point where the belt itself passes between oppositely moving rollers. Since amorphous metal ribbons have become a significant commercial product, devices for continuous fabrication are being developed all the time. (See, e.g. many papers in Steeb and Warlimont 1985, Suzuki 1982.)

As with amorphous semiconductors there is the question of whether the characteristics of a sample depend on its mode of preparation and, as the result is a solid out of thermodynamic equilibrium, the answer must in general be yes. The properties may then alter after fabrication; indeed many properties can be observed to change with time in samples held at a constant annealing temperature. For example, over some hundreds of hours, lengths may contract slightly, Young's modulus determined by sound velocity may increase a few per cent, viscosity measured by creep may rise, internal stresses fall, and so on. Electrical and magnetic properties change also. The

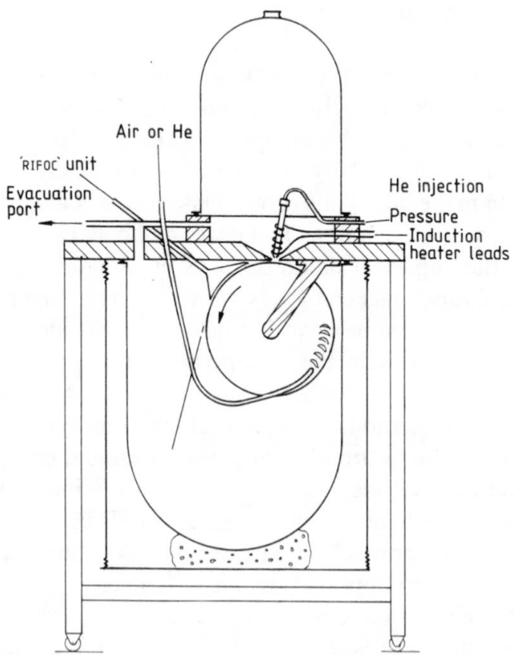

Figure 12.1 Schematic diagram of melt-spinning apparatus (from Pavuna 1981). RIFOC—ribbon fly-off control unit for secondary cooling and ribbon trajectory.

final value attained by a property depends on the annealing temperature. Presumably adjustments, some of an irreversible kind, proceed in the nearest-neighbour distances, altering both topological and chemical arrangement and acting to reduce overall disorder. Looked at from a free-volume point of view (§10.4), the free volume changes in amount and distribution. With very accurate diffraction studies made to large Q, such changes may be detectable in the radial distribution functions (Srolovitz *et al* 1981). Changes of properties with time must always be a consideration in technical applications and, if not allowed for, might vitiate comparisons of measurements made of samples of different age.

When a liquid alloy is quenched, vitrification competes with crystallisation and possible crystalline outcomes are homogeneous solid solutions or mixtures of two or more solutions or intermetallic compounds. As more and more alloys were vitrified it became clear that some alloys form glasses more readily than others and that some composition ranges are particularly favourable. Ready glass formation, or good glass-forming ability (GFA), means that the minimum rate of cooling, R_c, required to form a glass is relatively low, say less than about 10^5 K s^{-1}.

It was observed that the range of ready glass formation often embraced deep eutectic points (figure 12.2(*a*)). The low value of the liquidus temperature, T_L, at a eutectic point indicates a comparatively high stability of the liquid phase relative to the solid at that composition. The glass transition temperature, T_G, may be regarded as a measure of the stability of the glass: the higher T_G, the more stable the glass is. T_G is relatively insensitive to composition in a binary alloy but T_L frequently depends strongly on concentration. The reduced glass transition temperature, $T_{RG} \equiv T_G/T_L$, is often used as an index of glass-forming ability and has high values, say greater than about 0.6, at compositions where stable glasses (high T_G) can form near

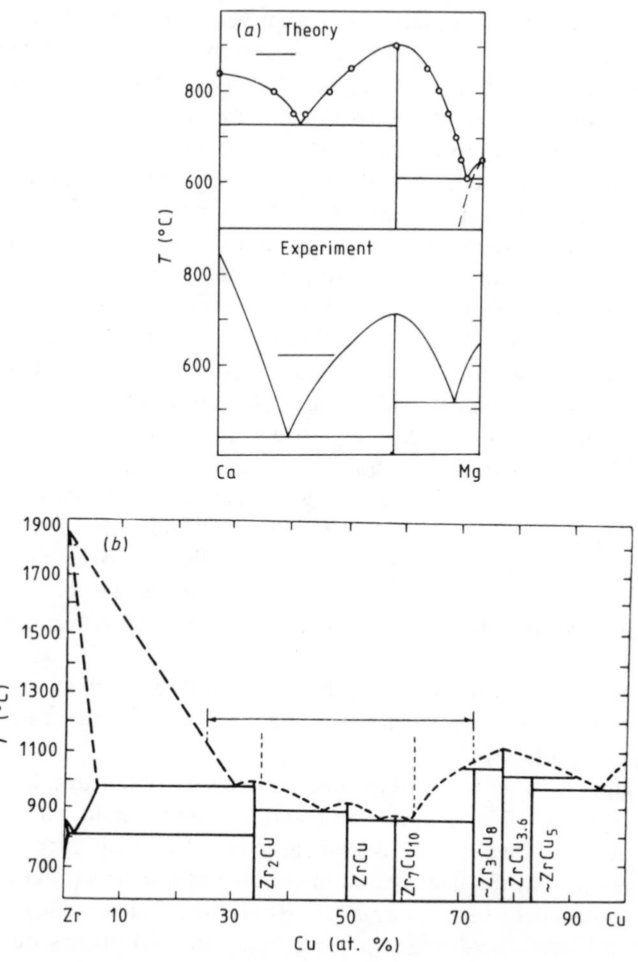

Figure 12.2 Equilibrium phase diagrams of (*a*) Ca–Mg. (After Hafner (1983).) (*b*) Cu–Zr, with eutectic points. Glass-forming ranges are shown by the horizontal lines above the liquidus. (From Giessen and Whang 1980a,b.)

deep eutectic points or, more generally, where T_L is low (figure 12.2(*b*)). T_G is not always known or easy to measure; sometimes the crystallisation temperature of the glass, T_{CR}, is used as an alternative index of glass stability.

An interesting and pertinent question to raise is: for a given material at a specified temperature how long does it take for a melt to nucleate and grow crystals to the extent that a very a small volume fraction, say 10^{-6}, is crystalline? This time would be expected to be very long at high temperatures where the liquid is stable and very long at low temperatures where the atomic mobility controlling nucleation and growth from the supercooled melt is very low. At some intermediate temperature the degree of supercooling will be optimum for crystallisation. These qualitative ideas can be expressed by time–temperature transformation (TTT) curves such as the schematic one shown in figure 12.3(*a*) and the others shown in 12.3(*b*) (Davies 1976). Calculating TTT curves is an exercise in homogeneous nucleation theory which we shall omit here; it involves hypotheses about both the probability of spontaneous formation of a nucleus and the rate of its growth. Diffusion coefficients, interfacial energy and free-energy differences between melt and crystal, as well as $T_L - T$, are all part of the input to the calculation and it is difficult to avoid numerous approximations. Once TTT curves are established however, it becomes possible to understand that R_c is a rate of cooling giving a vitrification trajectory which just grazes the vertex of the TTT curve (figure 12.3(*a*)). If, notionally, T_G is fixed and T_L varied, then the R_c could be calculated from a series of TTT curves corresponding to different values of T_{RG} because different T_L would require different inputs to the TTT calculation. As might be expected, R_c falls through several orders of magnitude as T_{RG} varies from about 0.2 to about 0.7 (Davies 1976).

Apart from the presence of eutectics, a number of semiempirical criteria emerged for anticipating ready glass formation and some of these will be described. They are not necessary, still less sufficient, conditions for ready glass formation but they are possible contributory factors that seem to favour vitrification in some cases (see, e.g., Sommer in Steeb and Warlimont 1985).

When a range of composition embraces several relatively complex crystal structures, the liquidus may stay low and the glass-forming range be prolonged (figure 12.2(*b*)).

In an alloy A_xB_{1-x} a size difference may be very significant. Many glasses form with A = a noble or transition metal, B = a metalloid, and $x \simeq 0.75$ to $\simeq 0.85$. Much-studied glasses like $Pd_{80}Si_{20}$, $Fe_{80}B_{20}$, $Ni_{20}P_{80}$ are examples. It was suggested that the dense random packing of the larger (A) component left interstices into which the smaller (B) atoms fitted—an idea compatible with $x \sim 0.8$. This may occur in some easily formed glasses but there are many counterexamples. Instead of, or as well as, size differences some chemical attraction between A and B may well help both to form and to stabilise glasses. When this is so, chemical short-range order (CSRO—see §§1.7, 2.7)

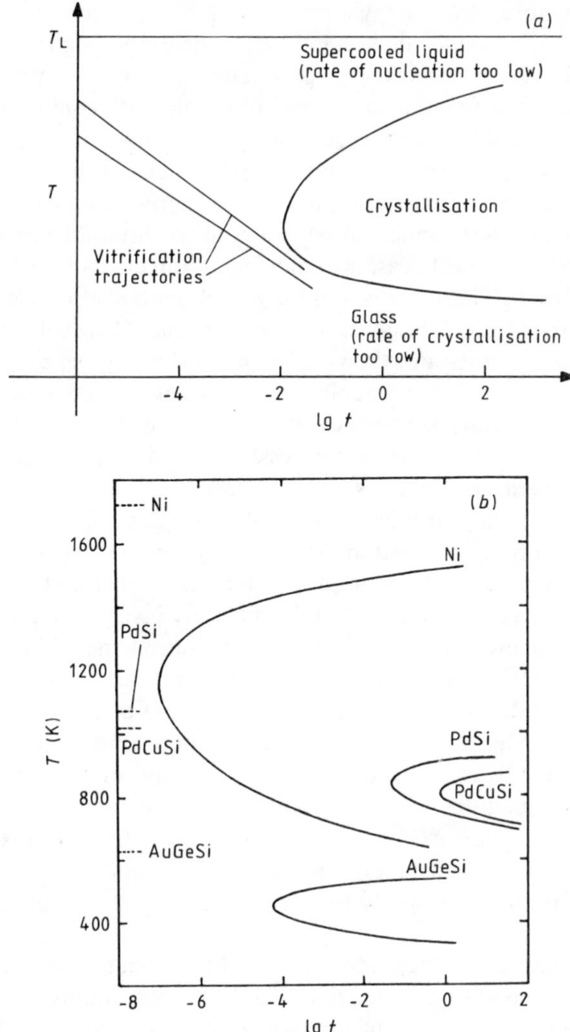

Figure 12.3 Examples of TTT curves: (*a*), schematic; (*b*), derived for the systems shown for a 10^{-6} volume fraction of crystallisation (from Davies 1976). Broken lines show the liquidus temperature for various compounds.

should be detectable and this is often the case; examples are discussed in the next section.

The conduction electrons play such an important role in metals that an electronic criterion for glass formation might be looked for. It was offered by Nagel and Tauc (1975) who argued as follows on the basis of NFE theory. If Q_p is the position of the first peak of $S(Q)$, then $2k_F \simeq Q_p$ is a strong-scattering condition corresponding as nearly as may be to Bragg scattering in an ordered structure. The energy $E_F \simeq \hbar^2(Q_p/2)^2(2m)^{-1}$ can therefore be assumed to be in a pseudogap with low $g(E_F)$ corresponding to the energy gap in a crystal (§9.10). However, $2k_F$ is determined by the valence electron density which is a function of composition and the requirement $2k_F = Q_p$ gives compositions for metal–metalloid systems about equal to those found for ready glass formation (e.g., $Pd_{80}Si_{20}$). The idea is that when E_F is in a pseudogap, and when structural fluctuations representing incipient crystallisation detract from the spherical symmetry in the liquid, the energy-lowering condition $2k_F = Q_p$ will cease to hold for all directions and an increase in energy will result. The liquid is thus metastable with respect to crystallisation and glasses tend to form easily as temperature falls. This explanation appears to work in some cases and the validity of considering the electronic energy cannot be doubted. But the dubiousness of using NFE concepts for systems with transition metals has been touched on before (§6.16) and evidence will be given below that E_F does not necessarily lie in a pseudogap.

It would be reasonable to assume that electron density, size differences and chemical affinity are all influences on glass formation. So is it possible to start from first principles and understand why certain ranges of composition are conducive to easy glass formation? Expressed thermodynamically, why are the free energies of the crystalline alternatives not so much lower than that of the supercooled liquid that crsytallisation inevitably occurs? The theory of metals given in Chapter 6 ought to be able to answer this at least for simple metals and that it could do so was first demonstrated by Hafner (1980) for Ca–Mg. It is an aspect of the general problem of calculating alloy phase diagrams from first principles and this requires an evaluation of ΔG, ΔH or ΔF as functions of composition (see §6.9 for the definitions of quantities like ΔG). Once $\Delta G(c)$ is found for various temperatures, there is a routine procedure in metallurgy for deducing the phase diagram from the principle that phases in equilibrium have equal chemical potentials (Hume-Rothery *et al* 1952).

Suppose that screened pseudopotentials have been constructed for the two metals at a given alloy composition and the corresponding electron density. As shown in §6.7 effective pair potentials follow from this. A number of techniques are then available for computing the structure and thermodynamic functions of the liquid alloy and these have been introduced before, namely, MD and MC (§§4.1, 4.2), GB, WCA and other variation and perturbation methods (§§5.8, 6.9). Such computations, based for example on a hard-sphere reference system, can lead *inter alia* to $\Delta G(c, T)$, $\Delta S(c, T)$ and $S_{ij}(Q)$ (§6.9). In the particular case of liquid Ca–Mg the results are in very reasonable agreement with experiment. There are, in addition, two approaches to the structure of the glassy state: the liquid computations could be repeated for $T \ll T_L$; or a cluster relaxation calculation could be done starting with a DRP cluster and relaxing it using the pair potentials (see §§2.5, 8.5). These two approaches lead to very similar, but not identical, results for

$g_{ij}(r)$, the difference showing up particularly in the second peak which is split in the cluster derivation (as often in experiments) but not in the other. A reasonable approximation to $\Delta G(c, T)$ etc, can therefore be derived for the glass by treating it as a supercooled liquid. It remains to consider the crystalline phases.

As the phase diagram of Ca–Mg shows (figure 12.2(*a*)), the compound relevant to glass formation is $CaMg_2$. With carefully constructed pseudopotentials it is possible to calculate the total energy (both volume and structure terms; equation (6.31)) for any assumed crystal structure and lattice parameters, and to vary both to identify the ground state. For the energetically preferred structure and lattice constants, the values of ΔU and ΔV could be found. At finite temperatures the contribution of lattice vibrations to the thermodynamic properties must be added in and can be estimated from a Debye model. In principle, many conceivable crystal structures at different compositions could be investigated this way to discover what intermetallic compounds would exist but, in Ca–Mg, other possibilities, such as Ca_2Mg, turn out to be relatively unstable and irrelevant to the present considerations. For a range of compositions in which some Ca is in equilibrium with some $CaMg_2$, the free energy is the sum of the two contributions. Conceivably, however, a solid solution might have formed instead, and $\Delta G(c)$ for this can also be computed from the pseudopotentials: in practice in the Ca–Mg system, the hypothetical solid solution turns out to have a high positive energy of formation and therefore to be energetically unfavourable compared with a mixture of Ca and $CaMg_2$.

What has just been described is the structure in outline of a very extensive calculation of which details are given by Hafner (1980, 1983). One outcome is the calculated phase diagram in figure 12.2(*a*) which is in qualitative agreement with the measured one. But because ΔG of supercooled liquids is also calculated, the range (if any) of composition where ΔG (supercooled liquid) is close to ΔG ($Ca + CaMg_2$ mixture) can be identified. We expect this to be the range in which—if rates of cooling are favourable—the supercooled liquid is likely to vitrify before crystallisation occurs. This range is shown in figure 12.4 where the ready-glass-forming range reveals itself near the eutetic. This establishes that, subject to the numerous approximations of a first-principles calculation, a rational basis exists for understanding why ready-glass-forming regions exist where they do. The electronic ingredients of pseudopotential theory, which lead to interionic potentials like those of §§4.5 and 6.7, automatically imply size differences (positions of ion–ion potential minima), chemical interactions (screening charges, charge transfer) and valence electron densities which all collectively determine the run of $\Delta G(c)$ curves and so ultimately the glass-forming ability. No doubt it will be difficult to demonstrate this in detail for more complicated glasses especially those with transition-metal constituents; Ca–Mg is a relatively simple case used here to illustrate the possibilities of first-principles explanations.

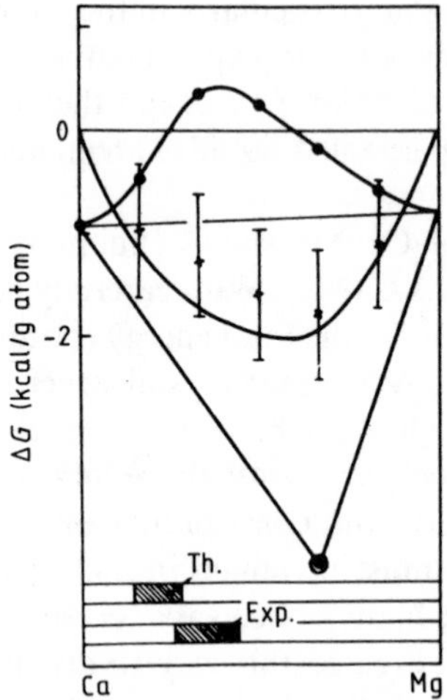

Figure 12.4 ΔG (in kcal per gm atom) calculated for Ca–Mg at 420 °C. Crosses with error bars show the theoretical ΔG of the supercooled liquid or glass and the curve through them is a possible interpolation. Full curves refer to a hypothetical solid solution. Sloping straight lines are linear interpolations of ΔG between the end points, the lowest end point being $CaMg_2$. The hatched regions indicate glass formation. (From Hafner 1980.)

12.2 Structure and vibrational spectrum

In the structural spread between dense random packing of hard-spheres (DRPHS) and covalent networks with low coordination (CRN), metallic glasses lie towards the former. The investigation of their LRO and SRO by diffraction, EXAFS and other techniques, and its representation by modelling and computation, are not essentially different from what has been described in earlier chapters either in general terms (Chapters 2, 3, 4 and 8) or in relation to insulating or semiconducting glasses (Chapters 10 and 11). Indeed metallic glasses have been taken as examples in §§2.5, 2.6, 2.7, 3.11, 3.13 and 8.5. This section therefore simply supplements what has gone before by a small number of additional examples. The primary problems are to obtain partial pair distribution functions, or equivalent information, and some insight into SRO. Good quality diffraction experiments, though difficult, will yield S_{NN}, S_{Nc}, S_{cc} or the S_{ij}. Ingenious use of null elements or zero alloys (§3.11), the combination of neutron work with x-ray work, and judicious choices of composition and concentration, can all lead to separate partial structure factors.

In an extensive study of melt-spun CuTi glasses, Sakata *et al* (see Suzuki (1982) p 330) demonstrated interesting relations between chemical short-range order (CSRO) and certain measures of stability in the glass and also established that CSRO in the liquid exists before vitrification. In this particular system, S_{Nc} is negligible and the difference between the values of the $|f(Q)|^2$ for x-rays and the $|b|^2$ for neutrons is large enough to give two simultaneous intensity equations soluble for S_{cc} and S_{NN} (see equation (3.32)). Coordination numbers, nearest-neighbour distances and SRO parameters, η, all follow from this for each of seven compositions. $\eta = \eta_{AB} = \eta_{BA}$ as defined in §2.7, and a positive value shows that CSRO exists and unlike neighbours are preferred. Figure 12.5 shows η as a function of composition and also its strong positive correlation with T_{CR} and T_{CR}/T_L. Here, T_{CR} is being used instead of T_G as a measure of glass stability and T_{CR}/T_L is thus a measure of glass-forming ability. It appears clearly that stability, glass-forming ability and CSRO go together. It would be somewhat surprising if the CSRO were completely absent in the liquid state and the experiment showed that it is there but that vitrification enhances η by a factor of at least two (Sakata *et al* 1981, Cowlam and Gardner 1984a).

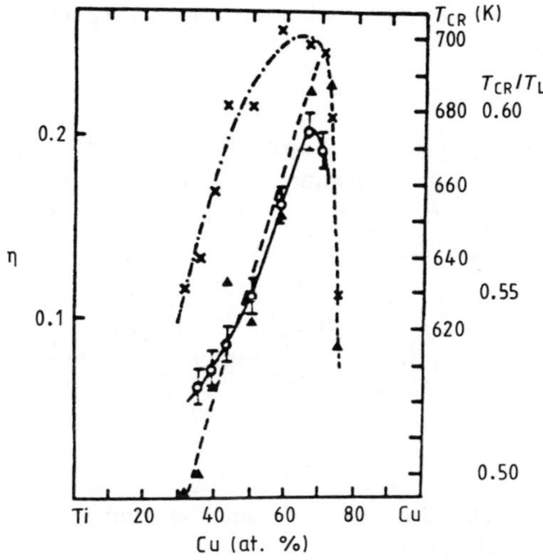

Figure 12.5 Chemical SRO parameter (η), T_{CR} and T_{CR}/T_L for some Cu–Ti glasses. Φ, η; —.×.—, T_{CR}; –▲––, T_{CR}/T_L. (From Cowlam *et al* in Suzuki 1982.)

Melt-spun metal–metalloid glasses are among the most widely studied amorphous metals and when they involve Ni it is possible to use ^{0}Ni which stands for the null isotopic mixture with zero coherent neutron scattering length. Isotopic substitution is possible for Fe, and neutron scattering can be supplemented by x-rays. Combining these possibilities, Lamparter *et al* derived partial structure factors for $Ni_{81}B_{19}$, $Fe_{80}B_{20}$, $Co_{81.5}B_{18.5}$ and inferred the $g_{ij}(r)$. Of the various results, figure 12.6 shows the $S_{ij}(Q)$ of the Fe–B

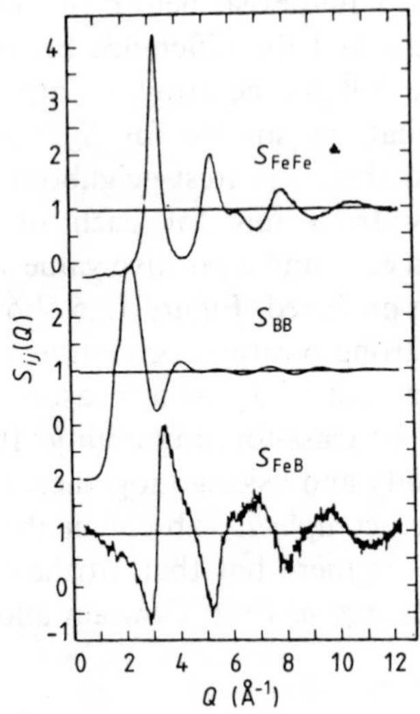

Figure 12.6 Partial structure factors of amorphous $Fe_{80}B_{20}$. (From Lamparter *et al* in Suzuki 1982.)

system. The resulting curve for $g_{FeB}(r)$ showed a strong narrow first peak implying a well defined Fe–B distance and therefore presumably a chemical bond. In the Ni–B system, $S_{B-B}(Q)$ is obtainable directly if ^{0}Ni is used and the corresponding $g_{BB}(r)$ had a split first peak indicating two separate B–B distances (Lamparter *et al*, in Suzuki (1982), p 343). Partial coordination numbers (see §2.4) are given in table 12.1 and the SRO parameter, $^0\eta_{AB}$ (see §2.7), for $Fe_{20}B_{80}$, is unity indicating complete chemical ordering. Studies of $Ni_{80}P_{20}$ give comparable but not identical results (see Steeb and Warlimont (1985) p 459).

Table 12.1 Partial coordination numbers in T–B systems.

	T	B
T = Fe	12.4	2.2
B	8.6	6.5
T = Co	12.7	1.5
B	6.6	—
T = Ni	10.8	2.2
B	9.3	3.6;3.7†

†Under the two separate peaks.

As a final example we take an R–T glass, namely, Dy_7Ni_3. It happens that both elements can be isotopically mixed to null form and the first 'double-null' experiment, by Wright *et al* (1985), illustrates further the power of the isotopic substitution method in cases where the isotopes are favourable. Dy–Ni has a eutectic point; Dy_7Ni_3 is near it and the glass can be melt spun. Figure 12.7 shows diffraction patterns taken at room temperature. In $Dy_7{}^0Ni_3$ the pattern is due to Dy–Dy distances which must also dominate the diffraction from the unsubstituted alloy because of the similarity of its pattern. Coherent neutron scattering being virtually absent from the double-null alloy, the ${}^0Dy_7{}^0Ni_3$ curve displays the magnetic scattering factor of Dy (compare figure 3.2). Magnetic ordering does not show itself at room temperature in the paramagnetic regime but peaks due to it show clearly, not only below the magnetic transition temperature of 38 ± 1 K, but also above this showing that short-range magnetic order persists up to about 150 K. More—such as inelastic scattering from the double-null alloy—can be expected from this type of investigation.

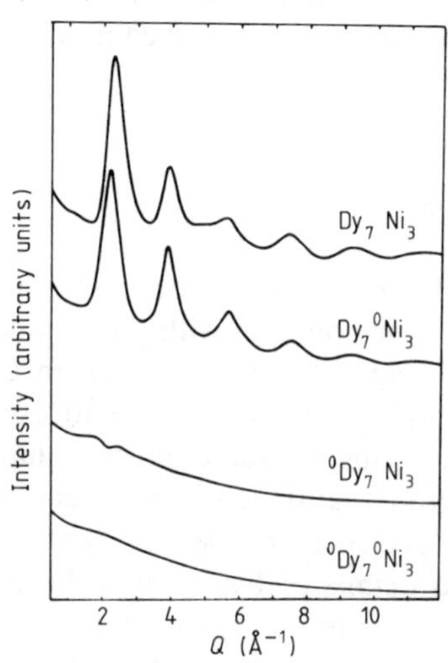

Figure 12.7 Diffraction intensity from amorphous Dy_7Ni_3 showing the use of null elements (from Wright *et al* 1984).

These three examples of diffraction and others in the literature show how important detailed studies are of the partial structure factors and how significant chemical ordering is. Hypotheses of random packing of spheres or small seeds (§2.6) are unlikely to lead to models that reproduce all the features that are emerging. Even with additional clues from EXAFS, NMR and Mössbauer experiments (Steeb and Warlimont 1985) the structural details of metallic glasses, as of other glasses, remain elusive (Cowlam and Gardner 1984a).

Computer modelling was described in §2.6 and it has been applied with some success to binary metallic glasses. The $g_{ij}(r)$ of $Fe_{80}B_{20}$, referred to above, were quite well reproduced by relaxing a DRPHS cluster of 1000 atoms using a truncated Lennard-Jones potential (Lewis and Harris in Wagner and Johnson (1984), p 547). But in general the initial DRP may prejudice the outcome making the relaxation method less reliable that an *ab initio* MD simulation (Grabow and Andersen 1985).

$S(Q, \omega)$, and the vibrational states of a metallic glass, were briefly discussed in §8.5 and figure 8.11, and have been reviewed by Suck and Rudin (in Beck and Güntherodt 1983). In general, the interpretation of neutron inelastic scattering from metallic glasses is not easy. For $\hbar\omega > 10$ meV there is a close resemblance between $S(Q, \omega)$ for a glass and for a polycrystalline metal of which the density and SRO are similar. As with infrared spectroscopy of non-metallic glass (§8.6), the inference seems to be that the SRO determines the main features of the thermal motion. This does not apply at $\hbar\omega < 10$ meV where there is greater intensity in the scattering from the glass which disappears on crystallisation. This implies low-energy modes which are characteristic of the amorphous structure and they may be localised and connected with the two-level system involved in low-temperature tunnelling processes (§§10.7, 12.3).

12.3 Electronic spectrum

The problems of calculating and measuring $g(E)$ and related quantities in metallic glasses have a lot in common with those in liquid metals and amorphous semiconductors, and the methods given in Chapters 7 and 11 are available. In §7.4, amorphous Fe was taken to exemplify the use of the recursion method with a cluster. Some more examples follow below. The point has been made before that the gross features of $g(E)$ are determined by the SRO; in general, the presence or absence of LRO determines the existence or sharpness of peaked details or band edges. Because of the difficulty of density of states calculations in non-periodic systems there is a tendency to find ways of keeping as close as possible to crystal calculations while incorporating some features of the SRO of the glassy state. Photoelectron spectroscopy and optical properties are obvious candidates for the experimental studies.

Before considering UPS it is interesting to enquire whether clues to $g(E_F)$ can be found in the specific heat (C) or magnetic susceptibility (χ). Neither are very illuminating. The specific heat of non-magnetic metallic glasses at low temperatures can be represented by an equation of the form used for insulating glasses (equation (10.5)) but the interpretation is not the same. Assuming the T^3-term can be understood as a Debye phonon contribution, the linear term, γT, is due mainly to the electronic specific heat and γ may be written

$$\gamma_{\text{obs}} = \gamma_{\text{el}}(1 + \lambda_{\text{ph}}) + \gamma_{\text{gl}}. \tag{12.1}$$

Here $\gamma_{\text{el}} = (\pi^2/3)k_B^2 g(E_F)$ and λ_{ph} is an enhancement factor from electron–phonon coupling. γ_{gl} is a contribution similar to that discussed in §10.7, for it appears that a two-level system operates to give a linear specific heat anomaly in metallic as in other glasses. However, $\gamma_{\text{gl}}/\gamma_{\text{el}}$ might be as low as 10^{-2} so the properties of γ_{gl} are difficult to disentangle. Nevertheless γ_{gl} was discovered in glassy Zr–Pd by Graebner *et al* in 1977 and some of the other phenomena associated with two-level systems in §10.7 have also been found in metallic glasses (Black in Güntherodt and Beck (1981) and Lasjaunias *et al* (1986)). $\gamma_{\text{obs}}/\gamma_{\text{FE}}$ varies from about 1 to about 2.5 but even if γ_{gl} is negligible, an independent value of λ_{ph} is needed to find $g(E_F)$ from γ_{obs}. λ_{ph} is referred to again in §12.5.

Similarly, it is not easy to infer $g(E_F)$ from χ. In glasses of simple metals very often $\chi < 0$ and the subtraction of the diamagnetic contribution of the ion cores leaves an electronic part which is usually significantly greater than χ_{FE} by a factor of two or more (Mizutani 1983). Part of this is presumably an enhancement due to electron–electron interactions (Nozières and Pines 1966).

In the previous section the structure of Cu–Ti was discussed. The UPS spectrum and reflectivity of a representative composition are shown in figure 12.8. By comparison with the spectra of pure polycrystals—also in figure 12.8(*a*)—and from a general knowledge of transition-metal states, we infer that the peak in the glass spectrum near 3 eV consists mostly of Cu 3d states and the peak near E_F of Ti 3d states. The Drude optical formulae (§6.15) served well for NFE liquid metals in the infrared but here, with two pronounced peaks in $g(E)$, it is not surprising that the reflectivity departs radically from the Drude prediction and shows evidence of transitions upward from the peaks (figure 12.8(*b*)). The authors of this work (Lapka *et al* 1985) were able to show that if $g(E)$ is parametrised to have the form of the UPS spectrum then equation (7.43*a*) for the real part of σ, inserted into equations (6.61) and (6.62) lead via ε_R and ε_I to the observed reflectivity. They also computed $g(E)$ for a crystalline sample having the FCC structure of CuAu. As is often the case, the results were in qualitative agreement with the UPS spectrum (see figure 12.8(*a*)), quantitative agreement being too much to expect without using the real SRO of the glass.

One way of incorporating the latter while still retaining some of the relative simplicity of lattice calculations was used by Khanna *et al* (1985) for the glass Ni–P. Any use of the CPA–KKR for substitutional alloys (§7.2) or the EMA–KKR for liquid or amorphous metals (§7.3) requires the construction of muffin-tin potentials. The potential inside a muffin-tin sphere is contributed in part by the neighbours and will therefore depend on where the neighbours are. To make the point for a disordered pure metal we could

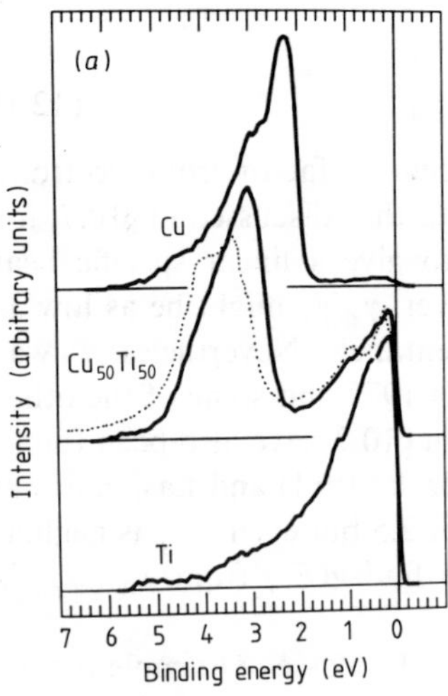

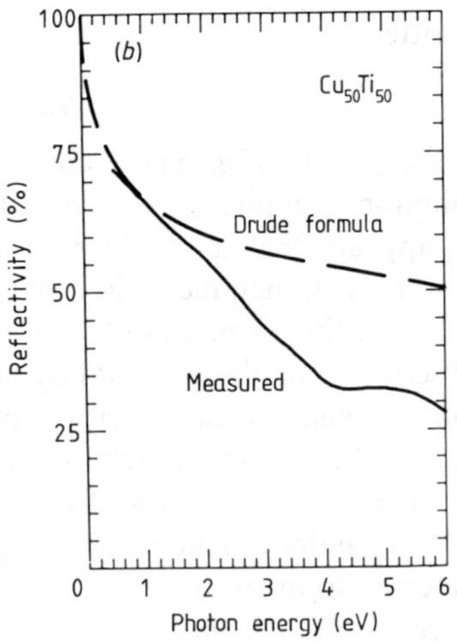

Figure 12.8 (*a*) The UPS observations on splat-cooled amorphous $Cu_{50}Ti_{50}$. The broken curve is calculated—see text. The Cu and Ti curves are for polycrystalline elements. The excitation energy was 21.2 eV. The zero of energy is E_F. (From Lapka *et al* 1985.) (*b*) Optical reflectivity of $Cu_{50}Ti_{50}$ glass.

write the charge density inside a muffin-tin sphere as its own density, ρ_a, plus a correction from the neighbours thus:

$$\rho(\boldsymbol{r}) = \rho_a(\boldsymbol{r}) + n_0 \int g(r)\rho_a(\boldsymbol{r} - \boldsymbol{r}')\,d\boldsymbol{r}'$$

where n_0 is the number density. This could be generalised to include the $g_{ij}(r)$ in an alloy and it makes clear that the potential depends on the SRO. This brings up yet again a problem posed by disorder: strictly every atom has a different neighbourhood and therefore a different potential, and using a mean potential derived from $g(r)$ involves the approximation of averaging at the beginning rather than at the end of the calculation. Khanna *et al* argued that the influence of SRO is more important on the potential than on the subsequent calculation of $g(E)$ and that one can therefore use the CPA instead of the more complicated EMA. It certainly appears that $g(E)$ calculated for the glass $Cu_{60}Zr_{40}$ reproduces the observed photoemission spectrum fairly well whether the EMA or CPA is used. Applying the CPA–KKR to a FCC random binary alloy of $Ni_{74}P_{26}$, with the number density of the glass, leads to figure 12.9. Experimental $g_{ij}(r)$ were used to compute the muffin-tin potential.

$g(E)$ for the glass has many interesting features. There is the general removal of peaks characteristic of crystalline Ni and the appearance of a bulge of s states of P at -0.4 Ryd. The p states of P hybridise with Ni states to increase $g(E)$ substantially above the bottom of the pure Ni band. $g(E_F)$

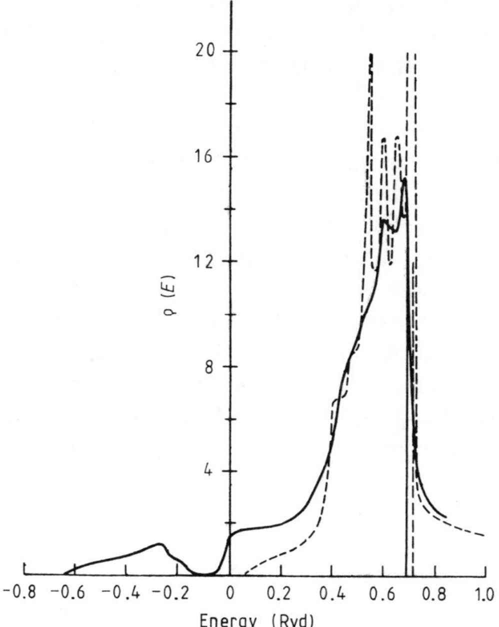

Figure 12.9 Density of states in Ni (broken curve) and $Ni_{74}P_{26}$ glass (full curve). Vertical lines indicate E_F. (After Khanna *et al* 1985.)

is reduced, but E_F remains near the top of the d band which does not fill up.

It is valuable to relate these results to various observations and to the hypothesis of Nagel and Tauc about glass formation (§12.1). E_F does not lie in a pseudogap and does not in fact change much as P enters, so there are always holes in the d band. Photoemission confirms both this and the increase in the total bandwidth. Both the Knight shift and the electronic specific heat are proportional to $g(E_F)$ and are shown by experiments to fall by factors of about two across the glass-forming range from about 15 to about 26 at. % of P. The calculation gives a decrease by about 1.4 which is much nearer to the observed factor than it would be if the P electrons simply filled up the Ni band and then pushed E_F upwards into the Ni s–p band where $g(E)$ is lower. The latter shift is what was supposed to happen on the rigid-band hypothesis with the consequence that k_F would increase with added P until $2k_F \sim Q_p$ in the glass-forming range as required by the Nagel and Tauc hypothesis. Although this appears not to happen, the extension of the s–p band downwards relative to the d band has the same effect of increasing k_F (Khanna *et al* 1985).

A number of other methods have been tried for calculating $g(E)$ in alloy glasses. Several are in Wagner and Johnson (1984) which also contains a review by Cyrot-Lackman. We refer to one other example: Fujiwara's calculation for Fe–P and Fe–B, which are both much studied glass-forming alloys (Fujiwara 1982).

The muffin-tin concept is used again but this time the structure is given by a relaxed DRP model of about 1500 spheres with diameters adjusted to give RDF in agreement with the observed ones. Inside a Wigner–Seitz cell in a crystal it is possible to conceive electron wavefunctions made up by linear superposition of muffin-tin orbitals derived from the cell, and from all its neighbours, in such a way that the sum is a solution of Schrödinger's equation inside the cell. This can be implemented numerically and works well in band structure calculations in crystals. On replacing the Wigner–Seitz cell with a sphere of the same volume, a simplification results; and by distributing the spheres according to the DRP model the method becomes applicable to glasses. The actual computation is performed by the recursion technique

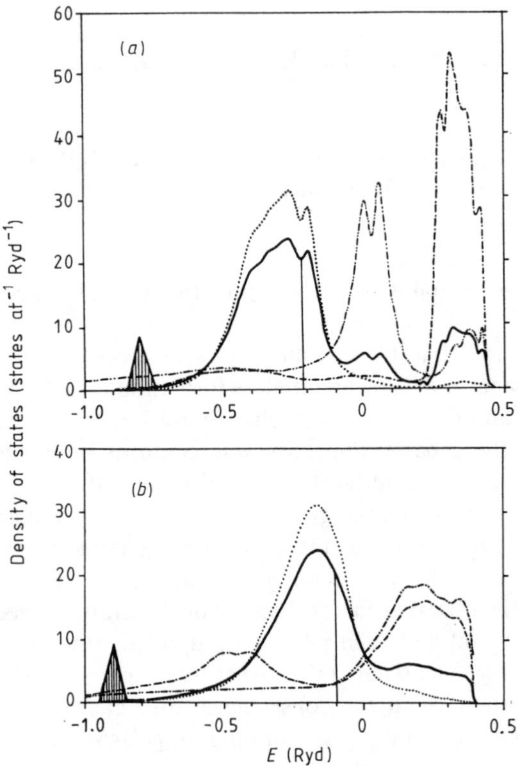

Figure 12.10 $g(E)$ in (*a*) $Fe_{75.2}B_{24.8}$ and (*b*) $Fe_{75.7}P_{24.3}$. Full curve, the glass; dotted curve, Fe 3d states; ..—.., Fe 4s states; chain curve, P 3p or B 2p states. The shaded triangle shows the position and weight of P 3s and B 2s states. (From Fujiwara 1982.)

(§7.4) and the details are given in Fujiwara (1982). The results have much in common with those quoted above for Ni–P. Figure 12.10 shows how the B or P s states appear at the low end of a broadened density of states. E_F remains near the top of the d band and in both Fe–B and Fe–P, $g(E_F)$ falls as the non-metal is added. The $g(E)$ is in qualitative agreement with UPS and XPS spectra.

These examples show that, in metallic glasses as in liquid metals and amorphous semiconductors, density of states calculations have become reasonably reliable and realistic in spite of the simplifications which are inevitable for modelling the structure and for making the computation tractable.

12.4 Electron transport properties I: simple metal alloys

Suppose that in the absence of experimental results a theory was required to predict the resistivity, $\rho(T)$, at constant volume, of an amorphous metallic element. In a pure perfect crystal the resistivity is known to derive from the electron–phonon interaction; scattering from the ions in their lattice positions is ignored because it leads to Bragg reflections and the band structure, not to resistance. This means that only the vibrational part of the dynamic structure factor is used for calculating the electron scattering, the part leading to Bragg scattering being subtracted out; thus the resistance vanishes at absolute zero. This does not make a good starting point for an amorphous metal because the topological disorder must involve scattering even if, as at 0 K, there are no thermal vibrations. In §6.12 the Ziman theory of resistance in liquid metals was set out. This uses the static structure factor, $S(Q)$, which certainly takes the topological disorder into account but also seems unsuitable for the present purpose because phonons would presumably play some part in the resistance of an amorphous metal. In §6.16 it was pointed out that Ziman's formula is actually the high-temperature limit of a more general theory in which $S(Q)$ is replaced by

$$S^{\rho}(Q) \equiv \int_{-\infty}^{+\infty} S(Q, \omega) \frac{\hbar\omega}{k_B T} n(\omega)\, d\omega \qquad (12.2a)$$

where

$$n(\omega) \equiv [\exp(\hbar\omega/k_B T) - 1]^{-1}.$$

This means that from equation (6.59)

$$\rho = \frac{3\pi}{e^2\hbar v_F^2}\left(\frac{N}{V}\right) \langle\langle S^{\rho}(Q)|u(Q)|^2\rangle\rangle. \qquad (12.2b)$$

$S^{\rho}(Q)$ is sometimes called the resistivity structure factor and includes both the static topological disorder and the vibrations; it looks the appropriate quantity for glassy metal resistance. Even so generalised, the NFE theory is only a diffraction theory suitable for weak scattering or low-resistivity systems and would probably only apply, if at all, to glasses of simple metals. Further, the partial resistivity structure factors, $S^{\rho}_{ij}(Q)$, would be required for alloys where chemical, as well as topological, disorder prevails.

This poses very formidable theoretical problems. The calculation of $S(Q, \omega)$ from first principles was referred to in §8.5 where an example was given to show that it can be done. But a number of theories of resistivity have been published which avoid having to do this by adopting approximations for $S^{\rho}_{ij}(Q)$. These result from simplifying assumptions about $S(Q, \omega)$, namely, that it represents plane-wave phonon propagation with a linear dispersion law like sound waves; or, alternatively, that every atom vibrates independently. These represent the extremes of coherent and incoherent thermal motion (see, e.g., §8.3). A detailed derivation of these approximate $S^{\rho}_{ij}(Q)$ will not be made here but they will be quoted to show what they, and consequently the resistance, depend on.

Before doing this we should pause to ask if it is reasonable to expect a NFE theory to apply to metallic glasses at all. With liquid metals we saw that one source of confidence was the free-electron value of R_H (§6.14), but that this could not always be relied on even for seemingly simple alloys. The matter has been quite thoroughly investigated for binary and ternary glasses of simple metals and it appears that $R_H \simeq R_H^{FE}$ in very many cases although, as might be expected from §6.14, not with Bi–Pb alloys (Mizutani 1983). Save for such exceptions, $2k_F$ may be inferred from $2k_F = 1.139 \times 10^{-3} (-R_H)^{-1/3} \text{Å}^{-1}$ where R_H is in $\text{m}^3\ \text{A}^{-1}\ \text{s}^{-1}$. $2k_F$ is of course an important input to the generalised Ziman theory. Alloys with $R_H \simeq R_H^{FE}$ are also of relatively low resistivity on the whole.

Returning now to $S(Q, \omega)$ and $S^{\rho}(Q)$ we quote from the work of Jäckle and Frobose via that of Hafner and Philipp (1984). For the plane-wave model of $S(Q, \omega)$, and in the case that one-phonon processes are adequate,

$$S^{\rho}_{ij}(Q) = \exp[-(W_i + W_j)]\left[S^{(\text{AL})}_{ij}(Q) + \frac{\hbar^2 Q^2}{k_B T (M_i M_j)^{1/2}} \times \left(c_s n(c_s Q)(n(c_s Q) + 1) + \frac{1}{(2\pi)^3 n_0} \int n(c_s k)(n(c_s k) + 1) S^{(\text{AL})}_{ij}(\boldsymbol{k} + \boldsymbol{Q})\, \mathrm{d}^3\boldsymbol{k} \right)\right] \quad (12.3a)$$

where M_i is the ion mass, c_s is the sound velocity and n_0 is the mean ion number density. In this expression, $\exp(-W_i)$ is the Debye–Waller factor for type-i atoms. This factor determines the reduction in intensity of Bragg peaks due to thermal vibrations during x-ray diffraction from a crystal, and reappears here in the one-phonon scattering of electrons (Ashcroft and Mermin 1976). W_i is a function of T and Q, namely,

$$W_i(Q, T) = \frac{\hbar^2 Q^2}{2M_i} \int_0^\infty \omega^{-1} g_i(\omega)(n(\omega) + \tfrac{1}{2})\, \mathrm{d}\omega \qquad (12.3b)$$

where $g_i(\omega)$ is the partial density of vibrational states. We note further that the static partial structure factors have been given the Ashcroft–Langreth form of equation (3.36). The sound velocity is assumed to be the same for both longitudinal and transverse modes—a simplifying assumption to help evaluate $S(Q, \omega)$.

For the incoherent vibration model,

$$S_{ij}^{\rho}(Q) = \exp[-(W_i + W_j)]\left(S_{ij}^{(\mathrm{AL})}(Q) + \frac{\hbar^2 Q^2 \delta_{ij}}{2k_{\mathrm{B}}T(M_i M_j)^{1/2}} \times \int_0^\infty n(\omega)(n(\omega)+1) g_i(\omega)\, \mathrm{d}\omega \right). \qquad (12.3c)$$

Equations (12.2) and (12.3) enable $\rho(T)$ to be calculated if the $S_{ij}^{(\mathrm{AL})}(Q)$ and $u_i(Q)$ are known. The $u_i(Q)$ are chosen from pseudopotential theory and lead to ion–ion potentials (§6.7). The latter can be used to determine the equilibrium density and the $S_{ij}^{(\mathrm{AL})}(Q)$ by a cluster relaxation method like that outlined in §8.5. The $g_i(\omega)$ are also required and the recursion method is available for this. This somewhat formidable programme of calculation has been carried out for $Ca_{70}Mg_{30}$ (Hafner and Philipp 1984) and for $Mg_{70}Zn_{30}$ (Hafner 1985). It amounts to a calculation of ρ from first principles and some results are given in table 12.2 and figure 12.11.

Table 12.2 Residual resistivities ($\mu\Omega$ cm).

	Experimental	Theoretical
$Ca_{70}Mg_{30}$	43.7	35.6
$Mg_{70}Zn_{30}$	43 to 57	43.6

The diagrams illustrate several points of general interest. First, the temperature coefficient of resistance (TCR) is positive in one and negative in the other. Liquid Zn also has a negative TCR and an explanation was given in §6.13 in terms of the proximity of $2k_{\mathrm{F}}$ to Q_{p}, and the temperature variation of the static structure factor. This kind of explanation is still available from the first term of equation (12.3). But the second or phonon term gives a positive TCR. In $Ca_{70}Mg_{30}$ the latter dominates and in $Mg_{70}Zn_{30}$ the former does. This can be shown in detail by examining the Q- and T-dependences of the factors in the integrand of equation (12.2*b*). It is clear that with three or more S_{ij} and two or more pseudopotentials and $g_i(\omega)$, there is scope for

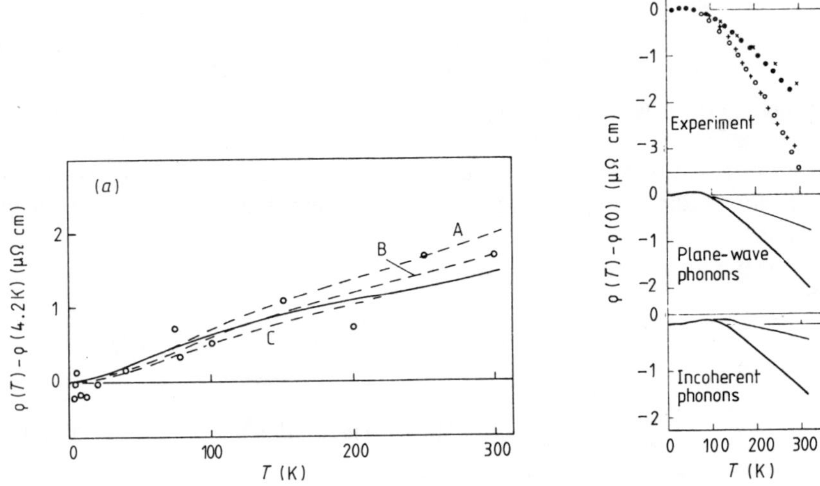

Figure 12.11 Experimental and calculated resistivities for: (*a*), $Ca_{70}Mg_{30}$; (*b*), $Mg_{70}Zn_{30}$. In (*a*) the full curve is theoretical and the points experimental; the broken curves show the effect of different Debye temperatures (A, 230 K; B, 250 K; C, 270 K). In (*b*) the thin line is isochoric and the thick line isobaric. The observed points are from four different experiments. (After Hafner and Philipp (1984) and Hafner (1985). For more detail, see these papers.)

much detailed behaviour and variation from glass to glass. This is further shown by the ability of the theory to reproduce the maxima and minima in ρ that are observed for $T < 60$ K in $Mg_{70}Zn_{30}$ (Hafner 1985).

$(\partial\rho/\partial T)_V$ and $(\partial\rho/\partial T)_p$ are different. The difference is not always great but is measurable in, say, liquid alkali metals near their melting points and is enormous in the expanded liquid metals of §11.12. Experiments are usually done at constant pressure so the theory should allow for thermal expansion. This is done in figure 12.11(*b*) and is obviously necessary.

Calculations of this type are no doubt improving but it seems clear that the generalised Faber–Ziman theory is quite capable of explaining $\rho(T)$ in glassy alloys of simple metals if pursued with sufficient attention to all the inputs to the equation. The negative TCR of $Mg_{70}Zn_{30}$ is by no means unusual, indeed dozens of simple metal glasses have it and in most cases $2k_F/Q_p \simeq 1$, strongly suggesting that the first term of equation (12.3) is dominating the temperature dependence giving an effect similar to that in divalent liquid metals (Mizutani 1983).

Although the NFE theory provides a formula for the thermoelectric power (§6.12), it is difficult to fit it to the data. This is partly because of the sensitivity of the theoretical value to the uncertain $u_i(2k_F)$ and $S_{ij}(2k_F)$ and also possibly to electron–phonon enhancement effects not included in the theory (see below). In the accumulated data, both positive and negative α and $\mathrm{d}\alpha/\mathrm{d}T$ are found (Mizutani 1983).

12.5 Electron transport properties II: non-simple cases

The two simple glasses mentioned above and a large number of others have ρ at 300 K well below 100 $\mu\Omega$ cm. Many other glasses containing non-simple metals have much higher resistivities and are consistent with interesting generalisations made by Mooij (1973). One of these says that the TCR correlates strongly with $\rho(300)$, being positive for $\rho(300) \lessgtr 150\ \mu\Omega$ cm and negative for higher ρ. This applies to crystalline or amorphous alloys and to liquids, with few exceptions. The other Mooij generalisation is that, irrespective of the low-temperature resistivity, alloys tend at high temperature towards a saturation value of $\rho \sim 150\ \mu\Omega$ cm. Some evidence bearing on this point is shown in figure 12.12.

To see that considering non-simple metals means relinquishing the NFE model, we have only to look at R_H which is sometimes positive and is, in any case, hard to compare with a free-electron value because of the difficulty of saying how many electrons are 'free'. An example is the Hall effect in melt-spun Cu–Zr, studied by Gallagher *et al* (1983). The resistivity at 300 K is typically about 180 $\mu\Omega$ cm and the TCR negative. Ageing the samples over many months greatly alters the TCR and R_H, apparently because of progressive surface oxidation leading to crystallisation. The reproducible results from new samples are added to others on liquid alloys from Künzi and Güntherodt (1980) in figure 12.13. Pure Cu has $R_H = R_H^{FE}$ but the sign change occurs at about 20 at. % of transition or rare-earth metal. R_H in Cu–Zr varies very little with temperature from 0 to about 250 K which is normal.

$R_H > 0$ remains hard to account for however. The need is for acceptable, preferably rigorous, expressions for σ_{xx} and $\sigma_{xy}(H)$ in equation (7.46), and a way of evaluating them with wavefunctions which are hybrids of s and d states in alloys containing transition metals. Both of these requirements seem not to have emerged fully from the formative stage at the time of writing but, given certain assumptions plausible for glasses, Morgan and collaborators have argued that R_H depends only on electron properties at E_F, that

$$R_H \propto (\mathrm{d}g(E)/\mathrm{d}E)_{E_F}$$

and that the latter can give $R_H > 0$ in regions of anomalous dispersion where $(\partial E/\partial k) < 0$ (see figure 7.5). On this basis R_H and σ can be computed reasonably well for CuZr alloys (Morgan and Howson 1985, Howson and Morgan 1985, Nguyen-Manh *et al* 1986).

To explain the existence of high resistivities some application of the Kubo equations will surely be required. In §7.7 it was pointed out that a Kubo

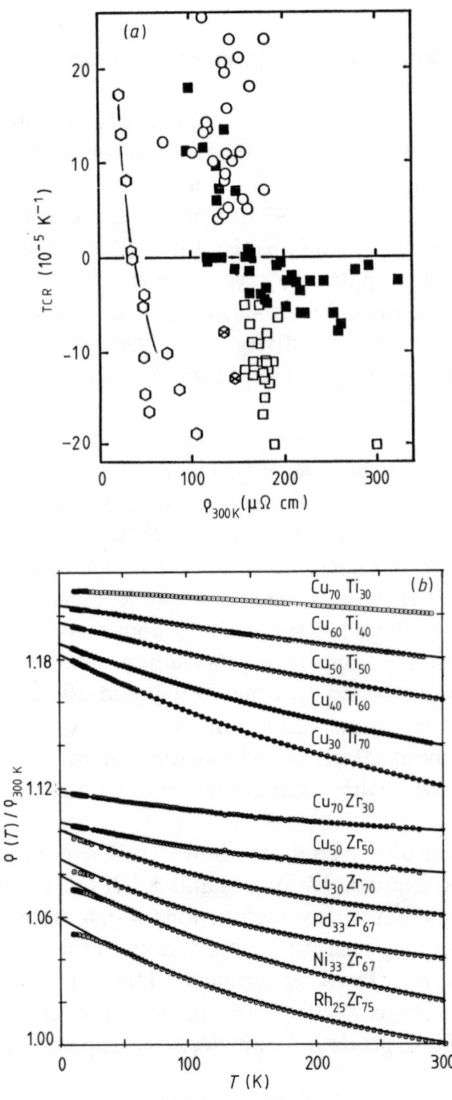

Figure 12.12 (*a*) Correlations of TCR with resistivity at 300 K. ○, ferromagnetic; ■, metal/metalloid alloys (paramagnetic); □, metal–metal alloys (paramagnetic); ⊗, ⬡, weakly paramagnetic or diamagnetic with/without $\rho \propto (1 - AT^2)$ at low T. (*b*) Examples of negative TCR in N–T and T–T alloys. The curves are vertically displaced by 0.02. (After Mizutani in Steeb and Warlimont (1986). For more detail see Mizutani (1983).)

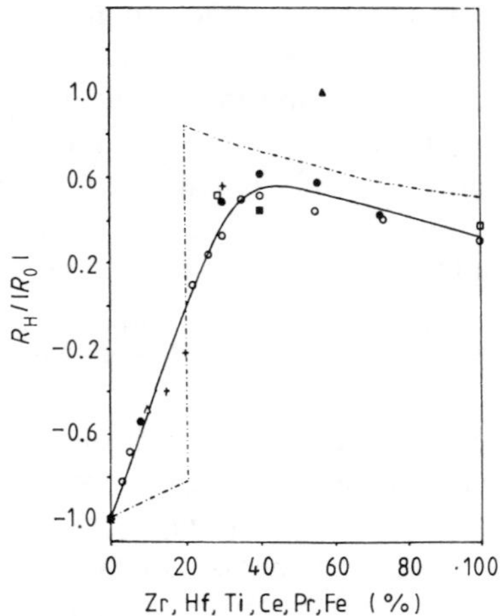

Figure 12.13 Onset of positive Hall coefficients in disordered alloys. ●, Cu–Zr; ■, Cu–Hf; ▲, Cu–Ti; ○, Cu–Ce (liq); □, Cu–Pr (liq); △, Ag–Pr (liq); +, Au–Fe (liq). (From Gallagher *et al* 1983.)

formula for diffusivity, applied to a cluster, had been successfully used with the recursion method to calculate ρ in disordered Fe. In fact a good account of ρ in liquid V, Cr, Mn, Fe, Co and Ni (mostly $\rho > 100\ \mu\Omega$ cm) has been given by this method and there is no reason in principle why it should not be used for amorphous metals (Ballentine 1985). It would be valuable in addition to have an argument which—like the NFE model in its context—gives some intuitive understanding of how small or negative temperature coefficients of resistance can arise in high-resistance amorphous alloys, and some work by Schirmacher and colleagues is interesting in this respect. Using arguments for which the original papers must be consulted, Belitz and Schirmacher inferred from the Kubo formulae that

$$\sigma = e^2\left(\frac{n/m}{M_0 + M_\mathrm{T}} + L_0 + L_\mathrm{T}\right).$$

The M-terms show σ controlled by scattering and the L-terms represent σ enhanced by hopping or tunnelling. The suffix 0 refers to elastic scattering or to tunnelling processes subject to static disorder; the suffix T is for inelastic scattering and tunnelling enhanced by dynamic disorder. The latter includes both coherent or phonon modes and incoherent or diffusive motion (Schirmacher 1985). In the expressions for the M and L it is not surprising to find $S(Q)$ and $S^\rho(Q)$ in the static and dynamic parts. The behaviour depends on the relative sizes of the terms. For metals of low resistivity the formula reduces to equation 12.2(b), but for more disorder and stronger interactions when the mean-free path is comparable with the de Broglie wavelength

$2\pi/k_F$, the L-terms begin to take effect and can give a conductivity which is enhanced by thermal motion leading to an overall negative TCR. The substantial amount of structural and dynamic input data needed for applying the theory to a real alloy is rarely, if ever, available.

The thermoelectric power usually requires careful interpretation and it was soon suspected that equation (6.49) for the diffusion thermopower was inadequate for amorphous alloys. We rewrite it here as

$$\frac{\alpha_{\text{th}}}{T} = \frac{-\pi^2 k_B^2}{3|e|E_F}\xi \qquad \xi \equiv \left(\frac{\partial \ln \sigma(E)}{\partial \ln E}\right)_{E_F} \tag{12.4}$$

One correction that need *not* be applied is for phonon drag. This arises from the non-equilibrium phonon distribution, or streaming by phonons in a temperature gradient, which is important in crystals but may be ignored in glasses because the disorder scatters phonons strongly and keeps them essentially in equilibrium (Jäckle 1980). Electron–phonon interaction is important nonetheless because it requires electrons to be treated as quasiparticles with effective mass m^* such that $m^*/m = 1 + \lambda_{\text{ph}}(T)$ where $\lambda_{\text{ph}}(T)$ is called the mass enhancement and depends on the electron–phonon coupling constant. This does not affect the conductivity but its energy dependence multiplies ξ by $1 + \lambda_{\text{ph}}(T)$ (Jäckle 1980).

The suggestion that $\lambda_{\text{ph}}(T)$ was important for metallic glasses was taken up experimentally by Gallagher and theoretically by Kaiser in 1981. Some experimental results for Cu–Zr, Cu–Ti and other alloys had the form shown in figure 12.14 which is from later work (Gallagher and Hickey 1985, see also Kaiser 1982). The argument was that ξ should vary very little, if at all, with temperature because magnetic scattering was absent, and because the small TCR counteracted any large changes with temperature in the relative importance of s and d electron contributions, or of competing scattering mechanisms. The observed failure of α/T to remain constant below about 200 K was therefore assigned to $\lambda_{\text{ph}}(T)$. The phenomenon has since been detected in many more glasses and the explanation essentially confirmed, though not without further examination which revealed several more corrections. For example, α_{th} should be replaced by $\alpha_{\text{th}}(1 + AT^{1/2})$ because of electron–electron interactions; $\lambda_{\text{ph}}(T)$ should be given a constant increment λ_{sf} due to spin fluctuations; and α/T acquires an increment due to the effect of electron–phonon interactions on electron velocities and relaxation times—an effect not included in m^*. Altogether

$$\frac{\alpha}{T} = \frac{\alpha_{\text{th}}}{T}(1 + AT^{1/2})(1 + \lambda_{\text{ph}}(T) + \lambda_{\text{sf}}) + (2\gamma_1 + \gamma_2)\lambda_{\text{ph}}(T)$$

where λ_{sf} and the constants γ_1, γ_2 are given by the theories. It appears however that the $AT^{1/2}$ and λ_{sf} terms have very little influence and that the observed λ_{ph} $(T = 0)$, about 0.4 to 1.0, agree well with those obtained independently from superconducting transition temperatures (Gallagher and

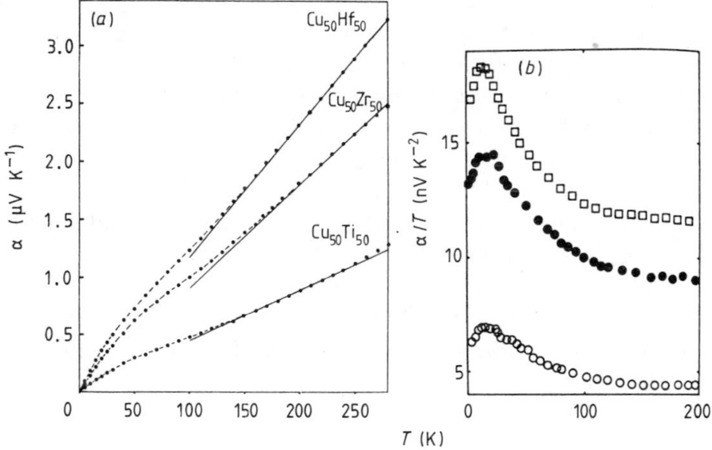

Figure 12.14 α and α/T in three glasses (from Gallagher and Hickey 1985).

Hickey 1985, Kaiser 1982). It remains a problem whether the observed $\lambda_{\text{ph}}(T)$ agrees with the theory of electron–phonon interaction which involves a coupling constant of uncertain frequency dependence (Naugle *et al* 1985).

There is still the question of the absolute value of ξ, for which a theory of $\sigma(E)$ is needed. Equations (6.60) from the NFE theory, or, for transition metals, its extended version (§6.16), are a possibility. Or, in a theory of Mott in which current is carried by s and p electrons but limited by scattering into empty d-band states, ξ depends on $(\text{d} \ln g(E)/\text{d} \ln E)_{E_{\text{F}}}$ and its sign therefore varies with that of $(\text{d}\, g(E)/\text{d}E)_{E_{\text{F}}}$. Neither of these models fits all the facts, and the extended Ziman theory certainly looks unsuited to non-simple glassy alloys. Since there is evidence for substantial current carrying by d electrons in some metals (§7.7), a full theory of α should include this (Gallagher and Greig 1982).

In §7.7, '$2k_{\text{F}}$ scattering' was introduced and §9.11 brought a logarithmic temperature dependence of conductivity into the weak localisation regime in 2-D. Since about 1980 these ideas have been imported into the theory of high-resistivity metallic glasses of which the negative TCR has presented a puzzle. This is an alternative approach to the one above by Schirmacher and Belitz and the relation between the two methods is not altogether clear. For simplicity let us assume that in the absence of $2k_{\text{F}}$ scattering, the conductivity would have the Boltzmann value, equation (6.52), which can easily be manipulated to read

$$\sigma_{\text{B}}(T) = \sigma_{\text{B}}(0) + \sigma_{\text{B}}^{(T)} = \frac{1}{3\pi^2}\left(\frac{e^2}{\hbar}\right)(k_{\text{F}} l_{\text{e}})^2 \left(\frac{1}{l_{\text{e}}} - \frac{1}{l_{\text{i}}(T)}\right). \tag{12.5}$$

Here, l_{e} and $l_{\text{i}}(T)$ are the elastic and inelastic mean-free paths as in §9.11.

$d\sigma_B(T)/dT$ is always negative because $l_i(T)$ falls with rising T. A correction would be necessary if the scattering were so great that $g(E_F)$ was seriously affected but we ignore this (Mott and Kaveh 1985).

A regime in which $2k_F$ scattering is present, but is not sufficient to cause localisation, is sometimes said to show 'localisation effects' or 'incipient localisation'. The theory in 2-D leads to equation (9.18) and, in 3-D, σ_B is corrected to $\sigma_B(T) + \sigma_L(T) = \sigma_B(T) + \sigma_L(0) + \sigma_L^{(T)}$. The correction σ_L is obtainable either from a multiple-scattering calculation (Kawabata 1982, see also Lee and Ramakrishnan 1985) or from a method in which electron wavefunctions, believed to combine components both extended and decreasing with a power law, are related to electron diffusion and then used in the Kubo–Greenwood equation of §7.7 (Mott and Kaveh 1985). In either case the formulae are

$$\sigma_L(0) = \frac{-C\sigma_B(0)}{(k_F l_e)^2} = \frac{-C}{3\pi^2}\left(\frac{e^2}{\hbar}\right)\frac{1}{l_e} \tag{12.6a}$$

$$\sigma_L^{(T)} = \frac{-C\sigma_B^{(T)}}{(k_F l_e)^2} + \frac{C\sigma_B l_e}{(k_F l_e)^2 L_i(T)} \simeq \frac{C}{3\pi^2}\left(\frac{e^2}{\hbar}\right)\frac{1}{L_i(T)} \tag{12.6b}$$

where $L_i(T) \simeq (l_i(T) l_e)^{1/2}$, the inelastic diffusion length (see also §9.11), and C is a constant of order unity. The last approximation uses $l_e \ll l_i$ and $L_i \ll l_i$, which are reasonable in disordered glasses.

Concentrating on the temperature-dependent part, $\sigma_B^{(T)} + \sigma_L^{(T)}$, we see it is approximately

$$\frac{1}{3\pi^2}\left(\frac{e^2}{\hbar}\right)\left(\frac{C}{L_i(T)} - \frac{(k_F l_e)^2}{l_i}\right)$$

and in high-resistivity glasses where l_e is very small—perhaps only about a nearest-neighbour distance—only the first term matters.

In connection with the first relevant experiments in glasses, Howson pointed out that the temperature dependence of σ would therefore depend on that of $l_i(T)$. Electron–phonon scattering is dominant and the theory of this shows that $l_i(T) \propto T^{-2}$, whence $\sigma \propto T$ for $T < \theta_D$; and $l_i(T) \propto T^{-1}$ whence $\sigma \propto T^{1/2}$ for $T > \theta_D$. Both give a negative TCR. This has been confirmed experimentally for a number of glasses—see figure 12.15.

As with localisation in 2-D (§§9.11, 9.12), the incipient localisation in 3-D can be affected by electron–electron interactions. There seems not to be an intuitive and transparent way of seeing what the effect is. Detailed calculations show that $g(E)$ is altered, up or down, by a correction proportional to $|E - E_F|^{1/2}$ and that a consequence is that σ is corrected by a term proportional to $T^{1/2}$. This term can be written (Al'tshuler and Aronov 1979)

$$\sigma_I \simeq \frac{1.3e^2}{4\pi^2\hbar}\left(\tfrac{4}{3} - \tfrac{3}{2}\tilde{F}\right)\left(\frac{k_B T}{2\hbar D}\right)^{1/2} \tag{12.7}$$

where the suffix I indicates 'interaction', D is the electron diffusion coefficient and $\tilde{F}$ was introduced in §9.11.

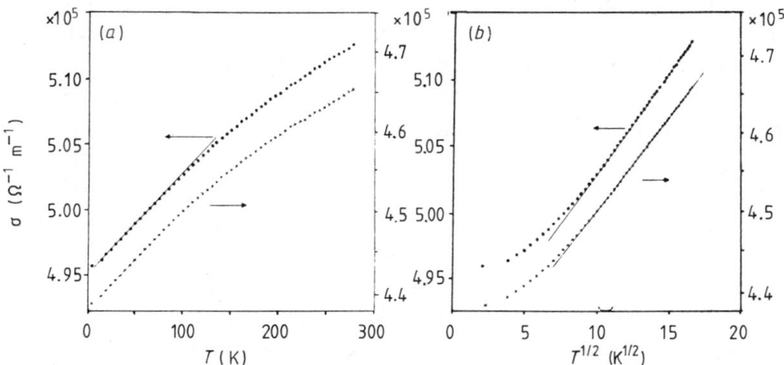

Figure 12.15 Temperature variation of σ in $Ti_{50}Be_{40}Zr_{10}$ (×) and $Cu_{50}Ti_{50}$ (●) glasses, showing the T- and $T^{1/2}$-dependences (from Howson 1984).

This effect is distinct from the $T^{1/2}$-dependence in σ_L and has been detected *below* the linear region of σ_L, say at $T \lesssim 10$ K. Figure 12.16 shows this and, as T rises from a few K to about 300 K, regimes of $T^{1/2}$-, T- and $T^{1/2}$-dependences succeed each other and conductivity extrema may well separate the regimes, giving a complex temperature dependence overall (Cochrane and Strom-Olsen 1984, Howson and Greig 1984).

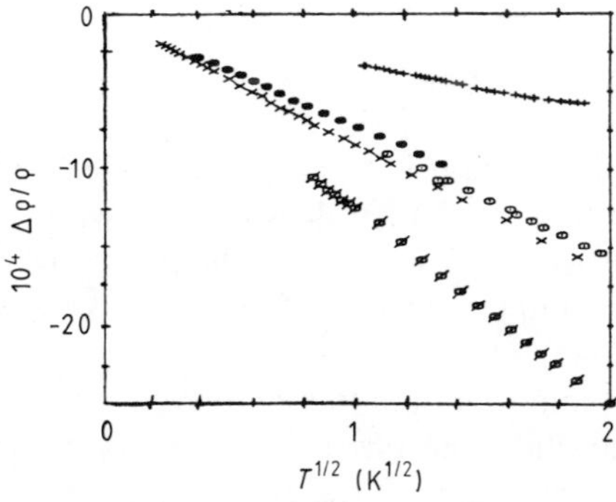

Figure 12.16 $T^{1/2}$-dependence of ρ at low temperatures (from Cochrane and Strom-Olsen 1984).

Recalling §9.11 again, we expect a magnetic field to produce complex effects in σ. There are at least four sources of this: cyclotron orbits introduce a new length into the problem and affect localisation thereby; the orbits, and also the Zeeman splitting of spin states, affect the electron–electron interaction term; these are the effects symbolised by the $\Delta\sigma(H)$ terms in equation (9.20). The spin–orbit (SO) interaction must also be allowed for and, in addition, if the material is superconducting, fluctuations in its thermodynamic variables, occurring at $T \sim T_c$, cause observable changes in electron transport (and other) properties (Skocpol and Tinkham 1975) which can be further modified by the magnetic field. By choosing appropriate samples, one or more of these effects can be virtually eliminated and a magnetoresistance experiment used to check the others. Thus Bieri *et al* (1984) observed positive magnetoresistance in $Cu_{57}Zr_{43}$ glass and interpreted it as the effect of H on localisation and superconductivity fluctuations. Poon *et al* (1985) show that a similar property exists in Lu-based glasses and argue that it is almost entirely due to the effect of H on localisation, superconductivity being absent and other sources largely eliminated by a high spin–orbit effect.

As first suggested by Howson and Greig (1983), the magnetoresistance in alloy glasses is a valuable indicator of localisation and interaction effects in 3-D. There are considerable difficulties in both experiment (quite small $\Delta\sigma(H)$ at very low T) and theory (many effects interfering with each other). Nevertheless at the time of writing valuable progress is elucidating not only the properties of metallic glasses but *ipso facto* the general phenomena of localisation and interaction characteristic of many disordered systems (Hickey *et al* 1986, Olivier *et al* 1986).

In this and the previous section, the electron transport properties of metallic glasses have been illustrated by examples. There are many more. In particular, magnetic impurities and ferromagnetism cause characteristic phenomena like the Kondo effect and the anomalous Hall effect but these will not be pursued here (see, e.g., Luborsky 1983).

12.6 Other important properties

The object of this final section is to draw the reader's attention to a few important properties that the chapter does not deal with. There is a growing number of comprehensive reviews on metallic glasses where these subjects can be pursued (Beck and Güntherodt 1983, Luborsky 1983, Moorjani and Coey 1984, Egami 1984).

Probably the most important omitted property is the magnetism of those alloys containing transition and rare-earth elements. Its importance partly derives from the technical value of the glasses, especially magnetically soft ones, for applications which include transformer cores, shielding and recording devices. Apart from this, the already profound problems of magnetism in crystals are intensified by disorder except that those arising from anisotropy are largely removed.

The itinerant electron model of the overall average properties of ferromagnets, such as saturation magnetisation, Curie temperature and spin wave dispersion, requires a knowledge of electronic quantities including $g(E)$, the bandwidth and the electron–electron interaction energy of two electrons. Calculations such as those of §12.3 are therefore pertinent and disorder will alter the magnetic properties as a consequence of smoothing the structure of $g(E)$ without changing the behaviour in principle. But it is rarely, if ever, possible to compare directly crystalline and amorphous forms of a ferromagnetic element and the process of alloying to form a glass—such as introducing B into Fe—itself alters the magnetism by electronic changes such as hybridisation and charge transfer between impurity and host. If crystalline and glassy alloys with similar composition and SRO can be compared, however, the disorder does not normally affect the ferromagnetism radically.

This is not in general true for other forms of magnetic order. Where the lowest energy of a pair of moments is achieved by antiparallel alignment (antiferromagnetism), a ring of interacting moments will be frustrated in the attempt to realise consistent antiparallelism if the ring is odd-numbered. Frustration also occurs when two or more neighbours of one atom with antiferromagnetic coupling are also neighbours of each other; the simplest case is the triangular lattice in 2-D. Topological disorder might create such situations and profoundly alter the energetics. Topological disorder is not a necessary condition for frustration however; it can occur in the square Ising lattice when there is random mixture of ferromagnetic and antiferromagnetic couplings, with or without a random dilution of magnetic sites by non-magnetic ones; it is impossible then to satisfy simultaneously all the competing coupling requirements and the ground-state energy rises over that of an ordered system. This is a rich source of highly interesting and challenging problems in magnetism.

It is difficult to pursue such questions without a preliminary account of magnetic theory in general and, for this reason, the reader will probably want to enter the very large literature through references already quoted above. From a valuable reference (Moorjani and Coey 1984) we take pictorial illustrations of magnetic ordering possibilities in topologically disordered arrays of local moments. The distinction between figures 12.17(*a*) and (*b*) is between systems with one kind of atom with a specified moment and magnetic interaction and systems with two kinds. To these we add in figure 12.17(*c*) a sequence of possibilities opened up as the atomic percentage of random magnetic impurities increases in a non-magnetic crystal lattice. The magnetic couplings are the various forms of exchange interaction which include the direct one, which was the original explanation of ferromagnetic couplings of spins, and the indirect one of Rudermann–Kittel–Kasuya–Yoshida (RKKY). In the latter, the electron gas, magnetically polarised by the moment of one ion, transmits the magnetic effect to a distant ion through an interaction which falls off as r^{-3} and oscillates in sign with wavelength $2\pi/2k_F$. The sign oscillations imply different couplings between pairs of different separations and can be a source of frustration in disordered arrays of moments. In the spin glass regime this results in the freezing-in of moments with disordered orientation at low temperatures. Spin glasses remain an intense focus of scientific interest (Chowdhury 1986).

Like the magnetic properties, but to a lesser extent, the mechanical properties of amorphous metals have also promised practical applications: they are possible constituents of strong composite structural materials for exam-

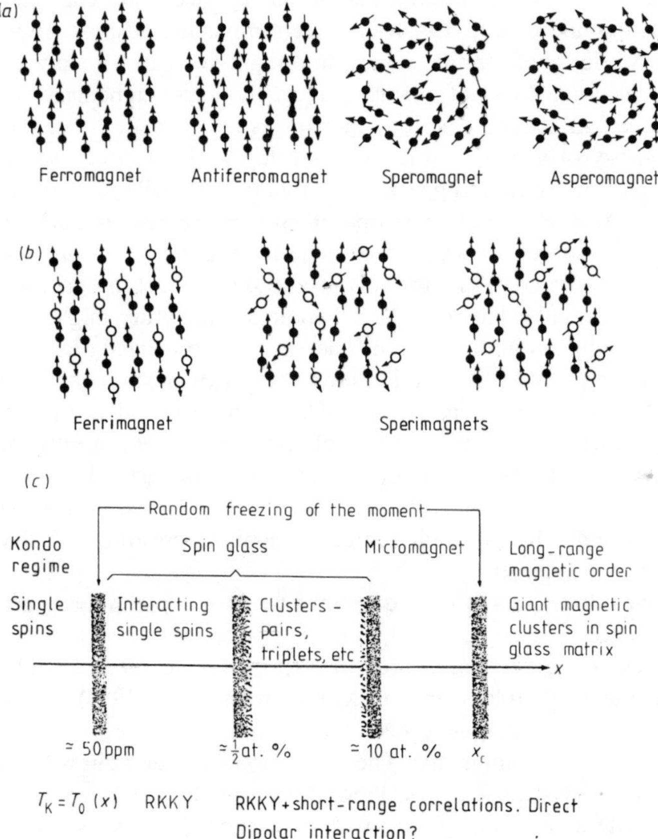

Figure 12.17 (*a*), (*b*) Schematic diagrams of various orientational ordering possibilities. Prefixes spero, speri, refer to random orientation systems with net magnetic moments. Asperomagnets have no net moment. (*c*) Various regimes at different magnetic impurity concentrations, x. The dominant interactions are shown below the diagram. x_c is the percolation threshold, e.g. about 15.5 at. % Fe in Au, at which direct interactions link up throughout the system. (After Moorjani and Coey 1984.)

ple. Their nature as glasses out of thermodynamic equilibrium complicates their mechanical properties both in theory and experiment. However, unless magnetisation and magnetoelastic effect intervene, the elasticity of metallic glasses is in general isotropic and there are only two independent properties, say the bulk and shear moduli. The former (K) is similar to, the latter (G) is about 30 % lower than, their values in the crystal. At least for simple metals, an equation of state ought to be derivable from the pseudopotential theory of ion–ion interactions in which the total energy, E, has volume and structure-dependent terms (§6.6). The bulk modulus is proportional to $d^2E/d\varepsilon^2$, where ε is the strain, and it was calculated successfully in this way for amorphous $Mg_{70}Zn_{30}$ by Hafner (see Suzuki (1982), p 1311). For shear strain at constant volume, a computation of $d^2E/d\varepsilon^2$ was originally performed by Weaire *et al* (1971) using pair potentials in a model glass of about 100 atoms. It appeared that the shear modulus was reduced relative to its value in the crystal by the adjustments of configuration to which the glass can resort in order to release partially its shear stress.

Of course there is a great variety of responses to stress (σ_{ij}) that metallic glasses can show. The elastic response which is virtually instantaneous is homogeneous and, if the stress is prolonged and also low (say $\sigma_{ij}/G < 0.01$), homogeneous creep can be detected. For sufficiently low stress the Newtonian viscosity, $\eta = \sigma_{ij}/\dot{\varepsilon}$, can be measured and it falls as the temperature rises. More complicated stress–strain relations occur at higher stresses (see, e.g., Spaepen and Taub in Luborsky (1983)).

Flow is not necessarily homogeneous. At high stresses (say $\sigma_{ij}/G > 0.02$), glasses subjected to tensile testing or cold working show local plastic deformations. In a tensile test the latter occur in thin shear bands making an angle with the direction of tensile force and the bands are detectable at the surface. They are the sites of very large strains leading ultimately to fracture. Metallic glasses do not in general show work hardening. This inhomogeneous behaviour is not very sensitive to temperature (Luborsky 1983). In crystal physics it would be natural to discuss plastic flow in terms of dislocations. It is obvious that dislocations defined by displacement of planes of atoms are not appropriate to a glass, but not obvious that the dislocation concept it therefore of no use at all. In crystals, dislocations are associated with characteristic stress distributions and similar ones could be notionally introduced into a glass by the kind of cuts and displacements shown in figure 2.9. This has been done in computer simulations but it is not established that mobile line defects play any role in plastic flow comparable with that in crystals (Beck and Güntherodt 1983, Luborsky 1983). Metallic glasses can show strengths up to the ideal limit set by the cohesive forces.

It was mentioned earlier that structural relaxation occurs during annealing or ageing. This affects all the properties and particularly some of the mechanical ones. For a rational investigation of the temperature dependence of some property P it is desirable that the structure should remain the same, statistically at least, during the measurements. It is found that P changes with time, t, during annealing but that $|P^{-1}\mathrm{d}P/\mathrm{d}t|$ gradually decreases. If P reaches a value P_0 at t_0 and temperature T_0, and thereafter changes by only a few per cent per day, it is possible to reduce T_0 to T_1 for an interval τ during which $P(T_1)$ is measured. If P_0 is regained on returning to T_0, it indicates that the structure did not change significantly during τ. This could be repeated for T_2, T_3 . . . , and $P(T)$ is then attributable to a particular structure and the measurement is said to be isoconfigurational. This is a means of effectively evading the relaxation effects in order to separate the variables.

But the relaxation phenomenon is of great interest in its own right, viscosity and diffusion being particularly sensitive to it. Atomic diffusion, like plastic flow, is not so easy to envisage in glasses as in crystals because the vacancy and its jumps are not well defined. Interstitial jumping of small atoms, notably hydrogen, occur in metallic glasses, but whatever the diffusion mechanisms of large atoms are, the diffusion coefficient generally obeys the Arrhenius law $D = D_0 \exp(-Q/T)$. Since Q does not depend on T, it seems that there is not a large spread of activation energies, and therefore presumably not a large variety of displacive motions, in spite of the fact that no two jumping environments are identical (see, e.g., Cantor in Steeb and Warlimont 1985). At the time of writing it is probably true to say that the atomic mechanisms involved in the relaxation process are not well enough understood to relate the relaxation of one property with that of another.

'At the time of writing' is a phrase used not simply in relation to metallic glasses but quite often throughout the book—which may therefore reasonably be concluded by the remark that, although much of the text has dealt with well established or fundamental ideas, some of it has presaged future developments in which the reader may well be taking part.

References

Al'tshuler B L and Aronov A C 1979 *Sov. Phys.–JETP* **50** 968

Aschroft N W and Mermin N D 1976 *Solid State Physics* (New York: Holt, Rinehard and Winston)

Ballentine L E 1985 *Proc. 5th Int. Conf. on Rapidly Quenched Metals* ed S Steeb and H Warlimont (Amsterdam: North-Holland) p 981

Ballentine L E and Kolář M 1986 *J. Phys. C: Solid State Phys.* **19** 981

Beck H and Güntherodt H-J (ed) 1983 *Glassy Metals II* (Berlin: Springer)

Belitz D and Schirmacher W 1984 *Proc. 5th Int. Conf. on Liquid and Amorphous Metals* (1984 *J. Non-Cryst. Solids* **61/62**) 1073

Bieri J B, Fert A, Creuzot G and Ousset J C 1984 *Solid State Commun.* **49** 849

Blatter A and von Allman M 1985 *Phys. Rev. Lett.* **54** 2103

Chowdhury D 1986 *Spin Glasses and other Frustrated Systems* (New York: Wiley)

Cochrane R W and Strom-Olsen J O 1984 *Phys. Rev.* B **29** 1088

Cowlam N and Gardner P P 1984a *J. Phys. F: Met. Phys.* **14** 1789
—— 1984b *Proc. 5th Int. Conf. on Liquid and Amorphous Metals* (1984 *J. Non. Cryst. Solids* **61**/**62**)
Cyrot-Lackmann F and Desré P (ed) 1980 *Proc. 4th. Int. Conf. on Liquid and Amorphous Metals* (*J. Physique* **41** C8)
Davies H A 1976 *Phys. Chem. Glasses* **17** 159
Egami T 1984 *Rep. Prog. Phys.* **47** 1601
Fujiwara T 1982 *J. Phys. F: Met. Phys.* **12** 661
Gallagher B L and Greig D 1982 *J. Phys. F: Met. Phys.* **12** 1721
Gallagher B L, Greig D, Howson M A and Croxon A A M 1983 *J. Phys. F: Met. Phys.* **13** 119
Gallagher B L and Hickey B J 1985 *J. Phys. F: Met. Phys.* **15** 911
Giessen B C and Whang S 1980a *Proc. 4th Int. Conf. on Liquid and Amorphous Metals* ed F Cyrot-Lackmann and P Desre (*J. Physique Coll.* **41** C8 95)
—— 1980b *Phys. Rev.* **21** 406
Grabow M H and Andersen H C 1985 *J. Non-Cryst. Solids* **75** 225
Güntherodt H-J and Beck H (ed) 1981 *Glassy Metals I* (Berlin: Springer)
Hafner J 1980 *Phys. Rev.* B **21** 406
—— 1983 *Phys. Rev.* B **28** 1734
—— 1985 *J. Non-Cryst. Solids* **69** 325
Hafner J and Philipp A 1984 *J. Phys. F: Met. Phys.* **14** 1685
Hickey B J, Gallagher B L and Howson M A 1986 *J. Phys. F: Met. Phys.* **16** L13
Howson M A 1984 *J. Phys. F: Met. Phys.* **14** L25
Howson M A and Greig D 1983 *J. Phys F: Met. Phys.* **13** L155
—— 1984 *Phys. Rev.* B **30** 4805
Howson M A and Morgan G J 1985 *Phil. Mag.* B **51** 439
Hume-Rothery W, Christian J W and Pearson W B 1952 *Metallurgical Equilibrium Diagrams* (London: Chapman and Hall)
Jäckle J 1980 *J. Phys. F: Met. Phys.* **10** L43
Kaiser A B 1982 *J. Phys. F: Met. Phys.* **12** L223
Kawabata A 1982 *Solid State Commun.* **38** 823
Khanna S N, Ibrahim A K, McKnight S W and Bansil A 1985 *Solid State Commun.* **55** 223
Künzi H-U and Güntherodt H-J 1980 *The Hall Effect and its Applications* (New York: Plenum)
Lapka R, Oelhafen P, Gubler U M and Güntherodt H-J 1985 *Phys. Rev.* B **31** 7734
Lasjaunias J C, Zougmore F and Béthoux O 1986 *Solid State Commun.* **60** 35
Lee P A and Ramakrishnan T V 1985 *Rev. Mod. Phys.* **57** 287
Luborsky F E (ed) 1983 *Amorphous Metallic Alloys* (London: Butterworths)
Mizutani U 1983 *Prog. Mater. Sci.* **28** 97
Mooij J H 1973 *Phys. Status Solidi* A **17** 521
Moorjani K and Coey J M D 1984 *Magnetic Glasses* (Amsterdam: Elsevier)
Morgan G J and Howson M A 1985 *J. Phys. C: Solid State Phys.* **18** 4327
Mott N F and Kaveh M 1985 *Adv. Phys.* **34** 329
Nagel S R and Tauc J 1975 *Phys. Rev. Lett.* **35** 380
Naugle D G, Delgado R, Armbrüster H, Tsai C L, Johnson W L and Williams A R 1985 *J. Phys. F: Met. Phys.* **15** 2189
Nguyen-Manh D, Mayou D, Morgan G J and Pasturel A 1987 *J. Phys. F: Met. Phys.* **17** 1309

Nozières P and Pines D 1966 *The Theory of Quantum Liquids* (New York: Benjamin)

Olivier M, Ström-Olsen J O, Altounian Z, Cochrane R W and Trudeau M 1986 *Phys. Rev.* B **33** 2799

Pavuna D 1981 *PhD Thesis* University of Leeds

Poon S J, Wong A M and Drehman A J 1985 *Phys. Rev.* B **31** 1668

Sakata M, Cowlam N and Davies H A 1981 *J. Phys. F: Met. Phys.* **11** L157

Schirmacher W 1985 *Proc. 5th Int. Conf. on Rapidly Quenched Metals* ed S Steeb and H Warlimont (Amsterdam: North-Holland) p 995

Skocpol W J and Tinkham M 1975 *Rep. Prog. Phys.* **38** 1049

Steeb S and Warlimont H (ed) 1985 *Proc. 5th Int. Conf. on Rapidly Quenched Metals* (Amsterdam: North-Holland)

Srolovitz D, Egami T and Vitek V 1981 *Phys. Rev.* B **24** 6936

Suzuki K (ed) 1982 *Proc. 4th Int. Conf. on Rapidly Quenched Metals* (Sendai: Japanese Institute of Metals)

Wagner C N J and Johnson W L (ed) 1984 *Proc. 5th Int. Conf. on Liquid and Amorphous Metals* (*J. Non-Cryst. Solids* **61/62**)

Weaire D, Ashby M F, Logan J and Weins M J 1971 *Acta Metall.* **19** 779

Wright A C, Hannon A C, Clare A G, Sinclair R N, Johnson W L, Atzman M and Mangin P 1985 *J. Physique Coll.* **46** C8 299

Wright A C, Hannon A C, Sinclair R N, Johnson W L and Atzman M 1984 *J. Phys. F: Met. Phys.* **14** L201

INDEX

Absorption, 64
electromagnetic, 161, 313, 320, 326, 355
ultrasonic, 289, 292
Acoustic modes, 198, 211, 222
Activation energy, 330–3
Activity, 148
Admittance, 191
Ag, 136
Al, 136, 158
Alloys
compound-forming, 341
glassy, 361
hard-sphere binary, 165
liquid, 147, 157–9, 341
order–disorder transformation 4, 5, 17
scattering from, 61, 67, 74
Amorphous semiconductors, 295
band structure, 320–8
defects, 302–8
electron transport, 328–40
preparation, 287–99
structure, 299–302
see also Doping, Gap states, etc
Amorphous solids, 7
Anderson model, 244
Anderson transition, 250, 254, 255
Annealing, 361, 391
Anomalous dispersion, 67
Antibonding orbital, 306, 322, 325
Antiferromagnetism, 22
Ar, 28, 51, 70, 94, 101, 127, 204, 209
Ashcroft–Langreth structure factors, 78
As–S, 301
As–Se, 326, 338
Atomic scattering factor, 57
Au, 160
Autocorrelation function *see* specific entries: density-, velocity- etc

B_2O_3, 284, 384
Ba, 176
$BaCl_2$, 72
Backbone, 235
Ball-and-spoke models, 38, 219, 279
Band gap *see* Energy gap
Bernal model, 25
Bhatia–Thornton structure factors, 75, 345, 368
Bi, 155, 160, 172, 186, 209, 361
Boltzmann equation, 153, 217
Bond angles, 38, 277, 282, 322
Bonding orbital, 305, 321
Born–Green–Yvon equation, 105, 110, 115, 124
Boroxol ring, 279, 284
Bravais lattice, 6
Bremsstrahlung isochromat spectroscopy, 324
Bridge function, 117
Bridging oxygen, 285, 297
Brillouin scattering, 213, 223
Brownian motion, 49, 213
Bulk modulus, 391

CaMg, 220, 366, 379
CCl_4, 69
Carnahan and Starling formulae, 91, 165
Cayley tree, 349, 355
Cell theory of liquids, 125
Cellular disorder, 9
Chalcogenide glasses, 13, 287, 297, 299, 326, 331, 338, 341
Charge transfer, 348, 349
Chemical
disorder, 9
potential, 147, 153, 190, 258
short-range order, 9, 342, 364, 368
thermodynamic model, 148, 343

Chemical (*cont*)
  vapour deposition, 299
Closure problem, 105
Cluster integrals, 106
Clusters, 179
Coherent potential approximation, 172–5, 179, 188, 193, 323, 373
Collective motion, 198, 205, 208, 210, 216, 221
Compound-forming alloys, 341
Compressibility, 92
  equation of state, 103, 115, 117
Compton scattering, 64
Computer glasses, 97, 222, 273
Conductance, 247–9, 259–61
Conduction electrons, 132
Conductivity
  electrical, 154, 156, 163, 190–5, 236, 259–63, 328–40, 348, 351, 377–88
  thermal, 154, 160, 190, 287
Configurational entropy, 10, 41
Continuous random network, 38, 276, 279, 296, 300
Coordination number, 33, 36
Core states, 133
Correlation energy, 144, 302, 306, 312, 315
Critical phenomena, 21, 50, 95, 117, 350–2
Cross section
  x-ray, 56
  neutron, 201
Crystallisation temperature, 364
Cs, 150, 165, 351, 354–6
Cs–Au, 348–50
Cu, 155, 381
Cu–Au, 189
Cu–Ni, 174, 188
Cu–Ti, 368, 373, 382
Cu–Zr, 374, 381, 388
Curie temperature, 21, 389

Dangling bond, 302–5, 308–12, 318, 323, 340
Debye heat capacity, 286, 367
Debye–Waller factor, 378
Deep-level transient spectroscopy, 317
Defect, 7, 302, 305
  see also Dangling bond, Dislocation etc
Dense random packing, 25, 220, 366, 368, 372, 376
Density
  autocorrelation function, 49, 58, 199
  expansions, 105
  function, 31
  of electron states, 133, 170, 179–83, 185–7, 242, 354, 355, 372–7
    local, 170, 180–3
  see also Optical density of states, Gap states, Tail states etc
Depletion hole, 144
Dielectric screening function, 137
Diffusion
  coefficient, 88, 190, 194, 208, 392
  length, 251
  -limited aggregation, 51
Direct correlation function, 111, 114, 119, 152, 345
Disclination, 44
Dislocation, 44, 391
Dispersion relation, 205, 210, 221–3
Distribution function, 30
Doping, 297, 303, 333
Double-null experiment, 371
Drude equations, 162, 353, 373
Dynamical structure factor, 163, 201–4, 216, 221, 344, 372, 378

Effective mass of electrons, 162, 164, 184, 384
Effective medium, 173
  approximation, 178, 185, 193, 373
Ehrenfest equations, 270
Electrochemical potential, 153, 258
Electron
  gas, 132, 140–4
  spin resonance, 300, 305, 308, 317
  transport, 153–65, 189–95, 328–40, 342, 351, 377–88
Electron–electron interaction, 137, 252, 254, 386
  see also Exchange and correlation
Electron–ion potential, 105, 133–7

Electron–phonon interaction, 309, 373, 384
Empty-core potential, 136
Energy gap, 320–8, 352
Energy transfer, 63, 163
Ensembles, 6, 89
  canonical, 83, 89
  grand canonical, 78, 102
Entropy, 9, 47, 92, 148
  electron gas, 146
  of mixing, 11, 148, 163, 344
  see also Configurational entropy
Equation of motion method, 183
Eutectics, 363
Exchange and correlation, 138, 144
Expanded liquid metals, 350
EXP approximation, 119
Extended state, 239
Extended x-ray absorption fine structure, 80, 282, 286, 300, 301
Extended Ziman theory, 164, 189, 385

Fe, 179, 183, 194, 383
Fermi glass, 250
Fermi surface, 154
  blurring of, 165, 171, 174, 188
Ferromagnetism, 18, 389
Fictive temperature, 271
Field effect, 314
  transistor, 253
Flory model, 149
Fluctuations, 74, 77, 212, 213–15, 224, 354, 388
Form factor, 136
Fractal, 47, 223
  dimension, 48, 50
Free volume, 125, 274, 290, 362

Ga, 155, 361
Gap states, 249, 297, 302, 305, 314–20, 333
Generalised Faber–Ziman theory, 163, 377
Gibbs–Bogoliubov inequality, 124, 146, 150
Gibbs' free energy, 147
  of mixing, 147, 366–8
Glass, 13, 266
  low-temperature properties, 286–93
  structure, 43, 279–86, *see also* Computer glasses, Metallic glass, Spin glass etc
  formation, 361–8, 375, *see also* Computer glasses, Glass transition
  -forming ability, 362, 369
  transition, 13, 97, 267–73
Glow discharge, 298
Green
  function, 169–72, 176–84, 193, 219, 348
  operator, 168

Hall effect, 155, 160, 194, 237, 335, 351, 378, 381
Hard-sphere model, 25, 89, 90, 110, 146, 152, 178, 180, 220
  binary formulae, 165
Heat capacity, *see* Specific heat
Hg, 155, 351, 355
High-temperature expansion
  of free energy, 121
Hopping, 257, 333, 383
  nearest-neighbour, 261
  variable-range, 262
Hubbard model, 256
Hybridisation, 164, 381
Hydrodynamic limit, 212
Hydrogenation of a-Si, 299, 300, 323
Hypernetted chain, 114–18, 129

Icosahedral symmetry, 17, 25
Ideal solution, 148
Imperfections, 7
Impurity band, 256
In, 160
Incipient localisation, 386
Incoherent scattering, 62, 203, 204, 210
Inelastic scattering, 62
  electron, 247, 251, 386–8
  neutron, 199–211, 224
Infrared spectra, 223, 301
Interatomic potential *see* Pair potential and specific cases, e.g., Lennard-Jones
Intermediate-range order, 8, 43, 281

Intermolecular potential, 100
  see also Pair potential
Interstitial hole, 29
Inverse participation ratio, 239
Inversion layer, 252, 263
Ion cores, 132–7
Ion–ion potential, 144, 152
Ising model, 17–22
Isoconfigurational measurements, 392
Isomorphous substitution, 68
Isothermal capacitance transient spectroscopy, 316
Isotopic substitution, 68, 224, 369

Joffe–Regel criterion, 239

K, 68
k-Conservation rule, 186, 325
Kauzmann paradox, 274
Kinetic energy correlation function, 99
KKR method, 174, 177, 188, 374
Knight shift, 185, 351, 355
Koch curve, 47, 49
Kr, 207
Kramers–Kronig relations, 161
Kronig–Penney model, 242
Kubo equations, 190, 381
Kubo–Greenwood equation, 192, 193, 329, 386

La, 183, 194
Langevin equation, 213
Laser glazing, 273, 361
Lattice animals, 230
Lattice gas, 22
Lennard-Jones
  glass, 99, 222
  model, 93, 108, 121, 124, 127, 209, 222
Li, 155, 165
Li–Pb, 342
Lifetime
  electron, 318, 337
  hole, 318, 337
  liquid complexes, 344
Light-induced ESR, 312
Linear chain
  masses, 217
  potentials, 242
Linear-response theory, 190, 202, 216
Liquid crystals, 7
Liquids,
  metallic, 131, 341, 350
  thermal motion in, 205, 211
  semiconducting, 295, 341, 351
  simple, 100
Local density of states *see* Density of states
Localisation
  Anderson, 238, 244, 255
  incipient, 386
  length, 238, 245, 246–9, 250, 259
  one-dimensional, 242, 248, 255
  scaling, 246
  two-dimensional, 245, 248, 251–4
  three-dimensional, 248, 255, 354
  of vibrations, 240
  weak, 248, 251, 253
Long-range
  order, 8
  oscillations, 95, 138, 145
Lorenz number, 160
LRO-II potential, 96, 107, 124

Magic-angle spinning, 283
Magnetic susceptibility, 373
Magnetic properties
  glasses, 388
Magnetoresistance, 252, 253, 388
Mayer function, 106
Mean free path
  electron, 154, 165, 172, 239, 251, 386
  phonon, 288
Mean spherical approximation, 118, 346
Mean-square displacement, 88, 198, 207
Melt spinning, 361
Memory function, 214–16
Metal–insulator transition, 237, 250, 255–7, 348, 350, 352–6
Metallic glass, 360
Metalloid, 360
Meyer–Neldel rule, 334
Mg, 254
Mg–Zn, 210, 221, 379, 391
Minimum metallic conductivity, 251, 330

Mobility
conductivity, 330
drift, 337
Hall, 237, 335
microscopic, 329
Mobility edge, 240, 249, 256, 325
Mobility gap, 249, 328, 353
Mode coupling, 217, 249, 276
Model potential, 135
Modified hypernetted chain, 118, 124
Molecular dynamics, 87, 97, 206–10, 215, 372
Molten salts, 69, 224
Moments *see* Scattering law
Momentum transfer, 55
Monte Carlo method, 28, 83, 108, 123, 180, 260
Mooij generalisations, 381
Mössbauer
scattering, 223
spectrum, 80
Mott
transition, 256, 354
g-factor, 194
Muffin-tin potential, 164, 174, 373, 376
Multiple scattering, 64, 164, 174, 189
Multiple trapping model, 337–40

Na, 68, 95, 107, 128, 136, 211
NaCl, 69
Ne, 209
Nearly-free-electron model, 132, 140–52, 366, 378
of electron transport, 156–63, 189
Network modifiers, 285
Neutron scattering, 62, 199–211, 368–71
Neutron spectrometers, 66
Ni, 72, 375, 383
Ni–B, 370
Ni–P, 373
Ni–Ti, 72
Nuclear magnetic resonance, 224, 283, 286, 300
see also Knight shift
Nuclear quadrupole relaxation, 224
Nucleation, 97, 364
Null elements, 72, 369–71

Off-diagonal disorder, 245, 354
One-component plasma, 152
Onsager equations, 190
Optical density of states, 186
Optical gap, 320
Optical properties, 161
amorphous semiconductors, 320, 324–8
liquid metals, 162, 353
Optimised random-phase approximation, 119, 147, 151
Order, 1
definition, 3
multiple, 4
orientational, 9
structural, 6
Order parameter, 5, 11, 19, 40, 349, 369
Order–disorder transformation, 5, 17, 189
Ordering potential, 19, 345
Ordering rule, 1
Ornstein–Zernike function, 112
see also Direct correlation function
Orthogonalisation hole, 144
Orthogonalised plane waves, 133
Overlap integral, 177

Packing fraction, 92, 115, 116, 165, 346
Pair correlation function, 103, 111, 345
time-dependent, 199–204
Pair potential, 60, 90, 93, 96, 100, 119, 391
from $g(r)$, 124
see also special cases—Lennard-Jones, ion–ion etc
Pair distribution function, 32, 60–1, 65, 92, 94, 96, 103, 118, 126
time-dependent, 199–204
Partial
coordination numbers, 36, 40, 370
molar quantities, 147
pair distribution functions, 35, 68, 345, 369
radial distribution function, 35, 61
structure factors, 61, 67, 74, 79, 369
Participation ratio, 239
Particle current autocorrelation function, 215

Partition function, 18, 19, 84, 102, 126
Passivation, 302, 323
Pb, 136, 155, 160, 209
Percolation, 227
  effective-medium theory, 236, 354
  expanded metals, 354
  fractals, 52
  glass transition, 276, 278
  indices, 232, 235
  lattice, 229
  rigidity, 278
Percus–Yevick equation, 114–18, 165
Periodic boundary conditions, 85, 239
Permittivity, 161
Penrose tiling, 14
Perturbation method
  nearly-free-electron, 134, 140
  thermodynamic, 121
Phase
  diagrams, 363, 367
  transition, 17, 93, 97, 273
Phonon drag, 384
Photoconductivity, 338
Photodarkening, 341
Photodissolution, 341
Photoemission, 162, 186, 188, 323, 373
Photoinduced absorption, 339
Photoluminescence, 308
Photothermal deflection spectroscopy, 320
Plastic crystals, 210
Polarisation, 64, 204
Polk model, 34, 38, 224
Polytope, 43
Positron annihilation, 186
Potential of mean force, 104, 124
Prigogine–Defay ratio, 270
Protein dynamics, 223
P–Se, 301
Pseudopotential, 133–46, 220, 367
Pseudowavefunction, 134
Pseudoions, 345
Pseudoatom, 137
Pseudogap, 250, 353, 366

Quantum interference, 195, 386
Quasicrystalline approximation, 178
Quasicrystals, 14, 46
Quasi-elastic peak, 205
Quasiperiodicity, 16
Quench echoes, 97
Quenching, 9, 12, 97, 271, 361

Radial distribution function, 33
Radiative recombination, 313
Radical planes, 29
Raman spectra, 224, 284, 301
Random-phase model, 193, 250, 335
Random walk, 50, 241
Rayleigh scattering, 212
Rb, 204–8, 216, 355
Rectilinear diameter, 352
Recursion method, 179, 221, 323
Reduced glass transition temperature, 363
Reference system, 120, 123, 152
Reflectivity, 162, 353, 373
Refractive index, 161
Regular assemblies, 18
Regular solution, 149
Relaxation, 37, 220, 266–73, 366, 372, 391
  time
    electronic, 153, 251
    configurational, 266, 269–73, 289
    Maxwell, 213, 215, 269
Resistivity *see* Conductivity
Resistivity structure factor, 377
Rigidity percolation, 278
Ring statistics, 38, 281, 322
RKKY interaction, 389
Rule, $8 - N$, 303

Sb, 155, 160
Scaling, 48, 246
Scattering, 55
  electron, 153, 155, 160, 164, 251
  neutron *see* Neutron scattering
  $2k_F$, 195, 385
  x-ray, 56, 282
  see also special cases—Brillouin, Rayleigh etc
Scattering amplitude, 55
Scattering law, 202, 216
  moments, 204
  symmetrised, 204

Scattering length, 62, 202
Schafli notation, 43
Schottky anomaly, 291
Screening, 133
  see also Dielectric screening function
Se, 301, 356
Self-energy, 176, 183
Self-correlation function, 203, 210–11
Self-similarity, 48, 50
SF_6, 210
Short-range order, 8
  see also Chemical SRO, Order parameter etc
Sierpinski gasket, 48, 50
Si, amorphous, 224, 296
  band structure, 320
  defects, 302
  electron transport, 328–40
  ESR, luminescence, 308
  gap states, 314
  preparation, 297
  structure, 299
  tail states, 325
  transient methods, 336
SiO_2, 38, 73, 219, 282–6, 287, 292
Significant structures, Eyring theory, 125
Simple metal, 133
Single-site approximation, 174, 178
Skew scattering, 160
Sn, 155
Space-charge-limited current, 320
Space–time correlation function, 199–202
Specific heat, 22, 268, 272, 292, 373
  low-temperature (linear), 287, 291, 292, 373
Spectral
  function, 170–5, 184
  operator, 169
Spin glasses, 22, 390
Spin–orbit interaction, 160, 195, 252, 254, 388
Splat cooling, 273, 361
Square-well potential, 118, 121
Sr, 176
Stability, of glass structure, 276
Staebler–Wronski effect, 340
Stokes's shift, 310
Structure factor, 49, 57, 60, 65, 115, 127, 136, 140, 152, 175, 202, 369
  binary systems, 61, 74
  long-wavelength limit, 76, 77, 115, 148, 272
  resistivity, 377
  see also Dynamical structure factor
Superconductivity fluctuations, 388
Supercooling, 97, 266, 364, 366
Superposition approximation, 105–10, 178

T-, *t*-matrix, 173, 189, 195
$T^{-1/4}$ hopping, 262, 333
Tail states, 249, 312, 325, 328, 334, 338
Te, 354
Tessellation, 14
Thermodynamic consistency, 117
Thermodynamics
  glasses, 266, 269, 367
  liquid metals, 146–52, 343, 347
Thermoelectric power, 154, 157–9, 190, 237, 330–4, 342, 348, 351, 380, 384
Three-body forces, 101, 207
Ti, 69
$TiCl_4$, 69
Tight-binding approximation, 174, 176, 179, 180, 348
Tiling, 14
Time-of-flight method, 337
Time–temperature transformation curve, 364
Tl, 202
Topological disorder, 9, 354
Topological network, 280
Total pair distribution function, 67
Transfer integral, 177, 245, 250
Transient methods, 132
  in semiconductors, 336–40
Transition-metal–metalloid glasses, 364, 370, 374
Trapping, of electrons, 337–40
Triplet distribution function, 32, 79, 103, 105
Tunnelling, 291
Two-level system, 47, 289–93, 372, 373

Ultrasonic attenuation, 289, 292
Umklapp processes, 221
Urbach edge, 326, 328, 353

Valence alternation pair, 306
Van Hove correlation function *see* Space–time correlation function
V, 65, 383
Variational methods, 123
Velocity autocorrelation function, 88, 207, 213–15
Virial
  coefficients, 91, 105
  equation, 88, 90, 117
Virtual-crystal approximation, 172, 173
Viscoelastic approximation, 213–16
Viscosity, 268, 391
Vitreous solid, 13
  see also Glass
Vogel–Fulcher equation, 269
Voronoi polyhedron, 27

Weaire–Thorpe model, 321
Weak localisation, *see* Localisation
Weeks, Chandler and Anderson method, 121, 147, 151
Wiedemann–Franz law, 154

X-ray photoemission spectroscopy, 187, 323

Zero alloy, 72
Ziman formula, 157, 377
  modifications of, 163
Zn, 136